Advanced Topics of

Thin-Walled Structures

Advanced Topics of

Thin-Walled Structures

Haim Abramovich

Technion, Israel Institute of Technology, Haifa, Israel

World Scientific

NEW JERSEY • LONDON • SINGAPORE • BEIJING • SHANGHAI • HONG KONG • TAIPEI • CHENNAI • TOKYO

Published by

World Scientific Publishing Co. Pte. Ltd.
5 Toh Tuck Link, Singapore 596224
USA office: 27 Warren Street, Suite 401-402, Hackensack, NJ 07601
UK office: 57 Shelton Street, Covent Garden, London WC2H 9HE

Library of Congress Cataloging-in-Publication Data
Names: Abramovits, Ḥayim, Dr., author.
Title: Advanced topics of thin-walled structures / Haim Abramovich,
Technion, Israel Institute of Technology, Haifa, Israel.
Other titles: Thin-walled structures
Description: Singapore ; Hackensack, NJ ; London : World Scientific, [2021] |
Includes bibliographical references and index.
Identifiers: LCCN 2020052511 (print) | LCCN 2020052512 (ebook) |
ISBN 9789811214240 (hardcover) | ISBN 9789811214257 (ebook for institutions) |
ISBN 9789811214264 (ebook for individuals)
Subjects: LCSH: Thin-walled structures. | Composite materials. | Airframes.
Classification: LCC TA660.T5 A23 2021 (print) | LCC TA660.T5 (ebook) | DDC 624.1/77--dc23
LC record available at https://lccn.loc.gov/2020052511
LC ebook record available at https://lccn.loc.gov/2020052512

British Library Cataloguing-in-Publication Data
A catalogue record for this book is available from the British Library.

For any available supplementary material, please visit
https://www.worldscientific.com/worldscibooks/10.1142/11663#t=suppl

Desk Editors: Vishnu Mohan/Steven Patt

Typeset by Stallion Press
Email: enquiries@stallionpress.com

Preface

Advanced Topics of Thin-Walled Structures aims at dissemination of advanced engineering topics using isotropic and composite materials as well as smart materials like piezoelectric and shape memory alloys (SMA). This book was written to enable students and scholars a good understanding of new emerging technologies using composite materials, mainly in the aerospace sector as well as smart intelligent structures. It can serve as an introductory book for graduate students wishing to study about composite made structures, enabling them to acquire the necessary physical and mathematical tools to understand the various aspects of composite-based structures as compared to isotropic ones. It also enables the reader to understand the new emerging field of intelligent based thin-walled structures.

The book is based on the various studies and investigations performed during the years by the author on metal , composite structures, piezoelectric actuations and sensing and shape memory alloys.

The present book aims at providing a textbook to a graduate or graduate–undergraduate course on thin-walled structures. It also can serve as a reference for engineers and scientists working in industry or academia.

The book starts with an extended introductory on thin-walled structures, characterizing aerospace and aeronautical structures. An introductory to elasticity topics and equations of motion is also presented to enable the understanding of structural analysis of aerospace-aeronautical structures. Chapter 2 is devoted to laminated composite materials, and presents the classical lamination theory, first-order shear deformation theory and advanced higher order theories. The chapter aims at providing mathematical and physical insight of composite materials.

The first two chapters are therefore suitable for an introduction on aluminum and/or composite materials for aerospace/aeronautical structures to the graduate or undergraduate students.

The third chapter displays various issues for the buckling for columns, plates and shells using isotropic and laminated composite materials. A complementary chapter, Chapter 4, presents the vibrational aspects for the structures presented in the previous chapter.

Chapter 5 is devoted to an important topic, the Vibration Correlation Technique (VCT) which is a non-destructive method to predict buckling of *in-situ* thin-walled structures. The chapter includes the theory behind the technique and many experimental studies for which the method was successfully applied. The buckling and vibrational behavior of circular and annular flat plates is presented and discussed in Chapter 6. Another important topic in the analysis of thin-walled structures, the dynamic buckling of structural components due to pulse loading, is described and presented in detail in Chapter 7, highlighting both analytical and experimental aspects of the subject.

The next two chapters of the book, Chapters 8 and 9, are devoted to smart intelligent materials, piezoelectric and shape memory alloys, respectively, and offers the reader basic knowledge to understand and calculate structures with those smart materials.

Chapter 10 includes first steps and basic knowledge of the use and applications of fiber optics, a technology with large applications in thin-walled structure.

The last chapter of the book (Chapter 11) introduces a new topic: nanobeams, presenting the theory behind it and how to calculate its static and dynamic behavior.

I wish that this book, together with other books and the huge numbers of references existing in the literature, would enable readers to become familiar with the treatment of thin-walled structures, and how to investigate and calculate their structural behavior.

I want to thank, my wife, Dorit and my children, Chen, Oz, Shir and Or for their support, understanding, love and devotion. Their continuous support throughout the writing period enabled the publishing of the present book.

Haim Abramovich
Haifa, October 2020

About the Author

Haim Abramovich is a retired professor from the Department of Aerospace, Technion, Israel Institute of Technology, IIT, Haifa, Israel. He has widely published many papers in the field of thin-walled structures, buckling and vibration, damage sensitivity of laminated composite structures, space structures, and smart structures.

Contents

Chapter 1

Thin-Walled Structures

1.1. Introduction

The present chapter aims at introducing the reader to basic thin-walled structures used in aerospace structures. To be able to analyze the aerospace structures, basic elasticity equations will be derived and presented. The following chapters will present additional topics and equations regarding the behavior of thin-walled structures and how to analyze them for design.

1.2. Thin-Walled Structures-Definition and Examples

Prior of presenting basic thin-walled structures, one has to introduce the so-called "aerospace vehicles". Vehicles such as an airplane, helicopter, glider or any autonomous device, capable of flying in the earth atmosphere are *aircrafts*, while those machines capable of flying in the space, like a satellite and space station are named *spacecraft*.

The aerospace structures must provide operational demands and safety within a minimum weight. These structures are made of thin load skins, frames, stiffeners and spars, all made of high strength and stiffness materials to comply with the minimum weight criterion.

To be able to design and calculate those structures, three basic structures, namely beams (or column for buckling and rods for tension), plates and shells have to be analyzed and their behavior under various type of loads be understood (see Fig. 1.1). As can be seen, a beam would be defined as a one-dimensional (1D) structure ($h, b \ll L$), a thin plate is considered as a two-dimensional (2D) structure ($a, b \gg t$), while a cylindrical shell forms a three-dimensional (3D) thin-walled structure.

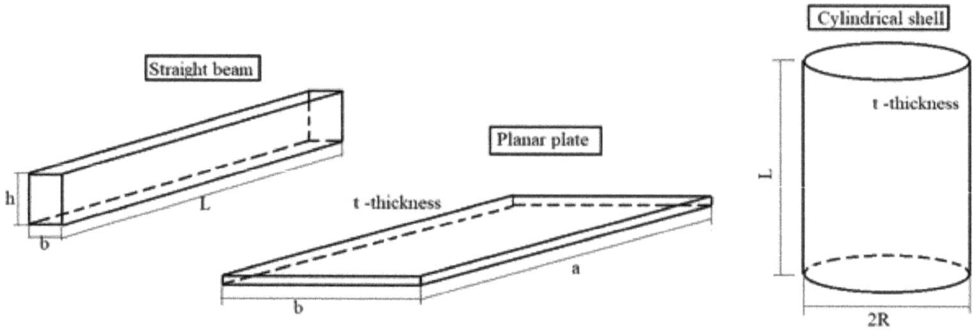

Fig. 1.1. Schematic drawings of the three basic structures used in aerospace structures: a straight beam, a planar plate and a cylindrical shell.

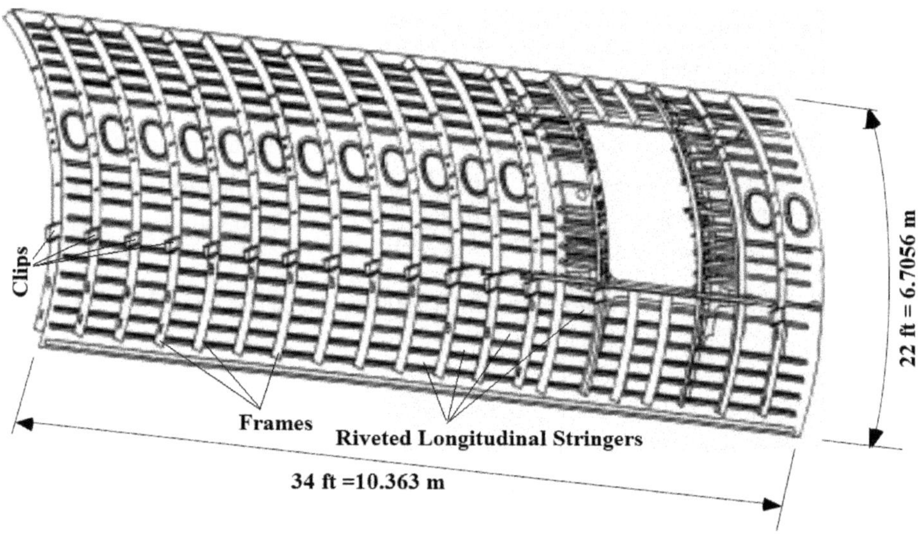

Fig. 1.2. Typical metal skin aircraft fuselage assembly (from NASA CR4730, [3]).

A typical metal skin aircraft fuselage assembly is presented in Fig. 1.2, depicting the frames, their clips, and the longitudinal riveted stringers, which actually are plates and beams. The wings would be made of an assembly of spars giving the aerodynamic shape to the wing, ribs connecting the spars, and all covered by skin panels.

The topic of aerospace structures and how to analyze and calculate their response to static and dynamic loads, had been widely addressed in the literature. Typical references can be found in [1–12].

1.3. Fundamental Issues of Elasticity

1.3.1. *Stresses, strains and rigid-body rotations*

The elasticity theory presents the fundamental expressions enabling the calculation of the structural behavior of basic and complicate components of aerospace structures. The basic topics will be next presented, while more in-depth notions can be found in [13–15].

For a 3D structure, the components of the mechanical stress (defined as force per area, with units N/m^2 or Pa) are presented in Fig. 1.3 for a cubic point in a structural continuum using a Cartesian coordinate system.

The expressions carrying only one index, like $\sigma_x, \sigma_y, \sigma_z$, are defined as normal stresses acting in the directions, x, y and z, respectively. Positive values will mean tension stress, while a negative value will stand for compression stress. The other expressions $\tau_{xz}, \tau_{zx}, \tau_{yz}, \tau_{zy}, \tau_{xy}, \tau_{yx}$ are shear stresses, where the first index denotes the normal of the surface on which the shear stress acts, while the second index denotes the direction of the stress. One should note that due to equilibrium, the following identities hold for the shear stresses

$$\tau_{xz} = \tau_{zx}, \quad \tau_{yz} = \tau_{zy}, \quad \tau_{xy} = \tau_{yx}. \tag{1.1}$$

The stress tensor will have the following form:

$$\begin{bmatrix} \sigma_{xx} & \tau_{xy} & \tau_{xz} \\ \tau_{yx} & \sigma_{yy} & \tau_{yz} \\ \tau_{zx} & \tau_{zy} & \sigma_{zz} \end{bmatrix} = \begin{bmatrix} \sigma_x & \tau_{xy} & \tau_{xz} \\ \tau_{xy} & \sigma_y & \tau_{yz} \\ \tau_{xz} & \tau_{yz} & \sigma_z \end{bmatrix}. \tag{1.2}$$

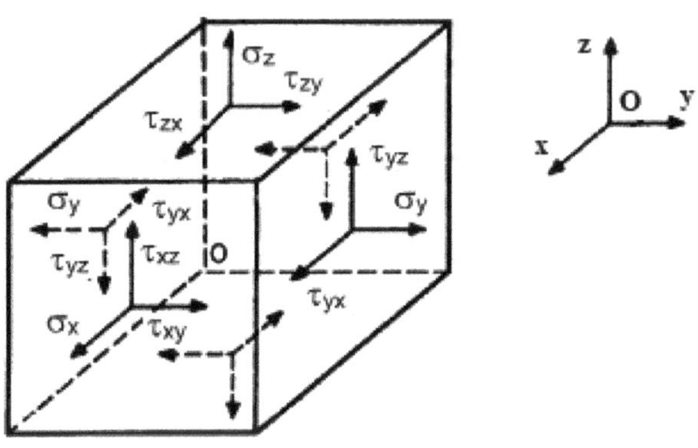

Fig. 1.3. Stresses in a cartesian coordinate system.

Table 1.1. Cylindrical and spherical coordinates.

	Cylindrical coordinates	Spherical coordinates
Axes	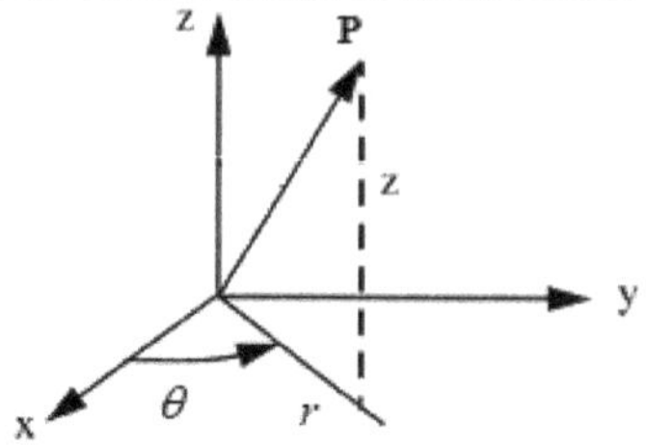	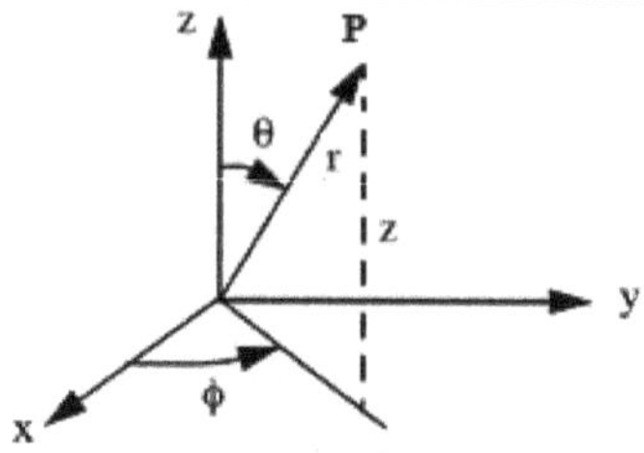
Relations	$x = r\cos\theta$ $y = r\sin\theta$ $\tan\theta = y/x$ $r = \sqrt{x^2+y^2}$	$x = r\sin\theta\cos\phi$ $y = r\sin\theta\sin\phi$ $z = r\cos\theta$ $r = \sqrt{x^2+y^2+z^2}$ $\tan\phi = y/x$ $\tan\theta = (\sqrt{x^2+y^2})/z$
Stresses	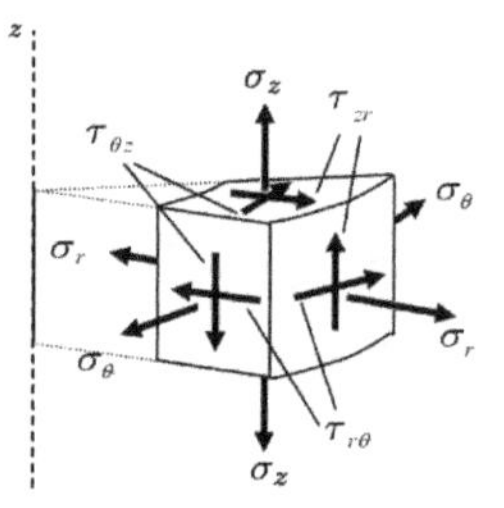	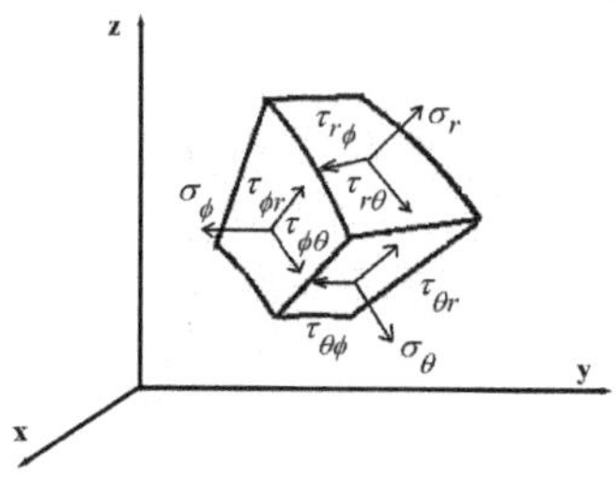
Stress tensor	$\begin{bmatrix} \sigma_r & \tau_{r\theta} & \tau_{rz} \\ \tau_{\theta r} & \sigma_\theta & \tau_{\theta z} \\ \tau_{zr} & \tau_{z\theta} & \sigma_z \end{bmatrix}$	$\begin{bmatrix} \sigma_r & \tau_{r\theta} & \tau_{r\phi} \\ \tau_{\theta r} & \sigma_\theta & \tau_{\theta\phi} \\ \tau_{\phi r} & \tau_{\phi\theta} & \sigma_\phi \end{bmatrix}$
Normal stresses	$\sigma_r, \sigma_\theta, \sigma_z$	$\sigma_r, \sigma_\theta, \sigma_\phi$
Shear stresses	$\tau_{r\theta}, \tau_{rz}, \tau_{\theta r}, \tau_{\theta z}, \tau_{zr}, \tau_{z\theta}$	$\tau_{r\theta}, \tau_{r\phi}, \tau_{\theta r}, \tau_{\theta\phi}, \tau_{\phi r}, \tau_{\phi\theta}$
Identities	$\tau_{r\theta} = \tau_{\theta r}, \tau_{rz} = \tau_{zr}, \tau_{\theta z} = \tau_{z\theta}$	$\tau_{r\theta} = \tau_{\theta r}, \tau_{r\phi} = \tau_{\phi r}, \tau_{\theta\phi} = \tau_{\phi\theta}$

Table 1.1 presents two other systems of coordinates: cylindrical and spherical ones.

It contains details of the axes, the relations between the cartesian and cylindrical and/or spherical coordinates, the stress tensor and its properties.

Another system of coordinates, the polar one, has only two coordinates (r, θ) and therefore is a 2D planar system, with only three stresses, $\sigma_r, \sigma_\theta, \tau_{r\theta} = \tau_{\theta r}$.

The associated strains and rotations for the various assumed displacements are presented in Table 1.2 for the three systems of coordinates: cartesian, cylindrical and spherical.

Table 1.2. Strains and rotations in cartesian, cylindrical and spherical coordinates.

	Cartesian coordinates	Cylindrical coordinates	Spherical coordinates
Displacements	u, v, w in x, y, z directions	u_r, u_θ, u_z in $r,\ \theta,\ z$ directions	u_r, u_θ, u_ϕ in $r,\ \theta,\ \phi$ directions
Normal strains	$\varepsilon_x = \frac{\partial u}{\partial x}$ $\varepsilon_y = \frac{\partial v}{\partial y}$ $\varepsilon_z = \frac{\partial w}{\partial z}$	$\varepsilon_r = \frac{\partial u_r}{\partial r}$ $\varepsilon_\theta = \frac{u_r}{r} + \frac{1}{r}\frac{\partial u_\theta}{\partial \theta}$ $\varepsilon_z = \frac{\partial u_z}{\partial z}$	$\varepsilon_r = \frac{\partial u_r}{\partial r}$ $\varepsilon_\theta = \frac{u_r}{r} + \frac{1}{r}\frac{\partial u_\theta}{\partial \theta}$ $\varepsilon_\phi = \frac{1}{r\sin\theta}\frac{\partial u_\phi}{\partial \phi} + \frac{u_r}{r} + \frac{u_\theta \cot\theta}{r}$
Shear strains	$\gamma_{xy} = \frac{\partial u}{\partial y} + \frac{\partial v}{\partial x}$ $\gamma_{yz} = \frac{\partial v}{\partial z} + \frac{\partial w}{\partial y}$ $\gamma_{xz} = \frac{\partial u}{\partial z} + \frac{\partial w}{\partial x}$	$\gamma_{r\theta} = \frac{\partial u_\theta}{\partial r} + \frac{1}{r}\frac{\partial u_r}{\partial \theta} - \frac{u_\theta}{r}$ $\gamma_{\theta z} = \frac{1}{r}\frac{\partial u_z}{\partial \theta} + \frac{\partial u_\theta}{\partial z}$ $\gamma_{zr} = \frac{\partial u_r}{\partial z} + \frac{\partial u_z}{\partial r}$	$\gamma_{r\theta} = r\frac{\partial}{\partial r}\left(\frac{u_\theta}{r}\right) + \frac{1}{r}\frac{\partial u_r}{\partial \theta}$ $= \frac{\partial u_\theta}{\partial r} - \frac{u_\theta}{r} + \frac{1}{r}\frac{\partial u_r}{\partial \theta}$ $\gamma_{\theta\phi} = \gamma_{\phi\theta} = \frac{\sin\theta}{r}\frac{\partial}{\partial \theta}\left(\frac{u_\phi}{\sin\theta}\right) + \frac{1}{r\sin\theta}\frac{\partial u_\theta}{\partial \phi}$ $= \frac{1}{r}\frac{\partial u_\phi}{\partial \theta} - \frac{u_\phi}{r}\cot\theta + \frac{1}{r\sin\theta}\frac{\partial u_\theta}{\partial \phi}$ $\gamma_{r\phi} = \frac{1}{r\sin\theta}\frac{\partial u_r}{\partial \phi} + r\frac{\partial}{\partial r}\left(\frac{u_\phi}{r}\right)$ $= \frac{1}{r\sin\theta}\frac{\partial u_r}{\partial \phi} + \frac{\partial u_\phi}{\partial r} - \frac{u_\phi}{r}$
Identities	$\gamma_{xy} = \gamma_{yx}$ $\gamma_{yz} = \gamma_{zy}$ $\gamma_{xz} = \gamma_{zx}$	$\gamma_{r\theta} = \gamma_{\theta r}$ $\gamma_{\theta z} = \gamma_{z\theta}$ $\gamma_{zr} = \gamma_{rz}$	$\gamma_{r\theta} = \gamma_{\theta r}$ $\gamma_{\theta\phi} = \gamma_{\phi\theta}$ $\gamma_{r\phi} = \gamma_{\phi r}$
Strain tensor	$\begin{bmatrix} \varepsilon_x & \varepsilon_{xy} & \varepsilon_{xz} \\ \varepsilon_{xy} & \varepsilon_y & \varepsilon_{yz} \\ \varepsilon_{xz} & \varepsilon_{yz} & \varepsilon_z \end{bmatrix} = \begin{bmatrix} \varepsilon_x & 0.5\gamma_{xy} & 0.5\gamma_{xz} \\ 0.5\gamma_{xy} & \varepsilon_y & 0.5\gamma_{yz} \\ 0.5\gamma_{xz} & 0.5\gamma_{yz} & \varepsilon_z \end{bmatrix}$	$\begin{bmatrix} \varepsilon_r & \varepsilon_{r\theta} & \varepsilon_{rz} \\ \varepsilon_{r\theta} & \varepsilon_\theta & \varepsilon_{\theta z} \\ \varepsilon_{rz} & \varepsilon_{\theta z} & \varepsilon_z \end{bmatrix} = \begin{bmatrix} \varepsilon_r & 0.5\gamma_{r\theta} & 0.5\gamma_{rz} \\ 0.5\gamma_{r\theta} & \varepsilon_\theta & 0.5\gamma_{\theta z} \\ 0.5\gamma_{rz} & 0.5\gamma_{\theta z} & \varepsilon_z \end{bmatrix}$	$\begin{bmatrix} \varepsilon_r & \varepsilon_{r\theta} & \varepsilon_{r\phi} \\ \varepsilon_{r\theta} & \varepsilon_\theta & \varepsilon_{\theta\phi} \\ \varepsilon_{r\phi} & \varepsilon_{\theta\phi} & \varepsilon_\phi \end{bmatrix} = \begin{bmatrix} \varepsilon_r & 0.5\gamma_{r\theta} & 0.5\gamma_{r\phi} \\ 0.5\gamma_{r\theta} & \varepsilon_y & 0.5\gamma_{\theta\phi} \\ 0.5\gamma_{r\phi} & 0.5\gamma_{\theta\phi} & \varepsilon_z \end{bmatrix}$

(Continued)

Table 1.2. (*Continued*)

	Cartesian coordinates	Cylindrical coordinates	Spherical coordinates
Rotations	$\omega_x = \frac{1}{2}\left(\frac{\partial w}{\partial y} - \frac{\partial v}{\partial z}\right)$ $\omega_y = \frac{1}{2}\left(\frac{\partial u}{\partial z} - \frac{\partial w}{\partial x}\right)$ $\omega_z = \frac{1}{2}\left(\frac{\partial v}{\partial x} - \frac{\partial u}{\partial y}\right)$ and $\omega_{zy} = -\omega_{yz} \equiv \omega_x$ $\omega_{xz} = -\omega_{zx} \equiv \omega_y$ $\omega_{yx} = -\omega_{xy} \equiv \omega_z$	$\omega_r = \frac{1}{2}\left(\frac{1}{r}\frac{\partial u_z}{\partial \theta} - \frac{\partial u_\theta}{\partial z}\right)$ $\omega_\theta = \frac{1}{2}\left(\frac{\partial u_r}{\partial z} - \frac{\partial u_z}{\partial r}\right)$ $\omega_z = \frac{1}{2r}\left[\frac{\partial}{\partial r}(r \cdot u_\theta) - \frac{\partial u_r}{\partial \theta}\right]$ and $\omega_{z\theta} = -\omega_{\theta z} \equiv \omega_r$ $\omega_{rz} = -\omega_{zr} \equiv \omega_\theta$ $\omega_{\theta r} = -\omega_{r\theta} \equiv \omega_z$	$\omega_r = \frac{1}{2r^2 \sin\theta}\left[\frac{\partial}{\partial \theta}(r \cdot u_\phi \cdot \sin\theta) - \frac{\partial}{\partial \phi}(r \cdot u_\theta)\right]$ $\omega_\theta = \frac{1}{2r \sin\theta}\left[\frac{\partial u_r}{\partial \phi} - \frac{\partial}{\partial r}(r \cdot u_\phi \cdot \sin\theta)\right]$ $\omega_\phi = \frac{1}{2r}\left[\frac{\partial}{\partial r}(r \cdot u_\theta) - \frac{\partial u_r}{\partial \theta}\right]$ and $\omega_{\phi\theta} = -\omega_{\theta\phi} \equiv \omega_r$ $\omega_{r\phi} = -\omega_{\phi r} \equiv \omega_\theta$ $\omega_{\theta r} = -\omega_{r\theta} \equiv \omega_\phi$

Table 1.3. Stress-strain expressions for an isotropic material.*

Cartesian coordinates (strains as a function of stresses)	$\varepsilon_x = \frac{1}{E}[\sigma_x - \nu(\sigma_y + \sigma_z)]$ $\varepsilon_y = \frac{1}{E}[\sigma_y - \nu(\sigma_x + \sigma_z)]$ $\varepsilon_z = \frac{1}{E}[\sigma_z - \nu(\sigma_x + \sigma_y)]$ $\gamma_{xy} = \frac{\tau_{xy}}{G}, \gamma_{yz} = \frac{\tau_{yz}}{G}, \gamma_{xz} = \frac{\tau_{xz}}{G}$
Cartesian coordinates (stresses as a function of strains)	$\sigma_x = \frac{E}{(\nu+1)(1-2\nu)}[(1-\nu)\varepsilon_x + \nu(\varepsilon_y + \varepsilon_z)]$ $\sigma_y = \frac{E}{(\nu+1)(1-2\nu)}[(1-\nu)\varepsilon_y + \nu(\varepsilon_x + \varepsilon_z)]$ $\sigma_z = \frac{E}{(\nu+1)(1-2\nu)}[(1-\nu)\varepsilon_z + \nu(\varepsilon_x + \varepsilon_y)]$ $\tau_{xy} = G\gamma_{xy}, \tau_{yz} = G\gamma_{yz}, \tau_{xz} = G\gamma_{xz}$
Cylindrical coordinates (strains as a function of stresses)	$\varepsilon_r = \frac{1}{E}[\sigma_r - \nu(\sigma_\theta + \sigma_z)]$ $\varepsilon_\theta = \frac{1}{E}[\sigma_\theta - \nu(\sigma_r + \sigma_z)]$ $\varepsilon_z = \frac{1}{E}[\sigma_z - \nu(\sigma_r + \sigma_\theta)]$ $\gamma_{r\theta} = \frac{\tau_{r\theta}}{G}, \gamma_{\theta z} = \frac{\tau_{\theta z}}{G}, \gamma_{rz} = \frac{\tau_{rz}}{G}$
Cylindrical coordinates (stresses as a function of strains)	$\sigma_r = \frac{E}{(\nu+1)(1-2\nu)}[(1-\nu)\varepsilon_r + \nu(\varepsilon_\theta + \varepsilon_z)]$ $\sigma_\theta = \frac{E}{(\nu+1)(1-2\nu)}[(1-\nu)\varepsilon_\theta + \nu(\varepsilon_r + \varepsilon_z)]$ $\sigma_z = \frac{E}{(\nu+1)(1-2\nu)}[(1-\nu)\varepsilon_z + \nu(\varepsilon_r + \varepsilon_\theta)]$ $\tau_{r\theta} = G\gamma_{r\theta}, \tau_{\theta z} = G\gamma_{\theta z}, \tau_{rz} = G\gamma_{rz}$
Spherical coordinates (strains as a function of stresses)	$\varepsilon_r = \frac{1}{E}[\sigma_r - \nu(\sigma_\theta + \sigma_\phi)]$ $\varepsilon_\theta = \frac{1}{E}[\sigma_\theta - \nu(\sigma_r + \sigma_\phi)]$ $\varepsilon_\phi = \frac{1}{E}[\sigma_\phi - \nu(\sigma_r + \sigma_\theta)]$ $\gamma_{r\theta} = \frac{\tau_{r\theta}}{G}, \gamma_{\theta\phi} = \frac{\tau_{\theta\phi}}{G}, \gamma_{r\phi} = \frac{\tau_{r\phi}}{G}$
Spherical coordinates (stresses as a function of strains)	$\sigma_r = \frac{E}{(\nu+1)(1-2\nu)}[(1-\nu)\varepsilon_r + \nu(\varepsilon_\theta + \varepsilon_\phi)]$ $\sigma_\theta = \frac{E}{(\nu+1)(1-2\nu)}[(1-\nu)\varepsilon_\theta + \nu(\varepsilon_r + \varepsilon_\phi)]$ $\sigma_\phi = \frac{E}{(\nu+1)(1-2\nu)}[(1-\nu)\varepsilon_\phi + \nu(\varepsilon_r + \varepsilon_\theta)]$ $\tau_{r\theta} = G\gamma_{r\theta}, \tau_{\theta\phi} = G\gamma_{\theta\phi}, \tau_{r\phi} = G\gamma_{r\phi}$

Notes: $^*E$, $G = \frac{E}{2(1+\nu)}$, ν are the Young's modulus, the shear modulus and the Poisson's ratio, respectively.

Once the stresses and strains had been defined for the three types of coordinates, one can apply the Hook's law for isotropic materials yielding the relations between the two. This is presented in Table 1.3 for the three systems of coordinates.

1.3.2. *Equilibrium and compatibility equations in elasticity*

Table 1.4 presents the equilibrium equations in the three systems of coordinates, expressed by stresses.

Table 1.4. Equilibrium equations for an isotropic material.

Cartesian coordinates*	$\frac{\partial \sigma_x}{\partial x} + \frac{\partial \tau_{xy}}{\partial y} + \frac{\partial \tau_{xz}}{\partial z} + \bar{X} = 0$ $\frac{\partial \tau_{xy}}{\partial x} + \frac{\partial \sigma_y}{\partial y} + \frac{\partial \tau_{yz}}{\partial z} + \bar{Y} = 0$ $\frac{\partial \tau_{xz}}{\partial x} + \frac{\partial \tau_{yz}}{\partial y} + \frac{\partial \sigma_z}{\partial z} + \bar{Z} = 0$
Cylindrical coordinates**	$\frac{\partial \sigma_r}{\partial r} + \frac{1}{r}\frac{\partial \tau_{r\theta}}{\partial \theta} + \frac{\partial \tau_{rz}}{\partial z} + \frac{\sigma_r - \sigma_\theta}{r} + \bar{R} = 0$ $\frac{\partial \tau_{r\theta}}{\partial r} + \frac{1}{r}\frac{\partial \sigma_\theta}{\partial \theta} + \frac{\partial \tau_{\theta z}}{\partial z} + \frac{2\tau_{r\theta}}{r} + \bar{\Theta} = 0$ $\frac{\partial \tau_{rz}}{\partial r} + \frac{1}{r}\frac{\partial \tau_{\theta z}}{\partial \theta} + \frac{\partial \sigma_z}{\partial z} + \frac{\tau_{rz}}{r} + \bar{Z} = 0$
Spherical coordinates***	$\frac{\partial \sigma_r}{\partial r} + \frac{1}{r}\frac{\partial \tau_{r\theta}}{\partial \theta} + \frac{1}{r \cdot \sin\theta}\frac{\partial \tau_{r\phi}}{\partial \phi} + \frac{1}{r}(2\sigma_r - \sigma_\theta - \sigma_\phi + \tau_{r\theta}\cot\theta) + \bar{R} = 0$ $\frac{\partial \tau_{r\theta}}{\partial r} + \frac{1}{r}\frac{\partial \sigma_\theta}{\partial \theta} + \frac{1}{r \cdot \sin\theta}\frac{\partial \tau_{\theta\phi}}{\partial \phi} + \frac{1}{r}[(\sigma_\theta - \sigma_\phi)\cot\theta + 3\tau_{r\theta}] + \bar{\Theta} = 0$ $\frac{\partial \tau_{r\phi}}{\partial r} + \frac{1}{r}\frac{\partial \tau_{\theta\phi}}{\partial \theta} + \frac{1}{r \cdot \sin\theta}\frac{\partial \sigma_\phi}{\partial \phi} + \frac{1}{r}(3\tau_{r\phi} + 2\tau_{\theta\phi}) + \bar{\Phi} = 0$

Notes: * $\bar{X}$, $\bar{Y}$, $\bar{Z}$ are body forces in the x, y, z directions, respectively.
** $\bar{R}$, $\bar{\Theta}$, $\bar{Z}$ are body forces in the r, θ, z directions, respectively.
*** $\bar{R}$, $\bar{\Theta}$, $\bar{\Phi}$ are body forces in the r, θ, ϕ directions, respectively.

Another way of presenting the equilibrium equations is the so-called Navier equations, which express the equilibrium equations using the assumed displacements, according to the chosen system of coordinates (see Table 1.5).

Note that the Lame's constants, appearing in Table 1.5, are defined as

$$G \equiv \frac{E}{2(1+\nu)} \qquad \lambda = \frac{\nu E}{(1+\nu)(1-2\nu)}, \tag{1.3}$$

where ν and E are the Poisson's ratio and the Young's modulus, respectively.

It is instructive to know the form of the Laplace operator, ∇^2, appearing in Table 1.5 for a cartesian coordinate system, for other system of coordinates. For a cylindrical system of coordinates, the Laplace operator would be

$$\nabla^2 \equiv \frac{\partial^2}{\partial r^2} + \frac{1}{r}\frac{\partial}{\partial r} + \frac{1}{r^2}\frac{\partial^2}{\partial \theta^2} + \frac{\partial^2}{\partial z^2}, \tag{1.4}$$

while for a spherical system of coordinates, the expression would be

$$\nabla^2 \equiv \frac{\partial^2}{\partial r^2} + \frac{2}{r}\frac{\partial}{\partial r} + \left(\frac{1}{r\sin(\phi)}\right)^2 \frac{\partial^2}{\partial \theta^2} + \frac{1}{r^2}\frac{\partial^2}{\partial \phi^2} + \frac{1}{r^2\tan(\phi)}\frac{\partial}{\partial \phi}. \tag{1.5}$$

Table 1.6 depicts the required compatibility equations to be fulfilled to ensure a unique solution for various systems of coordinates. These equations will answer the question how to get the displacements if the expressions for the strains are known and are presented in a compact form in Eq. (1.4):

$$\varepsilon_{ij,kl} + \varepsilon_{kl,ij} - \varepsilon_{ik,jl} - \varepsilon_{jl,ik} = 0, \quad \text{where } i,j,k,l = x,y,z. \tag{1.6}$$

Table 1.5. Equilibrium equations expressed using displacements (Navier equations) for an isotropic material.

Cartesian coordinates*	$(\lambda+G)\frac{\partial\varepsilon}{\partial x}+G\nabla^2 u+F_x=0$ where $\nabla^2\equiv\frac{\partial^2}{\partial x^2}+\frac{\partial^2}{\partial y^2}+\frac{\partial^2}{\partial z^2}$ $(\lambda+G)\frac{\partial\varepsilon}{\partial y}+G\nabla^2 v+F_y=0$ $\quad\varepsilon\equiv\varepsilon_x+\varepsilon_y+\varepsilon_z=\frac{\partial u}{\partial x}+\frac{\partial v}{\partial y}+\frac{\partial w}{\partial z}$ $(\lambda+G)\frac{\partial\varepsilon}{\partial z}+G\nabla^2 w+F_z=0$
Cylindrical coordinates**	$(\lambda+2G)\frac{\partial \mathrm{H}_\varepsilon}{\partial r}-2G(\frac{1}{r}\frac{\partial\rho_z}{\partial\theta}-\frac{\partial\rho_\theta}{\partial z})+F_r=0$ $(\lambda+2G)\frac{\partial \mathrm{H}_\varepsilon}{r\partial\theta}-2G(\frac{\partial\rho_r}{\partial z}-\frac{\partial\rho_z}{\partial r})+F_\theta=0$ $(\lambda+2G)\frac{\partial \mathrm{H}_\varepsilon}{\partial z}-2G(\frac{\partial(r\rho_\theta)}{\partial r}-\frac{\partial\rho_r}{\partial\theta})+F_z=0$ where $\mathrm{H}_\varepsilon\equiv\frac{1}{r}\frac{(ru_r)}{\partial r}+\frac{1}{r}\frac{\partial u_\theta}{\partial\theta}+\frac{\partial u_z}{\partial z}$ $\rho_r\equiv\frac{1}{2}[\frac{1}{r}\frac{\partial u_z}{\partial\theta}-\frac{\partial u_\theta}{\partial z}]$ $\rho_\theta\equiv\frac{1}{2}[\frac{\partial u_r}{\partial z}-\frac{\partial u_z}{\partial r}]$ $\rho_z\equiv\frac{1}{2}[\frac{\partial(ru_\theta)}{\partial r}-\frac{\partial u_r}{\partial\theta}]$
Spherical coordinates***	$(\lambda+2G)\frac{\partial\overline{\mathrm{H}}_\varepsilon}{\partial r}-\frac{2G}{r\sin\theta}[\frac{\partial(\overline{\rho}_\phi\sin\theta)}{\partial\theta}-\frac{\partial\overline{\rho}_\theta}{\partial\phi}]+\overline{F}_r=0$ $(\lambda+2G)\frac{\partial\overline{\mathrm{H}}_\varepsilon}{r\partial\theta}-\frac{2G}{r}[\frac{1}{\sin\theta}\frac{\partial(\overline{\rho}_r)}{\partial\phi}-\frac{\partial(r\overline{\rho}_\phi)}{\partial r}]+\overline{F}_\theta=0$ $\frac{(\lambda+2G)}{r\sin\theta}\frac{\partial\overline{\mathrm{H}}_\varepsilon}{\partial\phi}-\frac{2G}{r}[\frac{\partial(r\overline{\rho}_\theta)}{\partial r}-\frac{\partial(r\overline{\rho}_r)}{\partial\theta}]+\overline{F}_\phi=0$ where $\overline{\mathrm{H}}_\varepsilon\equiv\frac{1}{r^2}\frac{\partial(r^2u_r)}{\partial r}+\frac{1}{r\sin\theta}\frac{\partial(u_\theta\sin\theta)}{\partial\theta}+\frac{1}{r\sin\theta}\frac{\partial u_\phi}{\partial\phi}$ $\overline{\rho}_r\equiv\frac{1}{2r\sin\theta}[\frac{\partial(u_\phi\sin\theta)}{\partial\theta}-\frac{\partial u_\theta}{\partial\phi}]$ $\overline{\rho}_\theta\equiv\frac{1}{2}[\frac{1}{r\sin\theta}\frac{\partial u_r}{\partial\phi}-\frac{1}{r}\frac{\partial(ru_\phi)}{\partial r}]$ $\overline{\rho}_z\equiv\frac{1}{2r}[\frac{\partial(ru_\theta)}{\partial r}-\frac{\partial u_r}{\partial\theta}]$

Notes: $^*F_x\equiv\bar{X}$; $F_y\equiv\bar{Y}$; $F_z\equiv\bar{Z}$, $^{**}F_r\equiv\overline{R}$; $F_\theta\equiv\overline{\Theta}$; $F_z\equiv\bar{Z}$, $^{***}\overline{F}_r\equiv\overline{R}$; $\overline{F}_\theta\equiv\overline{\Theta}$; $\overline{F}_z\equiv\overline{\Phi}$.

The literature presents the compatibility equations in terms of stresses, naming them the Beltrami–Michell compatibility equations. Their compact form[a] is given as follows:

$$\sigma_{ij,kk}+\frac{1}{1+\nu}\sigma_{kk,uj}=-\frac{\nu}{1-\nu}F_{k,k}\delta_{ij}-F_{i,j}-F_{j,i}. \tag{1.7}$$

Detailed expressions for those equations are presented in Table 1.7 for the three systems of coordinates.

[a] δ_{ij} is the Kronecker delta defined as:

$$\delta_{ij}=\begin{cases}0 & \text{if} \quad i\neq j,\\ 1 & \text{if} \quad i=1.\end{cases}$$

Table 1.6. Compatibility equations for an isotropic material.*

Cartesian coordinates	$\frac{\partial^2 \varepsilon_x}{\partial y^2} + \frac{\partial^2 \varepsilon_y}{\partial x^2} = 2\frac{\partial^2 \varepsilon_{xy}}{\partial x \partial y}$ $\frac{\partial^2 \varepsilon_x}{\partial z^2} + \frac{\partial^2 \varepsilon_z}{\partial x^2} = 2\frac{\partial^2 \varepsilon_{xz}}{\partial x \partial z}$ $\frac{\partial^2 \varepsilon_y}{\partial z^2} + \frac{\partial^2 \varepsilon_z}{\partial y^2} = 2\frac{\partial^2 \varepsilon_{yz}}{\partial y \partial z}$ $\frac{\partial^2 \varepsilon_x}{\partial y \partial z} + \frac{\partial^2 \varepsilon_{yz}}{\partial x^2} = \frac{\partial^2 \varepsilon_{xz}}{\partial x \partial y} + \frac{\partial^2 \varepsilon_{xy}}{\partial x \partial z}$ $\frac{\partial^2 \varepsilon_y}{\partial x \partial z} + \frac{\partial^2 \varepsilon_{xz}}{\partial y^2} = \frac{\partial^2 \varepsilon_{yz}}{\partial x \partial y} + \frac{\partial^2 \varepsilon_{xy}}{\partial y \partial z}$ $\frac{\partial^2 \varepsilon_z}{\partial x \partial y} + \frac{\partial^2 \varepsilon_{xy}}{\partial z^2} = \frac{\partial^2 \varepsilon_{xz}}{\partial y \partial z} + \frac{\partial^2 \varepsilon_{yz}}{\partial x \partial z}$ where $\varepsilon_{xy} \equiv 0.5\gamma_{xy},\ \varepsilon_{xz} \equiv 0.5\gamma_{xz},\ \varepsilon_{yz} \equiv 0.5\gamma_{yz}$ $\varepsilon_{yx} \equiv 0.5\gamma_{yx},\ \varepsilon_{zx} \equiv 0.5\gamma_{zx},\ \varepsilon_{zy} \equiv 0.5\gamma_{zy}$
Cylindrical coordinates	$\frac{1}{r}\frac{\partial^2 \varepsilon_r}{\partial \theta^2} + \frac{\partial}{\partial r}\{r\frac{\partial \varepsilon_\theta}{\partial r} - (\varepsilon_r - \varepsilon_\theta)\} = \frac{\partial}{\partial \theta}\{\frac{\partial \gamma_{r\theta}}{\partial r} + \frac{\gamma_{r\theta}}{r}\}$ $\frac{1}{r^2}\frac{\partial^2 \varepsilon_z}{\partial \theta^2} + \frac{\partial^2 \varepsilon_\theta}{\partial z^2} + \frac{1}{r}\frac{\partial \varepsilon_z}{\partial r} = \frac{1}{r}\frac{\partial}{\partial z}\{\frac{\partial \gamma_{\theta z}}{\partial \theta} + \gamma_{rz}\}$ $\frac{\partial^2 \varepsilon_z}{\partial r^2} + \frac{\partial^2 \varepsilon_r}{\partial z^2} = \frac{\partial^2 \gamma_{rz}}{\partial r \partial z}$ $\frac{2}{r}\frac{\partial^2 \varepsilon_r}{\partial \theta \partial z} = \frac{1}{r}\frac{\partial}{\partial r}\{r\frac{\partial \gamma_{r\theta}}{\partial z} - \gamma_{\theta z}\} + \frac{\partial}{\partial r}\{\frac{1}{r}\frac{\partial \gamma_{rz}}{\partial \theta} - \frac{\partial \gamma_{\theta z}}{\partial r}\} + \frac{1}{r}\frac{\partial \gamma_{r\theta}}{\partial z} + \frac{\gamma_{\theta z}}{r^2}$ $2\frac{\partial}{\partial z}\{\frac{\partial \varepsilon_\theta}{\partial r} - (\frac{\varepsilon_r - \varepsilon_\theta}{r})\} = \frac{1}{r}\frac{\partial}{\partial \theta}\{\frac{\partial \gamma_{\theta z}}{\partial r} - \frac{1}{r}\frac{\partial \gamma_{rz}}{\partial \theta} + \frac{\partial \gamma_{r\theta}}{\partial z}\} + \frac{1}{r^2}\frac{\partial \gamma_{\theta z}}{\partial \theta}$ $\frac{2}{r}\frac{\partial}{\partial \theta}\{\frac{\partial \varepsilon_z}{\partial r} - \frac{\varepsilon_z}{r}\} = \frac{\partial}{\partial z}\{\frac{\partial \gamma_{\theta z}}{\partial r} + \frac{1}{r}\frac{\partial \gamma_{rz}}{\partial \theta} - \frac{\partial \gamma_{r\theta}}{\partial z} - \frac{\gamma_{\theta z}}{r}\}$ where $\gamma_{r\theta} \equiv 2\varepsilon_{r\theta}, \gamma_{rz} \equiv 2\varepsilon_{rz}, \gamma_{\theta z} \equiv 2\varepsilon_{\theta z}$ $\gamma_{\theta r} \equiv 2\varepsilon_{\theta r}, \gamma_{zr} \equiv 2\varepsilon_{zr}, \gamma_{z\theta} \equiv 2\varepsilon_{z\theta}$
Spherical coordinates	$\frac{\partial^2 \varepsilon_r}{\partial \theta^2} + \frac{\partial^2 \varepsilon_\theta}{\partial r^2} - 2\frac{\partial^2 \varepsilon_{r\theta}}{\partial r \partial \theta} = 0$ $\frac{\partial^2 \varepsilon_\theta}{\partial \phi^2} + \frac{\partial^2 \varepsilon_\phi}{\partial \theta^2} - 2\frac{\partial^2 \varepsilon_{\theta\phi}}{\partial \theta \partial \phi} = 0$ $\frac{\partial^2 \varepsilon_\phi}{\partial r^2} + \frac{\partial^2 \varepsilon_r}{\partial \phi^2} - 2\frac{\partial^2 \varepsilon_{\phi r}}{\partial \phi \partial r} = 0$ $\frac{\partial^2 \varepsilon_r}{\partial \theta \partial \phi} + \frac{\partial^2 \varepsilon_{\theta\phi}}{\partial r^2} - \frac{\partial^2 \varepsilon_{r\theta}}{\partial r \partial \phi} - \frac{\partial^2 \varepsilon_{r\phi}}{\partial r \partial \theta} = 0$ $\frac{\partial^2 \varepsilon_\theta}{\partial \phi \partial r} + \frac{\partial^2 \varepsilon_{\phi r}}{\partial \theta^2} - \frac{\partial^2 \varepsilon_{\theta\phi}}{\partial \theta \partial r} - \frac{\partial^2 \varepsilon_{\theta r}}{\partial \theta \partial \phi} = 0$ $\frac{\partial^2 \varepsilon_\phi}{\partial r \partial \theta} + \frac{\partial^2 \varepsilon_{r\theta}}{\partial \phi^2} - \frac{\partial^2 \varepsilon_{\phi r}}{\partial \phi \partial \theta} - \frac{\partial^2 \varepsilon_{\phi\theta}}{\partial \phi \partial r} = 0$ where $\varepsilon_{r\theta} \equiv 0.5\gamma_{r\theta},\ \varepsilon_{r\phi} \equiv 0.5\gamma_{r\phi},\ \varepsilon_{\theta\phi} \equiv 0.5\gamma_{\theta\phi}$ $\varepsilon_{\theta r} \equiv 0.5\gamma_{\theta r},\ \varepsilon_{\phi r} \equiv 0.5\gamma_{\phi r},\ \varepsilon_{\phi\theta} \equiv 0.5\gamma_{\phi\theta}$

1.3.3. *Two-dimensional representations: Plane stress and plane strain*

Often three-dimensional elasticity problems can be reduced to only two-dimensional representations, thus enabling easier solutions. These cases are named *plane stress* and/or *plane strain* problems.

Table 1.7. Compatibility equations for an isotropic material: the Beltrami–Michell equations.

Cartesian coordinates	$\nabla^2\sigma_x + \frac{1}{1+\nu}\frac{\partial^2\Lambda}{\partial x^2} = -\frac{1}{1-\nu}[\frac{\partial F_x}{\partial x} + \frac{\partial F_y}{\partial y} + \frac{\partial F_z}{\partial z}] - 2\frac{\partial F_x}{\partial x}$ $\nabla^2\sigma_y + \frac{1}{1+\nu}\frac{\partial^2\Lambda}{\partial y^2} = -\frac{1}{1-\nu}[\frac{\partial F_x}{\partial x} + \frac{\partial F_y}{\partial y} + \frac{\partial F_z}{\partial z}] - 2\frac{\partial F_y}{\partial y}$ $\nabla^2\sigma_z + \frac{1}{1+\nu}\frac{\partial^2\Lambda}{\partial z^2} = -\frac{1}{1-\nu}[\frac{\partial F_x}{\partial x} + \frac{\partial F_y}{\partial y} + \frac{\partial F_z}{\partial z}] - 2\frac{\partial F_z}{\partial z}$ $\nabla^2\tau_{yz} + \frac{1}{1+\nu}\frac{\partial^2\Lambda}{\partial y\partial z} = -(\frac{\partial F_z}{\partial y} + \frac{\partial F_y}{\partial z})$ $\nabla^2\tau_{zx} + \frac{1}{1+\nu}\frac{\partial^2\Lambda}{\partial z\partial x} = -(\frac{\partial F_x}{\partial z} + \frac{\partial F_z}{\partial x})$ $\nabla^2\tau_{xy} + \frac{1}{1+\nu}\frac{\partial^2\Lambda}{\partial x\partial y} = -(\frac{\partial F_y}{\partial x} + \frac{\partial F_x}{\partial y})$ where $\Lambda \equiv \sigma_x + \sigma_y + \sigma_z$
Cylindrical coordinates	$\frac{1}{r}\frac{\partial}{\partial r}(r\frac{\partial\sigma_r}{\partial r}) + \frac{1}{r^2}\frac{\partial\sigma_r}{\partial\theta^2} + \frac{\partial^2\sigma_r}{\partial z^2} + \frac{1}{1+\nu}\frac{\partial^2\Omega}{\partial r^2} = -\frac{1}{1-\nu}\frac{1}{r}[\frac{\partial(rF_r)}{\partial r} + \frac{\partial F_\theta}{\partial\theta} + \frac{r\partial F_z}{\partial z}] - 2\frac{\partial F_r}{\partial r}$ $\frac{1}{r}\frac{\partial}{\partial r}(r\frac{\partial\sigma_\theta}{\partial r}) + \frac{1}{r^2}\frac{\partial\sigma_\theta}{\partial\theta^2} + \frac{\partial^2\sigma_\theta}{\partial z^2} + \frac{1}{1+\nu}\frac{\partial^2\Omega}{\partial\theta^2} = -\frac{1}{1-\nu}\frac{1}{r}[\frac{\partial(rF_r)}{\partial r} + \frac{\partial F_\theta}{\partial\theta} + \frac{r\partial F_z}{\partial z}] - 2\frac{\partial F_\theta}{\partial\theta}$ $\frac{1}{r}\frac{\partial}{\partial r}(r\frac{\partial\sigma_z}{\partial r}) + \frac{1}{r^2}\frac{\partial\sigma_z}{\partial\theta^2} + \frac{\partial^2\sigma_z}{\partial z^2} + \frac{1}{1+\nu}\frac{\partial^2\Omega}{\partial z^2} = -\frac{1}{1-\nu}\frac{1}{r}[\frac{\partial(rF_r)}{\partial r} + \frac{\partial F_\theta}{\partial\theta} + \frac{r\partial F_z}{\partial z}] - 2\frac{\partial F_z}{\partial z}$ $\frac{1}{r}\frac{\partial}{\partial r}(r\frac{\partial\sigma_{\theta z}}{\partial r}) + \frac{1}{r^2}\frac{\partial\sigma_{\theta z}}{\partial\theta^2} + \frac{\partial^2\sigma_{\theta z}}{\partial z^2} + \frac{1}{1+\nu}\frac{\partial^2\Omega}{\partial\theta\partial z} = -(\frac{\partial F_\theta}{\partial z} + \frac{\partial F_z}{\partial\theta})$ $\frac{1}{r}\frac{\partial}{\partial r}(r\frac{\partial\sigma_{rz}}{\partial r}) + \frac{1}{r^2}\frac{\partial\sigma_{rz}}{\partial\theta^2} + \frac{\partial^2\sigma_{rz}}{\partial z^2} + \frac{1}{1+\nu}\frac{\partial^2\Omega}{\partial r\partial z} = -(\frac{\partial F_z}{\partial r} + \frac{\partial F_r}{\partial z})$ $\frac{1}{r}\frac{\partial}{\partial r}(r\frac{\partial\sigma_{r\theta}}{\partial r}) + \frac{1}{r^2}\frac{\partial\sigma_{r\theta}}{\partial\theta^2} + \frac{\partial^2\sigma_{r\theta}}{\partial z^2} + \frac{1}{1+\nu}\frac{\partial^2\Omega}{\partial r\partial\theta} = -(\frac{\partial F_r}{\partial\theta} + \frac{\partial F_\theta}{\partial r})$ where $\Omega \equiv \sigma_r + \sigma_\theta + \sigma_z$
Spherical coordinates	$\frac{1}{r^2}\frac{\partial}{\partial r}\left(r^2\frac{\partial\sigma_r}{\partial r}\right) + \frac{1}{r^2\sin\theta}\frac{\partial}{\partial\theta}\left(\sin\theta\frac{\partial\sigma_r}{\partial\theta}\right) + \frac{1}{r^2\sin^2\theta}\frac{\partial^2\sigma_r}{\partial\phi^2} + \frac{1}{1+\nu}\frac{\partial^2\Psi}{\partial r^2}$ $= -\frac{1}{1-\nu}\frac{1}{r}\left[\frac{1}{r}\frac{\partial(r^2F_r)}{\partial r} + \frac{1}{\sin\theta}\frac{\partial}{\partial\theta}(\sin\theta\cdot F_\theta) + \frac{1}{\sin\theta}\frac{\partial F_\phi}{\partial\phi}\right] - 2\frac{\partial F_r}{\partial r}$ $\frac{1}{r^2}\frac{\partial}{\partial r}\left(r^2\frac{\partial\sigma_\theta}{\partial r}\right) + \frac{1}{r^2\sin\theta}\frac{\partial}{\partial\theta}\left(\sin\theta\frac{\partial\sigma_\theta}{\partial\theta}\right) + \frac{1}{r^2\sin^2\theta}\frac{\partial^2\sigma_\theta}{\partial\phi^2} + \frac{1}{1+\nu}\frac{\partial^2\Psi}{\partial\theta^2}$ $= -\frac{1}{1-\nu}\frac{1}{r}\left[\frac{1}{r}\frac{\partial(r^2F_r)}{\partial r} + \frac{1}{\sin\theta}\frac{\partial}{\partial\theta}(\sin\theta\cdot F_\theta) + \frac{1}{\sin\theta}\frac{\partial F_\phi}{\partial\phi}\right] - 2\frac{\partial F_\theta}{\partial\theta}$ $\frac{1}{r^2}\frac{\partial}{\partial r}\left(r^2\frac{\partial\sigma_\phi}{\partial r}\right) + \frac{1}{r^2\sin\theta}\frac{\partial}{\partial\theta}\left(\sin\theta\frac{\partial\sigma_\phi}{\partial\theta}\right) + \frac{1}{r^2\sin^2\theta}\frac{\partial^2\sigma_\phi}{\partial\phi^2} + \frac{1}{1+\nu}\frac{\partial^2\Psi}{\partial\phi^2}$ $= -\frac{1}{1-\nu}\frac{1}{r}\left[\frac{1}{r}\frac{\partial(r^2F_r)}{\partial r} + \frac{1}{\sin\theta}\frac{\partial}{\partial\theta}(\sin\theta\cdot F_\theta) + \frac{1}{\sin\theta}\frac{\partial F_\phi}{\partial\phi}\right] - 2\frac{\partial F_\phi}{\partial\phi}$ $\frac{1}{r^2}\frac{\partial}{\partial r}\left(r^2\frac{\partial\sigma_{\theta\phi}}{\partial r}\right) + \frac{1}{r^2\sin\theta}\frac{\partial}{\partial\theta}\left(\sin\theta\frac{\partial\sigma_{\theta\phi}}{\partial\theta}\right) + \frac{1}{r^2\sin^2\theta}\frac{\partial^2\sigma_{\theta\phi}}{\partial\phi^2} + \frac{1}{1+\nu}\frac{\partial^2\Psi}{\partial\theta\partial\phi}$ $= -\left(\frac{\partial F_\theta}{\partial\phi} + \frac{\partial F_\phi}{\partial\theta}\right)$ $\frac{1}{r^2}\frac{\partial}{\partial r}\left(r^2\frac{\partial\sigma_{r\phi}}{\partial r}\right) + \frac{1}{r^2\sin\theta}\frac{\partial}{\partial\theta}\left(\sin\theta\frac{\partial\sigma_{r\phi}}{\partial\theta}\right) + \frac{1}{r^2\sin^2\theta}\frac{\partial^2\sigma_{r\phi}}{\partial\phi^2} + \frac{1}{1+\nu}\frac{\partial^2\Psi}{\partial\theta\partial\phi}$ $= -\left(\frac{\partial F_r}{\partial\phi} + \frac{\partial F_\phi}{\partial r}\right)$ $\frac{1}{r^2}\frac{\partial}{\partial r}\left(r^2\frac{\partial\sigma_{r\theta}}{\partial r}\right) + \frac{1}{r^2\sin\theta}\frac{\partial}{\partial\theta}\left(\sin\theta\frac{\partial\sigma_{r\theta}}{\partial\theta}\right) + \frac{1}{r^2\sin^2\theta}\frac{\partial^2\sigma_{r\theta}}{\partial\phi^2} + \frac{1}{1+\nu}\frac{\partial^2\Psi}{\partial r\partial\theta}$ $= -\left(\frac{\partial F_r}{\partial\theta} + \frac{\partial F_\theta}{\partial r}\right)$ where $\Psi \equiv \sigma_r + \sigma_\theta + \sigma_\phi$

1.3.3.1. *Plane stress problems*

Three-dimensional cases for which one dimension is much smaller than the other two, like for the case of thin flat plates, can be approximated using the plane stress assumption. For these cases, the plate thickness would be in the z direction, with the loads being applied in the z direction. Then, under these assumptions, all stress components having a subscript z are assumed zero. The remaining stress components are then assumed to be functions of only x and y coordinates. This will transform the 3D equations, presented in Table 1.3 (for a Cartesian coordinate system) into the following simplified 2D form:

$$\begin{aligned}
\varepsilon_x &= \frac{1}{E}[\sigma_x - \nu\sigma_y], \quad \varepsilon_y = \frac{1}{E}[\sigma_y - \nu\sigma_x], \\
\varepsilon_z &= -\frac{1}{E}[\nu(\sigma_x + \sigma_y)], \\
\gamma_{xy} &= \frac{\tau_{xy}}{G}, \ \gamma_{yz} = 0, \ \gamma_{xz} = 0.
\end{aligned} \tag{1.8}$$

Accordingly, when using a cylindrical coordinates system, we will get the following equations, which are also known as a *polar* representation:

$$\begin{aligned}
\varepsilon_r &= \frac{1}{E}[\sigma_r - \nu\sigma_\theta], \quad \varepsilon_\theta = \frac{1}{E}[\sigma_\theta - \nu\sigma_r], \\
\varepsilon_z &= -\frac{1}{E}[\nu(\sigma_r + \sigma_\theta)], \\
\gamma_{r\theta} &= \frac{\tau_{r\theta}}{G}, \ \gamma_{\theta z} = 0, \ \gamma_{rz} = 0.
\end{aligned} \tag{1.9}$$

The inverse equations, namely expressing the stresses as a function of the strains, are shown to be[b] for a cartesian coordinate system as

$$\begin{aligned}
\sigma_x &= \frac{E}{(1-\nu^2)}[\varepsilon_x + \nu\varepsilon_y], \\
\sigma_y &= \frac{E}{(1-\nu^2)}[\varepsilon_y + \nu\varepsilon_x], \\
\sigma_z &= 0 = [(1-\nu)\varepsilon_z + \nu(\varepsilon_x + \varepsilon_y)] \\
\tau_{xy} &= G\gamma_{xy}, \ \tau_{yz} = 0, \ \tau_{xz} = 0,
\end{aligned} \tag{1.10}$$

while for a polar coordinate system

$$\begin{aligned}
\sigma_r &= \frac{E}{(1-\nu^2)}[\varepsilon_r + \nu\varepsilon_\theta], \\
\sigma_\theta &= \frac{E}{(1-\nu^2)}[\varepsilon_\theta + \nu\varepsilon_r], \\
\sigma_z &= 0 = [(1-\nu)\varepsilon_z + \nu(\varepsilon_r + \varepsilon_\theta)], \\
\tau_{r\theta} &= G\gamma_{r\theta}, \ \tau_{\theta z} = 0, \ \tau_{rz} = 0.
\end{aligned} \tag{1.11}$$

[b]One should note that the strain in the z direction is given by, $\varepsilon_z \neq 0$!

The strains compatibility equations, as presented in Table 1.6, will have the following compact form for a plane stress case for using a Cartesian coordinates system

$$\frac{\partial^2 \varepsilon_x}{\partial y^2} + \frac{\partial^2 \varepsilon_y}{\partial x^2} = 2\frac{\partial^2 \varepsilon_{xy}}{\partial x \partial y},$$
$$\frac{\partial^2 \varepsilon_z}{\partial x^2} = 0, \frac{\partial^2 \varepsilon_z}{\partial y^2} = 0, \frac{\partial^2 \varepsilon_z}{\partial x \partial y} = 0. \tag{1.12}$$

One should note that the last three equations in Eq. (1.12) might cause some difficulties for plane stress problems.

For a cylindrical coordinates system, or a polar coordinates system, the strains compatibility equations would be

$$\frac{1}{r}\frac{\partial^2 \varepsilon_r}{\partial \theta^2} + \frac{\partial}{\partial r}\left\{r\frac{\partial \varepsilon_\theta}{\partial r} - (\varepsilon_r - \varepsilon_\theta)\right\} = \frac{\partial}{\partial \theta}\left\{\frac{\partial \gamma_{r\theta}}{\partial r} + \frac{\gamma_{r\theta}}{r}\right\},$$
$$\frac{1}{r}\frac{\partial^2 \varepsilon_z}{\partial \theta^2} + \frac{\partial \varepsilon_z}{\partial r} = 0, \frac{\partial^2 \varepsilon_z}{\partial r^2} = 0, \frac{\partial}{\partial \theta}\left\{\frac{\partial \varepsilon_z}{\partial r} - \frac{\varepsilon_z}{r}\right\} = 0. \tag{1.13}$$

1.3.3.2. *Plane strain problems*

Three-dimensional problems for which one dimension is much larger than the other two, like in the case of a long cylindrical body or water dam type structures, the solution can be approximated using the plane strain assumption. For these cases, the larger dimension would be in the z direction, and for any x–y plane (perpendicular to z) it is assumed that the loads are independent of z. Then, under these assumptions, all strain components having a subscript z are assumed zero. The remaining strain components are then assumed functions of only x and y coordinates[c]. This will transform the 3D equations into the following simplified 2D form:

$$\varepsilon_x = \frac{1}{E}[\sigma_x - \nu(\sigma_y + \sigma_z)], \quad \varepsilon_y = \frac{1}{E}[\sigma_y - \nu(\sigma_z + \sigma_x)],$$
$$\varepsilon_z = 0 = \frac{1}{E}[\sigma_z - \nu(\sigma_x + \sigma_y)], \ \gamma_{xy} = \frac{\tau_{xy}}{G}, \ \gamma_{yz} = 0, \gamma_{xz} = 0. \tag{1.14}$$

Accordingly, for a cylindrical (polar) coordinates system, we will obtain

$$\varepsilon_r = \frac{1}{E}[\sigma_r - \nu(\sigma_\theta + \sigma_z)], \quad \varepsilon_\theta = \frac{1}{E}[\sigma_\theta - \nu(\sigma_z + \sigma_r)],$$
$$\varepsilon_z = 0 = -\frac{1}{E}[\sigma_z - \nu(\sigma_r + \sigma_\theta)], \ \gamma_{r\theta} = \frac{\tau_{r\theta}}{G}, \ \gamma_{\theta z} = 0, \ \gamma_{rz} = 0. \tag{1.15}$$

[c]Note that in this case $\sigma_z \neq 0$!

The inverse equations, namely the stresses as a function of the strains, have the following form for a Cartesian system of coordinates:

$$\sigma_x = \frac{E}{(1+\nu)(1-2\nu)}[(1-\nu)\varepsilon_x + \nu\varepsilon_y],$$
$$\sigma_y = \frac{E}{(1+\nu)(1-2\nu)}[(1-\nu)\varepsilon_y + \nu\varepsilon_x],$$
$$\sigma_z = \frac{E}{(1+\nu)(1-2\nu)}[\nu(\varepsilon_x + \varepsilon_y)],$$
$$\tau_{xy} = G\gamma_{xy}, \quad \tau_{yz} = 0, \quad \tau_{xz} = 0, \tag{1.16}$$

while for a polar system of coordinates the expressions are

$$\sigma_r = \frac{E}{(1+\nu)(1-2\nu)}[(1-\nu)\varepsilon_r + \nu\varepsilon_\theta],$$
$$\sigma_\theta = \frac{E}{(1+\nu)(1-2\nu)}[(1-\nu)\varepsilon_\theta + \nu\varepsilon_r],$$
$$\sigma_z = \frac{E}{(1+\nu)(1-2\nu)}[\nu(\varepsilon_r + \varepsilon_\theta)],$$
$$\tau_{r\theta} = G\gamma_{r\theta}, \quad \tau_{\theta z} = 0, \quad \tau_{rz} = 0. \tag{1.17}$$

The strains compatibility equations will have the following compact form for the plane strain case for a cartesian system:

$$\frac{\partial^2 \varepsilon_x}{\partial y^2} + \frac{\partial^2 \varepsilon_y}{\partial x^2} = 2\frac{\partial^2 \varepsilon_{xy}}{\partial x \partial y},$$
$$\frac{\partial^2 \varepsilon_z}{\partial x^2} = 0, \frac{\partial^2 \varepsilon_z}{\partial y^2} = 0, \frac{\partial^2 \varepsilon_z}{\partial x \partial y} = 0. \tag{1.18}$$

Note that as for the plane stress case, the last three equations in Eq. (1.18) might complicate the solutions for plane strain problems. For a cylindrical coordinates system, the strain compatibility equations would be

$$\frac{1}{r}\frac{\partial^2 \varepsilon_r}{\partial \theta^2} + \frac{\partial}{\partial r}\left\{r\frac{\partial \varepsilon_\theta}{\partial r} - (\varepsilon_r - \varepsilon_\theta)\right\} = \frac{\partial}{\partial \theta}\left\{\frac{\partial \gamma_{r\theta}}{\partial r} + \frac{\gamma_{r\theta}}{r}\right\},$$
$$\frac{1}{r}\frac{\partial^2 \varepsilon_z}{\partial \theta^2} + \frac{\partial \varepsilon_z}{\partial r} = 0, \frac{\partial^2 \varepsilon_z}{\partial r^2} = 0, \frac{\partial}{\partial \theta}\left\{\frac{\partial \varepsilon_z}{\partial r} - \frac{\varepsilon_z}{r}\right\} = 0. \tag{1.19}$$

1.3.4. *The Airy function* $\phi(x, y)$

Sir George Airy suggested in 1862[d] one of the most efficient ways to solve two-dimensional problems in elasticity using a new variable, known as the Airy stress function, $\phi(x, y)$. This new variable leads to a new differential equation, which can be shown to be solvable in a much easier way compared to the solution of the equations of equilibrium, using the Navier's equations. This leads to the following expressions for the three planar stresses (in a Cartesian coordinates system):

$$\sigma_x(x,y) \equiv \frac{\partial^2 \phi(x,y)}{\partial y^2}, \sigma_y(x,y) \equiv \frac{\partial^2 \phi(x,y)}{\partial x^2}, \tau_{xy}(x,y) \equiv -\frac{\partial^2 \phi(x,y)}{\partial x \partial y}. \tag{1.20}$$

Having the stresses expressed by the Airy function, $\phi(x, y)$, we can substitute Eq. (1.20) into Eq. (1.14) yielding the compatibility equation for a state of plane strain, with no body forces, where the unknown is the Airy function.

$$\frac{\partial^4 \phi(x,y)}{\partial x^4} + 2\frac{\partial^4 \phi(x,y)}{\partial x^2 \partial y^2} + \frac{\partial^4 \phi(x,y)}{\partial y^4} \equiv \nabla^4 \phi(x,y) = 0. \tag{1.21}$$

Equation (1.21) is also known as the biharmonic equation, or $\nabla^4\phi(x, y) = 0$.

For a cylindrical coordinates system, the biharmonic equation has the following form:

$$\begin{aligned}\nabla^4 \phi(r,\theta) &\equiv \nabla^2[\nabla^2 \phi(r,\theta)] \\ &= \left(\frac{\partial^2}{\partial r^2} + \frac{1}{r}\frac{\partial}{\partial r} + \frac{1}{r^2}\frac{\partial^2}{\partial \theta^2}\right) \\ &\quad \times \left(\frac{\partial^2 \phi(r,\theta)}{\partial r^2} + \frac{1}{r}\frac{\partial \phi(r,\theta)}{\partial r} + \frac{1}{r^2}\frac{\partial^2 \phi(r,\theta)}{\partial \theta^2}\right) = 0,\end{aligned} \tag{1.22}$$

where the stress components are defined as

$$\begin{aligned}\sigma_r(r,\theta) &\equiv \frac{1}{r}\frac{\partial \phi(r,\theta)}{\partial r} + \frac{1}{r^2}\frac{\partial^2 \phi(r,\theta)}{\partial \theta^2}, \\ \sigma_\theta(r,\theta) &\equiv \frac{\partial^2 \phi(r,\theta)}{\partial r^2}, \\ \tau_{r\theta}(r,\theta) &\equiv -\frac{1}{r}\frac{\partial^2 \phi(r,\theta)}{\partial r \partial \theta} + \frac{1}{r^2}\frac{\partial \phi(r,\theta)}{\partial \theta}.\end{aligned} \tag{1.23}$$

Often, there are problems in which the applied loads have an axisymmetric distribution, namely there is no dependence on the θ variable, and all

[d]Sir George Biddell Airy (1801–1892), English astronomer and mathematician, (https://en.wikipedia.org/wiki/George_Biddell_Airy).

derivatives with respect to it would vanish. This results in the following stress field, $\sigma_r(r), \sigma_\theta(r)$ and $\sigma_{r\theta}(r) = 0$. For the axisymmetric stress distribution, the biharmonic function (Eq. (1.22)) reduces to

$$\nabla^4\phi(r) \equiv \left(\frac{\partial^2}{\partial r^2} + \frac{1}{r}\frac{\partial}{\partial r}\right)\left(\frac{\partial^2\phi(r)}{\partial r^2} + \frac{1}{r}\frac{\partial\phi(r)}{\partial r}\right)$$
$$= \frac{d^4\phi(r)}{dr^4} + \frac{2}{r}\frac{d^3\phi(r)}{dr^3} - \frac{1}{r^2}\frac{d^2\phi(r)}{dr^2} + \frac{1}{r^3}\frac{d\phi(r)}{dr} = 0. \quad (1.24)$$

The solution for Eq. (1.24) has the following form:

$$\phi(r) = A_1 + A_2 \log r + A_3 r^2 + A_4 r^2 \log r \quad (1.25)$$

and according to Eq. (1.23) the expressions for the stresses are

$$\sigma_r(r) \equiv \frac{1}{r}\frac{d\phi(r)}{dr} = \frac{A_2}{r^2} + 2A_3 + A_4(2\log r + 1),$$
$$\sigma_\theta(r) \equiv \frac{d^2\phi(r,\theta)}{dr^2} = -\frac{A_2}{r^2} + 2A_3 + A_4(2\log r + 3),$$
$$\tau_{r\theta}(r,\theta) \equiv 0, \quad (1.26)$$

where A_1, A_2, A_3 and A_4 are constants to be determined from the boundary conditions of the given problem.

For the case of plane stress, it can be shown that for the case of no body forces with symmetrically distributed applied loads, the Airy stress function $\phi(x,y)$ has the following form:

$$\phi(x,y) \equiv \psi(x,y) - \frac{\nu z^2}{2(1+\nu)r^2}\nabla^2\psi(x,y), \quad (1.27)$$

where $\psi(x,y)$ is the solution of the following biharmonic function $\nabla^4\psi(x,y)$. Assuming the condition for the plane strain, namely z is a very small quantity leads to $\phi(x,y) \approx \psi(x,y)$, which makes also the plane stress cases being solved by the biharmonic equation presented above.

In conclusion, the use of the Airy stress function representation reduces the problem of solving the stresses for an elastic body to that of finding a solution for the biharmonic partial differential equation, $\nabla^4\phi(x,y) = 0$, whose derivatives would satisfy certain boundary conditions, according to the posed problem.

1.3.5. *Thermal field in elasticity*

All the previous equations were derived assuming the temperature of the given structure is constant without any changes. However, for an unconstrained body, changing the uniform temperature, either by heating or by cooling, would cause the body to expand or contact, leading to normal strains. Preventing the expansion or the contraction would give rise to thermal stresses inside the structure.

To be able to take this effect into consideration, one must redefine the stress–strain relationships, superimposing the thermal strains on to the mechanical ones. Assuming the change in temperature is given spatially by the expression $T(x,\ y)$, the corresponding change in the length would be written by

$$\Delta L = \alpha L T\,(x, y), \tag{1.28}$$

where α is the linear one-dimensional thermal expansion coefficient. The associated thermal strain is then defined as

$$\varepsilon_t = \frac{\Delta L}{L} = \alpha T\,(x, y). \tag{1.29}$$

Adding the thermal strains to the mechanical one (see, for example, Eq. (1.9)) leads for the plane stress case to

$$\varepsilon_x = \frac{1}{E}[\sigma_x - \nu\sigma_y] + \alpha T, \varepsilon_y = \frac{1}{E}[\sigma_y - \nu\sigma_x] + \alpha T, \gamma_{xy} = \frac{\tau_{xy}}{G}. \tag{1.30}$$

Expressing the stresses as a function of the strains leads to

$$\sigma_x = \frac{E}{(1-\nu^2)}[\varepsilon_x + \nu\varepsilon_y] - \frac{E\alpha T}{(1-\nu)}, \tag{1.31}$$

$$\sigma_y = \frac{E}{(1-\nu^2)}[\varepsilon_y + \nu\varepsilon_x] - \frac{E\alpha T}{(1-\nu)}, \tag{1.32}$$

$$\tau_{xy} = G\gamma_{xy}.$$

Similar expressions can be obtained also for the plane strain case.

The biharmonic equation in the presence of a temperature field $T(x, y)$ can be shown to have the following form:

$$\nabla^4\phi(x, y) + \alpha E \nabla^2 T(x, y) = 0. \tag{1.33}$$

Equation (1.33) is true for both the plane stress and plane strains, provided the body forces can be assumed to be negligible.

References

[1] Bruhn, E.F., *Analysis and Design of Flight Vehicle Structures*, Jacobs Publication, 1973, 650 p.

[2] Niu, M.C.-Y., *Airframe Structural Design — Practical Design Information and Data on Aircraft Structures*, CONMILIT Press Ltd., 1988, 607 p.

[3] Polland, D.R., Finn, S.R., Griess, K.H., Hafenrichter, J.L., Hanson, C.T., Ilsewicz, L.B., Metschan, S.L., Scholz, D.B. and Smith, P.J., Global cost and weight evaluation of fuselage panel design concepts, NASA CR4730, April 1997, 319 p.

[4] Walker, T.H., Minguet, P.J., Flynn, B.W., Carbery, D.J., Swanson, G.D. and Ilsewicz, L.B., Advanced technology composite fuselage-structural performance, NASA CR4732, April 1997, 101 p.

[5] Flynn, B.W., Bodine, J.B., Dopker, B., Finn, S.R., Griess, K.H., Hanson, C.T., Harris, C.G., Nelson, K.M., Walker, T.H., Kennedy, T.C. and Nahan, M.F., Advanced technology composite fuselage-repair and damage assessment supporting maintenance, NASA CR4733, April 1997, 154 p.

[6] Willden, K.S., Harris, C.G., Flynn, B.W., Gessel, M.G., Scholz, D.B., Stawski, S. and Winson, V., Advanced technology composite fuselage-manufacturing, NASA CR4735, April 1997, 189 p.

[7] Megson, T.H.G., *Aircraft Structures for Engineering Students*, 4th Edition, Butterworth-Heinemann (an imprint of Elsevier Ltd.), 2007, 804 p.

[8] Gran B., *Bruhn Errata — A Companion to Analysis of Flight Vehicle Structures*, GRAN Corporation, 2nd Edition, 2008, 309 p.

[9] Weisshaar, T.A., *Aerospace Structures — An Introduction to Fundamental Problems*, Purdue University, USA, 28th of July, 2011, 197 p.

[10] Kalanchiam, M. and Chinnasamy, M., Advantages of composite materials in aircraft structures, *Int. J. Aerospace Mech. Eng.*, 6(11), 2012, pp. 2428–2432.

[11] Birch, H., Aerospace materials-changing planes, *Chem. World*, Oct. 2013, pp. 60–63.

[12] Abramovich, H., *Stability and Vibrations of Thin Walled Composite Structures*, Elsevier Ltd., 2017, 778 p.

[13] Timoshenko, S.P. and Goodier, J.N., *Theory of Elasticity*, 3rd Edition., International Student Edition, McGraw-Hill Book Company, 1970, McGraw-Hill Inc., 567 p.

[14] Chou, P.C. and Pagano, N.J., *ELASTICITY Tensor, Dyadic and Engineering Approaches*, Dover Publications, Inc., New York, 290 p.

[15] Carlucci, D., Payne, N. and Mehmedagic, I., Small strain compatibility conditions of an elastic solid in cylindrical coordinates, Technical Report ARDSM-TR-12001, U.S. Army Armament Research, Development and Engineering Center, Munitions Engineering Technology Center, Picatinny Arsenal, New Jersey, USA, April 2013, 14 p.

Chapter 2

Laminated Composite Materials

2.1. Introduction

Laminated composite materials are composite materials manufactured in layers. The common definition for a composite material, usually made of two constituents, one being the fiber (the reinforcement) and the other the glue (the matrix), states that a combination of the two materials would result in better properties than those of the individual components when they are used alone. The main advantages of composite materials over other existing materials, like metal or plastics, are their high strength and stiffness, combined with low density, allowing for a weight reduction in the finished part. The various types of composites are usually referred in the literature, by a block diagram, as depicted in Fig. 2.1 In the present chapter, when talking about a composite material, we will restrict ourselves to only continuous fibers (reinforcements) being embedded into a carrying matrix formed from an adequate glue.

Examples of such continuous reinforcements include unidirectional, woven cloth and helical winding (see Fig. 2.2). Continuous-fiber composites are often made into laminates by stacking single sheets of continuous fibers in different orientations to obtain the desired strength and stiffness properties with fiber volumes as high as 60–70 percent. Fibers produce high-strength composites because of their small diameter; they contain far fewer defects (normally surface defects) compared to the material produced in bulk. On top of it, due to their small diameter the fibers are flexible and suitable for complicated manufacturing processes, such as small radii or weaving. Materials like glass, graphite, carbon or aramid are used to produce fibers (see typical properties in Table 2.1). The today's usage of composite materials is mainly driven by the aerospace sector, with large percentage of the

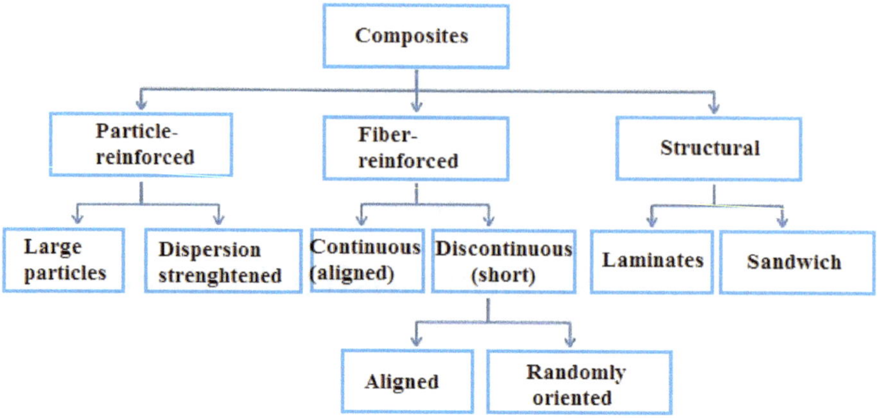

Fig. 2.1. Typical composite materials.

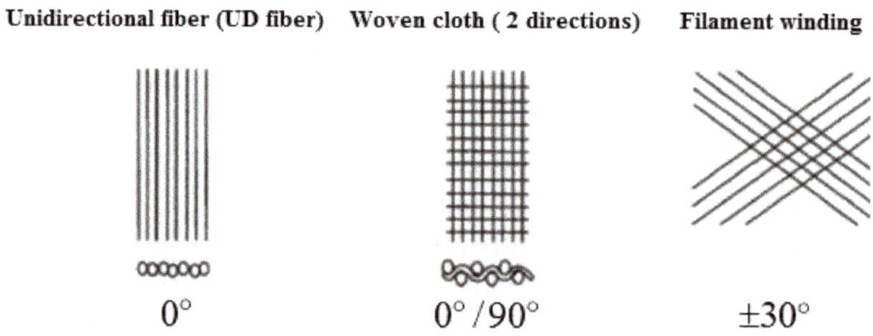

Fig. 2.2. Typical composite materials.

modern airplanes structures, like Boeing 787 or Airbus A380 (see Fig. 2.3), being manufactured from carbon, glass, aramid fibers. The main material for the matrix is a polymer, which has low strength and stiffness. The main functions of the matrix are to keep the fibers in the proper orientation and spacing and providing protection to the fiber from abrasion and the environment. In polymer matrix composites, the good and strong bond between the matrix and the reinforcement allows the matrix to transmit the outside loads from the matrix to the fibers through shear loading at the interface. Two types of polymer matrices are available: thermosets and thermoplastics. A thermoset starts as a low-viscosity resin that reacts and cures during processing, forming a solid. A thermoplastic is a high-viscosity resin that is processed by heating it above its melting temperature. Because a thermoset resin sets up and cures during processing, it cannot be reprocessed by reheating. A thermoplastic can be reheated above its melting temperature for additional processing.

Table 2.1. Typical properties of mostly used reinforced continuous fibers (from [1, 2]).

Material	Trade name	Density, ρ [kg/m^2]	Typical fiber diameter [μm]	Young's modulus, E [GPa]	Tensile strength [GPa]
$\alpha-Al_2O_3$ (Aluminum oxide)	FP (US)	3960	20	385	1.8
$Al_2O_3+SiO_2+B_2O_3$ (Mullite)	Nextel480 (USA)	3050	11	224	2.3
$Al_2O_3+SiO_2$ (Alumina-silica)	Altex (Japan)	3300	10–15	210	2.0
Boron (CVD* on tungsten)	VMC (Japan)	2600	140	410	4.0
Carbon (PAN** precursor)	T300 (Japan)	1800	7	230	3.5
Carbon (PAN** precursor)	T800 (Japan)	1800	5.5	295	5.6
Carbon (Pitch*** precursor)	Thorne IP755 (USA)	2060	10	517	2.1
SiC (+O) (Silicon carbide)	Nicalon (Japan)	2600	15	190	2.5–3.3
SiC (low O) (Silicon carbide)	Hi-Nicalon (Japan)	2740	14	270	2.8
SiC (+O+Ti) (Silicon carbide)	Tyranno (Japan)	2400	9	200	2.8
SiC (monofilament) (Silicon carbide)	Sigma	3100	100	400	3.5
E-glass (Silica)		2500	10	70	1.5–2.0
E-glass (Silica)		2500	10	70	1.5–2.0
Quartz (Silica)		2200	3–15	80	3.5
Aromatic polyamide	Kevlar 49 (USA)	1500	12	130	3.6
Polyethylene (UHMW)$^+$	Spectra 100 (USA)	970	38	175	3.0
High carbon steel	E.g. piano wire	7800	250	210	2.8
Aluminum	Electrical wire	2680	1670	75	0.27
Titanium	Wire	4700	250	115	0.434

*CVD = Chemical Vapor Deposition.
**PAN = PolyAcryoNitrile; about 90% of the carbon fibers produced are made from PAN.
***Pitch = A viscoelastic material that is composed of aromatic hydrocarbons. Pitch is produced via the distillation of carbon-based materials, such as plants, crude oil, and coal.
†UHMW = Ultra-high-molecular-weight *polyethylene* (or polyethene, the most common plastic produced in the world) is a subset of the thermoplastic polyethylene.

2.2. Unidirectional Composites

Unidirectional composites are usually composed of two constituents, the fiber and the matrix (which is the glue holding together the two components). Based on the rule of mixtures, one can calculate the properties of the unidirectional layer based on the properties of the fibers and the matrix and their volume fracture. The assumption to be made when applying the rule of mixtures is that the two constituents are bonded together and they behave like a single body. The major modulus (or the longitudinal modulus), E_{11}, of the layer can be written as:

$$E_{11} = E_f V_f + E_m V_m, \tag{2.1}$$

where E_f and E_m are the longitudinal moduli for the fibers and the matrix, respectively and V_f and V_m are their volume fractions[a].

The major Poisson's coefficient, ν_{12}, is given by

$$\nu_{12} = \nu_f V_f + \nu_m V_m, \tag{2.2}$$

where ν_f and ν_m are the longitudinal moduli for the fibers and the matrix, respectively.

One should note that the minor Poisson's coefficient, ν_{21}, will be calculated as follows:

$$\frac{\nu_{12}}{E_{11}} = \frac{\nu_{21}}{E_{22}} \Rightarrow \nu_{21} = \nu_{12}\frac{E_{22}}{E_{11}}. \tag{2.3}$$

The minor modulus (or the transverse modulus), E_{22}, of the layer is given by

$$\frac{1}{E_{22}} = \frac{V_f}{E_f} + \frac{V_m}{E_m} \Rightarrow E_{22} = \frac{E_m}{V_f\frac{E_m}{E_f} + V_m} = \frac{E_m}{V_f\frac{E_m}{E_f} + (1 - V_f)}. \tag{2.4}$$

The shear modulus of the layer, G_{12}, has the following form:

$$\frac{1}{G_{12}} = \frac{V_f}{G_f} + \frac{V_m}{G_m} \Rightarrow G_{12} = \frac{G_m}{V_f\frac{G_m}{G_f} + V_m} = \frac{G_m}{V_f\frac{G_m}{G_f} + (1 - V_f)} \tag{2.5}$$

where G_f and G_m are the shear moduli for the fibers and the matrix, respectively.

To be able to assess the differences between the properties of the fiber and to compare to those of the matrix, the reader is referred to Table 2.2 (from [1]).

[a]Note that $V_f + V_m = 1$.

Table 2.2. Typical properties of T300 carbon fibers and 914 epoxy resin.

Property	T300 carbon fibers	914 epoxy resin matrix
Young's modulus, E [GPa]	220	3.3
Shear modulus, G [GPa]	25	1.2
Poisson's ratio, ν	0.15	0.37

Table 2.3. Predictions of unidirectional composite properties by simple micromechanics models (adapted from [1]).

Equation	Relationship	Predicted values (moduli in [GPa])	Experimental values (moduli in [GPa])
(1.1)	$E_{11} = E_f V_f + E_m(1 - V_f)$	124.7	125.0
(1.4)	$\frac{1}{E_{22}} = \frac{V_f}{E_f} + \frac{(1-V_f)}{E_m}$	7.4	9.1
(1.5)	$\frac{1}{G_{12}} = \frac{V_f}{G_f} + \frac{(1-V_f)}{G_m}$	2.6	5.0
(1.2)	$\upsilon_{12} = \upsilon_f V_f + \upsilon_m(1 - V_f)$	0.25	0.34

As described in [1] the simple micromechanics model used in the rule of mixtures, predict well the values of the four variables, E_{11}, E_{22}, G_{12} and υ_{12}, as compared to experimental values, as can be seen in Table 2.3.

2.3. Properties of a Single Ply

A ply has two major dimensions and one, the thickness, very small as compared to the two major ones. Therefore, the 3D presentation of an orthotropic material will be simplified to a two-dimensional (2D) presentation (plane stress) by assuming that $\sigma_{33} = 0$ (see [1, 2]). This leads to a reduced compliance matrix, for the ply, in the following form:

$$\begin{Bmatrix} \varepsilon_{11} \\ \varepsilon_{22} \\ \gamma_{12} \end{Bmatrix} = \begin{bmatrix} \frac{1}{E_1} & -\frac{\nu_{21}}{E_2} & 0 \\ -\frac{\nu_{12}}{E_1} & \frac{1}{E_2} & 0 \\ 0 & 0 & \frac{1}{G_{12}} \end{bmatrix} \begin{Bmatrix} \sigma_{11} \\ \sigma_{22} \\ \sigma_{12} \end{Bmatrix}, \tag{2.6}$$

a third equation, for the strain in the thickness direction, ε_{33}, which is seldom used has the form

$$\varepsilon_{33} = -\frac{\gamma_{13}}{E_1}\sigma_{11} - \frac{\gamma_{23}}{E_2}\sigma_{22} \tag{2.7}$$

and the remaining two equations for the shear strains which are written as:

$$\begin{Bmatrix} \gamma_{23} \\ \gamma_{13} \end{Bmatrix} = \begin{bmatrix} \frac{1}{G_{23}} & 0 \\ 0 & \frac{1}{G_{13}} \end{bmatrix} \begin{Bmatrix} \sigma_{23} \\ \sigma_{13} \end{Bmatrix}. \tag{2.8}$$

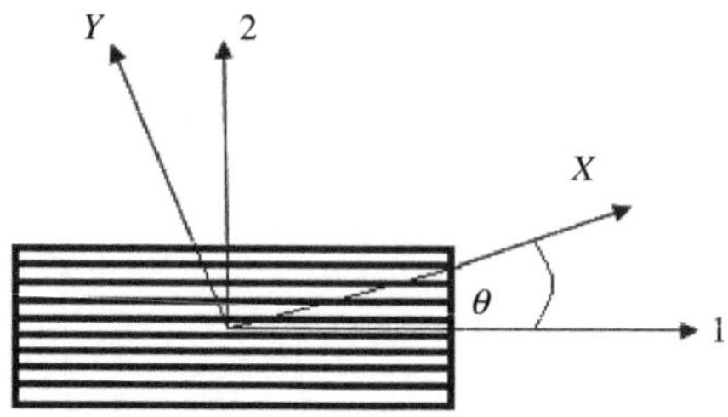

Fig. 2.4. Two coordinate systems: 1, 2 the ply orthotropic axes; X, Y arbitrary axes.

Calculation of the stresses as a function of strains would yield (by the use of Eqs. (2.6) and (2.8)):

$$\begin{Bmatrix} \sigma_{11} \\ \sigma_{22} \\ \sigma_{12} \end{Bmatrix} = \begin{bmatrix} Q_{11} & Q_{12} & 0 \\ Q_{21} & Q_{22} & 0 \\ 0 & 0 & Q_{66} \end{bmatrix} \begin{Bmatrix} \varepsilon_{11} \\ \varepsilon_{22} \\ \gamma_{12} \end{Bmatrix}$$

$$= \begin{bmatrix} \frac{E_1}{(1-\upsilon_{12}\upsilon_{21})} & \frac{\upsilon_{21}E_1}{(1-\upsilon_{12}\upsilon_{21})} & 0 \\ \frac{\upsilon_{12}E_2}{(1-\upsilon_{12}\upsilon_{21})} & \frac{E_2}{(1-\upsilon_{12}\upsilon_{21})} & 0 \\ 0 & 0 & G_{12} \end{bmatrix} \begin{Bmatrix} \varepsilon_{11} \\ \varepsilon_{22} \\ \gamma_{12} \end{Bmatrix}, \qquad (2.9)$$

with $Q_{12} = Q_{21}$ and $\upsilon_{12} \neq \upsilon_{21}$

$$\begin{Bmatrix} \sigma_{23} \\ \sigma_{13} \end{Bmatrix} = \begin{bmatrix} Q_{23} & 0 \\ 0 & Q_{13} \end{bmatrix} \begin{Bmatrix} \gamma_{23} \\ \gamma_{13} \end{Bmatrix} = \begin{bmatrix} G_{23} & 0 \\ 0 & G_{13} \end{bmatrix} \begin{Bmatrix} \gamma_{23} \\ \gamma_{13} \end{Bmatrix}. \qquad (2.10)$$

2.4. Transformation of Stresses and Strains

Consider the two coordinate systems, described in Fig. 2.4. The one with indexes 1 and 2 describes the ply orthotropic coordinate system, while the other one (x, y) is an arbitrary one, rotated at a given angle θ relative to the 1, 2 system. The transformation of the stresses and the strains from the 1, 2 coordinate system to the x, y coordinate system is done by multiplication of both the stresses and the strains at the ply level by the transformation matrix **T** as given by[b]:

[b]See, for example, Ref [1]: *Primer on Composite Materials: Analysis* by J. E. Ashton and J. C. Halpin, TECHNOMIC Publishing Co., Inc., Stanford, CT, 1969.

$$\begin{Bmatrix} \sigma_1 \\ \sigma_2 \\ \tau_{12} \end{Bmatrix}^k = [T] \begin{Bmatrix} \sigma_x \\ \sigma_y \\ \tau_{xy} \end{Bmatrix}^k, \tag{2.11}$$

$$\begin{Bmatrix} \varepsilon_1 \\ \varepsilon_2 \\ \frac{\gamma_{12}}{2} \end{Bmatrix}^k = [T] \begin{Bmatrix} \varepsilon_x \\ \varepsilon_y \\ \frac{\gamma_{xy}}{2} \end{Bmatrix}^k, \tag{2.12}$$

where k is the number of the ply, for which the transformation of strains and stresses is performed[c]. The transformation matrix T is given by

$$[T] = \begin{bmatrix} c^2 & s^2 & 2cs \\ s^2 & c^2 & -2cs \\ -cs & cs & c^2 - s^2 \end{bmatrix}, \quad \text{where } c \equiv \cos\theta,\ s \equiv \sin\theta. \tag{2.13}$$

To obtain the inverse of the matrix T, one needs simply to insert $-\theta$ instead of θ in Eq. (2.13) to yield

$$[T]^{-1} = [T(-\theta)] = \begin{bmatrix} c^2 & s^2 & -2cs \\ s^2 & c^2 & 2cs \\ cs & -cs & c^2 - s^2 \end{bmatrix}. \tag{2.14}$$

The ply (or lamina) strain–stress relationship transformed to the laminate references axis (x, y) is written as

$$\begin{Bmatrix} \sigma_1 \\ \sigma_2 \\ \tau_{12} \end{Bmatrix}^k = [T]^{-1} [Q]^k [T] \begin{Bmatrix} \varepsilon_x \\ \varepsilon_y \\ \gamma_{xy} \end{Bmatrix}^k, \quad \text{where } [Q]^k = \begin{bmatrix} Q_{11} & Q_{12} & 0 \\ Q_{12} & Q_{22} & 0 \\ 0 & 0 & 2Q_{66} \end{bmatrix}^k \tag{2.15}$$

and the expressions for Q_{11}, Q_{12}, Q_{22} and Q_{66} are given in Eq. (2.9). Performing the matrix multiplication in Eq. (2.15) yields:

$$\begin{Bmatrix} \sigma_1 \\ \sigma_2 \\ \tau_{12} \end{Bmatrix}^k = [\overline{Q}]^k \begin{Bmatrix} \varepsilon_x \\ \varepsilon_y \\ \gamma_{xy} \end{Bmatrix}^k \quad \text{where } [\overline{Q}]^k = \begin{bmatrix} \overline{Q_{11}} & \overline{Q_{12}} & \overline{Q_{16}} \\ \overline{Q_{12}} & \overline{Q_{22}} & \overline{Q_{26}} \\ \overline{Q_{16}} & \overline{Q_{26}} & \overline{Q_{66}} \end{bmatrix}^k \tag{2.16}$$

[c]Note that: $\sigma_{11} \equiv \sigma_1$, $\sigma_{22} \equiv \sigma_2$, $\varepsilon_{11} \equiv \varepsilon_1$, $\varepsilon_{22} \equiv \varepsilon_2$.

and

$$\begin{aligned}
\overline{Q_{11}} &= Q_{11}\cos^4\theta + 2(Q_{12} + 2Q_{66})\sin^2\theta\cos^2\theta + Q_{22}\sin^4\theta,\\
\overline{Q_{12}} &= (Q_{11} + Q_{22} - 4Q_{66})\sin^2\theta\cos^2\theta + Q_{12}(\sin^4\theta + \cos^4\theta),\\
\overline{Q_{22}} &= Q_{11}\sin^4\theta + 2(Q_{12} + 2Q_{66})\sin^2\theta\cos^2\theta + Q_{22}\cos^4\theta,\\
\overline{Q_{16}} &= (Q_{11} - Q_{12} - 2Q_{66})\sin\theta\cos^3\theta + (Q_{12} - Q_{22} + 2Q_{66})\sin^3\theta\cos\theta,\\
\overline{Q_{26}} &= (Q_{11} - Q_{12} - 2Q_{66})\sin^3\theta\cos\theta + (Q_{12} - Q_{22} + 2Q_{66})\sin\theta\cos^3\theta,\\
\overline{Q_{66}} &= (Q_{11} + Q_{22} - 2Q_{12} - 4Q_{66})\sin^2\theta\cos^2\theta + Q_{12}(\sin^4\theta + \cos^4\theta).
\end{aligned} \tag{2.17}$$

Another useful way of presenting the various terms of the matrix $[\overline{Q}]^k$ is the invariant procedure presented by Tsai and Pagano [2]:

$$\begin{aligned}
\overline{Q_{11}} &= U_1 + U_2\cos(2\theta) + U_3\cos(4\theta),\\
\overline{Q_{12}} &= U_4 - U_3\cos(4\theta),\\
\overline{Q_{22}} &= U_1 - U_2\cos(2\theta) + U_3\cos(4\theta),\\
\overline{Q_{16}} &= -\frac{1}{2}U_2\sin(2\theta) - U_3\sin(4\theta),\\
\overline{Q_{26}} &= -\frac{1}{2}U_2\sin(2\theta) + U_3\sin(4\theta),\\
\overline{Q_{66}} &= U_5 - U_3\cos(4\theta), \quad \text{where}\\
U_1 &= \frac{1}{8}[3Q_{11} + 3Q_{22} + 2Q_{12} + 4Q_{66}],\\
U_2 &= \frac{1}{2}[Q_{11} - Q_{22}],\\
U_3 &= \frac{1}{8}[Q_{11} + Q_{22} - 2Q_{12} - 4Q_{66}],\\
U_4 &= \frac{1}{8}[Q_{11} + Q_{22} + 6Q_{12} - 4Q_{66}],\\
U_5 &= \frac{1}{8}[Q_{11} + Q_{22} - 2Q_{12} + 4Q_{66}].
\end{aligned} \tag{2.18}$$

Note that the terms U_1, U_4 and U_5 are invariant to a rotation relative to the three axes (perpendicular to the 1–2 plane).

2.5. Classical Lamination Theory

Now we shall present the addition of the properties of each lamina to form a laminate, which is the structure to be investigated when we apply loads,

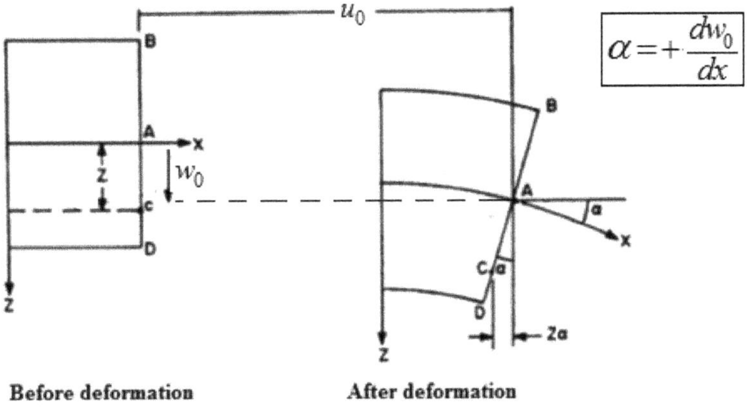

Fig. 2.5. The plate cross-section before and after the deformation (CLT approach).

using the classical lamination theory (CLT), which is based on Kirchhoff–Love plate theory (3–5).

Referring to Fig. 2.5, one can write the displacement in the x direction of a point at a z distance from the mid-plane as (where w is the displacement in the z direction)

$$u = u_0 - z\frac{\partial w}{\partial x}. \tag{2.19}$$

Similarly, the displacement in the y direction will be

$$v = v_0 - z\frac{\partial w}{\partial y}. \tag{2.20}$$

Then the strains (ε_x, ε_y and γ_{xy}) and the curvatures (κ_x, κ_y and κ_{xy}) can be written as

$$\begin{aligned}
\varepsilon_x &\equiv \frac{\partial u}{\partial x} = \frac{\partial u_0}{\partial x} - z\frac{\partial^2 w_0}{\partial x^2} = \varepsilon_x^0 + z\kappa_x,\\
\varepsilon_y &\equiv \frac{\partial v}{\partial y} = \frac{\partial v_0}{\partial y} - z\frac{\partial^2 w_0}{\partial y^2} = \varepsilon_y^0 + z\kappa_y,\\
\gamma_{xy} &\equiv \frac{\partial u}{\partial y} + \frac{\partial v}{\partial x} = \frac{\partial u_0}{\partial y} + \frac{\partial v_0}{\partial x} - 2z\frac{\partial^2 w}{\partial x \partial y} = \gamma_{xy}^0 + z\kappa_{xy},
\end{aligned} \tag{2.21}$$

where $\varepsilon_x^0, \varepsilon_y^0, \gamma_{xy}^0$ are the strains at the neutral plane. In matrix notation, Eq. (2.21) can be presented as

$$\begin{Bmatrix} \varepsilon_x \\ \varepsilon_y \\ \gamma_{xy} \end{Bmatrix} = \begin{Bmatrix} \varepsilon_x^0 \\ \varepsilon_y^0 \\ \gamma_{xy}^0 \end{Bmatrix} + z\begin{Bmatrix} \kappa_x \\ \kappa_y \\ \kappa_{xy} \end{Bmatrix} \Rightarrow \{\varepsilon\} = \{\varepsilon^0\} + z\{\kappa\}. \tag{2.22}$$

Then, the stresses at the lamina level is written as:

$$\{\sigma\}^k = [\overline{Q}]^k\{\varepsilon^0\} + z[\overline{Q}]^k[\kappa]. \tag{2.23}$$

Now we shall deal with force (N_x, N_y, N_{xy}) and moment (M_x, M_y, M_{xy}) resultants, per unit width (b in Fig. 2.6). Their definitions are given as (h is the total thickness of the laminate):

$$N_x \equiv \int_{-h/2}^{h/2} \sigma_x dz, N_y \equiv \int_{-h/2}^{h/2} \sigma_y dz, N_{xy} \equiv \int_{-h/2}^{h/2} \tau_{xy} dz,$$

$$M_x \equiv \int_{-h/2}^{h/2} \sigma_x z dz, M_y \equiv \int_{-h/2}^{h/2} \sigma_y z dz, M_{xy} \equiv \int_{-h/2}^{h/2} \tau_{xy} z dz. \tag{2.24}$$

Substituting the expressions of the stresses one obtains expressions for the force and moments resultants as a function of the strain on the mid plane, ε^0 and the curvature κ (see also Ref. [2]). The short written expressions are:

$$\begin{Bmatrix} \{N\} \\ \{M\} \end{Bmatrix} = \begin{bmatrix} [A] & [B] \\ [B] & [D] \end{bmatrix} \begin{Bmatrix} \{\varepsilon^0\} \\ \{\kappa\} \end{Bmatrix} \quad \text{or}$$

$$\begin{Bmatrix} \begin{Bmatrix} N_x \\ N_y \\ N_{xy} \end{Bmatrix} \\ \begin{Bmatrix} M_x \\ M_y \\ M_{xy} \end{Bmatrix} \end{Bmatrix} = \begin{bmatrix} \begin{bmatrix} A_{11} & A_{12} & A_{16} \\ A_{12} & A_{22} & A_{26} \\ A_{16} & A_{26} & A_{66} \end{bmatrix} & \begin{bmatrix} B_{11} & B_{12} & B_{16} \\ B_{12} & B_{22} & B_{26} \\ B_{16} & B_{26} & B_{66} \end{bmatrix} \\ \begin{bmatrix} B_{11} & B_{12} & B_{16} \\ B_{12} & B_{22} & B_{26} \\ B_{16} & B_{26} & B_{66} \end{bmatrix} & \begin{matrix} D_{11} & D_{12} & D_{16} \\ D_{12} & D_{22} & D_{26} \\ D_{16} & D_{26} & D_{66} \end{matrix} \end{bmatrix} \begin{Bmatrix} \begin{Bmatrix} \varepsilon_x^0 \\ \varepsilon_y^0 \\ \gamma_{xy}^0 \end{Bmatrix} \\ \begin{Bmatrix} \kappa_x \\ \kappa_y \\ \kappa_{xy} \end{Bmatrix} \end{Bmatrix}, \tag{2.25}$$

where the various constants are defined as:

$$A_{ij} \equiv \int_{-h/2}^{h/2} \overline{Q}_{ij}^k dz = \sum_{k=1}^{n} \overline{Q}_{ij}^k (h_k - h_{k-1})$$

$$B_{ij} \equiv \int_{-h/2}^{h/2} \overline{Q}_{ij}^k z dz = \frac{1}{2}\sum_{k=1}^{n} \overline{Q}_{ij}^k (h_k^2 - h_{k-1}^2)$$

$$D_{ij} \equiv \int_{-h/2}^{h/2} \overline{Q}_{ij}^k z^2 dz = \frac{1}{3}\sum_{k=1}^{n} \overline{Q}_{ij}^k (h_k^3 - h_{k-1}^3) \tag{2.26}$$

where $i, j = 1,1; 1,2; 2,2; 1,6; 2,6; 6,6.$

The way the sum is performed in Eq. (2.26) is according to the notations in Fig. 2.6. The passage from integral over the thickness of the laminate to the sum over the thickness is dictated by the fact that the individual plies

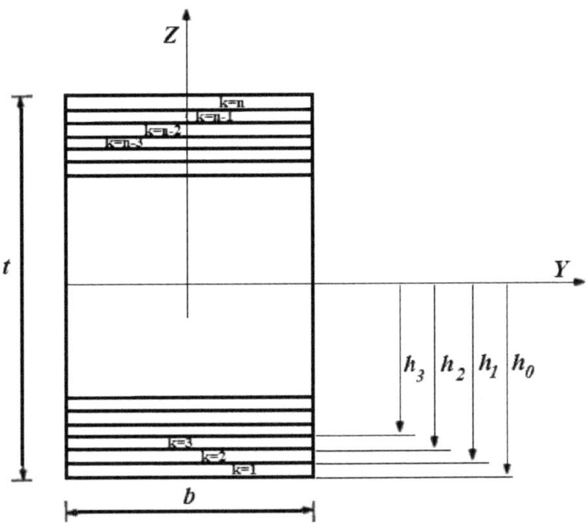

Fig. 2.6. Lamina notations within a given laminate.

are very thin and the properties within each lamina are assumed constant in the thickness direction.

Finally, the equations of motion for a general case, applied to a thin plate made of laminated composite plies, using the classical lamination theory (CLT), are given as follows (see [1, 2]):

$$\frac{\partial N_x}{\partial x} + \frac{\partial N_{xy}}{\partial y} = I_1 \frac{\partial^2 u_0}{\partial t^2} - I_2 \frac{\partial^2}{\partial t^2}\left(\frac{\partial w_0}{\partial x}\right),$$

$$\frac{\partial N_{xy}}{\partial x} + \frac{\partial N_y}{\partial y} = I_1 \frac{\partial^2 v_0}{\partial t^2} - I_2 \frac{\partial^2}{\partial t^2}\left(\frac{\partial w_0}{\partial y}\right),$$

$$\begin{aligned}&\frac{\partial^2 M_x}{\partial x^2} + 2\frac{\partial^2 M_{xy}}{\partial x \partial y} + \frac{\partial^2 M_y}{\partial y^2} + \frac{\partial}{\partial x}\left[N_{xx}\frac{\partial w_0}{\partial x} + N_{xy}\frac{\partial w_0}{\partial y}\right]\\ &\quad + \frac{\partial}{\partial y}\left[N_{yy}\frac{\partial w_0}{\partial y} + N_{xy}\frac{\partial w_0}{\partial x}\right]\\ &\quad = -p_z + I_1\frac{\partial^2 w_0}{\partial t^2} - I_3\frac{\partial^2}{\partial t^2}\left(\frac{\partial^2 w_0}{\partial x^2} + \frac{\partial^2 w_0}{\partial y^2}\right) + I_2\frac{\partial^2}{\partial t^2}\left(\frac{\partial w_0}{\partial x} + \frac{\partial w_0}{\partial y}\right),\end{aligned} \tag{2.27}$$

where p_z is the load per unit area in the z direction[d] and the subscript 0 represents the values at the mid plane of the cross-section. N represents the

[d]Note that the coordinate z is normally used for the thickness direction, while x and y coordinates define the plate area.

in-plane loads and the various moments of inertia, I_1, I_2 and I_3 are given by (ρ is the mass/unit length)

$$I_j = \int_{-h/2}^{h/2} \rho z^{j-1} dz; \; j = 1, 2, 3 \tag{2.28}$$

To obtain the equations for a beam, one can use Eq. (2.27), while all the derivations with respect with y are identically zero. This yields a one-dimensional equation in the following form:

$$\frac{\partial^2 M_x}{\partial x^2} + \frac{\partial}{\partial x}\left(N_{xx}\frac{\partial w_0}{\partial x}\right) = -p_z + I_1\frac{\partial^2 w}{\partial t^2} - I_3\frac{\partial^4 w}{\partial t^2 \partial x^2} + I_2\frac{\partial^3 w}{\partial t^2 \partial x}, \tag{2.29}$$

where N_x is the axial (in-plane, in the direction of the length of the beam) load. Remembering the relationship between transverse deflection, w, and the bending moment, we can rewrite Eq. (2.29) in terms of w only to yield

$$\begin{aligned} &-D_{11}\frac{\partial^2 w}{\partial x^2} = M_x \\ &\Rightarrow -\frac{\partial^2}{\partial x^2}\left(D_{11}\frac{\partial^2 w}{\partial x^2}\right) + \frac{\partial}{\partial x}\left(N_{xx}\frac{\partial w_0}{\partial x}\right) \\ &= -p_z + I_1\frac{\partial^2 w}{\partial t^2} - I_3\frac{\partial^4 w}{\partial t^2 \partial x^2} + I_2\frac{\partial^3 w}{\partial t^2 \partial x} \end{aligned} \tag{2.30}$$

with its associated boundary conditions:

$$\begin{aligned} &\text{Geometric: specify either } w \text{ or } \frac{\partial w}{\partial x}; \\ &\text{Natural: specify either } Q \equiv \frac{\partial M}{\partial x} \text{ or } M. \end{aligned} \tag{2.31}$$

Typical boundary conditions, normally used in the literature are in the following form:

$$\begin{aligned} &\text{Simply Supported: } w = 0 \text{ and } M = 0; \\ &\text{Clamped: } w = 0 \text{ and } \frac{\partial w}{\partial x} = 0; \\ &\text{Free: } Q \equiv \frac{\partial M}{\partial x} = 0 \text{ and } M = 0. \end{aligned} \tag{2.32}$$

The reader should be aware of thermal issues associated with the manufacturing of composite structures due to the differential thermal contraction during the post-curing phase and also as a consequence of any temperature changes during the service life of the structure. This issue is caused by the relatively small axial thermal expansion coefficient of the modern reinforcing fibers (for carbon fibers it is even slightly negative), while the resin matrix has a large thermal coefficient. When cooling from a typical curing temperature, like 140°C to room temperature, the fibers of the laminate composite will be in compression, while the matrix will show tension stresses [1]. Typical residual stresses due to thermal mismatch between the two components of the laminate are presented in Table 2.4.

Another important data for design is the experimental tension and compression strength as measured during various laboratory tests, as presented in Table 2.5 (see [1]).

Finally a table with a list of the main manufacturers of various composite materials is presented in Table 2.6.

Table 2.4. Typical thermal stresses in some common unidirectional composites (from [1]).

Matrix	Fiber	% Fiber volume, V_f	Temperature range ΔT [°K]	Fiber residual stress [MPa]	Matrix residual stress [MPa]
Epoxy (high T cure)	T300 carbon	65	120	−19	36
Epoxy (low T cure)	E glass	65	100	−15	28
Epoxy (low T cure)	Kevlar-49	65	100	−16	30
Borosilicate glass	T300 carbon	50	520	−93	93
CAS* glass-ceramic	Nicalon SiC	40	1000	−186	124

Note: *CAS = $CaO–Al_2O_3–SiO_2$.

Table 2.5. Typical experimental tension and compression strengths for common composite materials (from [1]).

Material	Lay-up	% Fiber volume, V_f	Tensile strength σ_t [GPa]	Compression strength σ_c [GPa]	Ratio $\sigma_c/\ \sigma_t$
GRP	ud	60	1.3	1.1	0.85
CFRP	ud	60	2.0	1.1	0.55
KFRP	ud	60	1.0	0.4	0.40
HTA/913 (CFRP)	$[(\pm45°, 0^\circ_2)_2]_S$	65	1.27	0.97	0.77
T800/924 (CFRP)	$[(\pm45°, 0^\circ_2)_2]_S$	65	1.42	0.90	0.63
T800/5245 (CFRP)	$[(\pm45°, 0^\circ_2)_2]_S$	65	1.67	0.88	0.53
SiC/CAS (CMC*)	ud	37	334	1360	4.07
SiC/CAS (CMC)	$[0°, 90°]_{3S}$	37	210	463	2.20

Note: *CMC = Ceramic matrix composites.

Table 2.6. List of the main manufacturers of various composite materials and resins.

Type of composite	Company	Company website
Thermoplastic composites	Milliken Tegris	tegris.milliken.com
Thermoplastic composites	Polystrand, Inc.	www.polystrand.com
Nonwoven fabrics (PolyWeb™) & foam	Wm. T. Burnett & Co.	www.williamtburnett.com
Thermoplastic composites	Schappe Techniques	www.schappe.com
Thermoplastic composites	TechFiber	www.fiber-tech.net
Thermoplastic composites	TenCate	www.tencate.com
Thermoplastic composites	Thercom	www.thercom.com
Thermoplastic composites	Vectorply	www.vectorply.com
Composite materials: resins and fibers	SF Composites	www.sf-composites.com
Formulation and manufacture of epoxy-based systems	SICOMIN	www.sicomin.com
Composite materials + resin, composite laminates	Lamiflex SPA	www.lamiflex.il
Composite materials + polyester	AMP Composite	www.amp-composite.il
Infusion, pultrusion, wet lay-up, prepreg, filament winding	Applied Poleramic Inc.	www.appliedpoleramic.com
Epoxy and Polyurethane	Endurance Technologies	www.epoxi.com
Composite materials	Gurit	www.gurit.com
Advanced thermoset resins	Huntsman Advanced Materials	www.huntsman.com/advanced_materials/a/Home
Advanced thermoset resins	Lattice Composites	www.latticecomposites.com
Kevlar	DuPont™ Kevlar®	www.dupont.com/products-and-services/fabrics-fibers-nonwovens/fibers/brands/kevlar.html

Table 2.6. (*Continued*)

Type of composite	Company	Company website
UHMWPE — ultra high molecular weight, high performance polyethylene material	DuPont™ Tenslyon™	www.dupont.com
Innegra™ HMPP (polypropylene), high performance fiber	Innegra Technologies	www.innegratech.com
Spread tow fabrics	TeXtreme	www.textreme.com/b2b
Adhesives and sealants	3M	solutions.3m.com
Prepreg and resins	Axiom Materials Inc.	www.axiommaterials.com
Fabrics, resins, composite materials	Barrday Advanced Materials Solutions	www.barrday.com
Carbon prepreg	Hankuk Carbon Co., Ltd.	www.hcarbon.com/eng/product/overview.asp
Carbon fibers and prepregs	Hexcel®	www.hexcel.com/Products/Industries/ICarbon-Fiber
Prepregs and compounds	Pacific Coast Composites	www.pccomposites.com
Prepregs and compounds	Quantum Composites	www.quantumcomposites.com

2.6. First-Order Shear Deformation Theory

Unlike isotropic materials, the ratio between the shear and the bending in orthotropic material, like laminated composite structures, is not negligible. The need to remove the somehow restricting assumptions from the classical lamination theory (CLT) which, as stated above, are based on Kirchhoff–Love plate theory [3–5], like the neglecting the influence of the shear strains and the fact that a plane before deformation remains plane after the deformation, led to derivation of more advanced bending theories for plates, like Mindlin theory of plates [6–8], which includes in-plane shear strains and is an extension of Kirchhoff–Love plate theory incorporating first-order shear effects.

Mindlin's theory assumes that there is a linear variation of displacement across the plate thickness but the plate thickness does not change during deformation. An additional assumption is that the normal stress through the thickness is ignored; an assumption which is also called the plane stress condition. The Mindlin theory is often called the first-order shear deformation theory of plates (FSDT) and its application to composite materials is

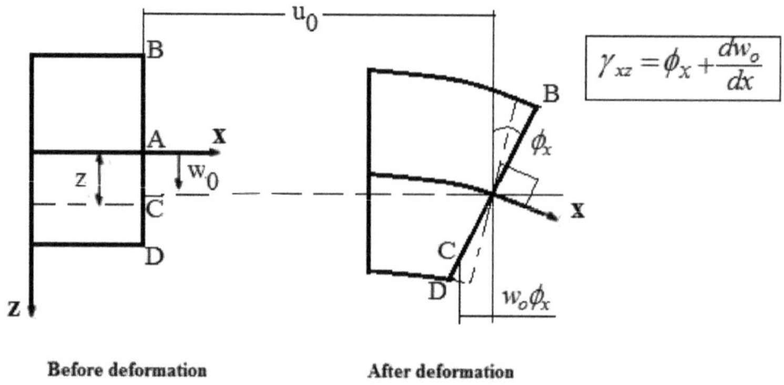

Fig. 2.7. The plate cross-section before and after the deformation (FSDT approach).

next presented. Under the assumptions and restrictions of Mindlin's theory (which has a similarity to Timoshenko's theory for beams [9–12]) the displacement field has five unknowns (u_0, v_0, w_0 — the displacements of the mid-plane in the x, y and z directions, respectively, and ϕ_x, ϕ_y — the rotations due to shear about x and y direction, respectively, and is given by (see also Fig. 2.7, which is similar to Fig. 2.5, but for FSDT approach)

$$u(x,y,z,t) = u_0(x,y,t) + z\phi_x(x,y,t),$$
$$v(x,y,z,t) = v_0(x,y,t) + z\phi_y(x,y,t),$$
$$w(x,y,z,t) = w_0(x,y,t). \tag{2.33}$$

The associated strains [13][e] assuming nonlinear terms are

$$\varepsilon_x = \frac{\partial u_0}{\partial x} + \frac{1}{2}\left(\frac{\partial w_0}{\partial x}\right)^2 + z\frac{\partial \phi_x}{\partial x},$$
$$\varepsilon_x = \frac{\partial v_0}{\partial y} + \frac{1}{2}\left(\frac{\partial w_0}{\partial y}\right)^2 + z\frac{\partial \phi_y}{\partial y},$$
$$\varepsilon_z = 0,\ \gamma_{xz} = \frac{\partial w_0}{\partial x} + \phi_x,\ \gamma_{yz} = \frac{\partial w_0}{\partial y} + \phi_y,$$
$$\gamma_{xy} = \left(\frac{\partial u_0}{\partial y} + \frac{\partial v_0}{\partial x} + \frac{\partial^2 w_0}{\partial x \partial y}\right) + z\left(\frac{\partial \phi_x}{\partial y} + \frac{\partial \phi_y}{\partial x}\right). \tag{2.34}$$

Multiplying Eq. (2.34) by the stiffness matrix $[\overline{Q}]$ and integrating through the thickness of the laminate, yields the force and moments resultants (as in

[e]Note that the assumption of constant shear strains across the height of the laminate is a rough approximation of the true strain distribution, which is at least quadratic through the thickness. However, although the rough approximation, the results of the application of Mindlin's plate theory presents very good results compared with experimental ones.

Eq. (2.23)), with two additional terms, the shear resultants, being defined by

$$\begin{Bmatrix} Q_x \\ Q_y \end{Bmatrix} = \kappa \int_{-h/2}^{h/2} \begin{Bmatrix} \tau_{xz} \\ \tau_{yz} \end{Bmatrix} dz, \tag{2.35}$$

where κ is called the shear correction coefficient and is defined by the ratio between the shear strain energies calculated by the actual shear distribution and the constant distribution assumed in the FSDT theory. The value of κ is taken as 5/6 for a rectangular cross-section[f].

The equations of motion will then have the following form:

$$\begin{aligned}
&\frac{\partial N_x}{\partial x} + \frac{\partial N_{xy}}{\partial y} = I_1 \frac{\partial^2 u_0}{\partial t^2} + I_2 \frac{\partial^2 \phi_x}{\partial t^2}, \\
&\frac{\partial N_{xy}}{\partial x} + \frac{\partial N_y}{\partial y} = I_1 \frac{\partial^2 v_0}{\partial t^2} + I_2 \frac{\partial^2 \phi_y}{\partial t^2}, \\
&\frac{\partial Q_x}{\partial x} + \frac{\partial Q_y}{\partial y} + \frac{\partial}{\partial x}\left[N_{xx}\frac{\partial w_0}{\partial x} + N_{xy}\frac{\partial w_0}{\partial y}\right] \\
&\quad + \frac{\partial}{\partial y}\left[N_{yy}\frac{\partial w_0}{\partial y} + N_{xy}\frac{\partial w_0}{\partial x}\right] = -p_z + I_1 \frac{\partial^2 w_0}{\partial t^2}, \\
&\frac{\partial M_x}{\partial x} + \frac{\partial M_{xy}}{\partial y} - Q_x = I_3 \frac{\partial^2 \phi_x}{\partial t^2} + I_1 \frac{\partial^2 u_0}{\partial t^2}, \\
&\frac{\partial M_{xy}}{\partial x} + \frac{\partial M_y}{\partial y} - Q_y = I_3 \frac{\partial^2 \phi_y}{\partial t^2} + I_1 \frac{\partial^2 v_0}{\partial t^2},
\end{aligned} \tag{2.36}$$

where p_z is the load per unit area in the z direction[g]. N_{xx}, N_{yy} and N_{xy} represent the in-plane loads and the various moments of inertia, I_1, I_2 and I_3 are given by (ρ is the mass/unit length) Eq. (2.28).

One should note that in addition to Eq. (2.25), which describe the resultants of the force and the moment as a function of the stiffness coefficients A_{ij}, B_{ij} and D_{ij}, the shear resultants Q_x and Q_y are defined as

$$\begin{Bmatrix} Q_y \\ Q_x \end{Bmatrix} = \kappa \begin{bmatrix} A_{44} & A_{45} \\ A_{45} & A_{55} \end{bmatrix} \begin{Bmatrix} \frac{\partial w_0}{\partial y} + \phi_y \\ \frac{\partial w_0}{\partial x} + \phi_x \end{Bmatrix}, \tag{2.37}$$

---

[f]The accurate value is $\kappa = \frac{10(1+\nu)}{12+11\nu}$ for a rectangular cross-section and $\kappa = \frac{6(1+\nu)}{7+6\nu}$ for a solid circular cross-section.

[g]Note that the coordinate z is normally used for the thickness direction, while x and y coordinates define the plate area.

where

$$A_{44} \equiv \kappa \int_{-h/2}^{h/2} \overline{Q}_{44}^{k} dz = \kappa \sum_{k=1}^{n} \overline{Q}_{44}^{k}(h_k - h_{k-1}),$$

$$A_{45} \equiv \kappa \int_{-h/2}^{h/2} \overline{Q}_{45}^{k} dz = \kappa \sum_{k=1}^{n} \overline{Q}_{45}^{k}(h_k - h_{k-1}),$$

$$A_{55} \equiv \kappa \int_{-h/2}^{h/2} \overline{Q}_{55}^{k} dz = \kappa \sum_{k=1}^{n} \overline{Q}_{55}^{k}(h_k - h_{k-1}), \tag{2.38}$$

and

$$\overline{Q}_{44} = Q_{44}\cos^2\theta + Q_{55}\sin^2\theta,$$

$$\overline{Q}_{45} = (Q_{55} - Q_{44})\cos\theta\sin\theta,$$

$$\overline{Q}_{55} = Q_{44}\sin^2\theta + Q_{55}\cos^2\theta,$$

where

$$Q_{44} = G_{23} \text{ and } Q_{55} = G_{13}. \tag{2.39}$$

Substituting the resultants defined in terms of the five unknown displacements $(u_0, v_0, w_0, \phi_x, \phi_y)$, we get [13] five differential equations for the five unknown displacements:

$$\begin{aligned}
&A_{11}\left[\frac{\partial^2 u_0}{\partial x^2} + \frac{\partial^3 w_0}{\partial x^3}\right] + A_{12}\left[\frac{\partial^2 v_0}{\partial x \partial y} + \frac{\partial^3 w_0}{\partial x \partial y^2}\right] \\
&+ A_{16}\left[2\frac{\partial^2 u_0}{\partial x \partial y} + \frac{\partial^2 v_0}{\partial x^2} + 3\frac{\partial^3 w_0}{\partial x^2 \partial y}\right] + A_{26}\left[\frac{\partial^2 v_0}{\partial y^2} + \frac{\partial^3 w_0}{\partial y^3}\right] \\
&+ A_{66}\left[\frac{\partial^2 u_0}{\partial y^2} + \frac{\partial^2 v_0}{\partial x \partial y} + 2\frac{\partial^3 w_0}{\partial x \partial y^2}\right] + B_{11}\frac{\partial^2 \phi_x}{\partial x^2} + B_{12}\frac{\partial^2 \phi_y}{\partial x \partial y} \\
&+ B_{16}\left[2\frac{\partial^2 \phi_x}{\partial x \partial y} + \frac{\partial^2 \phi_y}{\partial x^2}\right] + B_{26}\frac{\partial^2 \phi_y}{\partial y^2} \\
&+ B_{66}\left[\frac{\partial^2 \phi_x}{\partial y^2} + \frac{\partial^2 \phi_y}{\partial x \partial y}\right] = I_1\frac{\partial^2 u_0}{\partial t^2} + I_2\frac{\partial^2 \phi_x}{\partial t^2},
\end{aligned} \tag{2.40}$$

$$\begin{aligned}
&A_{22}\left[\frac{\partial^2 v_0}{\partial y^2} + \frac{\partial^3 w_0}{\partial y^3}\right] + A_{12}\left[\frac{\partial^2 u_0}{\partial x \partial y} + \frac{\partial^3 w_0}{\partial x^2 \partial y}\right] \\
&+ A_{16}\left[\frac{\partial^2 u_0}{\partial x^2} + \frac{\partial^3 w_0}{\partial x^3}\right] + A_{26}\left[\frac{\partial^2 u_0}{\partial y^2} + 2\frac{\partial^2 v_0}{\partial x \partial y} + 3\frac{\partial^3 w_0}{\partial x \partial y^2}\right] \\
&+ A_{66}\left[\frac{\partial^2 u_0}{\partial x \partial y} + \frac{\partial^2 v_0}{\partial x^2} + 2\frac{\partial^3 w_0}{\partial x^2 \partial y}\right] + B_{22}\frac{\partial^2 \phi_y}{\partial y^2} + B_{12}\frac{\partial^2 \phi_x}{\partial x \partial y}
\end{aligned}$$

$$+ B_{16}\frac{\partial^2 \phi_x}{\partial x^2} + B_{26}\left[\frac{\partial^2 \phi_x}{\partial y^2} + 2\frac{\partial^2 \phi_y}{\partial x \partial y}\right]$$

$$+ B_{66}\left[\frac{\partial^2 \phi_y}{\partial x^2} + \frac{\partial^2 \phi_x}{\partial x \partial y}\right] = I_1\frac{\partial^2 v_0}{\partial t^2} + I_2\frac{\partial^2 \phi_y}{\partial t^2}, \tag{2.41}$$

$$\kappa A_{55}\left[\frac{\partial^2 w_0}{\partial x^2} + \frac{\partial \varphi_x}{\partial x}\right] + \kappa A_{45}\left[2\frac{\partial^2 w_0}{\partial x \partial y} + \frac{\partial \varphi_y}{\partial x} + \frac{\partial \varphi_x}{\partial y}\right]$$

$$+ \kappa A_{44}\left[\frac{\partial^2 w_0}{\partial y^2} + \frac{\partial \varphi_y}{\partial y}\right]$$

$$+ \frac{\partial}{\partial x}\left[N_{xx}\frac{\partial w_0}{\partial x} + N_{xy}\frac{\partial w_0}{\partial y}\right] + \frac{\partial}{\partial y}\left[N_{yy}\frac{\partial w_0}{\partial y} + N_{xy}\frac{\partial w_0}{\partial x}\right]$$

$$= -p_z + I_1\frac{\partial^2 w_0}{\partial t^2}, \tag{2.42}$$

$$B_{11}\left[\frac{\partial^2 u_0}{\partial x^2} + \frac{\partial^3 w_0}{\partial x^3}\right] + B_{12}\left[\frac{\partial^2 v_0}{\partial x \partial y} + \frac{\partial^3 w_0}{\partial x \partial y^2}\right]$$

$$+ B_{16}\left[2\frac{\partial^2 u_0}{\partial x \partial y} + \frac{\partial^2 v_0}{\partial x^2} + 3\frac{\partial^3 w_0}{\partial x^2 \partial y}\right] + B_{26}\left[\frac{\partial^2 v_0}{\partial y^2} + \frac{\partial^3 w_0}{\partial y^3}\right]$$

$$+ B_{66}\left[\frac{\partial^2 u_0}{\partial y^2} + \frac{\partial^2 v_0}{\partial x \partial y} + 2\frac{\partial^3 w_0}{\partial x \partial y^2}\right] + D_{11}\frac{\partial^2 \phi_x}{\partial x^2} + D_{12}\frac{\partial^2 \phi_y}{\partial x \partial y}$$

$$+ D_{16}\left[2\frac{\partial^2 \phi_x}{\partial x \partial y} + \frac{\partial^2 \phi_y}{\partial x^2}\right] + D_{26}\frac{\partial^2 \phi_y}{\partial y^2} + D_{66}\left[\frac{\partial^2 \phi_x}{\partial y^2} + \frac{\partial^2 \phi_y}{\partial x \partial y}\right]$$

$$- \kappa A_{55}\left[\frac{\partial w_0}{\partial x} + \phi_x\right] - \kappa A_{45}\left[\frac{\partial w_0}{\partial y} + \phi_y\right] = I_3\frac{\partial^2 u_0}{\partial t^2} + I_3\frac{\partial^2 \phi_x}{\partial t^2}, \tag{2.43}$$

$$B_{22}\left[\frac{\partial^2 v_0}{\partial y^2} + \frac{\partial^3 w_0}{\partial y^3}\right] + B_{12}\left[\frac{\partial^2 u_0}{\partial x \partial y} + \frac{\partial^3 w_0}{\partial x^2 \partial y}\right]$$

$$+ B_{16}\left[\frac{\partial^2 u_0}{\partial x^2} + \frac{\partial^3 w_0}{\partial x^3}\right] + B_{26}\left[\frac{\partial^2 u_0}{\partial y^2} + 2\frac{\partial^2 v_0}{\partial x \partial y} + 3\frac{\partial^3 w_0}{\partial x \partial y^2}\right]$$

$$+ B_{66}\left[\frac{\partial^2 u_0}{\partial x \partial y} + \frac{\partial^2 v_0}{\partial x^2} + 2\frac{\partial^3 w_0}{\partial x^2 \partial y}\right] + D_{22}\frac{\partial^2 \phi_y}{\partial y^2} + D_{12}\frac{\partial^2 \phi_x}{\partial x \partial y}$$

$$+ D_{16}\frac{\partial^2 \phi_x}{\partial x^2} + D_{26}\left[\frac{\partial^2 \phi_x}{\partial y^2} + 2\frac{\partial^2 \phi_y}{\partial x \partial y}\right] + D_{66}\left[\frac{\partial^2 \phi_y}{\partial x^2} + \frac{\partial^2 \phi_x}{\partial x \partial y}\right]$$

$$- \kappa A_{44}\left[\frac{\partial w_0}{\partial y} + \phi_y\right] - \kappa A_{45}\left[\frac{\partial w_0}{\partial x} + \phi_x\right] = I_1\frac{\partial^2 v_0}{\partial t^2} + I_2\frac{\partial^2 \phi_y}{\partial t^2}. \tag{2.44}$$

To solve the five differential equations, ten boundary conditions should be supplied in the form of geometric and natural boundary conditions[h].

To obtain the equations of motion for a beam, using the first-order shear deformation theory, presented before for a plate, one should assume that all the derivations in the y direction should vanish, and v and φ_y should be identically zero. This yields for the general case, three coupled equations of motion having the following form (assuming constant properties along the beam):

$$\begin{aligned}
&A_{11}\frac{\partial^2 u_0}{\partial x^2} + B_{11}\frac{\partial^2 \phi_x}{\partial x^2} = I_1\frac{\partial^2 u_0}{\partial t^2} + I_2\frac{\partial^2 \phi_x}{\partial t^2},\\
&B_{11}\frac{\partial^2 u_0}{\partial x^2} + D_{11}\frac{\partial^2 \phi_x}{\partial x^2} - \kappa A_{55}\left[\frac{\partial w_0}{\partial x} + \phi_x\right] = I_3\frac{\partial^2 u_0}{\partial t^2} + I_3\frac{\partial^2 \phi_x}{\partial t^2},\\
&\kappa A_{55}\left[\frac{\partial^2 w_0}{\partial x^2} + \frac{\partial \varphi_x}{\partial x}\right] + N_{xx}\frac{\partial^2 w_0}{\partial x} = -p_z + I_1\frac{\partial^2 w_0}{\partial t^2}
\end{aligned} \tag{2.45}$$

with the following boundary conditions:

$$\begin{aligned}
&A_{11}\frac{\partial u_0}{\partial x} + B_{11}\frac{\partial \phi_x}{\partial x} = -N_{xx} \text{ or } u_0 = 0,\\
&A_{55}\left(\phi_x + \frac{\partial w_0}{\partial x}\right) - N_{xx}\frac{\partial w_0}{\partial x} = 0 \text{ or } w_0 = 0,\\
&B_{11}\frac{\partial u_0}{\partial x} + D_{11}\frac{\partial \phi_x}{\partial x} = 0 \text{ or } \phi_x = 0.
\end{aligned} \tag{2.46}$$

The reader is referred to Refs. [14–20], as typical sources for solving buckling and natural frequencies of beams using the FSDT approach.

2.7. Higher Order Theories

The need for higher order shear deformation theories, stem from the fact that although the FSDT approach present accurate results, it does not fulfil the shear free boundary conditions on the top and bottom sides of the beam or a plate. One should remember that the FSDT approach would need far less computer efforts as compared with any higher order theories, which are known as demanding large computer memory.

One of the higher order shear deformation theory was proposed by Reddy [13], and involves a third-order shear deformation theory (TSDT), which will be presented next.

[h]For further discussion about the types of boundary conditions to be imposed, the reader is referred to Ref. [13].

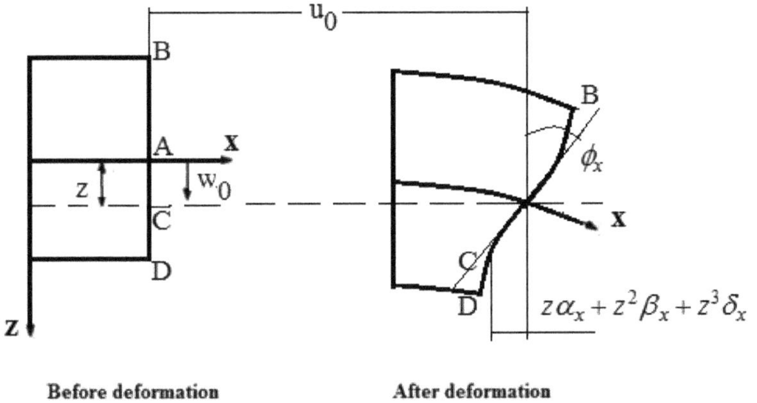

Fig. 2.8. The plate cross-section before and after the deformation (TSDT approach).

Based on the kinematics of the problem (Fig. 2.8) the displacement field has the following components:

$$u(x, y, z, t) = u_0(x, y, t) + z\alpha_x(x, y, t) + z^2\beta_x(x, y, t) + z^3\delta_x(x, y, t),$$
$$v(x, y, z, t) = v_0(x, y, t) + z\alpha_y(x, y, t) + z^2\beta_y(x, y, t) + z^3\delta_y(x, y, t),$$
$$w(x, y, z, t) = w_0(x, y, t). \tag{2.47}$$

The various terms in Eq. (2.47), like, α_x, α_y, β_x, β_y, δ_x and δ_y, are functions to be determined, having the following values at $z = 0$:

$$\alpha_x = \left[\frac{\partial u}{\partial z}\right], \alpha_y = \left[\frac{\partial v}{\partial z}\right], \beta_x = \frac{1}{2}\left[\frac{\partial^2 u}{\partial z^2}\right],$$
$$\beta_y = \frac{1}{2}\left[\frac{\partial^2 v}{\partial z^2}\right], \delta_x = \frac{1}{6}\left[\frac{\partial^3 u}{\partial z^3}\right], \delta_y = \frac{1}{6}\left[\frac{\partial^3 v}{\partial z^3}\right]. \tag{2.48}$$

Accordingly, nine independent unknowns are to be found from nine second-order partial differential equations. However, the number of independent unknowns can be reduced for certain boundary conditions, like traction free boundary conditions on the top and bottom layer of the plate [13], namely[i]

$$\tau_{xz}\left(x, y, +\frac{h}{2}, t\right) = 0, \quad \tau_{xz}\left(x, y, -\frac{h}{2}, t\right) = 0,$$
$$\tau_{yz}\left(x, y, +\frac{h}{2}, t\right) = 0, \quad \tau_{yz}\left(x, y, -\frac{h}{2}, t\right) = 0. \tag{2.49}$$

[i] h being the total height of the beam (or plate), while x-axis is along the beam's (or plate's) length, while y and z are perpendicular to it.

For arbitrary values of Q_{55}, Q_{45}, and Q_{44} (see also Eqs. (2.37)–(2.39)), one can show that the boundary equations (2.49) can be written in terms of the shear strains (see also [13]), using the following expressions:

$$\gamma_{xz}\left(x,y,+\frac{h}{2},t\right)=0,\gamma_{xz}\left(x,y,-\frac{h}{2},t\right)=0,$$
$$\gamma_{yz}\left(x,y,+\frac{h}{2},t\right)=0,\gamma_{yz}\left(x,y,-\frac{h}{2},t\right)=0. \quad (2.50)$$

Performing the necessary derivations, to express the shear strains using the assumed strain field (Eq. (2.47)), one gets

$$\gamma_{xz}\equiv\frac{\partial w}{\partial x}+\frac{\partial u}{\partial z}=\frac{\partial w_0}{\partial x}+\alpha_x+2z\beta_x+3z^2\delta_x,$$
$$\gamma_{yz}\equiv\frac{\partial w}{\partial y}+\frac{\partial v}{\partial z}=\frac{\partial w_0}{\partial y}+\alpha_y+2z\beta_y+3z^2\delta_y. \quad (2.51)$$

Substituting the boundary conditions (2.50) into (2.51), we get the following equations:

$$\text{at } z=+\frac{h}{2}:\gamma_{xz}=\frac{\partial w_0}{\partial x}+\alpha_x+h\beta_x+\frac{3h^2}{4}\delta_x=0,$$
$$\gamma_{yz}=\frac{\partial w_0}{\partial y}+\alpha_y+h\beta_y+\frac{3h^2}{4}\delta_y=0,$$
$$\text{at } z=-\frac{h}{2}:\gamma_{xz}=\frac{\partial w_0}{\partial x}+\alpha_x-h\beta_x+\frac{3h^2}{4}\delta_x=0,$$
$$\gamma_{yz}=\frac{\partial w_0}{\partial y}+\alpha_y-h\beta_y+\frac{3h^2}{4}\delta_y=0. \quad (2.52)$$

Solving Eq. (2.52) yields the following expressions:

$$\beta_x=0:\ \delta_x=-\frac{4}{3h^2}\left(\alpha_x+\frac{\partial w_o}{\partial x}\right),$$
$$\beta_y=0:\delta_y=-\frac{4}{3h^2}\left(\alpha_y+\frac{\partial w_o}{\partial y}\right). \quad (2.53)$$

Back substituting the expression (2.53) into (2.47), provides the strain field for the posed problem:

$$u(x,y,z,t)=u_0(x,y,t)+z\alpha_x(x,y,t)-z^3\frac{4}{3h^2}\left[\alpha_x(x,y,t)+\frac{\partial w_0(x,y,t)}{\partial x}\right],$$
$$v(x,y,z,t)=v_0(x,y,t)+z\alpha_y(x,y,t)-z^3\frac{4}{3h^2}\left[\alpha_y(x,y,t)+\frac{\partial w_0(x,y,t)}{\partial y}\right],$$
$$w(x,y,z,t)=w_0(x,y,t). \quad (2.54)$$

One should notice that the number of unknowns had been reduced to only 5, similar to the case of FSTD approach being applied to a rectangular plate.

For the case of small strains and moderate rotations, the strain–displacement relations can be shown as follows:

$$\varepsilon_{xx} = \frac{\partial u}{\partial x} + \frac{1}{2}\left(\frac{\partial w}{\partial x}\right)^2, \quad \varepsilon_{yy} = \frac{\partial v}{\partial y} + \frac{1}{2}\left(\frac{\partial w}{\partial y}\right)^2, \quad \varepsilon_{zz} = \frac{\partial w}{\partial z},$$
$$\varepsilon_{xy} = \frac{1}{2}\left(\frac{\partial u}{\partial y} + \frac{\partial v}{\partial x} + \frac{\partial^2 w}{\partial x \partial y}\right), \quad \varepsilon_{xz} = \frac{1}{2}\left(\frac{\partial u}{\partial z} + \frac{\partial w}{\partial x}\right), \quad \varepsilon_{yz} = \frac{1}{2}\left(\frac{\partial v}{\partial z} + \frac{\partial w}{\partial y}\right). \tag{2.55}$$

Substituting the displacement field, defined by Eq. (2.54), into the nonlinear strain–displacement equations, Eq. (2.55)) yields the following expressions:

$$\begin{Bmatrix} \varepsilon_{xx} \\ \varepsilon_{yy} \\ \gamma_{xy} \end{Bmatrix} = \begin{Bmatrix} \frac{\partial u_0}{\partial x} + \frac{1}{2}\left(\frac{\partial w_0}{\partial x}\right)^2 \\ \frac{\partial v_0}{\partial y} + \frac{1}{2}\left(\frac{\partial w_0}{\partial y}\right)^2 \\ \frac{\partial u_0}{\partial y} + \frac{\partial v_0}{\partial x} + \frac{\partial^2 w_0}{\partial x \partial y} \end{Bmatrix} + z \begin{Bmatrix} \frac{\partial \alpha_x}{\partial x} \\ \frac{\partial \alpha_y}{\partial y} \\ \frac{\partial \alpha_x}{\partial y} + \frac{\partial \alpha_y}{\partial x} \end{Bmatrix}$$
$$- z^3 \left(\frac{4}{3h^2}\right) \begin{Bmatrix} \frac{\partial \alpha_x}{\partial x} + \frac{\partial^2 w_0}{\partial x^2} \\ \frac{\partial \alpha_y}{\partial y} + \frac{\partial^2 w_0}{\partial y^2} \\ \frac{\partial \alpha_x}{\partial y} + \frac{\partial \alpha_y}{\partial x} + 2\frac{\partial^2 w_0}{\partial x \partial y} \end{Bmatrix},$$
$$\begin{Bmatrix} \gamma_{xz} \\ \gamma_{yz} \end{Bmatrix} = \begin{Bmatrix} \alpha_x + \frac{\partial w_0}{\partial x} \\ \alpha_y + \frac{\partial w_0}{\partial y} \end{Bmatrix} - z^2 \left(\frac{4}{h^2}\right) \begin{Bmatrix} \alpha_x + \frac{\partial w_0}{\partial x} \\ \alpha_y + \frac{\partial w_0}{\partial y} \end{Bmatrix} \quad \text{and } \varepsilon_{zz} = 0. \tag{2.56}$$

where

$$\gamma_{xy} \equiv 2\varepsilon_{xy}, \gamma_{xz} \equiv 2\varepsilon_{xz}, \gamma_{yz} \equiv 2\varepsilon_{yz}. \tag{2.57}$$

The stress resultant–strain relationship can be shown to have the following form (see also Ref. [13]):

$$\begin{Bmatrix} \{N\} \\ \{M\} \\ \{T\} \end{Bmatrix} = \begin{bmatrix} [A] & [B] & [E] \\ [B] & [D] & [F] \\ [E] & [F] & [G] \end{bmatrix} \begin{Bmatrix} \{\varepsilon^{(a)}\} \\ \{\varepsilon^{(b)}\} \\ \{\varepsilon^{(c)}\} \end{Bmatrix}, \tag{2.58}$$

$$\begin{Bmatrix} \{Q\} \\ \{S\} \end{Bmatrix} = \begin{bmatrix} [A] & [D] \\ [D] & [F] \end{bmatrix} \begin{Bmatrix} \{\gamma^{(a)}\} \\ \{\gamma^{(b)}\} \end{Bmatrix}, \tag{2.59}$$

where the terms $\{T\}$ and $\{S\}$ are higher order stress resultants, stemmed from the present third-order shear deformation theory. The various matrices terms for Eq. (2.58) are defined in the following way:

$$A_{ij} = \sum_{k=1}^{N} \int_{z_k}^{z_{k+1}} \bar{Q}_{ij} dz = \sum_{k=1}^{N} \bar{Q}_{ij}^{(k)} [z_{k+1} - z_k],$$

$$B_{ij} = \sum_{k=1}^{N} \int_{z_k}^{z_{k+1}} \bar{Q}_{ij} z dz = \frac{1}{2} \sum_{k=1}^{N} \bar{Q}_{ij}^{(k)} [(z_{k+1})^2 - (z_k)^2],$$

$$D_{ij} = \sum_{k=1}^{N} \int_{z_k}^{z_{k+1}} \bar{Q}_{ij} z^2 dz = \frac{1}{3} \sum_{k=1}^{N} \bar{Q}_{ij}^{(k)} [(z_{k+1})^3 - (z_k)^3], \text{ where } i,j = 1,2,6,$$

$$E_{ij} = \sum_{k=1}^{N} \int_{z_k}^{z_{k+1}} \bar{Q}_{ij} z^3 dz = \frac{1}{4} \sum_{k=1}^{N} \bar{Q}_{ij}^{(k)} [(z_{k+1})^4 - (z_k)^4],$$

$$F_{ij} = \sum_{k=1}^{N} \int_{z_k}^{z_{k+1}} \bar{Q}_{ij} z^4 dz = \frac{1}{5} \sum_{k=1}^{N} \bar{Q}_{ij}^{(k)} [(z_{k+1})^5 - (z_k)^5],$$

$$G_{ij} = \sum_{k=1}^{N} \int_{z_k}^{z_{k+1}} \bar{Q}_{ij} z^6 dz = \frac{1}{7} \sum_{k=1}^{N} \bar{Q}_{ij}^{(k)} [(z_{k+1})^7 - (z_k)^7], \tag{2.60}$$

while for Eq. (2.59) their definition is

$$A_{lm} = \sum_{k=1}^{N} \int_{z_k}^{z_{k+1}} \bar{Q}_{lm} dz = \sum_{k=1}^{N} \bar{Q}_{lm}^{(k)} [z_{k+1} - z_k],$$

$$D_{lm} = \sum_{k=1}^{N} \int_{z_k}^{z_{k+1}} \bar{Q}_{lm} z^2 dz = \frac{1}{3} \sum_{k=1}^{N} \bar{Q}_{lm}^{(k)} [(z_{k+1})^3 - (z_k)^3], \text{ where } l,m = 4,5.$$

$$F_{lm} = \sum_{k=1}^{N} \int_{z_k}^{z_{k+1}} \bar{Q}_{lm} z^4 dz = \frac{1}{5} \sum_{k=1}^{N} \bar{Q}_{lm}^{(k)} [(z_{k+1})^5 - (z_k)^5]. \tag{2.61}$$

Here

$$\{\varepsilon^{(a)}\} \equiv \left\{ \begin{array}{c} \frac{\partial u_0}{\partial x} + \frac{1}{2}\left(\frac{\partial w_0}{\partial x}\right)^2 \\ \frac{\partial v_0}{\partial y} + \frac{1}{2}\left(\frac{\partial w_0}{\partial y}\right)^2 \\ \frac{\partial u_0}{\partial y} + \frac{\partial v_0}{\partial x} + \frac{\partial^2 w_0}{\partial x \partial y} \end{array} \right\}, \quad \{\varepsilon^{(b)}\} \equiv \left\{ \begin{array}{c} \frac{\partial \alpha_x}{\partial x} \\ \frac{\partial \alpha_y}{\partial y} \\ \frac{\partial \alpha_x}{\partial y} + \frac{\partial \alpha_y}{\partial x} \end{array} \right\},$$

$$\{\varepsilon^{(c)}\} \equiv -\left(\frac{4}{3h^2}\right) \left\{ \begin{array}{c} \frac{\partial \alpha_x}{\partial x} + \frac{\partial^2 w_0}{\partial x^2} \\ \frac{\partial \alpha_y}{\partial y} + \frac{\partial^2 w_0}{\partial y^2} \\ \frac{\partial \alpha_x}{\partial y} + \frac{\partial \alpha_y}{\partial x} + 2\frac{\partial^2 w_0}{\partial x \partial y} \end{array} \right\},$$

$$\{\gamma^{(a)}\} \equiv \left\{ \begin{array}{c} \alpha_x + \frac{\partial w_0}{\partial x} \\ \alpha_y + \frac{\partial w_0}{\partial y} \end{array} \right\}, \quad \{\gamma^{(b)}\} \equiv -\left(\frac{4}{h^2}\right) \left\{ \begin{array}{c} \alpha_x + \frac{\partial w_0}{\partial x} \\ \alpha_y + \frac{\partial w_0}{\partial y} \end{array} \right\}. \tag{2.62}$$

Finally, the equations of motion, expressed in stress resultants, for a third-order shear deformation theory can be written as (see [13])

$$\frac{\partial N_{xx}}{\partial x}+\frac{\partial N_{xy}}{\partial y}=I_0\ddot{u}_0+\left(I_1-\frac{4}{3h^2}I_3\right)\ddot{\alpha}_x-\frac{4}{3h^2}I_3\frac{\partial \ddot{w}_0}{\partial x}, \tag{2.63a}$$

$$\frac{\partial N_{xy}}{\partial x}+\frac{\partial N_{yy}}{\partial y}=I_0\ddot{v}_0+\left(I_1-\frac{4}{3h^2}I_3\right)\ddot{\alpha}_y-\frac{4}{3h^2}I_3\frac{\partial \ddot{w}_0}{\partial y}, \tag{2.63b}$$

$$\begin{aligned}&\frac{\partial \bar{Q}_x}{\partial x}+\frac{\partial \bar{Q}_y}{\partial y}+\frac{\partial}{\partial x}\left(N_{xx}\frac{\partial w_0}{\partial x}+N_{xy}\frac{\partial w_0}{\partial y}\right)\\&\quad+\frac{\partial}{\partial y}\left(N_{xy}\frac{\partial w_0}{\partial y}+N_{yy}\frac{\partial w_0}{\partial y}\right)+\frac{4}{3h^2}\left(\frac{\partial^2 T_{xx}}{\partial x^2}+2\frac{\partial^2 T_{xy}}{\partial x\partial y}+\frac{\partial^2 T_{yy}}{\partial y^2}\right)+q\\&\quad=I_0\ddot{w}_0-\frac{16}{9h^4}I_6\left(\frac{\partial^2\ddot{w}_0}{\partial x^2}+\frac{\partial^2\ddot{w}_0}{\partial y^2}\right)+\frac{4}{3h^2}\left[I_3\left(\frac{\partial \ddot{w}_0}{\partial x}+\frac{\partial \ddot{w}_0}{\partial y}\right)\right.\\&\quad\left.+\left(I_4-\frac{4}{3h^2I_6}\right)\left(\frac{\partial\ddot{\alpha}_x}{\partial x}+\frac{\partial\ddot{\alpha}_y}{\partial y}\right)\right],\end{aligned} \tag{2.63c}$$

$$\begin{aligned}&\frac{\partial \bar{M}_{xx}}{\partial x}+\frac{\partial \bar{M}_{xy}}{\partial y}-\bar{Q}_x=\left(I_1-\frac{4}{3h^2}I_3\right)\ddot{u}_0+\left(I_2-\frac{8}{3h^2}I_4+\frac{16}{9h^4}I_6\right)\ddot{\alpha}_x\\&\quad-\frac{4}{3h^2}\left(I_4-\frac{4}{3h^2}I_6\right)\frac{\partial\ddot{w}_0}{\partial x},\end{aligned} \tag{2.63d}$$

$$\begin{aligned}&\frac{\partial \bar{M}_{xy}}{\partial x}+\frac{\partial \bar{M}_{yy}}{\partial y}-\bar{Q}_y=\left(I_1-\frac{4}{3h^2}I_3\right)\ddot{v}_0+\left(I_2-\frac{8}{3h^2}I_4+\frac{16}{9h^4}I_6\right)\ddot{\alpha}_y\\&\quad-\frac{4}{3h^2}\left(I_4-\frac{4}{3h^2}I_6\right)\frac{\partial\ddot{w}_0}{\partial y},\end{aligned} \tag{2.63e}$$

where

$$\begin{aligned}&I_0=\sum_{k=1}^{N}\int_{z_k}^{z_{k+1}}\rho^{(k)}dz,\quad I_1=\sum_{k=1}^{N}\int_{z_k}^{z_{k+1}}\rho^{(k)}zdz,\quad I_2=\sum_{k=1}^{N}\int_{z_k}^{z_{k+1}}\rho^{(k)}z^2dz,\\&I_3=\sum_{k=1}^{N}\int_{z_k}^{z_{k+1}}\rho^{(k)}z^3dz,\quad I_4=\sum_{k=1}^{N}\int_{z_k}^{z_{k+1}}\rho^{(k)}z^4dz,\\&I_5=\sum_{k=1}^{N}\int_{z_k}^{z_{k+1}}\rho^{(k)}z^5dz,\quad I_6=\sum_{k=1}^{N}\int_{z_k}^{z_{k+1}}\rho^{(k)}z^6dz,\end{aligned} \tag{2.64}$$

and

$$\begin{aligned}&\bar{M}_{xx}=M_{xx}-\frac{4}{3h^2}T_{xx},\bar{M}_{xy}=M_{xy}-\frac{4}{3h^2}T_{xy},\bar{M}_{yy}=M_{yy}-\frac{4}{3h^2}T_{yy},\\&\bar{Q}_x=Q_4-\frac{4}{h^2}S_4,\bar{Q}_y=Q_5-\frac{4}{h^2}S_5.\end{aligned} \tag{2.65}$$

Solution of Eqs. (2.63a)–(2.63e), will lead to the finding of the five unknowns, namely the displacements u_0, v_0, w_0 and the rotations α_x, α_y. Reddy [13] presents simplified results of the third-order theory, by neglecting the higher order terms (T_{xx}, T_{xy}, T_{yy}), while keeping the other higher order terms (S_x, S_y), however, the resulting theory comes out to be inconsistent in the energy sense.

A simpler high-order theory, a second-order shear deformation, was developed and presented in [21] by Khdeir and Reddy. They propose the following displacement field:

$$\begin{aligned} u(x, y, z, t) &= u_0(x, y, t) + z\alpha_x(x, y, t) + z^2\beta_x(x, y, t), \\ v(x, y, z, t) &= v_0(x, y, t) + z\alpha_y(x, y, t) + z^2\beta_y(x, y, t), \\ w(x, y, z, t) &= w_0(x, y, t). \end{aligned} \tag{2.66}$$

For the case of linear strains, the strain–displacement relations can be written as

$$\begin{aligned} \varepsilon_{xx} &\equiv \frac{\partial u}{\partial x} = \frac{\partial u_0}{\partial x} + z\frac{\partial \alpha_x}{\partial x} + z^2\frac{\partial \beta_x}{\partial x}, \\ \varepsilon_{yy} &\equiv \frac{\partial v_0}{\partial y} + z\frac{\partial \alpha_y}{\partial y} + z^2\frac{\partial \beta_y}{\partial y}, \quad \varepsilon_{zz} \equiv \frac{\partial w}{\partial z} = 0, \\ \varepsilon_{xy} &\equiv \frac{1}{2}\left(\frac{\partial u}{\partial y} + \frac{\partial v}{\partial x}\right) = \frac{1}{2}\left[\left(\frac{\partial u_0}{\partial y} + \frac{\partial v_0}{\partial x}\right) + z\left(\frac{\partial \alpha_x}{\partial y} + \frac{\partial \alpha_y}{\partial x}\right)\right. \\ &\quad \left. + z^2\left(\frac{\partial \beta_x}{\partial y} + \frac{\partial \beta_y}{\partial x}\right)\right], \\ \varepsilon_{xz} &\equiv \frac{1}{2}\left(\frac{\partial u}{\partial z} + \frac{\partial w}{\partial x}\right) = \frac{1}{2}\left[\left(\alpha_x + \frac{\partial w_0}{\partial x}\right) + 2z\beta_x\right], \\ \varepsilon_{yz} &\equiv \frac{1}{2}\left(\frac{\partial v}{\partial z} + \frac{\partial w}{\partial y}\right) = \frac{1}{2}\left[\left(\alpha_y + \frac{\partial w_0}{\partial x}\right) + 2z\beta_y\right]. \end{aligned} \tag{2.67}$$

The seven equations of motion presented in stress-resultant terms, for the seven unknown displacements (u_0, v_0, w_0, α_x, α_y, β_x and β_y), assumed for the second-order shear deformation theory, can be written as (see also [21])

$$\frac{\partial N_{xx}}{\partial x} + \frac{\partial N_{xy}}{\partial y} = I_1\ddot{u}_0 + I_2\ddot{\alpha}_x + I_3\ddot{\alpha}_y, \tag{2.68a}$$

$$\frac{\partial N_{xy}}{\partial x} + \frac{\partial N_{yy}}{\partial y} = I_1\ddot{v}_0 + I_2\ddot{\beta}_x + I_3\ddot{\beta}_y, \tag{2.68b}$$

$$\frac{\partial Q_x}{\partial x} + \frac{\partial Q_y}{\partial y} + q = I_1\ddot{w}_0, \tag{2.68c}$$

$$\frac{\partial M_{xx}}{\partial x} + \frac{\partial M_{xy}}{\partial y} - Q_x = I_2\ddot{u}_0 + I_3\ddot{\alpha}_x + I_4\ddot{\alpha}_y, \tag{2.68d}$$

$$\frac{\partial M_{xy}}{\partial x} + \frac{\partial M_{yy}}{\partial y} - Q_y = I_2\ddot{v}_0 + I_3\ddot{\beta}_x + I_4\ddot{\beta}_y, \tag{2.68e}$$

$$\frac{\partial T_{xx}}{\partial x} + \frac{\partial T_{xy}}{\partial y} - 2S_x = I_3\ddot{u}_0 + I_4\ddot{\alpha}_x + I_5\ddot{\alpha}_y, \tag{2.68f}$$

$$\frac{\partial T_{xy}}{\partial x} + \frac{\partial T_{yy}}{\partial y} - 2S_y = I_3\ddot{v}_0 + I_4\ddot{\beta}_x + I_5\ddot{\beta}_y, \tag{2.68g}$$

where

$$I_1 = \sum_{k=1}^{N}\int_{z_k}^{z_{k+1}} \rho^{(k)} dz, I_2 = \sum_{k=1}^{N}\int_{z_k}^{z_{k+1}} \rho^{(k)} z dz, I_3 = \sum_{k=1}^{N}\int_{z_k}^{z_{k+1}} \rho^{(k)} z^2 dz;$$

$$I_4 = \sum_{k=1}^{N}\int_{z_k}^{z_{k+1}} \rho^{(k)} z^3 dz, I_5 = \sum_{k=1}^{N}\int_{z_k}^{z_{k+1}} \rho^{(k)} z^4 dz. \tag{2.69}$$

The stress resultant–strain relationships, based on the above second-order displacement field, can be written as

$$\begin{Bmatrix} \{N\} \\ \{M\} \\ \{T\} \end{Bmatrix} = \begin{bmatrix} [A] & [B] & [D] \\ [B] & [D] & [E] \\ [D] & [E] & [F] \end{bmatrix} \begin{Bmatrix} \{\varepsilon^{(a)}\} \\ \{\varepsilon^{(b)}\} \\ \{\varepsilon^{(c)}\} \end{Bmatrix}, \tag{2.70}$$

$$\begin{Bmatrix} \{Q\} \\ \{S\} \end{Bmatrix} = \begin{bmatrix} [A] & [B] \\ [B] & [D] \end{bmatrix} \begin{Bmatrix} \{\gamma^{(a)}\} \\ \{\gamma^{(b)}\} \end{Bmatrix}, \tag{2.71}$$

while for Eq. (2.70), the components are defined as

$$A_{ij} = \sum_{k=1}^{N}\int_{z_k}^{z_{k+1}} \bar{Q}_{ij} dz = \sum_{k=1}^{N} \bar{Q}_{ij}^{(k)} [z_{k+1} - z_k],$$

$$B_{ij} = \sum_{k=1}^{N}\int_{z_k}^{z_{k+1}} \bar{Q}_{ij} z dz = \frac{1}{2}\sum_{k=1}^{N} \bar{Q}_{ij}^{(k)} [(z_{k+1})^2 - (z_k)^2],$$

$$D_{ij} = \sum_{k=1}^{N}\int_{z_k}^{z_{k+1}} \bar{Q}_{ij} z^2 dz = \frac{1}{3}\sum_{k=1}^{N} \bar{Q}_{ij}^{(k)} [(z_{k+1})^3 - (z_k)^3], \text{ where } i,j = 1,2,6,$$

$$E_{ij} = \sum_{k=1}^{N}\int_{z_k}^{z_{k+1}} \bar{Q}_{ij} z^3 dz = \frac{1}{4}\sum_{k=1}^{N} \bar{Q}_{ij}^{(k)} [(z_{k+1})^4 - (z_k)^4],$$

$$F_{ij} = \sum_{k=1}^{N}\int_{z_k}^{z_{k+1}} \bar{Q}_{ij} z^4 dz = \frac{1}{5}\sum_{k=1}^{N} \bar{Q}_{ij}^{(k)} [(z_{k+1})^5 - (z_k)^5], \tag{2.72}$$

while for (Eq. (2.71)) their definition is

$$A_{lm} = \sum_{k=1}^{N} \int_{z_k}^{z_{k+1}} \bar{Q}_{lm} dz = \sum_{k=1}^{N} \bar{Q}_{lm}^{(k)} [z_{k+1} - z_k],$$

$$B_{lm} = \sum_{k=1}^{N} \int_{z_k}^{z_{k+1}} \bar{Q}_{lm} z dz = \frac{1}{2} \sum_{k=1}^{N} \bar{Q}_{lm}^{(k)} [(z_{k+1})^2 - (z_k)^2], \text{ where } l, m = 4, 5.$$

$$D_{lm} = \sum_{k=1}^{N} \int_{z_k}^{z_{k+1}} \bar{Q}_{lm} z^2 dz = \frac{1}{3} \sum_{k=1}^{N} \bar{Q}_{lm}^{(k)} [(z_{k+1})^3 - (z_k)^3], \tag{2.73}$$

Here

$$\{\varepsilon^{(a)}\} \equiv \begin{Bmatrix} \frac{\partial u_0}{\partial x} \\ \frac{\partial v_0}{\partial y} \\ \frac{\partial u_0}{\partial y} + \frac{\partial v_0}{\partial x} \end{Bmatrix}, \{\varepsilon^{(b)}\} \equiv \begin{Bmatrix} \frac{\partial \alpha_x}{\partial x} \\ \frac{\partial \alpha_y}{\partial y} \\ \frac{\partial \alpha_x}{\partial y} + \frac{\partial \alpha_y}{\partial x} \end{Bmatrix}, \{\varepsilon^{(c)}\} \equiv \begin{Bmatrix} \frac{\partial \beta_x}{\partial x} \\ \frac{\partial \beta_y}{\partial y} \\ \frac{\partial \beta_x}{\partial y} + \frac{\partial \beta_y}{\partial x} \end{Bmatrix},$$

$$\{\gamma^{(a)}\} \equiv \begin{Bmatrix} \alpha_x + \frac{\partial w_0}{\partial x} \\ \alpha_y + \frac{\partial w_0}{\partial y} \end{Bmatrix}, \{\gamma^{(b)}\} \equiv 2 \begin{Bmatrix} \beta_x \\ \beta_y \end{Bmatrix} \tag{2.74}$$

and

$$\gamma_{xy} \equiv 2\varepsilon_{xy}, \quad \gamma_{xz} \equiv 2\varepsilon_{xz}, \quad \gamma_{yz} \equiv 2\varepsilon_{yz}. \tag{2.75}$$

The equations of motion (2.68a)–(2.68g) can be expressed using the seven unknown displacements to yield (see [21])

$$A_{11}\frac{\partial^2 u_0}{\partial x^2} + A_{66}\frac{\partial^2 u_0}{\partial y^2} + (A_{12} + A_{66})\frac{\partial^2 v_0}{\partial x \partial y} + 2B_{16}\frac{\partial^2 \alpha_x}{\partial x \partial y} + B_{16}\frac{\partial^2 \beta_x}{\partial x^2},$$
$$B_{26}\frac{\partial^2 \beta_x}{\partial y^2} + D_{11}\frac{\partial^2 \alpha_y}{\partial x^2} + D_{66}\frac{\partial^2 \alpha_y}{\partial y^2} + (D_{12} + D_{66})\frac{\partial^2 \beta_y}{\partial x \partial y} = I_1 \ddot{u}_0 + I_3 \ddot{\alpha}_y, \tag{2.76a}$$

$$A_{22}\frac{\partial^2 v_0}{\partial y^2} + A_{66}\frac{\partial^2 v_0}{\partial x^2} + (A_{12} + A_{66})\frac{\partial^2 u_0}{\partial x \partial y} + B_{16}\frac{\partial^2 \alpha_x}{\partial x^2} + B_{26}\frac{\partial^2 \alpha_x}{\partial y^2},$$
$$2B_{26}\frac{\partial^2 \beta_x}{\partial x \partial y} + D_{22}\frac{\partial^2 \beta_y}{\partial y^2} + D_{66}\frac{\partial^2 \beta_y}{\partial x^2} + (D_{12} + D_{66})\frac{\partial^2 \alpha_y}{\partial x \partial y} = I_1 \ddot{v}_0 + I_3 \ddot{\beta}_y, \tag{2.76b}$$

$$A_{55}\left(\frac{\partial \alpha_x}{\partial x}+\frac{\partial^2 w_0}{\partial x^2}\right)+A_{44}\left(\frac{\partial \beta_x}{\partial y}+\frac{\partial^2 w_0}{\partial y^2}\right)$$
$$+2B_{45}\left(\frac{\partial \beta_y}{\partial x}+\frac{\partial \alpha_y}{\partial y}\right)+q=I_1\ddot{w}_0, \tag{2.76c}$$

$$2B_{16}\frac{\partial^2 u_0}{\partial x\partial y}+B_{16}\frac{\partial^2 v_0}{\partial x^2}+B_{26}\frac{\partial^2 v_0}{\partial y^2}+D_{11}\frac{\partial^2 \alpha_x}{\partial x^2}+D_{66}\frac{\partial^2 \alpha_x}{\partial y^2}$$
$$+(D_{12}+D_{66})\frac{\partial^2 \beta_x}{\partial x\partial y}+2E_{16}\frac{\partial^2 \alpha_y}{\partial x\partial y}+E_{16}\frac{\partial^2 \beta_y}{\partial x^2}+E_{26}\frac{\partial^2 \beta_y}{\partial y^2}$$
$$-A_{55}\left(\alpha_x+\frac{\partial w_0}{\partial x}\right)-2B_{45}\beta_y=I_3\ddot{\alpha}_x, \tag{2.76d}$$

$$2B_{26}\frac{\partial^2 v_0}{\partial x\partial y}+B_{16}\frac{\partial^2 u_0}{\partial x^2}+B_{26}\frac{\partial^2 u_0}{\partial y^2}+D_{22}\frac{\partial^2 \beta_x}{\partial y^2}+D_{66}\frac{\partial^2 \beta_x}{\partial x^2}$$
$$+(D_{12}+D_{66})\frac{\partial^2 \alpha_x}{\partial x\partial y}+E_{26}\frac{\partial^2 \alpha_y}{\partial y^2}+2E_{26}\frac{\partial^2 \beta_y}{\partial x\partial y}+E_{16}\frac{\partial^2 \alpha_y}{\partial x^2}$$
$$-A_{44}\left(\alpha_x+\frac{\partial w_0}{\partial y}\right)-2B_{45}\alpha_y=I_3\ddot{\beta}_x, \tag{2.76e}$$

$$D_{11}\frac{\partial^2 u_0}{\partial x^2}+D_{66}\frac{\partial^2 u_0}{\partial y^2}+(D_{12}+D_{66})\frac{\partial^2 v_0}{\partial x\partial y}+2E_{16}\frac{\partial^2 \alpha_x}{\partial x\partial y}$$
$$+E_{16}\frac{\partial^2 \beta_x}{\partial x^2}+E_{26}\frac{\partial^2 \beta_x}{\partial y^2}+F_{11}\frac{\partial^2 \alpha_y}{\partial x^2}+F_{66}\frac{\partial^2 \alpha_y}{\partial y^2}$$
$$+(F_{12}+F_{66})\frac{\partial^2 \beta_y}{\partial x\partial y}-2B_{45}\left(\beta_x+\frac{\partial w_0}{\partial y}\right)-4D_{55}\alpha_y=I_3\ddot{u}_0+I_5\ddot{\alpha}_y, \tag{2.76f}$$

$$D_{22}\frac{\partial^2 v_0}{\partial y^2}+D_{66}\frac{\partial^2 v_0}{\partial x^2}+(D_{12}+D_{66})\frac{\partial^2 u_0}{\partial x\partial y}+2E_{26}\frac{\partial^2 \beta_x}{\partial x\partial y}$$
$$+E_{16}\frac{\partial^2 \alpha_x}{\partial x^2}+E_{26}\frac{\partial^2 \alpha_x}{\partial y^2}+F_{221}\frac{\partial^2 \beta_y}{\partial y^2}$$
$$+F_{66}\frac{\partial^2 \beta_y}{\partial x^2}+(F_{12}+F_{66})\frac{\partial^2 \alpha_y}{\partial x\partial y}$$
$$-2B_{45}\left(\alpha_x+\frac{\partial w_0}{\partial x}\right)-4D_{44}\beta_y=I_3\ddot{v}_0+I_5\ddot{\beta}_y. \tag{2.76g}$$

The solution in [21] is given for a case of an antisymmetric angle-ply laminated rectangular plate with two opposite edges having simply supported boundary conditions (in the x direction) while the other two (in the y direction) may have any arbitrary combinations of free, clamped or simply supported edge conditions. The natural frequencies of such a plate are calculated using a generalized Lèvy type solution, namely selecting the assumed deflections and rotations to fulfill in x direction, the simply supported boundary conditions (these function will be dependent only on x), while in the perpendicular direction, having arbitrary boundary conditions, the functions will be dependent only on y. These functions (in the y direction) will be now the seven unknowns of the problem, after fulfilment of the prescribed boundary conditions in the y direction.

Another approach for using high-order approximation is known in the literature as "the Zig-Zag" theory, as can be found in [22–35]. This approach is best summarized and reviewed by Carrera in [25]. The idea behind this approach is to describe a piecewise continuous displacement field for basic elements like beams, plates and shells in the thickness direction and fulfill interlaminar continuity of the transverse stresses at each layer interface. One should note that in the literature the piecewise assumption of the stresses and the displacement fields are often referred as Zig-Zag (ZZ) and Interlaminar Continuity (IC), respectively, while the theories applying both assumptions are named as Zig-Zag theories. As Carerra [25] pointed out in his review, Lekhnitskii [36] was the first to suggest a theory for multilayered structures, followed by other fundamental contributors like Ambartsumian [37, 38] and Reisner [39, 40].

References

[1] Harris, B., *Engineering Composite Materials*, The Institute of Materials, London, UK, 1999, 193 p.
[2] Jones, R.M., *Mechanics of Composite Materials*, 2nd Edition, Taylor & Francis, Philadelphia, 1999, 519 p.
[3] Love, A.E.H., On the small free vibrations and deformations of elastic shells, *Philos. Trans. Royal Soc. (London)*, 1888, Série A, No. 17 pp. 491–549.
[4] Reddy, J. N., *Theory and Analysis of Elastic Plates and Shells*, CRC Press, Taylor & Francis, 2007.
[5] Timoshenko, S. and Woinowsky-Krieger, S., *Theory of Plates and Shells*, McGraw-Hill New York, 1959.
[6] Mindlin, R.D., Influence of rotatory inertia and shear on flexural motions of isotropic, elastic plates, *ASME Journal of Applied Mechanics*, Vol. 18, 1951, pp. 31–38.
[7] Reddy, J.N., *Theory and Analysis of Elastic Plates*, Taylor & Francis, Philadelphia, 1999.
[8] Lim, G.T. and Reddy, J.N., On canonical bending relationships for plates, *Int. J. Solids Struct.*, Vol. 40, 2003, pp. 3039–3067.

[9] Timoshenko, S.P., On the correction factor for shear of the differential equation for transverse vibrations of bars of uniform cross-section, *Philos. Mag.*, 1921, 744 p.
[10] Timoshenko, S.P., On the transverse vibrations of bars of uniform cross-section, *Philos. Maga.*, 1922, 125 p.
[11] Rosinger, H.E. and Ritchie, I.G., On Timoshenko's correction for shear in vibrating isotropic beams, *J. Phys. D: Appl. Phys.*, Vol. 10, 1977, pp. 1461–1466.
[12] Timoshenko, S.P. and Gere, J.M., *Mechanics of Materials*, Van Nostrand Reinhold Co., 1972.
[13] Reddy, J.N., *Mechanics of Laminated Composite Plates and Shells*, 2nd Edition, CRC Press LLC, 2004.
[14] Chandrashekhara, K., Krishnamurthy, K. and Roy, S., Free vibration of composite beams including rotary inertia and shear deformation, *Composite Struct.*, Vol. 14, Issue 4, 1990, pp. 269–279.
[15] Singh, G. and Venkateswara, R., Analysis of the nonlinear vibrations of unsymmetrically laminated composite beams, *AAIA J.*, Vol. 29, No. 10, 1991.
[16] Abramovich, H. and Livshits, A., Free vibrations of non-symmetric cross-ply laminated composite beams, *J. Sound Vibration*, Volume 176, Issue 5, 6 October 1994, pp. 597–612.
[17] Abramovich, H., Eisenberger, M. and Shulepov, O., Vibrations and buckling of non-symmetric laminated composite beams via the exact element method, *AAIA J.*, 1995, http://arc.aiaa.org/doi/abs/10.2514/6.1995-1459.
[18] Abramovich, H., Eisenberger, M. and Shulepov, O., Vibrations and buckling of cross-ply nonsymmetric laminated composite beams, *AAIA J.*, Vol. 34, No. 5, 1996.
[19] Li, J., Wu, Z., Kong, X., Li, X. and Wu, W., Comparison of various shear deformation theories for free vibration of laminated composite beams with general lay-ups, *Composite Struct.*, Vol. 108, 2014, pp. 767–778.
[20] Abadi, M.M. and Daneshmehr, A.R., An investigation of modified couple stress theory in buckling analysis of micro composite laminated Euler–Bernoulli and Timoshenko beams, *Int. J. Eng. Sci.*, Vol. 75, 2014, pp. 40–53.
[21] Khdeir, A.A. and Reddy, J.N., Free vibrations of laminated composite plates using second-order shear deformation theory. *Composite Struct.*, Vol. 71, Issue 6, 1999, pp. 617–626.
[22] Li, X. and Liu, D., Zigzag theory for composite laminates, *AIAA J.*, Vol. 33, No. 6, 1995, pp. 1163–1165.
[23] Cho, Y.B. and Averill, R.C., First-order zig-zag sublaminate plate theory and finite element model for laminated composite and sandwich panels, *Composite Struct.*, Vol. 50, 2000, pp. 1–15.
[24] Fares, M.E. and Elmarghany, M. Kh., A refined zigzag nonlinear first-order shear deformation theory of composite laminated plates, *Composite Struct.*, Vol. 82, 2008, pp. 71–83.
[25] Carrera, E., Historical review of zig-zag theories for multilayered plates and shells, *Appl. Mech. Rev.*, Vol. 56, No. 3, 2003, pp. 287–308.
[26] Demasi, L., ∞^6 mixed plate theories based on the generalized unified formulation. Part I: Governing equations, *Composite Struct.*, Vol. 87, 2009, pp. 1–11.
[27] Demasi, L., ∞^6 mixed plate theories based on the generalized unified formulation. Part II: Layerwise theories, *Composite Struct.*, Vol. 87, 2009, pp. 12–22.
[28] Demasi, L., ∞^6 mixed plate theories based on the generalized unified formulation. Part III: Advanced mixed high order shear deformation theories, *Composite Struct.*, Vol. 87, 2009, pp. 183–194.

[29] Demasi, L., ∞^6 mixed plate theories based on the generalized unified formulation. Part IV: Zig-zag theories, *Composite Struct.*, Vol. 87, 2009, pp. 195–205.

[30] Demasi, L., ∞^6 mixed plate theories based on the generalized unified formulation. Part V: Results, *Composite Struct.*, Vol. 88, 2009, pp. 1–16.

[31] Sahoo, R. and Singh, B.N., A new inverse hyperbolic zigzag theory for the static analysis of composite and sandwich plates, *Composite Struct.*, Vol. 105, 2013, pp. 385–397.

[32] Tessler, A., Refined zigzag theory for homogenous, laminated composite, and sandwich plates: A homogenous limit methodology for zigzag function selection, NASA/TP-2010-216214, March 2010.

[33] Barut, A., Madenci, E. and Tessler, A., C^0-continuous triangular plate element for laminated composite and sandwich plates using the {2,2}-refined zigzag theory, *Composite Struct.*, Vol. 106, 2013, pp. 835–853.

[34] Sahoo, R. and Singh, B.N., A new trigonometric zigzag theory for buckling and free vibration analysis of laminated composite and sandwich plates, *Composite Struct.*, Vol. 117, 2014, pp. 316–332.

[35] Sahoo, R. and Singh, B.N., A new trigonometric zigzag theory for static analysis of laminated composite and sandwich plates, *Aerospace Sci. Technol.*, Vol. 35, 2014, pp. 15–28.

[36] Lekhnitskii, S.G., Strength calculation of composite beams, *Vestnik Inzhen I Tekhnikov*, No. 9, 1935.

[37] Ambartsumian, S.A., Analysis of two-layer orthotropic shells, *Investiia Akad Nauk SSSR, Ot Tekh Nauk*, No. 7, 1957.

[38] Ambartsumian, S.A., Two analysis method for two-layer orthotropic shells, *Izv An Arm SSR Seiya Fiz-Matem Nauk* X(2), 1957.

[39] Reissner, E., On a certain mixed variational theory and a proposed application, *Int. J. Numer. Methods Eng.* Vol. 20, 1984, pp. 1366–1368.

[40] Reissner, E., On a mixed variational theorem and on a shear deformable plate theory, *Int. J. Numer. Methods Eng.* Vol. 23, 1986, pp. 193–198.

Chapter 3

Buckling of Thin-Walled Structures

3.1. Introduction

This chapter will present the behavior of thin-walled structures like columns, plates and shells. The thin-walled structures can be either isotropic or orthotropic and their critical loads will be analytically evaluated for applicable cases. Finally, some approximate energy methods will also be highlighted, enabling the calculation of buckling loads for cases where no closed-form solutions are available.

3.2. Buckling of Columns

3.2.1. *Euler buckling*

To find the buckling[a] load of a column, having the length L, and stiffness EI[b] resting on simply supported, or hinged-hinged[c] boundary conditions and being compressed by a compressive load P (see Fig. 3.1) one needs to write the moment at a chosen axial coordinate, x, and equate it to the known expression of EId^2z/dx^2. One should note that in Fig. 3.1, the column is described after its deformation, in an exaggerated manner. The result is (see

[a] "*Buckling*" phenomenon has various definitions. One of those definitions *describes* its as the state of a compressed column, when slight increase in the compressive load would induce sudden large lateral (unbounded) deformation of structure, whereas before that compressive load (critical load), the column had experienced little, if any, lateral deformation.
[b] E = Young's Modulus, I = moment of Inertial of the cross-section (for a rectangular cross, having a height of h and width of $b, I = bh^3/12$).
[c] In some of the books and manuscripts, these boundary conditions are referred as *pinned–pinned* boundary conditions.

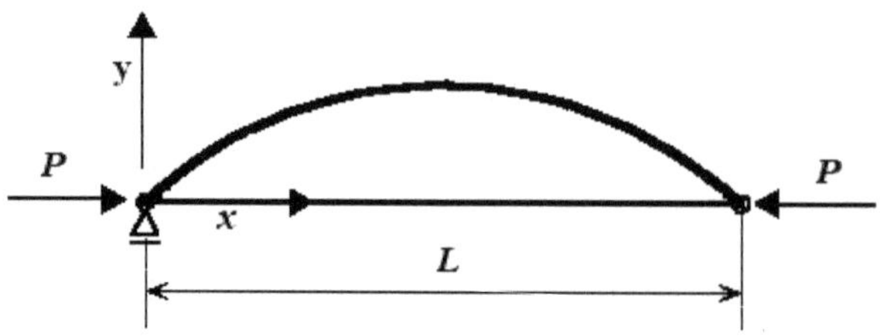

Fig. 3.1. A hinged–hinged column under compressive loads P.

[1, 8]) as follows:

$$EI\frac{d^2y(x)}{dx^2} = -Py(x). \tag{3.1}$$

Rearranging Eq. (3.1) and dividing by EI, we obtain

$$\frac{d^2y(x)}{dx^2} + k^2y(x) = 0, \quad k^2 = \frac{P}{EI}. \tag{3.2}$$

The general solution for Eq. (3.2) is given by

$$y(x) = A\sin(kx) + B\cos(kx), \tag{3.3}$$

while A and B are constants to be determined from the boundary conditions of the problem. For the hinged–hinged column case, the boundary conditions would require the lateral displacement and the moment to vanish at its both ends. One should note that for the equation of motion displayed by Eq. (3.1) the natural boundary conditions, namely the moment is inherently fulfilled. Therefore, we are left with only two geometric boundary conditions which lead to

$$\begin{aligned}
&y(0) = 0 \Rightarrow B = 0,\\
&y(L) = 0 \Rightarrow A\sin(kL) = 0,\\
&\quad A \neq 0 \text{ and } \sin(kL) = 0 \Rightarrow kL = n\pi,\ n = 1, 2, 3, \ldots.
\end{aligned} \tag{3.4}$$

The mathematical result from Eq. (3.4) is that the calculation of the buckling of a column leads to an eigenvalue problem. Substituting the expression for k (see Eq. (3.2)) and taking $n = 1$ leads to the expression for what it is called the *Euler buckling load*:

$$k^2 \equiv \frac{P}{EI} = \left(\frac{n\pi}{L}\right)^2 \Rightarrow P = \left(\frac{n\pi}{L}\right)^2 EI,\ P_{n=1} \equiv P_{\text{Euler}} = \frac{\pi^2 EI}{L^2}. \tag{3.5}$$

For other boundary conditions (beside hinged–hinged), incorporating both geometrical and natural boundary conditions, the equation of motion would have the following form[d]:

$$\frac{d^4y(x)}{dx^4} + k^2\frac{d^2z}{dx^2}(x) = 0,\ k^2 = \frac{P}{EI}. \tag{3.6}$$

The general solution for Eq. (3.6) is then given by

$$y(x) = A\sin(kx) + B\cos(kx) + Cx + D. \tag{3.7}$$

Application of various boundary conditions (see Table 3.1 for the most used one) would provide the relevant buckling load and its associated mode. For an enhanced understanding of the buckling phenomenon, which is a stability issue, the reader is referred to [1, 8–12].

3.2.2. *Rankin–Gordon formula*

The buckling formulas derived in the previous section dealt with slender columns. Slenderness ratio of a column (λ) is defined as the ratio of the effective length of a column (L_{eff} — see last row in Table 3.1) and the least radius of gyration (r) about the axis under consideration. Consequently, this can be expressed as

$$\lambda \equiv \frac{L_{\text{eff}}}{r},\ r = \sqrt{\frac{I}{A}}, \tag{3.8}$$

where A is the cross-section area of the column and I is its moment of inertia.

The Euler buckling formula, developed in the previous section, is not suitable for short stubby columns. There are a few formulas to be used for relatively low slenderness ratio, with the Johnson's parabola [13] and the Rankine–Gordon[e] semi-empirical formula, being the most known ones.

Johnson's parabola [13] is given by the following expression (see also Fig. 3.2):

$$\sigma \equiv \sigma_y - \frac{1}{E}\left(\frac{\sigma_y}{2\pi}\right)^2\left(\frac{L_{\text{eff}}}{r}\right)^2 = \sigma_y - \frac{1}{E}\left(\frac{\sigma_y}{2\pi}\right)^2\lambda^2, \tag{3.9}$$

where σ_y is the yield stress for a given material.

Rankine–Godman semi-empirical formula can be written as

$$\frac{1}{P_{\text{R-G}}} \equiv \frac{1}{P_{\text{Euler}}} + \frac{1}{P_{\text{Crush}}}, \tag{3.10}$$

where $P_{\text{Crush}} \equiv \sigma_y \cdot A$ is the crushing load or yield point load in compression for a given material. Equation (3.10) can be rearranged to show the stress

---

[d]The theory leading to Eq. (3.6) is usually denoted in the literature as the *Bernoulli–Euler* theory.
[e]William John Macquorn Rankine (1820–1872) and Perry Hugesworth Gordon (1894–1966).

Table 3.1. Buckling factors and associated modes.

Hinged-hinged	Clamped-clamped	Clamped-free	Hinged-guided	Clamped-guided	Clamped-hinged
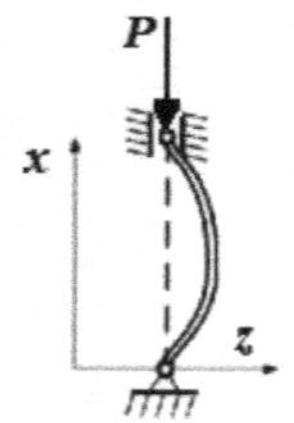	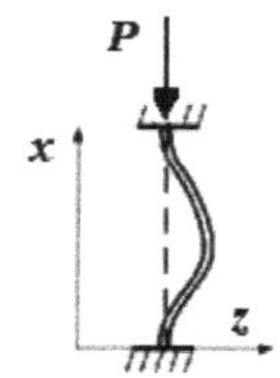	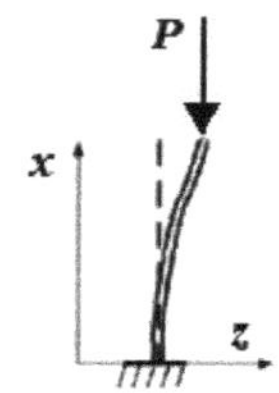	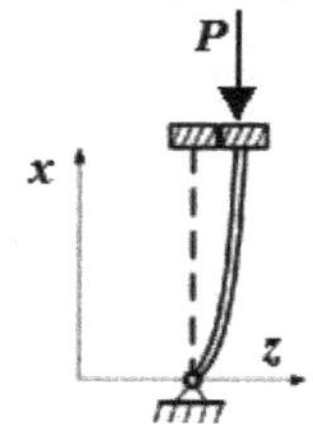	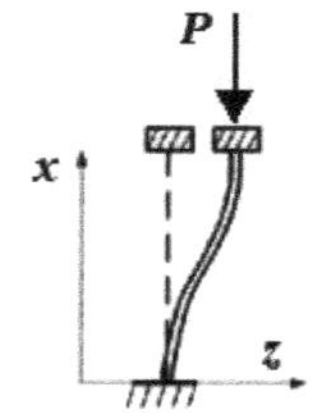	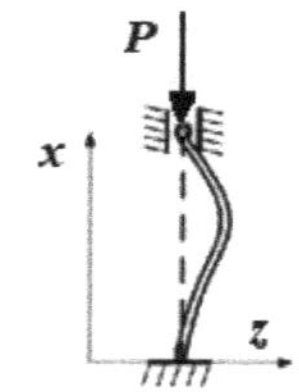
$y(0) = y(L) = 0$ $y''(0) = y''(L) = 0$	$y(0) = y(L) = 0$ $y'(0) = y'(L) = 0$	$y(0) = y'(0) = 0$ $y''(L) = 0$ $y'''(L) + k^2 y'(L) = 0$	$y(0) = y''(0) = 0$ $y'(L) = 0$ $y'''(L) + k^2 y'(L) = 0$	$y(0) = y'(0) = 0$ $y'(L) = 0$ $y'''(L) + k^2 y'(L) = 0$	$y(0) = y(L) = 0$ $y'(0) = y''(L) = 0$
$K = 1$	$K = 4$	$K = 0.25$	$K = 0.25$	$K = 1$	$K = 2.045$
$y(x) = \sin\left(\frac{\pi x}{L}\right)$	$y(x) = 1 - \cos\left(\frac{2\pi x}{L}\right)$	$y(x) = 1 - \cos\left(\frac{\pi x}{2L}\right)$	$y(x) = 1 - \cos\left(\frac{\pi x}{2L}\right)$	$y(x) = 1 - \cos\left(\frac{\pi x}{2L}\right)$	$y(x)$ $= \sin(\alpha x)$ $+\alpha L\left[(1 - \cos\alpha x) - \frac{x}{L}\right]$ $\alpha = 1.4318\frac{\pi}{L}$
$L_{\text{eff}} = L$	$L_{\text{eff}} = 0.5L$	$L_{\text{eff}} = 2L$	$L_{\text{eff}} = 2L$	$L_{\text{eff}} = L$	$L_{\text{eff}} = 0.7L$

Note: $y'(x) \equiv \frac{dy(x)}{dx}; y''(x) \equiv \frac{d^2y(x)}{dx^2}; y'''(x) \equiv \frac{d^3y(x)}{dx^3}$, $P_{(\text{anyBC})} = KP_{(\text{Euler})} = K\frac{\pi^2 EI}{L^2}$, $0 < K \leq 4$; L_{eff} **(effective length) of a column** = the distance between two successive inflection points.

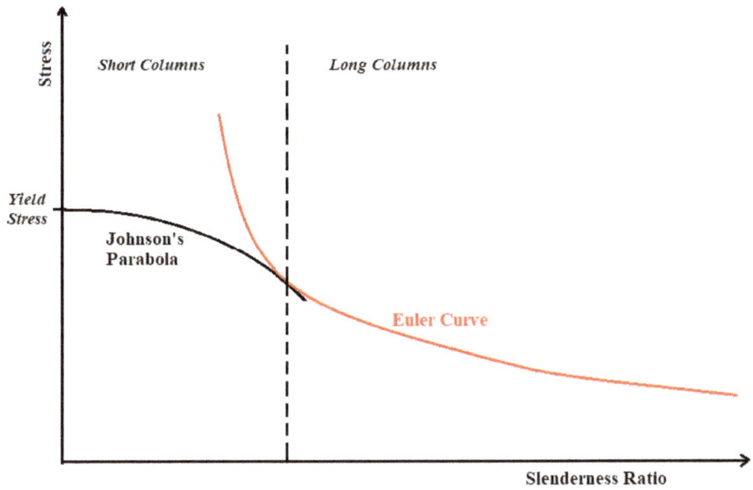

Fig. 3.2. Schematic diagram presenting Euler curve and Johnson's parabola.

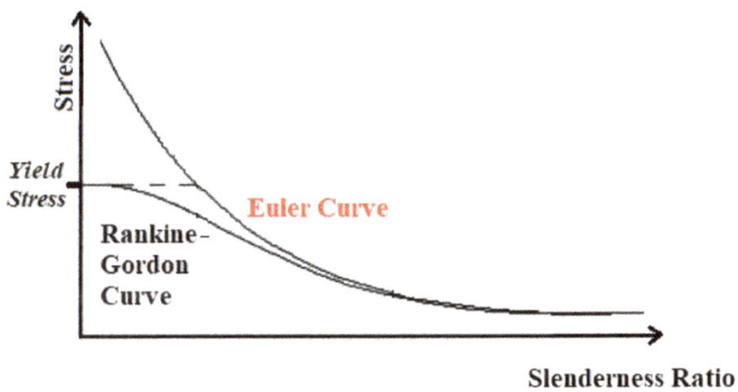

Fig. 3.3. Schematic diagram presenting Euler curve and Rankine–Gordon curve.

in the column according to Rankine–Godman approach to yield

$$\sigma_{R-G} = \frac{\sigma_y}{1 + \frac{\sigma_y}{\pi^2 E}\left(\frac{L_{\text{eff}}}{r}\right)^2} = \frac{\sigma_y}{1 + \frac{\sigma_y}{\pi^2 E}\lambda^2} \tag{3.11}$$

A schematic diagram presenting the Euler curve and the Rankine–Godman curve is shown in Fig. 3.3.

One can see from Fig. 3.3 that for short columns the stress predicted by the Rankine–Gordon curve is tending to the yield stress, while for long columns, both curves would coincide.

3.2.3. *Composite columns — CLT approach*

Buckling of laminated composite columns (see Chapter 2), which are considered as one-dimensional elements (see Fig. 3.4) can be derived for a general

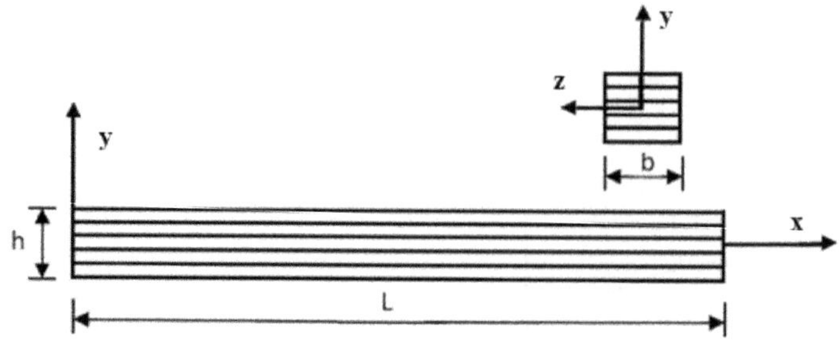

Fig. 3.4. The laminated composite column model.

laminate using the classical lamination theory (CLT), to yield the following equations of motion:

$$A_{11}\left(\frac{d^2u_0}{dx^2}+\frac{d^3w_0}{dx^3}\right)-B_{11}\frac{d^3w_0}{dx^3}=0, \tag{3.12}$$

$$-B_{11}\left(\frac{d^3u_0}{dx^3}+\frac{d^4w_0}{dx^4}\right)+D_{11}\frac{d^4w_0}{dx^4}+\overline{N}_{xx}\frac{d^2w_0}{dx^2}=0, \tag{3.13}$$

while u_0 and w_0 are the displacements in the x and y directions, respectively, A_{11}, B_{11} and D_{11} are the stretching, coupled stretching-bending and bending stiffness, respectively, and $\overline{N}_{xx}$ stands for the axial compression per unit width (b) and it is assumed to be independent of the coordinate x. The two coupled equations of motion (Eqs. (3.12) and (3.13)) can be decoupled to yield one single equation having the following form:

$$\left(D_{11}-\frac{B_{11}^2}{A_{11}}\right)\frac{d^4w_0}{dx^4}+\overline{N}_{xx}\frac{d^2w_0}{dx^2}=0 \tag{3.14}$$

or

$$\frac{d^4w_0}{dx^4}+\frac{\overline{N}_{xx}}{\left(D_{11}-\frac{B_{11}^2}{A_{11}}\right)}\frac{d^2w_0}{dx^2}=0 \tag{3.15}$$

3.2.3.1. *Symmetric laminate* ($B_{11}=0$)

For the case of symmetric laminate, $B_{11}=0$, where w_b being the buckling deflection, one obtains

$$\frac{d^4w_b}{dx^4}+\frac{P}{E_{xx}\cdot I_{yy}}\frac{d^2w_b}{dx^2}=0, \tag{3.16}$$

where

$$b\cdot D_{11}=E_{xx}\cdot I_{yy} \text{ and } b\cdot\overline{N}_{xx}=0 \tag{3.17}$$

One should note that Eq. (3.16) is exactly the Bernoulli–Euler equation of buckling for columns (see Eq. (3.6)).

Now we shall present the general solution for Eq. (3.16), applicable to the buckling of symmetric-laminated composite columns ($B_{11} = 0$). Rearranging Eq. (3.16) we obtain

$$\frac{d^4 w_b}{dx^4} + \lambda^2 \frac{d^2 w_b}{dx^2} = 0, \quad \lambda^2 = \frac{\overline{N}_{xx}}{D_{11}}. \tag{3.18}$$

The solution of Eq. (3.18) has the following general form (which is identical to Eq. (3.7) applied for Bernoulli–Euler columns):

$$w_b(x) = C_1 \sin(\lambda x) + C_2 \cos(\lambda x) + C_3 x + C_4, \tag{3.19}$$

where C_1, C_2, C_3 and C_4 are constant to be determined using four boundary conditions of a given problem. For example, assuming a simply supported column having a length L, one can write its boundary conditions as:

$$\begin{aligned} w_b(0) = 0;\ M_{yy}(0) = 0 \ &\Rightarrow \ \frac{d^2 w_b(0)}{dx^2} \equiv w_b''(0) = 0, \\ w_b(L) = 0;\ M_{yy}(L) = 0 \ &\Rightarrow \ \frac{d^2 w_b(L)}{dx^2} \equiv w_b''(L) = 0. \end{aligned} \tag{3.20}$$

Substituting the boundary conditions in Eq. (3.19) yields a set of four equations with four unknowns, which can be written in a 4×4 matrix while the right-hand sides of the four equations are identically zero. To obtain a unique solution, the determinant of the matrix must vanish. This leads to the following characteristic equation

$$\sin(\lambda L) = 0 \ \Rightarrow \ \lambda L = n\pi, n = 1, 2, 3, 4, \ldots. \tag{3.21}$$

The critical buckling load of the column (the lowest one, $n = 1$) will then be

$$(\overline{N}_{xx})_{cr} = \frac{\pi^2}{L^2} D_{11} \tag{3.22}$$

One should remember that the term in Eq. (3.22) should be multiplied by the width of the beam, b, to obtain P_{cr} [N]. One can see that the buckling load for a laminated symmetric composite column (Eq. (3.22)) is exactly the same as for the Bernoulli–Euler column presented in Eq. (3.5).

The column buckling shape (the eigenfunction) has the following form (this is obtained by back substituting the eigenvalue from Eq. (3.21) into the matrix form):

$$w_b(x) = C_1 \sin(\lambda x) = C_1 \sin\left(\frac{\pi x}{L}\right). \tag{3.23}$$

Table 3.2. Buckling of laminated composite columns: Out-of-plane boundary conditions — CLT approach.

No.	BC name	Out-of-plane boundary conditions $x = 0$	Out-of-plane boundary conditions $x = L$
1	SS–SS*	$w_b = 0;\ \frac{d^2w_b}{dx^2} = 0$	$w_b = 0;\ \frac{d^2w_b}{dx^2} = 0$
2	C–C**	$w_b = 0;\ \frac{dw_b}{dx} = 0$	$w_b = 0;\ \frac{dw_b}{dx} = 0$
3	C–F***	$w_b = 0;\ \frac{dw_b}{dx} = 0$	$\frac{d^2w_b}{dx^2} = 0;\ \frac{d^3w_b}{dx^3} + \lambda^2\frac{dw_b}{dx} = 0$
4	F–F	$\frac{d^2w_b}{dx^2} = 0;\ \frac{d^3w_b}{dx^3} + \lambda^2\frac{dw_b}{dx} = 0$	$\frac{d^2w_b}{dx^2} = 0;\ \frac{d^3w_b}{dx^3} + \lambda^2\frac{dw_b}{dx} = 0$
5	SS–C	$w_b = 0;\ \frac{d^2w_b}{dx^2} = 0$	$w_b = 0;\ \frac{dw_b}{dx} = 0$
6	SS–F	$w_b = 0;\ \frac{d^2w_b}{dx^2} = 0$	$\frac{d^2w_b}{dx^2} = 0;\ \frac{d^3w_b}{dx^3} + \lambda^2\frac{dw_b}{dx} = 0$
7	G****–F	$\frac{dw_b}{dx} = 0;\ \frac{d^3w_b}{dx^3} + \lambda^2\frac{dw_b}{dx} = 0$	$\frac{d^2w_b}{dx^2} = 0;\ \frac{d^3w_b}{dx^3} + \lambda^2\frac{dw_b}{dx} = 0$
8	G–SS	$\frac{dw_b}{dx} = 0;\ \frac{d^3w_b}{dx^3} + \lambda^2\frac{dw_b}{dx} = 0$	$w_b = 0;\ \frac{d^2w_b}{dx^2} = 0$
9	G–G	$\frac{dw_b}{dx} = 0;\ \frac{d^3w_b}{dx^3} + \lambda^2\frac{dw_b}{dx} = 0$	$\frac{dw_b}{dx} = 0;\ \frac{d^3w_b}{dx^3} + \lambda^2\frac{dw_b}{dx} = 0$
10	G–C	$\frac{dw_b}{dx} = 0;\ \frac{d^3w_b}{dx^3} + \lambda^2\frac{dw_b}{dx} = 0$	$w_b = 0;\ \frac{dw_b}{dx} = 0$

Note: *SS = Simply supported or hinged–hinged or pinned–pinned; **C = Clamped; ***F = Free; G**** = Guided.

Tables 3.2 and 3.3[f] present some of the most encountered column cases, while the schematic drawings of the various out-of-plane boundary conditions are presented in Fig. 3.5. One should note that only the out-of-plane boundary conditions can influence the critical buckling load and the shape at buckling (see also Appendix A at the end of the present chapter).

To conclude this section, a column with general boundary conditions is presented in Fig. 3.6.

Defining the following variables as

$$\overline{k}_1 = \frac{k_1}{\overline{D}_{11}};\quad \overline{k}_2 = \frac{k_2}{\overline{D}_{11}};\quad \overline{k}_{\theta_1} = \frac{k_{\theta_1}}{\overline{D}_{11}};\quad \overline{k}_{\theta_2} = \frac{k_{\theta_2}}{\overline{D}_{11}}, \tag{3.24}$$

where $\overline{D}_{11} \equiv D_{11} - \frac{B_{11}^2}{A_{11}}$ and k_1 and k_2 are linear springs, while $\overline{k}_{\theta_1}$ and $\overline{k}_{\theta_2}$ are torsion springs, the boundary conditions of the problem depicted in Fig. 3.6 are given by

$$\begin{aligned} &\frac{d^3w_b(0)}{dx^3} + \lambda^2\frac{dw_b(0)}{dx} = -\overline{k}_1 \cdot w_b(0), \quad \frac{d^2w_b(0)}{dx^2} = \overline{k}_{\theta_1}\frac{dw_b(0)}{dx}, \\ &\frac{d^3w_b(L)}{dx^3} + \lambda^2\frac{dw_b(L)}{dx} = \overline{k}_2 \cdot w_b(L), \quad \frac{d^2w_b(L)}{dx^2} = -\overline{k}_{\theta_2}\frac{dw_b(L)}{dx}. \end{aligned} \tag{3.25}$$

[f] Some of the boundary conditions and their relevant buckling loads are also presented in Table 3.1.

Table 3.3. Buckling loads and relevant buckling modes of laminated composite columns using CLT approach.

No.	Name	Characteristic equation	Critical load	Mode shape
1	SS–SS	$\sin(\lambda L) = 0$ $\lambda L = n\pi$	$(\overline{N}_{xx})_{cr} = \frac{\pi^2}{L^2} D_{11}$	$\sin(\frac{\pi x}{L})$
2	C–C	$\lambda L \sin(\lambda L) = 2[1 - \cos(\lambda L)]$ $\lambda L = 2\pi, 8.987, 4\pi, \ldots$	$(\overline{N}_{xx})_{cr} = 4\frac{\pi^2}{L^2} D_{11}$	$1 - \cos(\frac{2\pi x}{L})$
3	C–F	$\cos(\lambda L) = 0$ $\lambda L = (2n - 1)\pi/2$	$(\overline{N}_{xx})_{cr} = \frac{\pi^2}{4L^2} D_{11}$	$1 - \cos(\frac{\pi x}{2L})$
4	F–F	$\sin(\lambda L) = 0$ $\lambda L = n\pi$	$(\overline{N}_{xx})_{cr} = \frac{\pi^2}{L^2} D_{11}$	$\sin(\frac{\pi x}{L})$
5	SS–C	$\tan(\lambda L) = \lambda L$ $\lambda L = 1.430\pi, 2.459\pi, \ldots$	$(\overline{N}_{xx})_{cr} = 2.045\frac{\pi^2}{L^2} D_{11}$	$\sin(\alpha x) + \alpha L[1 - \cos(\alpha x) - \frac{x}{L}]$ while $\alpha = 1.4318\frac{\pi}{L}$
6	SS–F	$\sin(\lambda L) = 0$ $\lambda L = n\pi$	$(\overline{N}_{xx})_{cr} = \frac{\pi^2}{L^2} D_{11}$	$\sin(\frac{\pi x}{L})$
7	G–F	$\cos(\lambda L) = 0$ $\lambda L = (2n - 1)\pi/2$	$(\overline{N}_{xx})_{cr} = \frac{\pi^2}{4L^2} D_{11}$	$\cos(\frac{\pi x}{2L})$
8	G–SS	$\cos(\lambda L) = 0$ $\lambda L = (2n - 1)\pi/2$	$(\overline{N}_{xx})_{cr} = \frac{\pi^2}{4L^2} D_{11}$	$\cos(\frac{\pi x}{2L})$
9	G–G	$\sin(\lambda L) = 0$ $\lambda L = n\pi$	$(\overline{N}_{xx})_{cr} = \frac{\pi^2}{L^2} D_{11}$	$\cos(\frac{\pi x}{L})$
10	G–C	$\sin(\lambda L) = 0$ $\lambda L = n\pi$	$(\overline{N}_{xx})_{cr} = \frac{\pi^2}{L^2} D_{11}$	$1 - \cos(\frac{\pi x}{L})$

Application of the boundary conditions to the general solution presented by Eq. (3.19) yields the following characteristic equation (see details in [9]):

$$\begin{aligned}
&\{-(\overline{k}_1 + \overline{k}_2)\lambda^6 + [\overline{k}_{\theta_1} \cdot \overline{k}_{\theta_2}(\overline{k}_1 + \overline{k}_2) + \overline{k}_1 \cdot \overline{k}_2 \cdot L]\lambda^4 \\
&\quad + \overline{k}_{\theta_1} \cdot \overline{k}_{\theta_2}(\overline{k}_{\theta_1} + \overline{k}_{\theta_2} - \overline{k}_{\theta_1} \cdot \overline{k}_{\theta_2} \cdot L)\lambda^2\} \sin(\lambda L) \\
&\quad + \{(\overline{k}_1 + \overline{k}_2)(\overline{k}_{\theta_1} + \overline{k}_{\theta_2})\lambda^3 - \overline{k}_1 \cdot \overline{k}_2 \cdot L(\overline{k}_{\theta_1} + \overline{k}_{\theta_2})\lambda^3 \\
&\quad - 2 \cdot \overline{k}_1 \cdot \overline{k}_2 \cdot \overline{k}_{\theta_1} \cdot \overline{k}_{\theta_2} \cdot \lambda\} \cos(\lambda L) \\
&\quad + \overline{k}_1 \cdot \overline{k}_2 \cdot \overline{k}_{\theta_1} \cdot \overline{k}_{\theta_2} \cdot \lambda = 0.
\end{aligned} \tag{3.26}$$

Equation (3.26) can be used to solve not only problems involving boundary conditions with springs, but also classical boundary conditions. Letting the

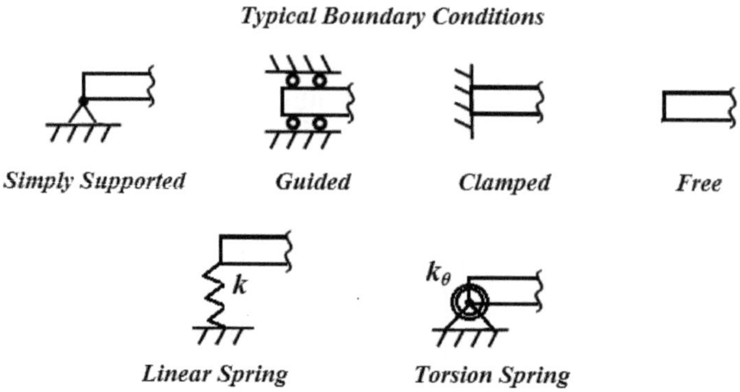

Fig. 3.5. Typical out-of-plane boundary conditions for laminated composite columns (and also isotropic columns) under axial compression.

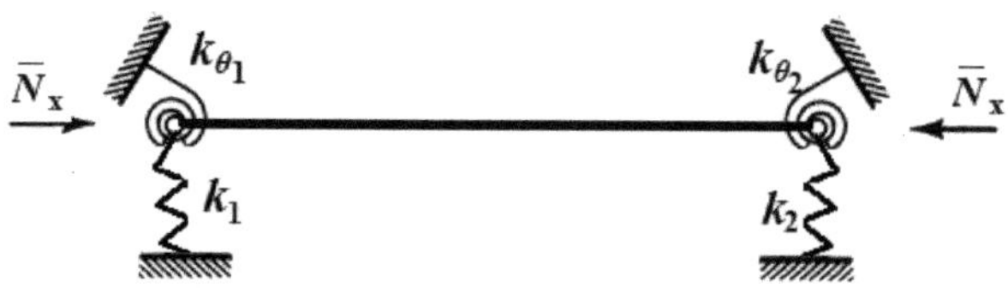

Fig. 3.6. A typical laminated composite column under axial compression having springs as boundary conditions.[g]

linear springs k_1 and k_2 tend to infinity (∞) while the torsional ones, $\overline{k}_{\theta_1}$ and $\overline{k}_{\theta_2}$, are set to zero, would lead to the classical hinged–hinged boundary conditions at both ends of the compressed beam. Similarly, setting all the springs to zero would yield free–free boundary conditions. Clamped boundary conditions can be obtained when both types of springs (linear and torsion) would tend to infinity.

3.2.4. *Non-symmetric laminate* ($B_{11} \neq 0$)

To solve the case of a non-symmetric laminate ($B_{11} \neq 0$), we can use Eq. (3.15). The general solution can be written as

$$w_0(x) = C_1 \sin(\lambda x) + C_2 \cos(\lambda x) + C_3 x + C_4, \tag{3.27}$$

where

$$\lambda^2 = \frac{\overline{N}_{xx}}{\left(D_{11} - \frac{B_{11}^2}{A_{11}}\right)}. \tag{3.28}$$

[g]The drawing presented can also be suitable for a Bernoulli–Euler column.

The in-plane displacement will then be given as (see Appendix A):

$$u_0(x) = C_5 \sin(\lambda x) + C_6 \cos(\lambda x) + C_7 x + C_8, \tag{3.29}$$

where

$$\begin{aligned} C_5 &= -\frac{B_{11}}{A_{11}} \cdot \lambda \cdot C_2, \\ C_6 &= +\frac{B_{11}}{A_{11}} \cdot \lambda \cdot C_1. \end{aligned} \tag{3.30}$$

Solving for a hinged–hinged column (or pinned–pinned, or simply supported at both ends), with the following boundary conditions, two for the in-plane and four for the out-of-plane (see also Appendix A):

$$w_0(0) = 0; M_{xx}(0) = 0 \Rightarrow -B_{11}\frac{du_0(0)}{dx} + D_{11}\frac{d^2 w_0(0)}{dx^2} = 0; \tag{3.31}$$

$$w_0(L) = 0; M_{xx}(L) = 0 \Rightarrow -B_{11}\frac{du_0(L)}{dx} + D_{11}\frac{d^2 w_0(L)}{dx^2} = 0. \tag{3.32}$$

$$u_0(0) = 0, \tag{3.33}$$

$$A_{11}\frac{du_0(L)}{dx} - B_{11}\frac{d^2 w_0(L)}{dx^2} = -\overline{N}_{xx} \tag{3.34}$$

yields the following eigenvalue:

$$\sin(\lambda L) = 0 \Rightarrow \lambda L = n\pi, n = 1, 2, 3, 4, \ldots, \tag{3.35}$$

which gives the critical buckling load per unit width having the following form:

$$\left(\overline{N}_{xx}\right)_{cr} = \frac{\pi^2}{L^2}\left(D_{11} - \frac{B_{11}^2}{A_{11}}\right). \tag{3.36}$$

The buckling shape will be like the symmetric case presented in Table 3.2, namely $\sin(\frac{\pi x}{L})$. Note that as expected, the buckling load for a symmetric laminate will be higher than a non-symmetric one, having the same number of layers. Tables 3.4 and 3.5 present the boundary conditions, for the non-symmetric case, while Table 3.6 gives the buckling loads and the associated buckling modes for the most used cases.

Table 3.4. Buckling of non-symmetric laminated composite columns: out-of-plane boundary conditions using CLT approach.

No.	Name	Out-of-plane boundary conditions $x = 0$	Out-of-plane boundary conditions $x = L$
1	SS–SS*	$w_0 = 0;\ -B_{11}\frac{du_0}{dx} + D_{11}\frac{d^2w_0}{dx^2} = 0$	$w_0 = 0;\ -B_{11}\frac{du_0}{dx} + D_{11}\frac{d^2w_0}{dx^2} = 0$
2	C–C**	$w_0 = 0;\ \frac{dw_0}{dx} = 0$	$w_0 = 0;\ \frac{dw_0}{dx} = 0$
3	C–F***	$w_0 = 0;\ \frac{dw_0}{dx} = 0$	$-B_{11}\frac{du_0}{dx} + D_{11}\frac{d^2w_0}{dx^2} = 0$ $-B_{11}\frac{d^2u_0}{dx^2} + D_{11}\frac{d^3w_0}{dx^3} + \overline{N}_{xx}\frac{dw_0}{dx} = 0$
4	F–F	$-B_{11}\frac{du_0}{dx} + D_{11}\frac{d^2w_0}{dx^2} = 0$ $-B_{11}\frac{d^2u_0}{dx^2} + D_{11}\frac{d^3w_0}{dx^3} + \overline{N}_{xx}\frac{dw_0}{dx} = 0$	$-B_{11}\frac{du_0}{dx} + D_{11}\frac{d^2w_0}{dx^2} = 0$ $-B_{11}\frac{d^2u_0}{dx^2} + D_{11}\frac{d^3w_0}{dx^3} + \overline{N}_{xx}\frac{dw_0}{dx} = 0$
5	SS–C	$w_0 = 0$ $-B_{11}\frac{du_0}{dx} + D_{11}\frac{d^2w_0}{dx^2} = 0$	$w_0 = 0;\ \frac{dw_0}{dx} = 0$
6	SS–F	$w_0 = 0;\ -B_{11}\frac{du_0}{dx} + D_{11}\frac{d^2w_0}{dx^2} = 0$	$-B_{11}\frac{du_0}{dx} + D_{11}\frac{d^2w_0}{dx^2} = 0$ $-B_{11}\frac{d^2u_0}{dx^2} + D_{11}\frac{d^3w_0}{dx^3} + \overline{N}_{xx}\frac{dw_0}{dx} = 0$
7	G***–F	$\frac{dw_0}{dx} = 0;$ $-B_{11}\frac{d^2u_0}{dx^2} + D_{11}\frac{d^3w_0}{dx^3} + \overline{N}_{xx}\frac{dw_0}{dx} = 0$	$-B_{11}\frac{du_0}{dx} + D_{11}\frac{d^2w_0}{dx^2} = 0$ $-B_{11}\frac{d^2u_0}{dx^2} + D_{11}\frac{d^3w_0}{dx^3} + \overline{N}_{xx}\frac{dw_0}{dx} = 0$
8	G–SS	$\frac{dw_0}{dx} = 0$ $-B_{11}\frac{d^2u_0}{dx^2} + D_{11}\frac{d^3w_0}{dx^3} + \overline{N}_{xx}\frac{dw_0}{dx} = 0$	$w_0 = 0;\ -B_{11}\frac{du_0}{dx} + D_{11}\frac{d^2w_0}{dx^2} = 0$
9	G–G	$\frac{dw_0}{dx} = 0$ $-B_{11}\frac{d^2u_0}{dx^2} + D_{11}\frac{d^3w_0}{dx^3} + \overline{N}_{xx}\frac{dw_0}{dx} = 0$	$\frac{dw_0}{dx} = 0$ $-B_{11}\frac{d^2u_0}{dx^2} + D_{11}\frac{d^3w_0}{dx^3} + \overline{N}_{xx}\frac{dw_0}{dx} = 0$
10	G–C	$\frac{dw_0}{dx} = 0$ $-B_{11}\frac{d^2u_0}{dx^2} + D_{11}\frac{d^3w_0}{dx^3} + \overline{N}_{xx}\frac{dw_0}{dx} = 0$	$w_0 = 0;\ \frac{dw_0}{dx} = 0$

Note: *SS = Simply Supported, or hinged–hinged or pinned–pinned; **C = Clamped; ***F = Free; G*** = Guided.

3.4. Buckling of Columns: FSDT Approach

Buckling of columns using the first-order shear deformation theory (FSDT) (see also [5, 8, 11, 14–16, 22–29]) can be obtained using the equations presented in Chapter 2[h]:

$$A_{11}\left(\frac{d^2u_0}{dx^2} + \frac{d^3w_b}{dx^3}\right) + B_{11}\frac{d^2\phi_x}{dx^2} = 0, \tag{3.37}$$

$$KA_{55}\left(\frac{d^2w_b}{dx^2} + \frac{d\phi_x}{dx}\right) - \overline{N}_{xx}\frac{d^2w_b}{dx^2} = 0, \tag{3.38}$$

[h] $w_o = w_0^p + w_b$ (where w_0^p is the prebuckling deflection and $w_b$being the buckling deflection).

Table 3.5. Buckling of non-symmetric laminated composite columns: in-plane boundary conditions using CLT approach.

No.	Name	In-plane boundary conditions	
		$x = 0$	$x = L$
1	SS–SS	$u_0 = 0$	$A_{11}\frac{du_0}{dx} - B_{11}\frac{d^2w_0}{dx^2} = -\overline{N}_{xx}$
2	C–C	$u_0 = 0$	$A_{11}\frac{du_0}{dx} - B_{11}\frac{d^2w_0}{dx^2} = -\overline{N}_{xx}$
3	C–F	$u_0 = 0$	$A_{11}\frac{du_0}{dx} - B_{11}\frac{d^2w_0}{dx^2} = -\overline{N}_{xx}$
4	F–F	$A_{11}\frac{du_0}{dx} - B_{11}\frac{d^2w_0}{dx^2} = -\overline{N}_{xx}$	$A_{11}\frac{du_0}{dx} - B_{11}\frac{d^2w_0}{dx^2} = -\overline{N}_{xx}$
5	SS–C	$u_0 = 0$	$A_{11}\frac{du_0}{dx} - B_{11}\frac{d^2w_0}{dx^2} = -\overline{N}_{xx}$
6	SS–F	$u_0 = 0$	$A_{11}\frac{du_0}{dx} - B_{11}\frac{d^2w_0}{dx^2} = -\overline{N}_{xx}$
7	G–F	$u_0 = 0$	$A_{11}\frac{du_0}{dx} - B_{11}\frac{d^2w_0}{dx^2} = -\overline{N}_{xx}$
8	G–SS	$u_0 = 0$	$A_{11}\frac{du_0}{dx} - B_{11}\frac{d^2w_0}{dx^2} = -\overline{N}_{xx}$
9	G–G	$u_0 = 0$	$A_{11}\frac{du_0}{dx} - B_{11}\frac{d^2w_0}{dx^2} = -\overline{N}_{xx}$
10	G–C	$u_0 = 0$	$A_{11}\frac{du_0}{dx} - B_{11}\frac{d^2w_0}{dx^2} = -\overline{N}_{xx}$

$$B_{11}\left(\frac{d^2u_0}{dx^2} + \frac{\partial^3 w_b}{dx^3}\right) + D_{11}\frac{d^2\phi_x}{dx^2} - KA_{55}\left(\frac{dw_b}{dx} + \phi_x\right) = 0. \quad (3.39)$$

Decoupling the equations leads to the following uncoupled ones[i] (see also [22, 23]):

$$\frac{d^4w_b}{dx^4} + \Gamma^2\frac{d^2w_b}{dx^2} = 0, \quad (3.40)$$

$$\frac{d^3\phi_x}{dx^3} + \Gamma^2\frac{d^2\phi_x}{dx} = 0, \quad (3.41)$$

$$\frac{d^4u_0}{dx^4} + \Gamma^2\frac{d^2u_0}{dx^2} = 0, \quad (3.42)$$

where

$$\Gamma^2 = \frac{\overline{N}_{xx}}{(D_{11} - \frac{B_{11}^2}{A_{11}})(1 - \frac{\overline{N}_{xx}}{KA_{55}})} \Rightarrow \overline{N}_{xx} = \frac{\Gamma^2(D_{11} - \frac{B_{11}^2}{A_{11}})}{1 + \frac{\Gamma^2(D_{11} - \frac{B_{11}^2}{A_{11}})}{KA_{55}}} \quad (3.43)$$

[i]Note that to uncouple the equations one needs to neglect high order terms like $\frac{d^3w_b}{dx^3}$.

Table 3.6. Buckling loads and relevant buckling modes of laminated composite columns using CLT approach.

No.	Name	Characteristic Equation	Critical buckling load per unit width	Mode shape
1	SS–SS	$\sin(\widehat{\lambda}L) = 0;\ \widehat{\lambda}L = n\pi$	$(\overline{N}_{xx})_{cr} = \frac{\pi^2}{L^2}\overline{D}_{11}^*$	$\sin(\frac{\pi x}{L})$
2	C–C	$\widehat{\lambda}L\sin(\widehat{\lambda}L) = 2[1-\cos(\widehat{\lambda}L)]$ $\widehat{\lambda}L = 2\pi, 8.987, 4\pi, \ldots.$	$(\overline{N}_{xx})_{cr} = 4\frac{\pi^2}{L^2}\overline{D}_{11}$	$1-\cos(\frac{2\pi x}{L})$
3	C–F	$\cos(\widehat{\lambda}L) = 0$ $\widehat{\lambda}L = (2n-1)\frac{\pi}{2}$	$(\overline{N}_{xx})_{cr} = \frac{\pi^2}{4L^2}\overline{D}_{11}$	$1-\cos(\frac{\pi x}{2L})$
4	F–F	$\sin(\widehat{\lambda}L) = 0;\ \widehat{\lambda}L = n\pi$	$(\overline{N}_{xx})_{cr} = \frac{\pi^2}{L^2}\overline{D}_{11}$	$\sin(\frac{\pi x}{L})$
5	SS–C	$\tan(\widehat{\lambda}L) = \widehat{\lambda}L$ $\widehat{\lambda}L = 1.430\pi, 2.459\pi, \ldots$	$(\overline{N}_{xx})_{cr} = 2.045\frac{\pi^2}{L^2}\overline{D}_{11}$	$\sin(\alpha x) + \alpha L[1-\cos(\alpha x) - \frac{x}{L}]$ while $\alpha = 1.4318\frac{\pi}{L}$
6	SS–F	$\sin(\widehat{\lambda}L) = 0$ $\widehat{\lambda}L = n\pi$	$(\overline{N}_{xx})_{cr} = \frac{\pi^2}{L^2}\overline{D}_{11}$	$\sin(\frac{\pi x}{L})$
7	G–F	$\cos(\widehat{\lambda}L) = 0;\ \widehat{\lambda}L = (2n-1)\frac{\pi}{2}$	$(\overline{N}_{xx})_{cr} = \frac{\pi^2}{4L^2}\overline{D}_{11}$	$\cos(\frac{\pi x}{2L})$
8	G–SS	$\cos(\widehat{\lambda}L) = 0$ $\widehat{\lambda}L = (2n-1)\pi/2$	$(\overline{N}_{xx})_{cr} = \frac{\pi^2}{4L^2}\overline{D}_{11}$	$\cos(\frac{\pi x}{2L})$
9	G–G	$\sin(\widehat{\lambda}L) = 0$ $\widehat{\lambda}L = n\pi$	$(\overline{N}_{xx})_{cr} = \frac{\pi^2}{L^2}\overline{D}_{11}$	$\cos(\frac{\pi x}{L})$
10	G–C	$\sin(\widehat{\lambda}L) = 0$ $\widehat{\lambda}L = n\pi$	$(\overline{N}_{xx})_{cr} = \frac{\pi^2}{L^2}\overline{D}_{11}$	$1-\cos(\frac{\pi x}{L})$

Note: $^*\overline{D}_{11} = (D_{11} - \frac{B_{11}^2}{A_{11}})$.

Accordingly, the solutions for Eq. (3.40)–(3.42) have a form similar to that presented before, namely

$$w_b(x) = C_1\sin(\Gamma x) + C_2\cos(\Gamma x) + C_3 x + C_4, \tag{3.44}$$

$$\phi_x(x) = C_5\sin(\Gamma x) + C_6\cos(\Gamma x) + C_7, \tag{3.45}$$

$$u_0(x) = C_8\sin(\Gamma x) + C_9\cos(\Gamma x) + C_{10}x + C_{11}. \tag{3.46}$$

One should note that the 11 constants ($C_1 - C_{11}$) are not independent. Their dependency among them can be obtained by back-substituting the solutions (Eqs. (3.44)–(3.46)) in the coupled equations, Eq. (3.37) and Eq. (3.39), to yield the following five relations:

$$A_{11}\cdot C_8 + B_{11}\cdot C_5 = 0, \tag{3.47}$$

$$A_{11}\cdot C_9 + B_{11}\cdot C_6 = 0, \tag{3.48}$$

$$-B_{11}\cdot C_8\cdot\widehat{\lambda}^2 - D_{11}\cdot C_5\cdot\widehat{\lambda}^2 - KA_{55}(-C_2\cdot\widehat{\lambda} + C_5) = 0, \tag{3.49}$$

$$-B_{11} \cdot C_9 \cdot \widehat{\lambda}^2 - D_{11} \cdot C_6 \cdot \widehat{\lambda}^2 - KA_{55}(C_1 \cdot \widehat{\lambda} + C_6) = 0, \tag{3.50}$$

$$-KA_{55} \cdot (C_3 + C_7) = 0 \Rightarrow C_3 + C_7 = 0. \tag{3.51}$$

These five relationships together with the six boundary conditions of the problem will provide the needed eleven equations with eleven unknowns, $C_1 - C_{11}$. The six boundary conditions of the problem are (see also Chapter 2):

$$A_{11}\frac{du_0}{dx} + B_{11}\frac{d\phi_x}{dx} = -\overline{N}_{xx} \text{ or } u_0 = 0, \tag{3.52}$$

$$KA_{55}\left(\frac{dw_b}{dx} + \phi_x\right) - \overline{N}_{xx}\frac{dw_b}{dx} = 0 \text{ or } w_b = 0, \tag{3.53}$$

$$B_{11}\frac{du_0}{dx} + D_{11}\frac{d\phi_x}{dx} = 0 \text{ or } \phi_x = 0. \tag{3.54}$$

Let us present the general case of a simply supported laminated composite beam having the following boundary conditions:

$$u_0(0) = 0; A_{11}\frac{du_0(L)}{dx} + B_{11}\frac{d\phi_x(L)}{dx} = -\overline{N}_{xx}. \tag{3.55}$$

$$w_b(0) = 0;\ w_b(L) = 0, \tag{3.56}$$

$$B_{11}\frac{du_0(0)}{dx} + D_{11}\frac{d\phi_x(0)}{dx} = 0;\ B_{11}\frac{du_0(L)}{dx} + D_{11}\frac{d\phi_x(L)}{dx} = 0. \tag{3.57}$$

Solving for the eleven unknowns, $C_1 - C_{11}$, we get the following results for the first four constants:

$$C_1 = \frac{B_{11}}{A_{11}}\frac{[1-\cos(\Gamma L)]}{\sin(\Gamma L)},\ C_2 = \frac{B_{11}}{A_{11}},\ C_3 = 0,\ C_4 = -\frac{B_{11}}{A_{11}}. \tag{3.58}$$

To obtain the buckling load, one should demand that $w_0(x) \to \infty$, leading to the same procedure applied for a general beam using the CLT approach. Observing the constant C_1, leads to the following eigenvalue solution

$$\sin(\Gamma L) = 0 \Rightarrow \Gamma L = n\pi,\ n = 1, 2, 3, 4, \ldots, \tag{3.59}$$

which leads to the critical buckling load per unit width having the following form:

$$(\overline{N}_{xx})_{cr} = \frac{(\frac{\pi}{L})^2(D_{11} - \frac{B_{11}^2}{A_{11}})}{1 + \frac{(\frac{\pi}{L})^2(D_{11} - \frac{B_{11}^2}{A_{11}})}{KA_{55}}} = \frac{\left[(\overline{N}_{xx})_{cr}\right]_{CLT}}{1 + \frac{\left[(\overline{N}_{xx})_{cr}\right]_{CLT}}{KA_{55}}}. \tag{3.60}$$

If one defines the following term

$$\mathrm{H} \equiv \frac{[(\overline{N}_{xx})_{cr}]_{CLT}}{KA_{55}}, \tag{3.61}$$

Table 3.7. Buckling loads and relevant buckling modes of laminated composite columns using FSDT approach.

No.	Name	$[(\overline{N}_{xx})_{cr}]_{CLT}$	Critical buckling load per unit width	Mode shape
1	SS–SS	$(\overline{N}_{xx})_{\text{cr}} = \frac{\pi^2}{L^2}\overline{D}_{11}^*$	$(\overline{N}_{xx})_{\text{cr}} = \frac{[(\overline{N}_{xx})_{cr}]_{CLT}}{1+\text{H}}^{**}$	$\sin(\frac{\pi x}{L})$
2	C–C	$(\overline{N}_{xx})_{cr} = 4\frac{\pi^2}{L^2}\overline{D}_{11}$	$(\overline{N}_{xx})_{cr} = \frac{[(\overline{N}_{xx})_{cr}]_{CLT}}{1+\text{H}}$	$1-\cos(\frac{2\pi x}{L})$
3	C–F	$(\overline{N}_{xx})_{cr} = \frac{\pi^2}{4L^2}\overline{D}_{11}$	$(\overline{N}_{xx})_{cr} = \frac{[(\overline{N}_{xx})_{cr}]_{CLT}}{1+\text{H}}$	$1-\cos(\frac{\pi x}{2L})$
4	F–F	$(\overline{N}_{xx})_{cr} = \frac{\pi^2}{L^2}\overline{D}_{11}$	$(\overline{N}_{xx})_{cr} = \frac{[(\overline{N}_{xx})_{cr}]_{CLT}}{1+\text{H}}$	$\sin(\frac{\pi x}{L})$
5	SS–C	$(\overline{N}_{xx})_{cr} = 2.045\frac{\pi^2}{L^2}\overline{D}_{11}$	$(\overline{N}_{xx})_{cr} = \frac{[(\overline{N}_{xx})_{cr}]_{CLT}}{1+\text{H}}$	$\sin(\alpha x) + \alpha L[1-\cos(\alpha x) - \frac{x}{L}]$ while $\alpha = 1.4318\frac{\pi}{L}$
6	SS–F	$(\overline{N}_{xx})_{cr} = \frac{\pi^2}{L^2}\overline{D}_{11}$	$(\overline{N}_{xx})_{cr} = \frac{[(\overline{N}_{xx})_{cr}]_{CLT}}{1+\text{H}}$	$\sin(\frac{\pi x}{L})$
7	G–F	$(\overline{N}_{xx})_{cr} = \frac{\pi^2}{4L^2}\overline{D}_{11}$	$(\overline{N}_{xx})_{cr} = \frac{[(\overline{N}_{xx})_{cr}]_{CLT}}{1+\text{H}}$	$\cos(\frac{\pi x}{2L})$
8	G–SS	$(\overline{N}_{xx})_{cr} = \frac{\pi^2}{4L^2}\overline{D}_{11}$	$(\overline{N}_{xx})_{cr} = \frac{[(\overline{N}_{xx})_{cr}]_{CLT}}{1+\text{H}}$	$\cos(\frac{\pi x}{2L})$
9	G–G	$(\overline{N}_{xx})_{cr} = \frac{\pi^2}{L^2}\overline{D}_{11}$	$(\overline{N}_{xx})_{cr} = \frac{[(\overline{N}_{xx})_{cr}]_{CLT}}{1+\text{H}}$	$\cos(\frac{\pi x}{L})$
10	G–C	$(\overline{N}_{xx})_{cr} = \frac{\pi^2}{L^2}\overline{D}_{11}$	$(\overline{N}_{xx})_{cr} = \frac{[(\overline{N}_{xx})_{cr}]_{CLT}}{1+\text{H}}$	$1-\cos(\frac{\pi x}{L})$

Note: $^*\overline{D}_{11} = \left(D_{11} - \frac{B_{11}^2}{A_{11}}\right)$; $^{**}\text{H} \equiv \frac{\left[(\overline{N}_{xx})_{cr}\right]_{CLT}}{KA_{55}}$.

then Eq. (3.60) can be written as

$$(\overline{N}_{xx})_{cr} = \frac{[(\overline{N}_{xx})_{cr}]_{CLT}}{1+\text{H}}. \tag{3.62}$$

The shape of the buckling will be $\sin(\frac{\pi x}{L})$ like for the symmetric case, calculated using the CLT approach. One should notice that the buckling load calculated using FSDT approach is lower than the one calculated according to CLT. Like before, a non-symmetric laminate would yield an even lower buckling load as compared with a symmetric one having the same layers. Some typical buckling loads are presented in Table 3.7.

3.5. Buckling of Plates

In contrast to the previous sections, which were dealing with a one-dimensional structure, like columns or beams, plates are considered to be two-dimensional entities, with the third dimension, the thickness, being much smaller than the other planar dimensions. Although the issue of stability of plates had been treated intensively in the literature (see, for example, [1–4, 6, 7, 10, 11, 17, 18, 20, 21, 30–34, 36–39]) few closed-form solutions are available. Most of the buckling solutions are based on approximate methods, like Rayleigh–Ritz, Galerkin–Bubnov or extended Kantorovich methods (which will be presented in Appendix B).

3.5.1. *Buckling of isotropic plates*

The results for isotropic plates to be presented in this section are based on the seminal work of Timoshenko and Gere [1]. The equation of motion for a thin isotropic plate (thickness, h) under in-plane loads (assuming the plate has some initial curvature due to those loads) is given by the following expression:

$$\frac{\partial^4 w(x,y)}{\partial x^4} + 2\frac{\partial^4 w(x,y)}{\partial x^2 \partial y^2} + \frac{\partial^4 w(x,y)}{\partial y^4} = \frac{1}{D}\left(N_x \frac{\partial^2 w(x,y)}{\partial x^2} + 2N_{xy}\frac{\partial^2 w(x,y)}{\partial x \partial y} + N_y \frac{\partial^2 w(x,y)}{\partial y^2}\right), \quad (3.63)$$

where N_x and N_y are compressive loads per unit length in the x and y directions of the plate (acting at it midplane), N_{xy} is the shear load per unit length and $D = \frac{Eh^3}{12(1-\nu^2)}$ is the bending rigidity of the plate, while E and ν are its Young's modulus and Poisson's ratio, respectively.

Let us present the solution for simply supported boundary conditions around all four sides of a thin plate having the length a, width b and thickness h, subjected to in-plane compression load N_x, while $N_y = N_{xy} = 0$, also known as the Navier method[j]. The out-of-plane displacement can be assumed to be

$$w(x,y) = \sum_{m=1}^{\infty}\sum_{n=1}^{\infty} \sin\frac{m\pi}{a} \sin\frac{n\pi}{b}. \quad (3.64)$$

Note that the double series, representing the lateral deflection, satisfy both Eq. (3.63) and the boundary conditions of the case, namely

$$w_0(x,0) = 0,\ w_0(x,b) = 0,\ w_0(0,y) = 0,\ w_0(a,y) = 0,$$
$$M_{xx}(0,y) = 0,\ M_{xx}(a,y) = 0,\ M_{yy}(x,0) = 0,\ M_{yy}(x,b) = 0, \quad (3.65)$$

where

$$M_{xx} = -D\left[\frac{\partial^2 w}{\partial x^2} + \nu\frac{\partial^2 w}{\partial y^2}\right];\ M_{yy} = -D\left[\frac{\partial^2 w}{\partial y^2} + \nu\frac{\partial^2 w}{\partial x^2}\right]. \quad (3.66)$$

[j]Claude-Louis Navier (1785–1836).

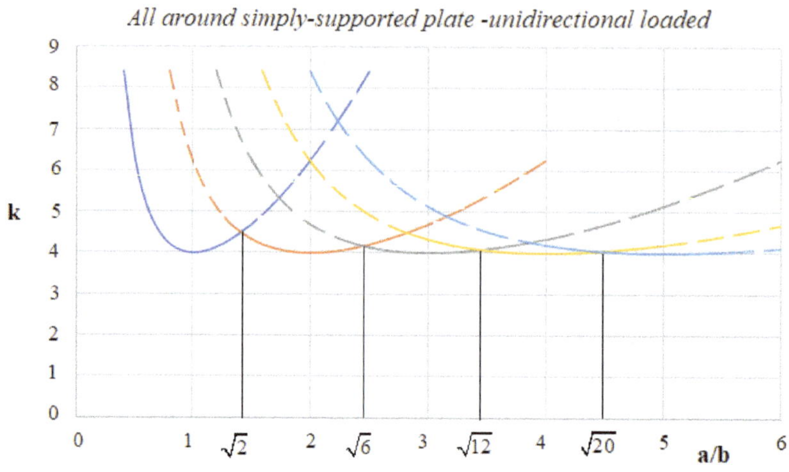

Fig. 3.7. All around simply supported thin plate loaded in the longitudinal direction-buckling factor, k, vs. aspect ratio, a/b.

Substituting Eq. (3.64) into Eq. (3.63), leads to the following expression:

$$N_x = \frac{\pi^2 a^2}{m^2} D \left[\left(\frac{m}{a}\right)^2 + \left(\frac{n}{b}\right)^2 \right]^2. \tag{3.67}$$

To obtain the critical load, one should find the smallest value of N_x in Eq. (3.67), which is easy to see that it will happen for $n = 1$, namely half wave in the width direction (perpendicular do the applied in-plane load N_x). Therefore, the critical buckling load can be written as

$$(N_x)_{cr} = \frac{\pi^2}{m^2 b^2} D \left[\left(\frac{b}{a}\right) m^2 + \left(\frac{a}{b}\right) \right]^2 = k \frac{\pi^2}{b^2} D$$

$$k = \frac{1}{m^2} \left[\left(\frac{b}{a}\right) m^2 + \left(\frac{a}{b}\right) \right]^2. \tag{3.68}$$

The variation of k with the aspect ratio a/b and the number of half waves in the longitudinal direction, m, is presented in Fig. 3.7. The minimal value of k is 4, and it occurs at an integer value of the aspect ratio, a/b, with the critical buckling waves following the aspect ratio, in the longitudinal direction.

Other cases of interest would include plates with two opposite sides on simply-supported boundary conditions, while the other two sides have any combination of boundary conditions, with the axial compression load being applied on the simply supported edges, also known as the Lévy method[k].

[k]Maurice Lévy (1838–1910), *Comptes Rendus*, vol. 129, 1899, pp. 535–539.

The general solution has the following form (see [1]):

$$w(x,y) = \sin\frac{m\pi}{a}[A_1 e^{-\alpha y} + A_2 e^{\alpha y} + A_3\cos(\beta y) + A_4 \sin(\beta y)],$$

where

$$\alpha = \sqrt{\left(\frac{m\pi}{a}\right)^2 + \sqrt{\frac{N_x}{D}}\left(\frac{m\pi}{a}\right)^2}, \beta \equiv \sqrt{-\left(\frac{m\pi}{a}\right)^2 + \sqrt{\frac{N_x}{D}}\left(\frac{m\pi}{a}\right)^2}. \quad (3.69)$$

The constants $A_1 - A_4$ are to be determined by application of the proper boundary conditions, leading to an eigenvalue problem which determines the critical buckling load (per unit width). Typical boundary condition for the unloaded edges of the plate are next given for clamped–clamped and free–free cases:

(a) Clamped–clamped

$$w(x,0) = 0; \quad \frac{\partial w(x,0)}{\partial x} = 0; \quad w(x,b) = 0; \quad \frac{\partial w(x,0)}{\partial x} = 0. \quad (3.70)$$

(b) Free–free

$$\frac{\partial^2 w(x,0)}{\partial y^2} + \nu\frac{\partial^2 w(x,0)}{\partial x^2} = 0; \quad \frac{\partial^3 w(x,0)}{\partial y^3} + (2-\nu)\frac{\partial^3 w(x,0)}{\partial x^2 \partial y} = 0;$$
$$\frac{\partial^2 w(x,b)}{\partial y^2} + \nu\frac{\partial^2 w(x,b)}{\partial x^2} = 0; \quad \frac{\partial^3 w(x,b)}{\partial y^3} + (2-\nu)\frac{\partial^3 w(x,b)}{\partial x^2 \partial y} = 0. \quad (3.71)$$

Typical results for the common boundary conditions are presented in Fig. 3.8.

Timoshenko and Gere [1] presented an expression for the Case E (see Fig. 6.8) for long plates, having the following form:

$$k_c = 0.456 + \left(\frac{b}{a}\right)^2. \quad (3.72)$$

More details about the buckling of flat isotropic plates, involving other in-plane loadings, bending and buckling of plates having elastic boundary conditions can be found in [1, 40].

To conclude this section, we shall present the case of a rectangular flat isotropic plate under pure shear ($N_x = N_y = 0$, $N_{xy} \neq 0$), which was treated in [1, 40–43] The equation of motion for this case is given by

$$\frac{\partial^4 w(x,y)}{\partial x^4} + 2\frac{\partial^4 w(x,y)}{\partial x^2 \partial y^2} + \frac{\partial^4 w(x,y)}{\partial y^4} = \frac{2N_{xy}}{D}\frac{\partial^2 w(x,y)}{\partial x \partial y}. \quad (3.73)$$

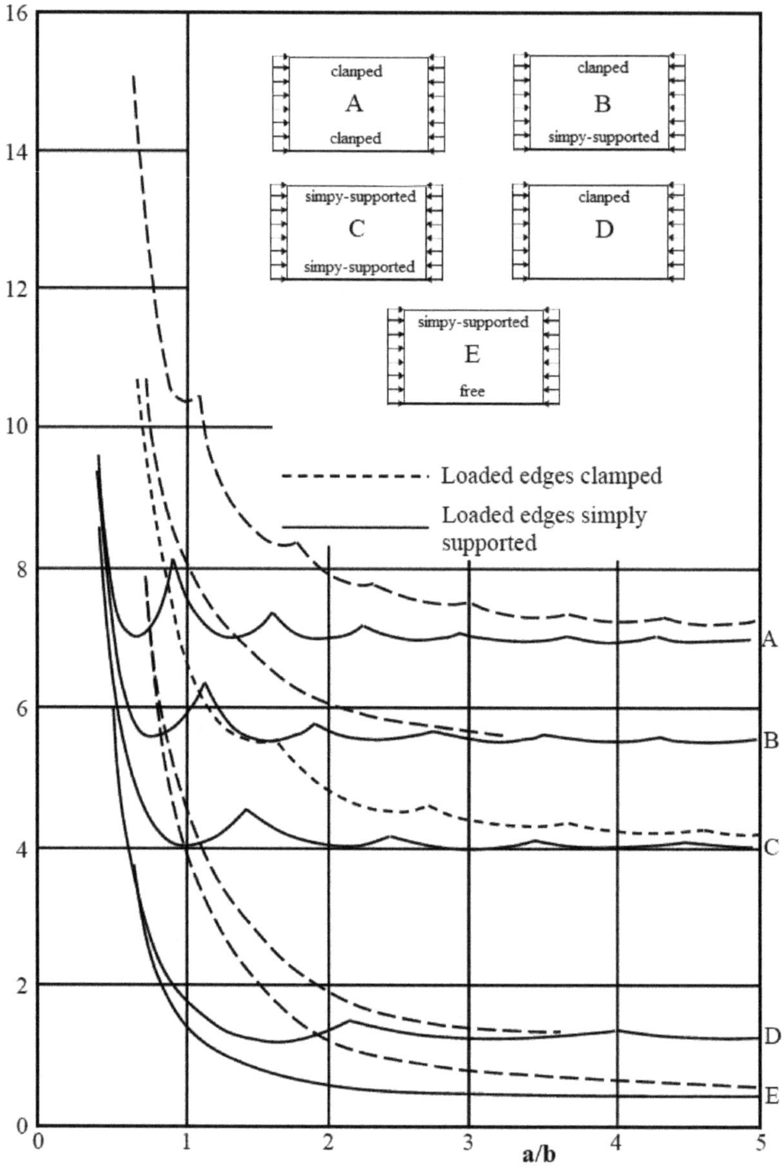

Fig. 3.8. Rectangular thin plates loaded in the longitudinal direction-buckling factor, k, vs. aspect ratio, a/b, for various boundary conditions (adapted from [40]).

Due to the mixed derivation on the right-hand side of Eq. (3.73), no trigonometric solution for the deflection, w, is possible. The approximate solutions had been obtained using energy methods (see [1, 42]). Figure 3.9 presents the buckling coefficient, k_s for a rectangular flat plate resting on simply-supported boundary conditions, under shear stresses along its perimeter.

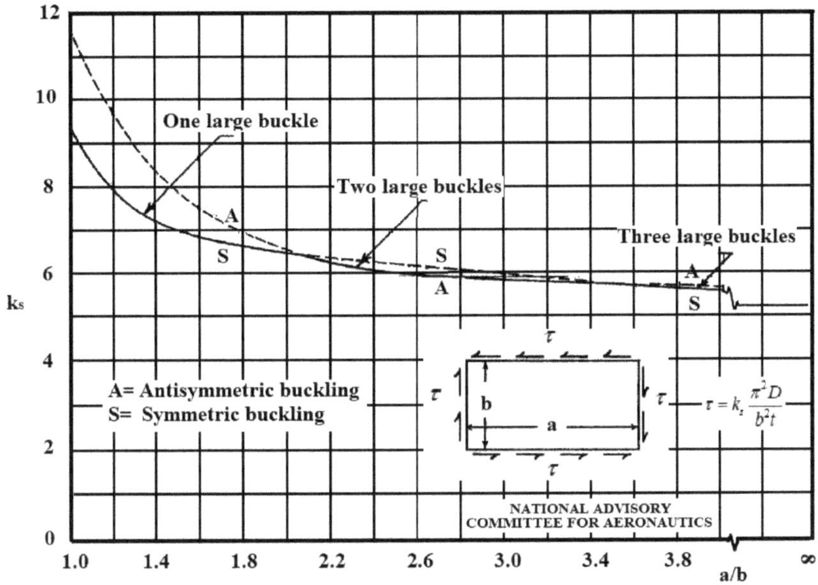

Fig. 3.9. Rectangular thin plates loaded in pure shear-buckling factor, k, vs. aspect ratio, a/b, for simply supported boundary conditions (adapted from [42]).

From Fig. 3.9 one can obtain the buckling coefficient for infinitely long flat plates on simply supported boundary conditions to yield

$$(\tau_{\mathrm{cr}})_{\mathrm{SS}} = k_s \frac{\pi^2 D}{b^2 h} = 5.35 \frac{\pi^2 D}{b^2 h}. \tag{3.74}$$

A similar expression is presented in [1] for infinitely long flat plates on clamped boundary conditions, namely

$$(\tau_{\mathrm{cr}})_{\mathrm{clamped}} = k_s \frac{\pi^2 D}{b^2 h} = 8.98 \frac{\pi^2 D}{b^2 h}. \tag{3.75}$$

In [1], a general expression for the shear buckling coefficient k_s is presented for rectangular plates on simply supported boundary conditions, starting with a ratio $a/b = 1$ up to long infinite one, $a/b \to \infty$,

$$k_s = 5.35 + 4\left(\frac{b}{a}\right)^2. \tag{3.76}$$

3.5.2. *Buckling of orthotropic plates*

Finding the buckling of orthotropic plates is more complicated as compared to the isotropic case and only few closed form solutions are available for certain boundary conditions and layups. Other rigorous plate solutions can be found based on Navier method which treats rectangular plates on simply supported boundary conditions all around it or the Lévy method suitable for plates with two opposite simply supported edges, while the other two can have any combination of boundary conditions. Approximate solutions can be obtained using Rayleigh–Ritz, Galerkin–Bubnov or extended Kantorovich methods, all being based on energy approaches.

The first case to be presented is also called *special orthotropic plates* for which the bending-stretching coupling terms B_{ij} and the bending-twisting coefficients D_{16} and D_{26} are set to zero. This leads to the following equation, assuming constant in-plane compressive loads per unit width in the x and y directions, $\overline{N}_{xx}$ and $\overline{N}_{yy}$, respectively and shear loads $\overline{N}_{xy}$:

$$D_{11}\frac{\partial^4 w_0}{\partial x^4} + 2(D_{12} + 2D_{66})\frac{\partial^4 w_0}{\partial x^2 \partial y^2} + D_{22}\frac{\partial^4 w_0}{\partial y^4}$$
$$= \overline{N}_{xx}\frac{\partial^2 w_0}{\partial x^2} + 2\overline{N}_{xy}\frac{\partial^2 w_0}{\partial x \partial y} + \overline{N}_{yy}\frac{\partial^2 w_0}{\partial y^2}. \tag{3.77}$$

Note that the simply supported boundary conditions presented in Eq. (6.65), for the isotropic case, still hold, however the expressions for the moments, per unit width are for the orthotropic case:

$$M_{xx} = -\left[D_{11}\frac{\partial^2 w_0}{\partial x^2} + D_{12}\frac{\partial^2 w_0}{\partial y^2}\right],$$
$$M_{yy} = -\left[D_{12}\frac{\partial^2 w_0}{\partial x^2} + D_{22}\frac{\partial^2 w_0}{\partial y^2}\right], M_{xy} = -2D_{66}\frac{\partial^2 w_0}{\partial x \partial y}. \tag{3.78}$$

For the clamped boundary conditions along the plate's perimeter, Eq. (3.70) presented for the isotropic case is valid also for the present orthotropic case. The issue of free boundary conditions on all four sides of the plate, presented in Eq. (3.71) for the isotropic case, is somehow more complicated and is presented as:

$$M_{xx}(0, y) = 0, M_{xx}(a, y) = 0, M_{yy}(x, 0) = 0, M_{yy}(x, b) = 0,$$
$$V_y(0, y) = 0, V_y(a, y) = 0, V_x(x, 0) = 0, V_y(x, b) = 0, \tag{3.79}$$

where

$$V_x = \frac{\partial M_{xx}}{\partial x} + \frac{\partial M_{xy}}{\partial y} + \frac{\partial}{\partial x}\left(\overline{N}_{xy}\frac{\partial w_0}{\partial y} - \overline{N}_{xx}\frac{\partial w_0}{\partial x}\right) + \frac{\partial M_{xy}}{\partial y},$$
$$V_y = \frac{\partial M_{yy}}{\partial y} + \frac{\partial M_{xy}}{\partial x} + \frac{\partial}{\partial y}\left(\overline{N}_{xy}\frac{\partial w_0}{\partial x} - \overline{N}_{yy}\frac{\partial w_0}{\partial y}\right) + \frac{\partial M_{xy}}{\partial x}. \tag{3.80}$$

Solving Eq. (3.77) for a rectangular plate loaded in the x direction only ($\overline{N}_{xy} = \overline{N}_{yy} = 0$) for all around simply supported boundary conditions

(Eqs. (3.65) and (3.78)) which can be shown to be simplified yielding the following expressions:

$$w_0(x,0) = \frac{\partial^2 w_0(x,0)}{\partial x^2} = 0, \ w_0(x,b) = \frac{\partial^2 w_0(x,b)}{\partial x^2} = 0,$$
$$w_0(0,y) = \frac{\partial^2 w_0(0,y)}{\partial y^2} = 0, \ w_0(a,y) = \frac{\partial^2 w_0(a,y)}{\partial y^2} = 0, \qquad (3.81)$$

and substituting the following out-of-plane deflection w_0, which satisfies the boundary conditions (Eq. (3.81)), into Eq. (3.77)

$$w_0 = A_{mn} \sin\frac{m\pi x}{a} \sin\frac{n\pi y}{b}, \ m,n = 1,2,3,\ldots, \qquad (3.82)$$

where A_{mn} is a small arbitrary amplitude coefficient, yields the following critical load per unit width (b being the width of the plate)

$$(\overline{N}_{xx})_{\text{cr}} = \pi^2 \left[D_{11}\left(\frac{m}{a}\right)^2 + 2\left(D_{12} + 2D_{66}\right)\left(\frac{n}{b}\right)^2 + D_{22}\left(\frac{n}{b}\right)^4\left(\frac{a}{m}\right)^2 \right]. \qquad (3.83)$$

Equation (3.83) is a function of both m and n, which are the number of half-waves in the x and y directions, respectively. It is clearly that the minimum can be obtained for $n = 1$. Rearranging Eq. (3.83) leads to the following expression:

$$\frac{\mathrm{K}_x}{\pi^2} \equiv \frac{\overline{N}_{xx}\cdot b^2}{\pi^2 D_{22}} = \left[\frac{D_{11}}{D_{22}}\left(\frac{b}{a}\right)^2 m^2 + 2\left(\frac{D_{12}}{D_{22}} + 2\frac{D_{66}}{D_{22}}\right) + \left(\frac{a}{b}\right)^2 \frac{1}{m^2}\right]. \qquad (3.84)$$

Equation (3.84) enables to find the lowest values for K_x as a function of the aspect ratio a/b and m for given values of D_{11}, D_{22}, D_{12} and D_{66}. Figure 3.10 presents the behavior of Eq. (3.84) for three values of D_{11}/D_{22} assuming $(D_{12} + 2D_{66})/D_{22} = 1$. Similar curves would be achieved for other values of those parameters.

One should note that the minimum value for each curve generated by Eq. (3.84) is found for a certain aspect ratio thus obtaining $\left(\mathrm{K}_x/\pi^2\right)_{\min}$

$$\frac{a}{b} = m\cdot\sqrt[4]{\frac{D_{11}}{D_{22}}} \Rightarrow \left(\frac{\mathrm{K}_x}{\pi^2}\right)_{\min} = 2\left(\sqrt{\frac{D_{11}}{D_{22}}} + \frac{D_{12}}{D_{22}} + 2\frac{D_{66}}{D_{22}}\right). \qquad (3.85)$$

The biaxial case, where only in-plane compressive forces ($\overline{N}_{xy} = 0$) are applied in the x and y directions of an all-around simply supported plate can also be solved, provided one can write $\kappa \equiv \overline{N}_{yy}/\overline{N}_{xx}$ (the y compression

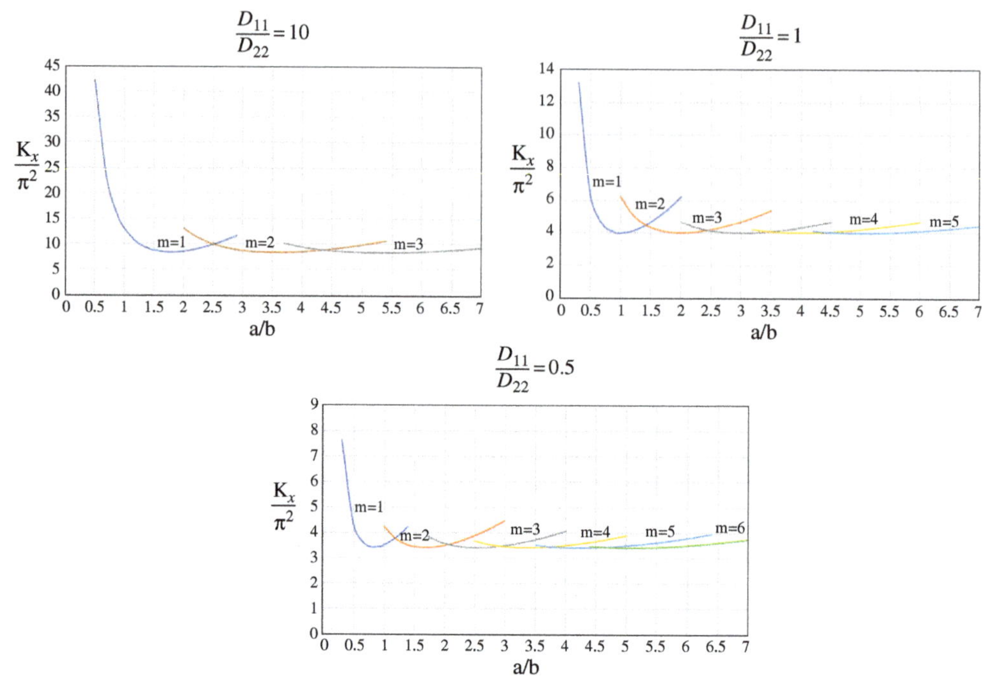

Fig. 3.10. Uniaxial buckling of a rectangular special orthotropic plate simply supported on all its sides for three values of D_{11}/D_{22} while $(D_{12} + 2D_{66})/D_{22} = 1$.

load is proportional to the x compression load). Then using Eqs. (3.77) and (3.82) one obtains the expression, for the critical buckling load, as

$$(\overline{N}_{xx})_{cr} = \pi^2 \frac{[D_{11}(\frac{m}{a})^2 + 2\,(D_{12} + 2D_{66})\,(\frac{n}{b})^2 + D_{22}(\frac{n}{b})^4(\frac{a}{m})^2]}{1 + \kappa(\frac{a}{b})^2(\frac{n}{m})^2} \tag{3.86}$$

yielding as before the value of (K_x/π^2) as a function of the aspect ratio (a/b), stiffness D_{11}, D_{12}, D_{22}, D_{66} and the half waves in the x and y plate directions, m and n, respectively,

$$\frac{\mathrm{K}_x}{\pi^2} = \frac{[D_{11}(\frac{m}{a})^2 + 2\,(D_{12} + 2D_{66})\,(\frac{n}{b})^2 + D_{22}(\frac{n}{b})^4(\frac{a}{m})^2]}{1 + \kappa(\frac{a}{b})^2(\frac{n}{m})^2}. \tag{3.87}$$

Note that in contrast to the isotropic case, the assumption of $n = 1$ for the special orthotropic case of a plate under biaxial loads will not necessary lead to the lowest value of the critical compression load, and therefore its lowest value will be found after calculating the critical buckling loads for a few combinations of m and n.

The case of a special orthotropic rectangular plate on simply supported boundary conditions under pure shear $(\overline{N}_{xy} \neq 0, \overline{N}_{xx} = 0, \overline{N}_{yy} = 0)$ case is more complicated than the previous two ones. The equation to be solved is given by

$$D_{11}\frac{\partial^4 w_0}{\partial x^4} + 2(D_{12} + 2D_{66})\frac{\partial^4 w_0}{\partial x^2 \partial y^2} + D_{22}\frac{\partial^4 w_0}{\partial y^4} = 2\overline{N}_{xy}\frac{\partial^2 w_0}{\partial x \partial y}. \tag{3.88}$$

There are no analytical solutions for this case, even for an all-around simply supported rectangular plate. Bergmann and Reisner [31] assumed an infinitely long plate in the x direction ($a/b \to \infty$) with the D_{11} bending rigidity being neglected. The exact solution for this case, where simply supported boundary conditions were assumed along the long edges, has the following form:

$$\frac{\overline{N}_{xy} \cdot b^2}{4\sqrt{[D_{22}(D_{12} + 2D_{66})]}} = 11.71. \tag{3.89}$$

Another solution for Eq. (3.88) was given by Leissa [30] having the following form:

$$w(x, y) = f(y)e^{i\kappa\frac{x}{b}}, \tag{3.90}$$

where $i = \sqrt{-1}$, b is the plate's width and κ is a wavelength constant to be determined. Substituting Eq. (3.90) into Eq. (3.88) yields:

$$\begin{aligned} &D_{11}\left(\frac{\kappa}{b}\right)^4 f(y) - 2\left(D_{12} + 2D_{66}\right)\left(\frac{\kappa}{b}\right)^2 \frac{d^2 f(y)}{dy^2} + D_{22}\frac{d^4 f(y)}{dy^4} \\ &- 2i\overline{N}_{xy}\left(\frac{\kappa}{b}\right)\frac{df(y)}{dy} = 0. \end{aligned} \tag{3.91}$$

The solution of Eq. (3.91) has the following form:

$$f(y) = A_1 e^{i\beta_1\frac{y}{b}} + A_2 e^{i\beta_2\frac{y}{b}} + A_3 e^{i\beta_3\frac{y}{b}} + A_4 e^{i\beta_4\frac{y}{b}}, \tag{3.92}$$

where β_1, β_2, β_3 and β_4 are the roots of the fourth-degree polynomial equation presented by Eq. (3.91). Then, the simply supported boundary conditions are substituted at $y = 0, b$ into Eq. (3.92), leading to a fourth-order characteristic determinant, with κ and $\overline{N}_{xy}$ being free parameters. It was shown in [30] that the following parameter, $\sqrt{D_{11}/D_{22}}/\left[(D_{12} + 2D_{66})/D_{22}\right]$, determines the solution.

When only two opposite edges of a plate are on simply supported boundary conditions, while the other two sides can be either clamped, free or any other combination of boundaries, the Lévy method can be applied to solve the buckling problem. Assuming the out-of-plane displacement has the following form:

$$w_0(x, y) = G_n(x)\sin\frac{n\pi y}{b},\ n = 1, 2, 3, \tag{3.93}$$

and substituting in Eq. (3.77), for constant in-plane loads, without shear type ones ($\overline{N}_{xx} \neq 0, \overline{N}_{yy} \neq 0, \overline{N}_{xy} = 0$) yields

$$D_{11}\frac{d^4G_n(x)}{dx^4} - \left[2(D_{12} + 2D_{66})\left(\frac{n\pi}{b}\right)^2 + \overline{N}_{xx}\right]$$

$$\frac{d^2G_n(x)}{dx^2} + \left(\frac{n\pi}{b}\right)^2\left[D_{22}\left(\frac{n\pi}{b}\right)^2 + \overline{N}_{yy}\right]G_n(x) = 0. \qquad (3.94)$$

The general solution for Eq. (3.94) can be assumed to have the following form:

$$G_n(x) = A_1\sinh(\lambda_1 x) + A_2\cosh(\lambda_1 x) + A_3\sin(\lambda_2 x) + A_4\cos(\lambda_2 x), \quad (3.95)$$

where λ_1 and λ_2 are the roots of the following characteristic equation:

$$D_{11}\lambda^4 - \left[2(D_{12} + 2D_{66})\left(\frac{n\pi}{b}\right)^2 + \overline{N}_{xx}\right]\lambda^2$$

$$+\left[D_{22}\left(\frac{n\pi}{b}\right)^4 + \overline{N}_{yy}\left(\frac{n\pi}{b}\right)^2\right] = 0. \qquad (3.96)$$

The roots of Eq. (3.96) are given by

$$(\lambda_1)^2 = \frac{1}{D_{11}}[B + \sqrt{B^2 - D_{11}C}],$$

$$(\lambda_2)^2 = \frac{1}{D_{11}}[-B + \sqrt{B^2 - D_{11}C}], \qquad (3.97)$$

where

$$B \equiv \left[(D_{12} + 2D_{66})\left(\frac{n\pi}{b}\right)^2 + \frac{\overline{N}_{xx}}{2}\right], \quad C \equiv \left[D_{22}\left(\frac{n\pi}{b}\right)^4 + \overline{N}_{yy}\left(\frac{n\pi}{b}\right)^2\right]. \qquad (3.98)$$

The constants $A_1 - A_4$ are to be determined using the appropriate boundary conditions at $x = 0$, a, yielding the buckling load of the plate.

3.6. Buckling of Shells

In contrast to the previous two sections, which were dealing with either one-dimensional structures, like columns or beams or two-dimensional structures, like plates, shells are considered to be three-dimensional entities, with the thickness, being much smaller than the other planar dimensions. Although the issue of stability of various types of shells had been treated intensively in the literature (see for example typical studies 44–65) very few closed form solutions are available. To understand the complexity of this thin-walled structure type we shall first present some types of the shells, their structural behavior and then the various theories existing in the literature.

3.6.1. *Introduction to shells*

Shells are thin-walled structures used as structural elements in aerospace, marine, civil, architecture and mechanical engineering. Typical examples are aircraft body, missiles, rockets, ships and submarines or water tanks, silos, piping systems and pressure vessels. They provide many advantages like high stiffness, high strength to weight ratio, a good efficiency of load carrying behavior and structural integrity.

Two classes of shells can be classified as: thin shells ($h/R \ll 1$[1]) and thick shells. From the engineering point of view, one can define for a thin shell the following inequality

$$\left(\frac{h}{R}\right)_{\max} \leq \frac{1}{20} \text{ or } \frac{1}{1000} \leq \left(\frac{h}{R}\right) \leq \frac{1}{20} \tag{3.99}$$

while thick shells their relative thickness lies in the range of

$$\left(\frac{h}{R}\right)_{\max} \geq \frac{1}{20}. \tag{3.100}$$

All shells have a curvature. The shape of a shell is classified according to its curvature leading to cylindrical (circular and non-circular), conical, spherical, ellipsoidal, paraboloid, toroidal and hyperbolic paraboloid shells as presented in Fig. 3.11 and Table 3.8. Another classification is done based on the material of the shell, leading to metal shells, composite shells or laminated composite shells. Depending on the shape of the shell, various loads can be applied: axial, lateral, a combination of lateral and axial loads, external and internal pressure, bending and torsion.

Figure 3.12 presents some of the shells being tested at the Aerospace Structures Laboratory, Technion by the author of this book.

The behavior of a circular cylindrical shell is schematically described in Fig. 3.13. It highlights the difference between a perfect ideal shell (λ_p) and a real imperfect shell (λ_L), with the bifurcation load (λ_C) as a reference.

As is depicted in Fig. 3.13, $\lambda_p > \lambda_L$ and therefore a great discrepancy between tests results and theoretical solutions exists for shells. Some of the factors attributed to this discrepancy are as follows:

(a) Boundary conditions;
(b) Initial geometric imperfections;
(c) Load eccentricity;
(d) Shell geometry (length and the ratio R/t);
(e) Plasticity and residual stresses from welding and rolling;

[1] h = shell thickness and R = shell radius.

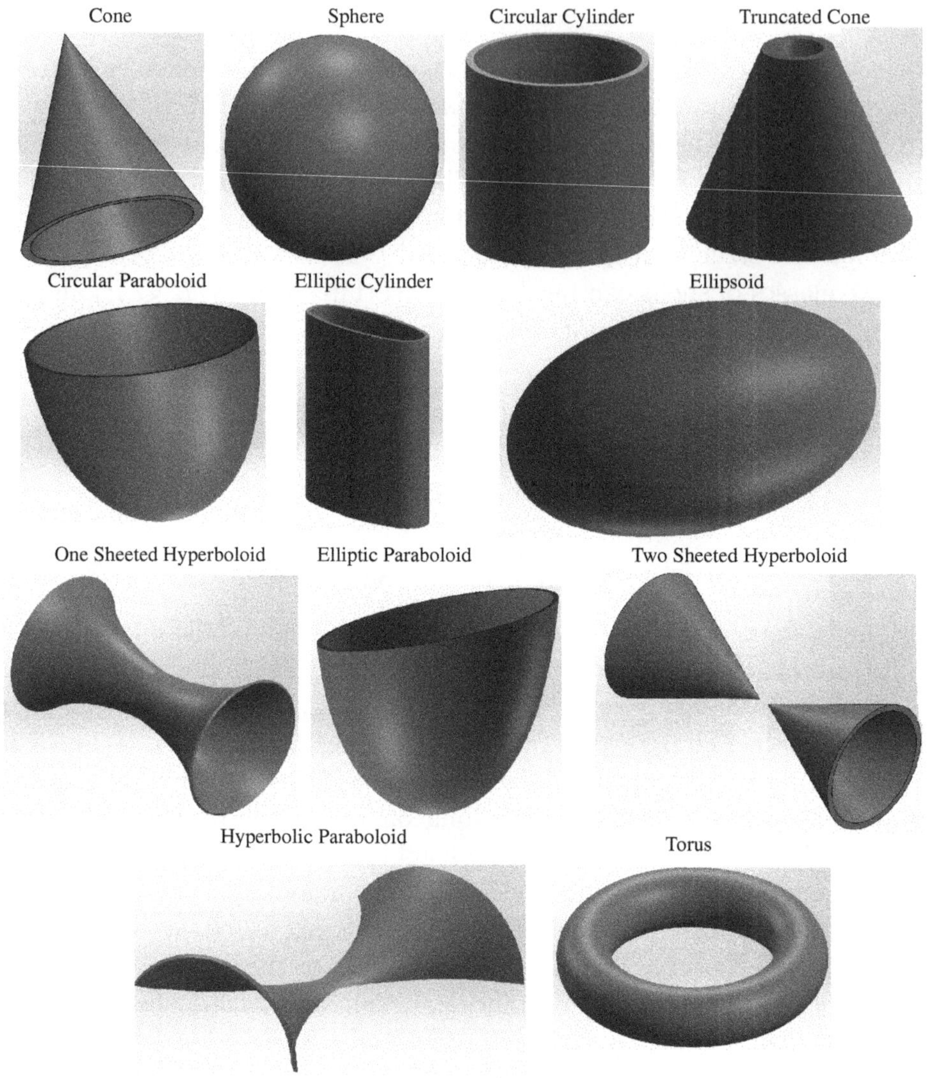

Fig. 3.11. Schematic shapes of shells (not to scale).

(f) Uneven load distribution;
(g) Variation of temperatures.

The first two factors, boundary conditions and initial geometric imperfections, are the most significant factors in reducing the buckling strength of axially compressed cylindrical shells.

This is further stressed in Fig. 3.14, where the experimental buckling loads for numerous tests are compared to the theoretical predicted buckling

Table 3.8. Mathematical expressions for various types of shells

Type of shell	Mathematical expression
Ellipsoid	$\frac{x^2}{a^2}+\frac{y^2}{b^2}+\frac{z^2}{c^2}=1$
Sphere	$x^2+y^2+z^2=a^2$
Cone	$\frac{x^2}{a^2}+\frac{y^2}{b^2}=\frac{z^2}{c^2}$
Cylinder with a circular cross-section	$x^2+y^2=a^2$
Cylinder with an elliptic cross-section	$\frac{x^2}{a^2}+\frac{y^2}{b^2}=1$
Hyperboloid of one sheet	$\frac{x^2}{a^2}+\frac{y^2}{b^2}-\frac{z^2}{c^2}=1$
Hyperboloid of two sheets	$-\frac{x^2}{a^2}-\frac{y^2}{b^2}+\frac{z^2}{c^2}=1$
Elliptic paraboloid	$\frac{x^2}{a^2}+\frac{y^2}{b^2}=\frac{z}{c}$
Circular paraboloid	$\frac{x^2}{a^2}+\frac{y^2}{a^2}=\frac{z}{c}$
Hyperbolic paraboloid	$\frac{x^2}{a^2}-\frac{y^2}{b^2}=\frac{z}{c}$
Torus*	$[R-\sqrt{x^2+y^2}]^2+z^2=r^2$

Note: *R is the radius from the center of the hole to the center of the tube, while r is the radius of the tube.

loads. According to the recommendations from NASA SP-8007 a curve is drawn to engulf from bellow all the experimental results leading to

$$F_{\text{design}} = F_{\text{theory}} \cdot \kappa, \text{ where}$$
$$\kappa \equiv \frac{P}{P_{CL}} = 1 - 0.901\left[1 - e^{-\frac{1}{16}\sqrt{\frac{R}{t}}}\right]. \tag{3.101}$$

Accordingly, this knock down factor, κ, depends on the ratio R/t and yields the following values:

$$\frac{P}{P_{CL}}\left(\frac{R}{t}=500\right) \equiv \kappa = 0.321,$$
$$\frac{P}{P_{CL}}\left(\frac{R}{t}=1000\right) \equiv \kappa = 0.223,$$
$$\frac{P}{P_{CL}}\left(\frac{R}{t}=1500\right) \equiv \kappa = 0.178,$$
$$\frac{P}{P_{CL}}\left(\frac{R}{t}=2000\right) \equiv \kappa = 0.153. \tag{3.102}$$

Therefore to be able to design a shell for axial compression the theoretical buckling load should multiplied by κ to yield the design load.

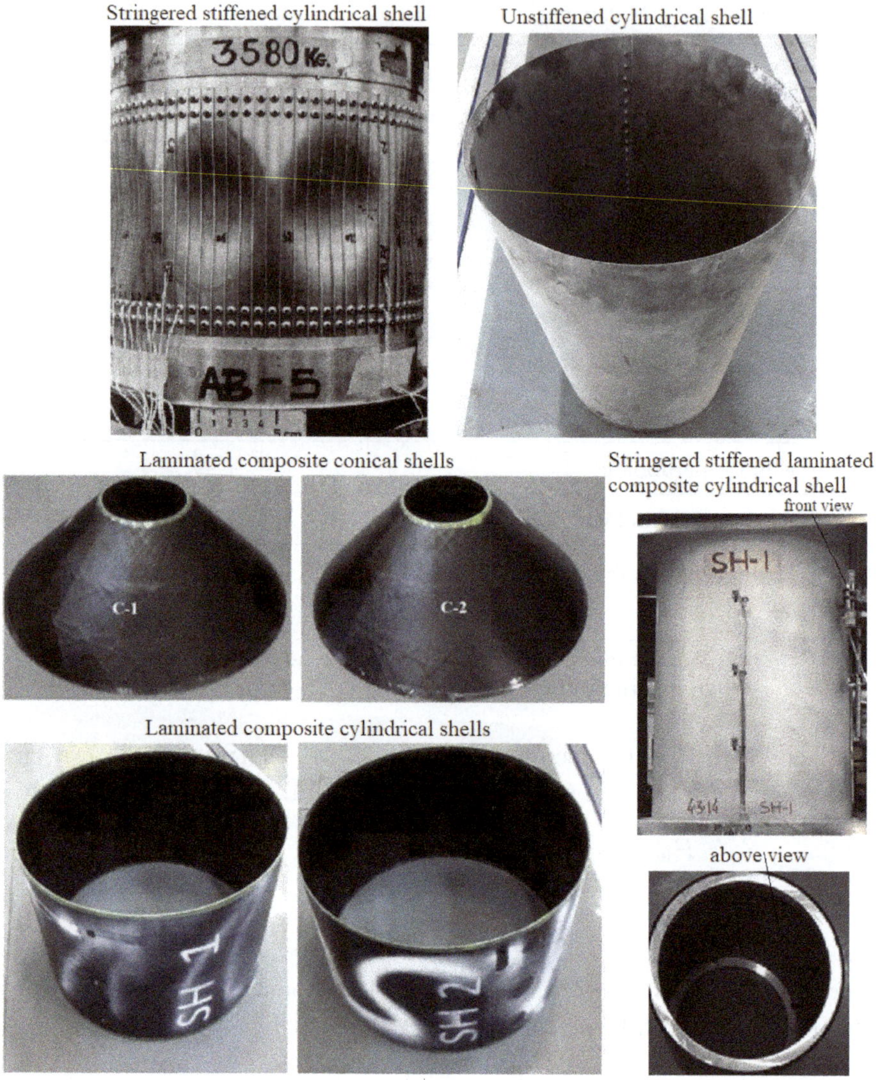

Fig. 3.12. Realized shells tested at the Technion, Aerospace Structures Laboratory [8].

3.6.2. *Buckling of isotropic cylindrical shells*

3.6.2.1. *Axial compression*

An approximate buckling equation for axially loaded cylinders (compression) [46] is

$$N_x = k_x \frac{\pi^2 D}{L^2}, \text{ where}$$

$$D = \frac{Et^3}{12(1 - \upsilon^2)}. \tag{3.103}$$

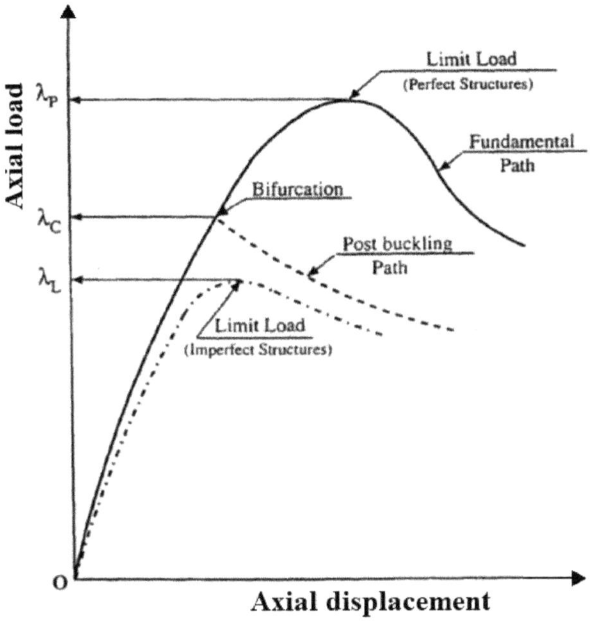

Fig. 3.13. Typical behavior of a compressed circular cylindrical shell.

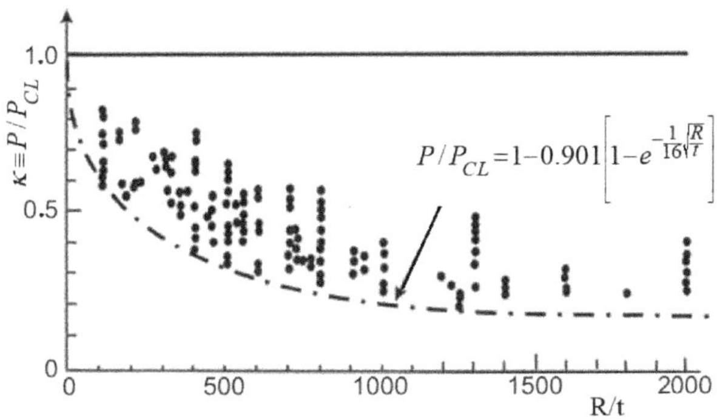

Fig. 3.14. The knock down factor κ vs. R/t for circular cylindrical shell based on data from [65].

Note that N_x is the axial compression load per unit shell circumference ($N_x = F_{\text{axial}}/2\pi R$), R is the shell radius, t is its thickness, E is the Young's modulus and υ is the Poisson's ratio.

The buckling factor k_x in Eq. (3.103) is defined [66] as

$$k_x = m^2(1+\beta^2)^2 + \frac{12}{\pi^4}\frac{\kappa^2 Z^2}{m^2(1+\beta^2)^2}, \text{ where}$$

$$\beta = \frac{nL}{\pi Rm} \quad Z = \frac{L^2}{Rt}\sqrt{1-\upsilon^2}. \tag{3.104}$$

It is claimed in [66] that for $\kappa Z > \frac{\sqrt{3}\pi^2}{6}$ (moderately long cylinders) the buckling factor can be approximated to

$$k_x = \frac{4\sqrt{3}}{\pi^2}\kappa Z. \tag{3.105}$$

Substituting Eq. (3.105) into Eq. (3.103), and remembering the expressions for D and Z we obtain the critical buckling stress as

$$\sigma_x = \sigma_{cr} = \frac{N_x}{t} = \frac{\kappa E}{\sqrt{3(1-\upsilon^2)}}\frac{t}{R}$$

$$\Rightarrow \sigma_{\text{cr}} = 0.608\kappa E\frac{t}{R} \text{ (for } \upsilon = 0.316). \tag{3.106}$$

The corresponding buckling load would then be

$$P_{cr} = 2\pi Rt \cdot \sigma_{\text{cr}} = 1.216\pi\kappa Et^2. \tag{3.107}$$

Note that for an isotropic unpressurized circular shell, the equation presented for axial compressed shell should be used with

$$\kappa = 1 - 0.731\left(1 - e^{-\frac{1}{16}\sqrt{\frac{R}{t}}}\right).$$

3.6.2.2. *External pressure*

When discussing the case of a cylindrical shell under external pressure, one has to differentiate between the term "lateral pressure" which means the external pressure is applied only on the curved wall of the shell and the term "hydrostatic pressure" for which the pressure is applied also to the ends of the cylinder. For the lateral pressure case, the load on the shell wall is given by

$$N_y = \sigma_y t = p \cdot R,\ N_x = 0, \tag{3.108}$$

while for the hydrostatic pressure the load on the cylinder wall would be

$$N_y = \sigma_y t = p \cdot R,\ N_x = \sigma_x t = \frac{p \cdot R}{2}, \tag{3.109}$$

where p is the pressure and R stands for the shell radius.

An approximate buckling equation for cylinders loaded by lateral pressure [46] is

$$N_y = k_y\frac{\pi^2 D}{L^2}. \tag{3.110}$$

The buckling factor k_y in Eq. (3.11) is defined [66] as

$$k_x \equiv \frac{p \cdot R \cdot L^2}{\pi^2 D} = \frac{1}{\beta^2}\left[(1+\beta^2)^2 + \frac{12}{\pi^4}\frac{\kappa^2 Z^2}{(1+\beta^2)^2}\right], \text{ where}$$

$$\beta = \frac{nL}{\pi Rm},\ Z = \frac{L^2}{Rt}\sqrt{1-\upsilon^2}. \tag{3.111}$$

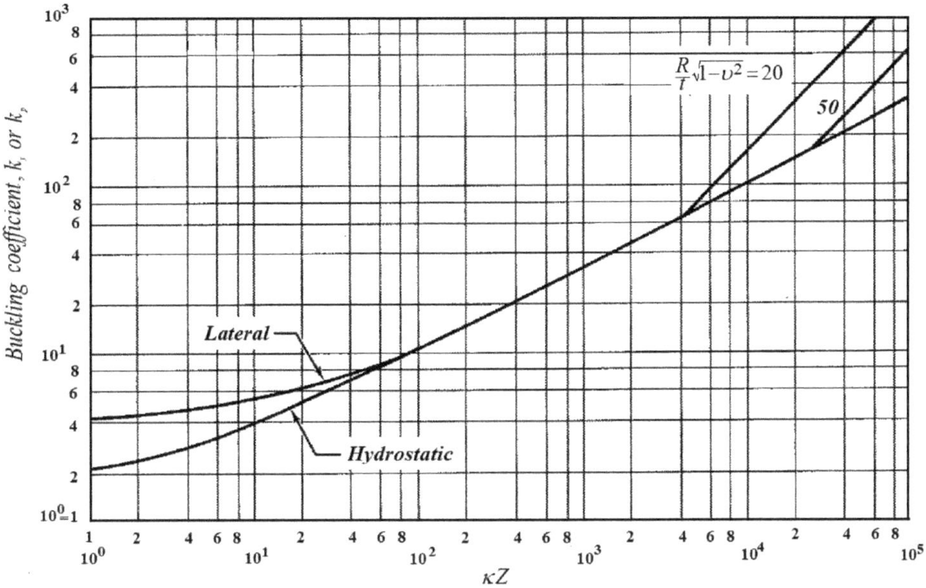

Fig. 3.15. Buckling coefficients for simply supported isotropic circular shell under external pressure (adapted from [66]).

For the hydrostatic pressure case, we will have similar expressions, namely

$$N_y = k_p \frac{\pi^2 D}{L^2}. \tag{3.112}$$

The buckling factor k_p is defined [66] as

$$k_p \equiv \frac{p \cdot R \cdot L^2}{\pi^2 D} = \frac{1}{(\beta^2 + 0.5)} \left[\left(1 + \beta^2\right)^2 + \frac{12}{\pi^4} \frac{\kappa^2 Z^2}{\left(1 + \beta^2\right)^2} \right]. \tag{3.113}$$

Figure 3.15 presents the variation of the minimal values of the buckling coefficients as a function of κZ. It was obtained by allowing β (the buckle aspect ratio) to be a continuous variable along the region of κZ and by differentiation with respect to β to obtain minimal values for the buckling coefficients.

As can be seen in Fig. 3.15, the curve has a straight-line, for $\kappa Z > 100$ and it can be approximated by the following equation [66]:

$$k_y = 1.04\sqrt{\kappa Z}. \tag{3.114}$$

This leads to the following expression for the critical pressure:

$$p_{\text{cr}} = \frac{0.855}{(1 - \upsilon^2)^{\frac{3}{4}}} \frac{E\sqrt{\kappa}}{\left(\frac{R}{t}\right)^{\frac{5}{2}} \left(\frac{L}{R}\right)}$$

$$\Rightarrow p_{\text{cr}}(\upsilon = 0.3) = 0.926 \frac{E\sqrt{\kappa}}{\left(\frac{R}{t}\right)^{\frac{5}{2}} \left(\frac{L}{R}\right)}. \tag{3.115}$$

For high values of κZ appearing in Fig. 3.15, the buckling coefficient would have the following form (corresponding to the buckling of the shell into an oval shape):

$$k_y = \frac{3}{\pi^2}\frac{\kappa Z}{(\frac{R}{t})\sqrt{1-\upsilon^2}}$$

$$\Rightarrow p_{cr} = \frac{\kappa E}{4(1-\upsilon^2)}\left(\frac{t}{R}\right)^3. \qquad (3.116)$$

Note that by changing the boundary conditions to no longitudinal movement and/or no rotation might increase the theoretical buckling pressure by 50% [66].

3.6.2.3. *Torsion*

Applying a torsion moment would induce shear stresses in the wall of the cylinder. The buckling coefficient k_{xy} vs. κZ is presented in Fig. 3.16. For the straight part of the curve $(50 < \kappa Z < 78(R/t)^2(1-\upsilon^2))$ in Fig. 3.16, we can write the following relationship the buckling coefficient and its associated critical shear stress:

$$k_{xy} = \frac{N_{xy}L^2}{\pi^2 D} = 0.85(\kappa Z)^{\frac{3}{4}}$$

$$\Rightarrow \tau_{xy} = \frac{0.747\,(\kappa)^{\frac{3}{4}}\,E}{(\frac{R}{t})^{\frac{5}{4}}\sqrt{(\frac{L}{R})}}. \qquad (3.117)$$

For cases in which $\kappa Z > 78(R/t)^2(1-\upsilon^2)$, the shell is expected to buckle having two circumferential half waves ($m = 2$) [66] and its corresponding buckling coefficient will be

$$k_{xy} = \frac{2\sqrt{2}\kappa Z}{\pi^2\sqrt{(\frac{R}{t})}\sqrt[4]{(1-\upsilon^2)}}$$

$$\Rightarrow \tau_{xy} = \frac{\kappa E}{3\sqrt{2}\sqrt[4]{(1-\upsilon^2)^3}}\left(\frac{t}{R}\right)^{\frac{3}{2}}. \qquad (3.118)$$

It is recommended for moderately long shells [66] to use the knockdown factor of

$$\kappa^{\frac{3}{4}} = 0.67 \Rightarrow \kappa = 0.67^{\frac{4}{3}} = 0.586. \qquad (3.119)$$

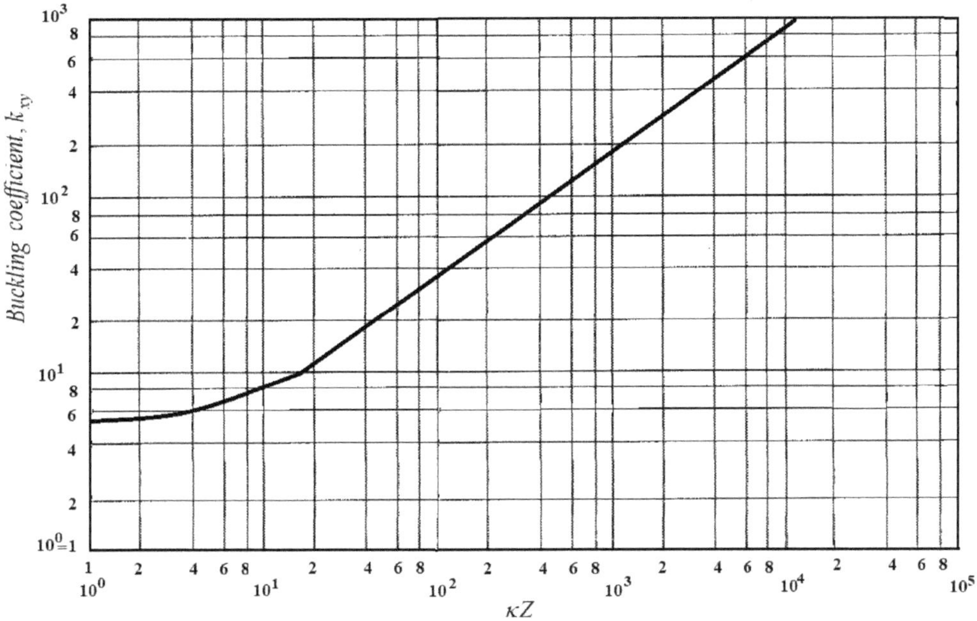

Fig. 3.16. Buckling coefficients for simply supported isotropic circular shell under torsion (adapted from [66]).

3.6.2.4. *Combined loadings*

In real life, the circular shell might be loaded by more than one type of loading. For those cases, the following equations should be used.

Combination of axial compression and bending:

$$\frac{\sigma_x}{\sigma_{x,cr}} + \frac{\sigma_b}{\sigma_{b,cr}} = 1, \tag{3.120}$$

where $\sigma_{x,cr}$ and $\sigma_{b,cr}$ are the critical stresses for axial compression and bending, respectively.

Combination of axial compression and external pressure:

$$\frac{\sigma_x}{\sigma_{x,cr}} + \frac{\sigma_p}{\sigma_{p,cr}} = 1, \tag{3.121}$$

where $\sigma_{x,cr}$ and $\sigma_{p,cr}$ are the critical stresses for axial compression and external pressure, respectively.

Combination of axial compression and torsion:

$$\frac{\sigma_x}{\sigma_{x,cr}} + \frac{\sigma_t}{\sigma_{t,cr}} = 1, \tag{3.122}$$

where $\sigma_{x,cr}$ and $\sigma_{t,cr}$ are the critical stresses for axial compression and torsion, respectively.

3.6.2.5. *Buckling of orthotropic cylindrical shells under axial compression*

Jones [67] presented in 1968 an expression to calculate the buckling load for an orthotropic shell in compression with more than four half circumferential waves ($n \geq 4$). Its form is

$$N_x = \left(\frac{L}{\pi m}\right)^2 \frac{\det(A)}{\det(\overline{A})}, \tag{3.123}$$

where

$$\det(A) = \begin{vmatrix} A_{11} & A_{12} & A_{13} \\ A_{21} & A_{22} & A_{23} \\ A_{31} & A_{32} & A_{33} \end{vmatrix}; \ \det(\overline{A}) = \begin{vmatrix} A_{11} & A_{12} \\ A_{21} & A_{22} \end{vmatrix}, \tag{3.124}$$

and the various members of the determinants are defined as

$$\begin{aligned}
A_{11} &= \widetilde{E}_x \left[\frac{m\pi}{L}\right]^2 + \widetilde{G}_{xy} \left[\frac{n}{R}\right]^2, \ A_{12} = A_{21} = [\widetilde{E}_x + \widetilde{G}_{xy}] \left[\frac{m\pi}{L}\right] \left[\frac{n}{R}\right], \\
A_{22} &= \widetilde{E}_y \left[\frac{n}{R}\right]^2 + \widetilde{G}_{xy} \left[\frac{m\pi}{L}\right]^2, \\
A_{31} &= A_{13} = [\widetilde{C}_{xy} + 2\widetilde{K}_{xy}] \left[\frac{n}{R}\right]^2 \left[\frac{m\pi}{L}\right] + \frac{\widetilde{E}_{xy}}{R} \left[\frac{m\pi}{L}\right] + \widetilde{C}_x \left[\frac{m\pi}{L}\right]^3, \\
A_{23} &= A_{32} = [\widetilde{C}_{xy} + 2\widetilde{K}_{xy}] \left[\frac{m\pi}{L}\right]^2 \left[\frac{n}{R}\right] + \frac{\widetilde{E}_y}{R} \left[\frac{n}{R}\right] + \widetilde{C}_y \left[\frac{n}{R}\right]^3, \\
A_{33} &= \widetilde{D}_x \left[\frac{m\pi}{L}\right]^4 + \widetilde{D}_{xy} \left[\frac{m\pi}{L}\right]^2 \left[\frac{n}{R}\right]^2 \\
&\quad + \widetilde{D}_y \left[\frac{n}{R}\right]^4 + \frac{\widetilde{E}_y}{R^2} + 2\frac{\widetilde{C}_y}{R} \left[\frac{n}{R}\right]^2 + 2\frac{\widetilde{C}_{xy}}{R} \left[\frac{m\pi}{L}\right]^2.
\end{aligned} \tag{3.125}$$

The various stiffness expressions in Eq. (3.125) are next presented.

The extensional stiffness of the shell's wall is written as

$$\widetilde{E}_x = \sum_{k=1}^{N} \left[\frac{E_x}{1 - \upsilon_x \upsilon_y}\right]_k t_k, \ \widetilde{E}_y = \sum_{k=1}^{N} \left[\frac{E_y}{1 - \upsilon_x \upsilon_y}\right]_k t_k,$$

$$\widetilde{E}_{xy} = \sum_{k=1}^{N} \left[\frac{\upsilon_y E_x}{1 - \upsilon_x \upsilon_y}\right]_k t_k = \sum_{k=1}^{N} \left[\frac{\upsilon_x E_y}{1 - \upsilon_x \upsilon_y}\right]_k t_k,$$

while for an isotropic shell

$$\widetilde{E}_x = \widetilde{E}_y = \frac{E \cdot t}{1 - \upsilon^2}, \ \widetilde{E}_{xy} = \frac{\upsilon \cdot \leq E \cdot t}{1 - \upsilon^2}. \tag{3.126}$$

The shear stiffness of the shell's wall is given as

$$\widetilde{G}_{xy} = \sum_{k=1}^{N} [G_{xy}]_k t_k$$

while for an isotropic shell

$$\widetilde{G}_{xy} = G \cdot t. \tag{3.127}$$

The bending stiffness (in the directions x and y and twisting) of the shell's wall (per unit wall width) is given as (see also Fig. 3.17)

$$\widetilde{D}_x = \sum_{k=1}^{N} \left[\frac{E_x}{1 - \upsilon_x \upsilon_y}\right]_k \left[\frac{t_k^3}{12} + t_k \overline{z}_k^2\right], \; \widetilde{D}_y = \sum_{k=1}^{N} \left[\frac{E_y}{1 - \upsilon_x \upsilon_y}\right]_k \left[\frac{t_k^3}{12} + t_k \overline{z}_k^2\right],$$

$$\widetilde{D}_{xy} = \sum_{k=1}^{N} \left[4G_{xy} + \frac{\upsilon_y E_x}{1 - \upsilon_x \upsilon_y} + \frac{\upsilon_x E_y}{1 - \upsilon_x \upsilon_y}\right]_k \left[\frac{t_k^3}{12} + t_k \overline{z}_k^2\right], \tag{3.128}$$

while for an isotropic shell

$$\widetilde{D}_x = \widetilde{D}_y = D, \; \widetilde{D}_{xy} = 2D.$$

The coupling stiffness of the shell's wall is given as (see also Fig. 3.17)

$$\widetilde{C}_x = \sum_{k=1}^{N} \left[\frac{E_x}{1 - \upsilon_x \upsilon_y}\right]_k t_k \overline{z}_k, \widetilde{C}_y = \sum_{k=1}^{N} \left[\frac{E_y}{1 - \upsilon_x \upsilon_y}\right]_k t_k \overline{z}_k,$$

$$\widetilde{C}_{xy} = \sum_{k=1}^{N} \left[\frac{\upsilon_y E_x}{1 - \upsilon_x \upsilon_y}\right]_k t_k \overline{z}_k = \sum_{k=1}^{N} \left[\frac{\upsilon_x E_y}{1 - \upsilon_x \upsilon_y}\right]_k t_k \overline{z}_k,$$

$$\widetilde{K}_{xy} = \sum_{k=1}^{N} [G_{xy}]_k t_k \overline{z}_k, \tag{3.129}$$

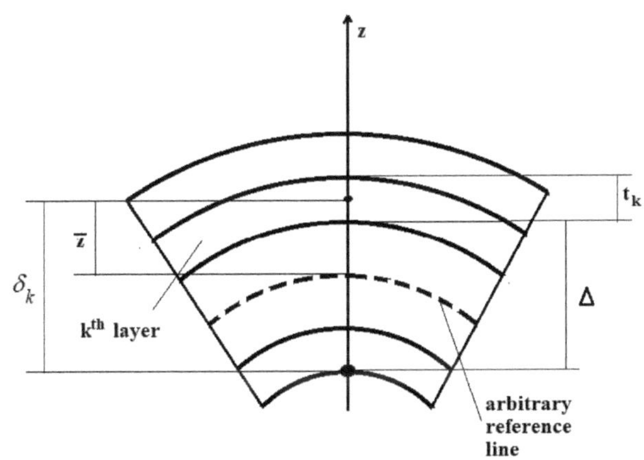

Fig. 3.17. The arrangement of the layers in an orthotropic shell wall (adapted from [66]).

where the subscript k refers to the properties (material and geometry) of the kth layer in an N-layered shell's wall.

Note that due to the large number of variables, no closed-form solution can be presented. For configuration for which the coupling coefficients can be neglected, the buckling load should be calculated using the following equation:

$$N_x = \frac{m^2\pi^2}{L^2}\widetilde{D}_x\left[1 + 2\frac{\widetilde{D}_{xy}}{\widetilde{D}_x}\beta^2 + \frac{\widetilde{D}_y}{\widetilde{D}_x}\beta^4\right] + \frac{\kappa^2 L^4}{\pi^4 m^2 \widetilde{D}_x R^2}\frac{\widetilde{E}_x\widetilde{E}_y - \widetilde{E}_{xy}^2}{\widetilde{E}_x + \left[\frac{\widetilde{E}_x\widetilde{E}_y - \widetilde{E}_{xy}^2}{\widetilde{G}_{xy}} - 2\widetilde{E}_{xy}\right]\beta^2 + \widetilde{E}_y\beta^4}. \tag{3.130}$$

The correlation factor should be taken as the same form used for isotropic shells, while the wall thickness should be replaced by the geometrical mean of the radii of gyrations for the axial and circumferential directions, namely

$$\kappa = 1 - 0.901\left[1 - e^{-\frac{1}{29.8}\sqrt{\frac{R}{\sqrt[4]{\frac{\widetilde{D}_x\widetilde{D}_y}{\widetilde{E}_x\widetilde{E}_y}}}}}\right]. \tag{3.131}$$

Note that for the bending case is similar to the previous case where the correlation factor should be taken as [66]

$$\kappa = 1 - 0.731\left[1 - e^{-\frac{1}{29.8}\sqrt{\frac{R}{\sqrt[4]{\frac{\widetilde{D}_x\widetilde{D}_y}{\widetilde{E}_x\widetilde{E}_y}}}}}\right]. \tag{3.132}$$

3.6.2.6. *Buckling of orthotropic cylindrical shells under external pressure*

For this case, the equation for the buckling under lateral pressure can be written as in Eq. (3.123) namely

$$p = \frac{R}{n^2}\frac{\det(A)}{\det(\overline{A})}, \tag{3.133}$$

where $\det(A)$ and $\det(\overline{A})$ are defined by Eqs. (3.124) and (3.125).

For the hydrostatic case, we have

$$p = \frac{R}{n^2 + 0.5(\frac{m\pi R}{L})^2}\frac{\det(A)}{\det(\overline{A})} \tag{3.134}$$

Note that, for the lateral pressure case, m would be unity while n would vary to yield a minimal critical pressure (but not less than $n = 2$) in Eq. (3.133).

For the hydrostatic pressure case, one should yield also the value of m in addition to n. For long shells, Eq. (3.133) is simplified to the following form:

$$p = \frac{3(\widetilde{D}_y - \frac{\widetilde{C}_y^2}{\widetilde{E}_y})}{R^3}. \tag{3.135}$$

3.6.2.7. *Buckling of orthotropic cylindrical shells under torsion*

Reference [66] reports the torsional buckling of orthotropic cylindrical shells, for negligible coupling of bending and extension. The expression is given as

$$M_{t,cr} \approx 21.75\widetilde{D}_y^{\frac{5}{8}} \frac{R^{\frac{5}{4}}}{\sqrt{L}} \left[\frac{\widetilde{E}_x\widetilde{E}_y - \widetilde{E}_{xy}^2}{\widetilde{E}_y}\right]^{\frac{3}{8}} \cdot \kappa \text{ with } \kappa = 0.67$$

$$\text{for } \sqrt{\frac{\widetilde{E}_x\widetilde{E}_y - \widetilde{E}_{xy}^2}{12\widetilde{E}_y\widetilde{D}_x}} \left[\frac{\widetilde{D}_y}{\widetilde{D}_x}\right]^{\frac{5}{6}} \frac{L^2}{R} \geq 500. \tag{3.136}$$

3.6.2.8. *Combined bending and axial compression of orthotropic cylindrical shells*

Due to a limited experimental data, Ref. [66] recommends the following expression:

$$\frac{\sigma_x}{\sigma_{x,cr}} + \frac{\sigma_b}{\sigma_{b,cr}} = 1. \tag{3.137}$$

References

[1] Timoshenko, S.P. and Gere, J.M., *Elastic Stability*, 2nd Edition, McGraw-Hill, New York, 1963, 541 p.

[2] Timoshenko, S.P. and Woinowsky-Krieger, S., *Theory of Plates and Shells*, McGraw-Hill, New York, 1959, 580 p.

[3] Reissner, E. and Stavsky, Y., Bending and stretching of certain types of heterogeneous aelotropic elastic plates, *J. Appl. Mech.*, Vol. 28(3), 1961, pp. 402–408.

[4] Dong, S.B., Pister, K.S. and Taylor, R.L., On the theory of laminated anisotropic shells and plates, *J. Aerospace Sci.*, Vol. 29(8), 1962, pp. 969–975.

[5] Whitney, J.M., Cylindrical bending of unsymmetrically laminated plates, *J. Composite Mater.*, Vol. 3(4), 1969, pp. 715–719.

[6] Whitney, J.M. and Leissa, A.W., Analysis of heterogeneous anisotropic plates, *J. Appl. Mech.*, Vol. 36(2), 1969, pp. 261–266,

[7] Reddy, J.N., *Mechanics of Laminated Composite Plates and Shells — Theory and Analysis*, 2nd Edition, CRC Press, 2004, 831p.

[8] Abramovich, H. (ed)., *Stability and Vibrations of Thin-Walled Composite Structures*, Woodhead Publishing Series in Composites Science and Engineering, Elsevier, Duxford, 758 p.

[9] Simitses, G.J., *An Introduction to the Elastic Stability of Structures*, Prentice-Hall, Inc., Englewood Cliffs, NJ, 1976, 253 p.
[10] Reddy, J.N., *Energy Principles and Variational Methods in Applied Mechanics*, 2nd Edition, John Wile & Sons, New York, 2002, 608 p.
[11] Brush, D.O. and Almroth, B.O., *Buckling of Bars, Plates and Shells*, McGraw-Hill International Student Edition, McGraw-Hill 1975, 379 p.
[12] Thompson, J.M.T. and Hunt, G.W., *Elastic Instability Phenomena*, John Wiley & Sons, 1984, 216 p.
[13] Johnson, J.B., Bryan, C.W., Turneaure, F.E. and Kinne, W.S., *The Theory and Practice of Modern Framed Structures — Designed for the Use of Schools and for Engineers in Professional Practice, Part III: Design,* John Wiley & Sons, New York, 1916, 565 p.
[14] Timoshenko, S.P., On the correction factor for shear of the differential equation for transverse vibrations of bars of uniform cross-section, *Philos. Maga.*, Vol. 41(25), 1921, pp. 744–746.
[15] Timoshenko, S.P., On the transverse vibrations of bars of uniform cross-section, *Philos. Maga.*, Vol. 43(253), 1922, pp. 125–131.
[16] Reissner, E., The effect of transverse shear deformation on the bending of elastic plates, *J. Appl. Mech.*, Vol. 12(2), 1945, pp. A69–A77.
[17] Mindlin, R. D., Influence of rotatory inertia and shear on flexural motions of isotropic, elastic plates, *ASME J. Appl. Mech.*, Vol. 18(1), 1951, pp. 31–38.
[18] Whitney, J.M., The effect of transverse shear deformation in the bending of laminated plates, *J. Composite Mater.*, Vol. 3(4), 1969, pp. 534–547.
[19] Whitney, J.M. and Pagano, N.J., Shear deformation in heterogeneous anisotropic plates, *J. Appl. Mech.*, Vol. 37(4), 1970, pp. 1031–1036.
[20] Whitney, J.M., Shear correction factors for orthotropic laminates under static load, *J. Appl. Mech.*, Vol. 40(1), 1973, pp. 302–304.
[21] Reissner, E., Note on the effect of transverse shear deformation in laminated anisotropic plates, *Computer Methods Appl. Mech. Eng.*, Vol. 20, 1979, pp. 203–209.
[22] Abramovich, H., Shear deformation and rotary inertia effects of vibrating composite beams, *Composite Struct.*, Vol. 20, 1992, pp. 165–173.
[23] Abramovich, H. and Hamburger, O., Vibration of a uniform cantilever Timoshenko beam with translational and rotational springs and with a tip mass, *J. Sound Vibration*, Vol. 154(1), 1992, pp. 67–80.
[24] Abramovich, H., A note on experimental investigation on a vibrating Timoshenko cantilever beam, *J. Sound Vibration*, Vol. 160(1), 1993, pp. 167–171.
[25] Abramovich, H. and Livshits, A., Dynamic behavior of cross-ply laminated beams with piezoelectric layers, *Composite Struct.*, Vol. 25(1-4), 1993, pp. 371–379.
[26] Abramovich, H. and Livshits, A., Free vibrations of non-symmetric cross-ply laminated composite beams, *J. Sound Vibration*, 176(5), 1994, pp. 597–612.
[27] Abramovich, H., Thermal buckling of cross-ply composite laminates using a first-order shear deformation theory, *Composites Struct.*, Vol. 28, 1994, pp. 201–213.
[28] Abramovich, H., Deflection control of laminated composite beams with piezoceramic layers — closed form solutions, *Composite Struct.*, Vol. 43, 1998, pp. 217–231.
[29] Abramovich, H., Eisenberger, M. and Shulepov, O., Vibrations and buckling of cross-ply non-symmetric laminated composite beams, *AIAA J.*, Vol. 34(5), 1996, pp. 1064–1069.
[30] Leissa, A.W., Buckling of laminated composite plates and shell panels, AFWAL-TR-85-3069, June 1985.
[31] Bergmann, S. and Reissner, H., Neuere probleme aus der flugzeugstatik uber die knickung von wellblechstreifen bei schubbeanspruchung, *Zeit. Flugtechnik Motorluftschiffahrt (Z.F.M.)*, Vol. 20(18), 1929, pp. 475–481.

[32] Jones, R.M., Buckling and vibration of unsymmetrically laminated cross-ply rectangular plates, *AIAA J.*, Vol. 11(12), 1973, pp. 1626–1632.
[33] Whitney, J.M., A study of the effects of coupling between bending and stretching on the mechanical behavior of layered anisotropic composite materials, Ph.D. Dissertation, Ohio State University, 1968. Also Tech. Rept. AFML-TR-68-330, 80 p., April 1969.
[34] Whitney, J.M. and Leissa, A.W., Analysis of heterogeneous anisotropic plates, *Trans. ASME, J. Appl. Mech.*, Vol. 36(2), 1969, pp. 261–266.
[35] Jones, R.M., Morgan, H.S. and Whitney, J.M., Buckling and vibration of antisymmetrically laminated angle-ply rectangular plates, *Trans. ASME J. Appl. Mech.*, Vol. 12, 1973, pp. 1143–1144.
[36] Whitney, J.M., Bending, vibrations, and buckling of laminated anisotropic rectangular plates, Wright Patterson AFBML, Ohio, Technical Rept. AFML-TR-70-75, 35 p. Aug. 1970.
[37] Whitney, J.M., The effect of boundary conditions of the response of laminated composites, *J. Composite Mater.*, Vol. 4, 1970, pp. 192–203.
[38] Khdeir, A.A., Stability of antisymmetric angle-ply laminated plates, *J. Eng. Mech.*, Vol. 115(5), 1989, pp. 952–962.
[39] Reddy, J.N., Khdeir, A.A. and Librescu, L., Lévy type solutions for symmetrically laminated rectangular plates using first order shear deformation theory, *J. Appl. Mech.*, Vol. 54, 1987, pp. 740–742.
[40] Gerard, G. and Becker, H., Handbook of structural stability, Part I — Buckling of flat plates, NACA TN 3781, 106 p.
[41] Johns, D.J., Shear buckling of isotropic and orthotropic plates — A review, Aeronautical Research Council, R&M No. 3677, London: Her Majesty's Stationery Office, UK, March 1972, 37 p.
[42] Stein, M. and Neff, J., Buckling stresses of simply supported rectangular flat plates in shear, NACA TN 1222, March 1947, 14 p.
[43] Batdorf, S.B. and Stein, M., Critical combinations of shear and direct stress for simply supported rectangular flat plates, NACA TN 1223, March 1947, 32 p.
[44] Donnell, L.H., Stability of thin-walled tubes under torsion, NACA Report 479, 1933, 27 p.
[45] Donnell, L.H., A new theory for the buckling of thin cylinders under axial compression and bending, AER-56-12, Transaction of the American Society of Mechanical Engineers, Aeronautical Engineering Division, Annual Meeting, New York, Dec. 3–7, 1934, pp. 796–806.
[46] Batdorf, S.B., A simplified method of elastic — stability analysis for thin cylindrical shells II — Modified equilibrium equation, NACA TN-1342, 1947, 36 p.
[47] Sanders, J.L. Jr., An improved first-approximation theory for thin shells, NASA TR R-24, 1959, 11 p.
[48] Weingarten, V.I., Morgan, E.J. and Seide, P., Elastic stability of thin-walled cylindrical and conical shells under axial compression, *AIAA J.* Vol. 3(3), 1965, pp. 500–505.
[49] Block, D.L., Card, M.F. and Mikulas, M.M. Jr., Buckling of eccentrically stiffened orthotropic cylinders, NASA TN D-2960, NASA Langley Research Center, Hampton, Virginia, Aug. 1965, 31 p.
[50] Jones, R.M. and Morgan H.S., Buckling and vibration of cross-ply laminated circular cylindrical shells, *AIAA J.*, Vol. 13(5), 1975, pp. 664–671.
[51] Simitses, G.J., Buckling of moderately thick laminated cylindrical shells — A review, Composites Part B 27B, 1359-8368(95)00013-5, 1996, pp. 581–587.
[52] Simitses, G.J., Sheinman, I. and Shaw, D., The accuracy of Donnell's equations for axially-loaded, imperfect orthotropic cylinders, *Computers Struct.*, Vol. 20(6), 1985, pp. 939–945.

[53] Simitses, G.J., Buckling and postbuckling of imperfect cylindrical shells — A review, *Appl. Mech. Rev.*, Vol. 39(10), 1986, pp. 1517–1524.
[54] Hutchinson, J., Buckling of imperfect cylindrical shells under axial compression and external pressure, *AIAA J.*, Vol. 3(10), 1965, pp. 1969–1970.
[55] Teng, J.G., Buckling of thin shells: Recent advances and trends, *Appl. Mech. Rev.*, Vol. 49(4), 1996, pp. 263–274.
[56] Hunt, G.W., Lord, G.J. and Peletier, M.A., Cylindrical shell buckling: A characterization of localization and periodicity, *Discrete Contin. Dynam. Syst. — Ser. B*, Vol. 3(4), 2003, pp. 505–518.
[57] Ventsel, E. and Krauthammer, T., *Thin Plates and Shells, Theory Analysis and Application*, Marcel Dekker, Inc. New York, 2001, 671 p.
[58] Aly, S.S.S., Buckling assessment of axially loaded cylindrical shells with random imperfections, Ph.D. thesis, IOWA State University, 1995, 154 p.
[59] Bushnell, D., Computerized buckling analysis of shells, AFWAL-TR-81-3049, Dec. 1981, 805 p.
[60] Godoy, L.A. and Sosa, E.M., Computational buckling analysis and practice, *Mećanica Computacional*, Vol. XXI, Idelsohn, S.R., Sonzogni, V.E. and Cardona A. (eds.), Santa Fe-Paraná, Argentina, 2002, pp. 1652–1667.
[61] Takano, A., Simple closed-form solution for the buckling of moderately thick anisotropic cylinders, *Trans. JSASS Aerospace Tech.*, Vol. 10, 2012, pp. 17–26.
[62] Lvov, G., Pupazescu, A., Beschetnikov, D. and Zaharia, M., Buckling analysis of a thin-walled cylindrical shell strengthened by fiber-reinforced polymers, *Materiale Plastic*, Vol. 52(1), 2015, pp. 28–31.
[63] Chung, S.W. and Hong, S.G., A comparison of membrane shell theories of hybrid anisotropic materials, *Eur. J. Eng. Technol.*, Vol. 4(5), 2016, pp. 83–106.
[64] Bushnell, D., Buckling of shells-pitfall for designers, *AIAA J.*, Vol. 19(9), 1981, pp. 1183–1226.
[65] Arbocz, J. and Starnes, J.H. Jr., Future directions and challenges in shell stability analysis, *Thin-Walled Struct.*, Vol. 40(9), 2002, pp. 729–754.
[66] Weingarten, V.I., Seide, P. and Peterson, J.P., Buckling of thin-walled circular cylinders, NASA SP-8007, Sept. 1965, revised version Aug. 1968, 49 p.
[67] Jones, R.M., Buckling of circular cylindrical shells with multiple orthotropic layers and eccentric stiffeners, *AIAA J.*, Vol. 6(12), 1968, pp. 2301–2305.

Appendix A: Non-symmetric Laminated Composite Beam — CLT Approach

The two coupled equations of motion for a non-symmetric laminated composite beam under compressive loads, based on the CLT approach are as follows:

$$A_{11}\frac{d^2u_0}{dx^2} - B_{11}\frac{d^3w_0}{dx^3} = 0, \tag{A.1}$$

$$-B_{11}\frac{d^3u_0}{dx^3} + D_{11}\frac{d^4w_0}{dx^4} + \overline{N}_{xx}\frac{d^2w_0}{dx^2} = 0. \tag{A.2}$$

Derivation with respect to x of Eq. (A.1) and substituting in Eq. (A.2) leads to the following decoupled equation:

$$-\frac{B_{11}}{A_{11}}\frac{d^4w_0}{dx^4}+D_{11}\frac{d^4w_0}{dx^4}+\overline{N}_{xx}\frac{d^2w_0}{dx^2}=0. \tag{A.3}$$

The solution of Eq. (A.3) has the following general form:

$$w_0(x)=C_1\sin(\lambda x)+C_2\cos(\lambda x)+C_3x+C_4, \tag{A.4}$$

where

$$\lambda^2=\frac{\overline{N}_{xx}}{\left(D_{11}-\frac{B_{11}^2}{A_{11}}\right)}. \tag{A.5}$$

Let us assume that the axial displacement $u0(x)$ has the same for as the lateral displacement (see, for example, [5, 18–29]), namely

$$u_0(\lambda x)=C_5\sin(\lambda x)+C_6\cos(\lambda x)+C_7x+C_8. \tag{A.6}$$

The four constants in Eq. (A.6) are connected to the four constants in Eq. (A.4). By substituting Eqs. (A.4) and (A.6) in Eq. (A.1) we get the following relationships:

$$C_5=-\frac{B_{11}}{A_{11}}\cdot\lambda\cdot C_2,$$

$$C_6=+\frac{B_{11}}{A_{11}}\cdot\lambda\cdot C_1. \tag{A.7}$$

The rest of the constants, C_1, C_2, C_3, C_4, C_7 and C_8 will be determined using the six boundary conditions of the problem. For the case of simply supported at both ends of the column, the out-of-plane boundary conditions have the following expressions:

$$w_0(0)=0, M_{xx}(0)=0\Rightarrow -B_{11}\frac{du_0(0)}{dx}+D_{11}\frac{d^2w_0(0)}{dx^2}=0, \tag{A.8}$$

$$w_0(L)=0, M_{xx}(L)=0\Rightarrow -B_{11}\frac{du_0(L)}{dx}+D_{11}\frac{d^2w_0(L)}{dx^2}=0, \tag{A.9}$$

and the in-plane boundary conditions can be written (assuming that at $x=0$ there is no axial displacement, while at the other end of the beam, at $x=L$, there is a compression load) as:

$$u_0(0)=0, \tag{A.10}$$

$$A_{11}\frac{du_0(L)}{dx}-B_{11}\frac{d^2w_0(L)}{dx^2}=-\overline{N}_{xx}. \tag{A.11}$$

Applying the four out-of-plane boundary conditions (Eqs. (A.8) and (A.9)) leads to the following four equations:

$$C_2 + C_4 = 0,$$

$$\frac{B_{11}^2}{A_{11}} \cdot \widehat{\lambda}^2 \cdot C_2 - B_{11} \cdot C_7 - D_{11} \cdot \widehat{\lambda}^2 \cdot C_2 = 0,$$

$$C_1 \sin(\widehat{\lambda} L) + C_2 \cos(\widehat{\lambda} L) + C_3 x + C_4 = 0,$$

$$-\frac{B_{11}^2}{A_{11}} \cdot \widehat{\lambda}^2 \cdot C_2 \cdot \cos(\widehat{\lambda} L) - \frac{B_{11}^2}{A_{11}} \cdot \widehat{\lambda}^2 \cdot C_1 \cdot \sin(\widehat{\lambda} L) + B_{11} \cdot C_7,$$

$$+D_{11} \cdot \widehat{\lambda}^2 \cdot C_1 \cdot \sin(\widehat{\lambda} L) + D_{11} \cdot \widehat{\lambda}^2 \cdot C_2 \cdot \cos(\widehat{\lambda} L) = 0. \tag{A.12}$$

Applying the two in-plane boundary conditions (Eqs. (A.10) and (A.11)) yields

$$\frac{B_{11}}{A_{11}} \cdot \widehat{\lambda} \cdot C_1 + C_8 = 0,$$

$$- B_{11} \cdot \widehat{\lambda}^2 \cdot C_2 \cdot \cos(\widehat{\lambda} L) - B_{11} \cdot \widehat{\lambda}^2 \cdot C_1 \cdot \sin(\widehat{\lambda} L) + A_{11} \cdot C_7,$$

$$+ B_{11} \cdot \widehat{\lambda}^2 \cdot C_1 \cdot \sin(\widehat{\lambda} L) + -B_{11} \cdot \widehat{\lambda}^2 \cdot C_2 \cdot \cos(\widehat{\lambda} L)$$

$$= -\overline{N}_{xx} \Rightarrow C_7 = -\frac{\overline{N}_{xx}}{A_{11}}. \tag{A.13}$$

Solving for the six constants we get the following results:

$$C_1 = \frac{B_{11}}{A_{11}} \frac{[\cos(\widehat{\lambda} L) - 1]}{\sin(\widehat{\lambda} L)}, \quad C_2 = \frac{B_{11}}{A_{11}}, \quad C_3 = 0,$$

$$C_4 = -\frac{B_{11}}{A_{11}}, \quad C_7 = -\frac{\overline{N}_{xx}}{A_{11}}, \quad C_8 = -\widehat{\lambda}\frac{B_{11}^2}{A_{11}^2} \frac{[\cos(\widehat{\lambda} L) - 1]}{\sin(\widehat{\lambda} L)}. \tag{A.14}$$

To obtain the buckling load, one should demand that $w_0(x) \to \infty$, namely finding when one of the constants would tend to infinity. Observing the constant C_1 leads to the following eigenvalue solution

$$\sin(\widehat{\lambda} L) = 0 \Rightarrow \widehat{\lambda} L = n\pi, n = 1, 2, 3, 4, \ldots, \tag{A.15}$$

which leads to the critical buckling load per unit width, b, having the following form:

$$(\overline{N}_{xx})_{cr} = \frac{\pi^2}{L^2} \left(D_{11} - \frac{B_{11}^2}{A_{11}} \right). \tag{A.16}$$

The shape of the buckling will be $\sin(\frac{\pi x}{L})$ like for the symmetric case.

One should note that the critical buckling load per unit width $(\overline{N}_{xx})_{cr}$ is influenced by the coupling coefficient B_{11}, presenting for non-symmetric cases, a buckling load below that obtained for the symmetric case, having the same number of laminates.

For the clamped–clamped case, the involved boundary conditions are as follows:

$$w_0(0) = 0, \quad \frac{dw_0(0)}{dx} = 0, \tag{A.17}$$

$$w_0(L) = 0, \quad \frac{dw_0(L)}{dx} = 0, \tag{A.18}$$

$$u_0(0) = 0, \quad A_{11}\frac{du_0(L)}{dx} - B_{11}\frac{d^2w_0(L)}{dx^2} = -\overline{N}_{xx}. \tag{A.19}$$

Applying the boundary conditions yields the following characteristic equation:

$$(\widehat{\lambda}L) \cdot \sin(\widehat{\lambda}L) = 2[1 - \cos(\widehat{\lambda}L)] \;\Rightarrow\; \widehat{\lambda}L = 2\pi, 8.987, 4\pi, \ldots \tag{A.20}$$

with $C_1 = C_3 = C_4 = C_8 = 0$, $C_2 \neq 0$ and $C_7 = -\frac{\overline{N}_{xx}}{A_{11}}$. The critical buckling load per unit width can then be written as

$$(\overline{N}_{xx})_{cr} = \frac{4\pi^2}{L^2}\left(D_{11} - \frac{B_{11}^2}{A_{11}}\right). \tag{A.21}$$

The buckling shape can be written as

$$w_0(x) = C_2\left[\cos\left(\frac{2\pi x}{L}\right) - 1\right]. \tag{A.22}$$

Appendix B: Energy Methods

Rayleigh–Ritz method: The Rayleigh–Ritz method is attributed to John William Strutt, 3rd Baron Rayleigh (Lord Rayleigh), a British scientist with many contributions to both experimental and theoretical physics and Walther Heinrich Ritz (2nd Feb. 1878–7th July 1909) a Swiss theoretical physicist.

The method asks for assuming a deflection shape, satisfying only the geometric boundary conditions[m] (like deflection or angle) having the following general form [1, 2]:

$$v(x) = \alpha_1 f_1(x) + \alpha_2 f_2(x) + \cdots + \alpha_n f_n(x), \tag{B.1}$$

where α_j is the unknown coefficient for the known function $f_j(x)'$ which satisfies the geometric boundary condition of the posed problem. Note that

[m] Natural boundary conditions are the boundary conditions for the moment and the shear force.

instead of assuming a polynomial series, one can any other functions, like trigonometric series satisfying the geometric boundary conditions.

The second step is to write the total energy as a function of the unknown coefficients

$$\pi(\alpha_1, \alpha_2, \ldots, \alpha_n) = U + T + V, \tag{B.2}$$

where U, V and T are the strain energy, the work of the external loads and the kinetic energy, respectively.

The third step is to apply the principle of minimum potential energy to determine the unknown coefficients α_j, namely

$$\frac{\partial \pi}{\partial \alpha_1} = \frac{\partial \pi}{\partial \alpha_2} = \cdots = \frac{\partial \pi}{\partial \alpha_n} = 0. \tag{B.3}$$

Equation (B.3) leads to n equations with n unknowns, α_j.

Example. Determining the buckling load of a cantilever beam using the Rayleigh–Ritz method.

The strain energy U and the work of the external force, V, are given by the following equations:

$$U = \int_0^L \frac{M^2}{2EI} dx = \frac{EI}{2} \int_0^L \left(\frac{d^2 v}{dx^2}\right)^2 dx,$$

$$V = -P_{cr}\delta = -\frac{P_{cr}}{2} \int_0^L \left(\frac{dv}{dx}\right)^2 dx,$$

$$\Rightarrow \pi = U + V = \frac{EI}{2} \int_0^L \left(\frac{d^2 v}{dx^2}\right)^2 dx - \frac{P_{cr}}{2} \int_0^L \left(\frac{dv}{dx}\right)^2 dx. \tag{B.4}$$

Assuming that the shape of the deflection is the shape of a cantilever beam, loaded at its free end (see Fig. B.1) we have

$$v(x) = \frac{v_o x^2}{2L^3}(3L - x), \tag{B.5}$$

where v_o is the free parameter to be determined.

Note that the expression in Eq. B.5 satisfies the geometric boundary conditions of the cantilever problem, namely

$$\begin{aligned} v(x = 0) &= 0, \\ \frac{dv(x = 0)}{dx} &= 0, \\ v(x = L) &= v_0, \end{aligned} \tag{B.6}$$

and the moment is zero at its free end, $M = d^2v/dx^2 = 0$.

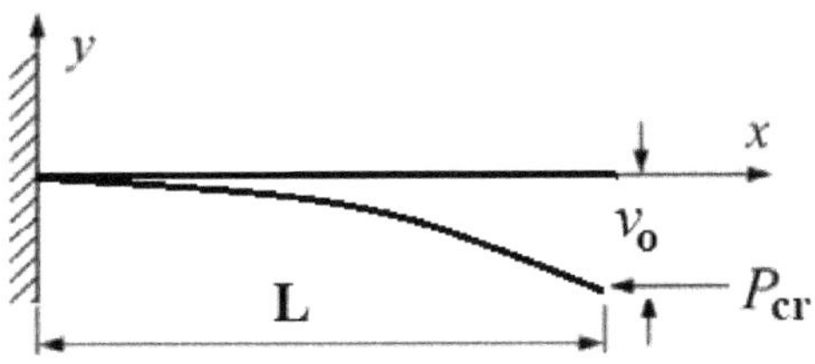

Fig. B.1. The shape of the buckled cantilever beam.

The bending moment at any point along the buckled beam is given by

$$M(x) = P_{cr}(v_o - v). \tag{B.7}$$

Substituting the expressions for the moment and the assumed shape into Eq. (B.4) yields

$$\pi = U + V = \frac{P_{cr}^2 v_o^2}{2EI} \int_0^L \left(1 - \frac{3x^2}{2L^2} + \frac{x^3}{2L^3}\right)^2 dx$$
$$- \frac{P_{cr}}{2} \int_0^L \left(\frac{3v_0}{2L^3}\right)^2 x^2(2L - x)^2 dx$$
$$\Rightarrow \pi = U + V = \frac{17}{35}\frac{P_{cr}^2 v_o^2 L}{2EI} - \frac{3}{5}P_{cr}\frac{v_o^2}{L}. \tag{B.8}$$

Minimizing the potential gives

$$\frac{\partial \pi}{\partial v_o} = \frac{\partial\,(U+V)}{\partial v_o} = \frac{17}{35}\frac{P_{cr}^2 v_o L}{EI} - \frac{6P_{cr}v_o}{5L} = 0$$
$$\Rightarrow P_{cr} = \frac{42EI}{17L^2} = 2.471\frac{EI}{L^2}. \tag{B.9}$$

The value of the buckling load using the Rayleigh–Ritz method, $P_{cr} = 2.471EI/L^2$ compares very well with the exact value $P_{cr} = \pi^2 EI/4L^2 = 2.467EI/L^2$.

For plates, the strain energy U, the work performed by the in-plane forces V and the kinetic energy are given as

$$U = \frac{D}{2}\iint_A \left\{\left(\frac{\partial^2 w}{\partial x^2} + \frac{\partial^2 w}{\partial y^2}\right)^2 + 2\,(1 - v)\left[\frac{\partial^2 w}{\partial x^2}\frac{\partial^2 w}{\partial y^2} - \left(\frac{\partial^2 w}{\partial x \partial y}\right)^2\right]\right\} dA,$$
$$V = -\frac{1}{2}\iint_A \left\{N_x\left(\frac{\partial w}{\partial x}\right)^2 + N_y\left(\frac{\partial w}{\partial y}\right)^2 + 2N_{xy}\left(\frac{\partial w}{\partial x}\frac{\partial w}{\partial y}\right)\right\} dA,$$
$$T = \frac{\rho t \omega^2}{2}\iint_A w^2 dA, \tag{B.10}$$

where $D = Et^3/[12(1 - \nu^2)]$, E is the Young's modulus, t is the thickness of the plate, ν is the Poisson's ratio, $w(x, y)$ stands for the deflection of the plate, ρ is the mass density per unit plate area and ω is the angular frequency of vibration.

Galerkin method: A Russian mathematician, Boris Grigoryevich Galerkin (1871–1945), first proposed the method, and therefore it carries its name [1, 3, 4]. It is a strong and useful method to approximate solutions of ordinary and partial differential equations, and is widely applied to solve wide range of problems in engineering. The method uses the governing equation for the posed problem and its boundary conditions and assumes a trial function, which satisfies the boundary conditions (both geometric and natural).

The first step is to pose the governing equation of the problem. For a plate under a lateral load, q, the governing equation would have the following form:

$$\nabla^4 w \equiv \frac{\partial^4 w}{\partial x^4} + 2\frac{\partial^4 w}{\partial x^2 \partial y^2} + \frac{\partial^4 w}{\partial y^4} = \frac{q}{D}. \tag{B.11}$$

Next, a deflection trial function is assumed which satisfies all boundary conditions of the problem.

The approximate solution for the plate case would be

$$\widetilde{w}\,(x, y) = \sum_{i=1}^{N} \alpha_i \nabla^4 \varphi_i\,(x, y), \tag{B.12}$$

where α_i are unknown constants and $\varphi_i(x, y)$ are linearly independent trial functions to satisfy the boundary condition of the given plate. For a plate, we have

$$\text{Clamped-edge } w(x,y)_{@x=a} = 0 \quad \left.\frac{\partial w\,(x,y)}{\partial x}\right|_{@x=a} = 0,$$

$$w(x,y)_{@y=b} = 0 \quad \left.\frac{\partial w(x,y)}{\partial y}\right|_{@y=b} = 0;$$

$$\text{Simply supported-edge } w(x,y)_{@x=a} = 0$$

$$M(x,y)_{@x=a} = -D\left(\frac{\partial^2 w}{\partial x^2} + \nu\frac{\partial^2 w}{\partial y^2}\right)_{@x=a}$$

$$= -D\left.\frac{\partial^2 w}{\partial x^2}\right|_{@x=a} = 0,$$

$$w(x,y)_{@y=b} = 0$$

$$M(x,y)_{@y=b} = -D\left(\frac{\partial^2 w}{\partial x^2} + \nu\frac{\partial^2 w}{\partial y^2}\right)_{@y=b}$$

$$= -D\left.\frac{\partial^2 w}{\partial y^2}\right|_{@y=b} = 0;$$

$$\text{Free}-\text{edge}\ M(x,y)_{@x=a} = -D\left(\frac{\partial^2 w}{\partial x^2}+\nu\frac{\partial^2 w}{\partial y^2}\right)_{@x=a} = 0,$$

$$V_{yz}(x,y)_{@x=a} = -D\left(\frac{\partial^3 w}{\partial x^3}+(2-\nu)\frac{\partial^3 w}{\partial x\partial y^2}\right)_{@x=a} = 0,$$

$$M(x,y)_{@y=b} = -D\left(\frac{\partial^2 w}{\partial x^2}+\nu\frac{\partial^2 w}{\partial y^2}\right)_{@y=b} = 0,$$

$$V_{yz}(x,y)_{@y=b} = -D\left(\frac{\partial^3 w}{\partial y^3}+(2-\nu)\frac{\partial^3 w}{\partial x^2\partial y}\right)_{@y=b} = 0. \tag{B.13}$$

The weighted integral will be:

$$\iint_A\left[\sum_{i=1}^{N}\alpha_i\nabla^4\varphi_i(x,y)-\frac{q(x,y)}{D}\right]\varphi_i(x,y)dxdy = 0$$

$$\Rightarrow \sum_{i=1}^{N}\alpha_i\iint_A[\nabla^4\varphi_i(x,y)\varphi_j(x,y)]dxdy = \frac{q(x,y)}{D}\iint_A\varphi_j\,(x,y)dxdy,\ \text{or} \tag{B.14}$$

$$\sum_{i=1}^{N}\alpha_i K_{ij} = \frac{q\,(x,y)}{D}F_j,\ \text{where}$$

$$F_j = \iint_A\varphi_j(x,y)dxdy,\ K_{ij} = \iint_A[\nabla^4\varphi_i(x,y)\varphi_j(x,y)]dxdy.$$

Example: Determining the deflection of a beam using the Galerkin method.

The beam has — any boundary conditions at its two ends and is loaded by a lateral load per unit length (L) of the beam, $q(x)$. The governing equation is written as

$$EIy^{IV} = q(x) \Rightarrow y^{IV} = \frac{q(x)}{EI}. \tag{B.15}$$

Assuming the approximate deflection of the beam, where the origin of the longitudinal coordinate, x, is at the mid-span of the beam, $\widetilde{y} = \alpha_i\varphi_i(x)$ and substituting in Eq. (B.15) leads to the residual $R(x)$ of the problem, namely

$$\widetilde{y}^{IV} - \frac{q(x)}{EI} \equiv R(x) \neq 0. \tag{B.16}$$

Minimizing the residual, we obtain

$$\int_{-\frac{L}{2}}^{\frac{L}{2}} \varphi_i(x)R(x)dx = 0, \quad i = 1, 2, 3, \ldots, n$$

$$\Rightarrow \int_{-\frac{L}{2}}^{\frac{L}{2}} \varphi_i(x)\left(\tilde{y}^{IV} - \frac{q(x)}{EI}\right)dx \tag{B.17}$$

$$= \int_{-\frac{L}{2}}^{\frac{L}{2}} \varphi_i(x)\left(\frac{\alpha_j d^4\varphi_j}{dx^4} - \frac{q(x)}{EI}\right)dx = 0, \quad i = 1, 2, 3, \ldots, n.$$

Equation (B.17) can be written in a matrix form, namely

$$[K]\{\alpha\} = \{P\}, \quad \text{where}$$

$$K_{ij} = \int_{-\frac{L}{2}}^{\frac{L}{2}} \varphi_i(x)\left(\frac{d^4\varphi_j}{dx^4}\right)dx = 0, \tag{B.18}$$

$$P_i = \int_{-\frac{L}{2}}^{\frac{L}{2}} \varphi_i(x)\frac{q(x)}{EI}dx.$$

Solution of the matrix equation will provide the constants α_i needed to define the deflection curve $\varphi_i(x)$, which was earlier assumed to be known. Note that a high number of terms in the deflection curve will lead to a better accurate solution. The deflection curve $\varphi_i(x)$ must satisfy the four boundary conditions of the beam.

Kantorovich method: The method carries the name of Leonid Vitaliyevich Kantorovich (1912–1986), a soviet mathematician and economist who shared the 1975 Nobel Prize for Economics with Tjalling Koopmans for their work on the optimal allocation of scares resources. The Kantorovich method is applied in the area of thin-walled 2D structures [5–11], by reducing the problem of a double integral to a single simple integral.

As presented by Kantorovich and Krylov [5], in what it is known in the literature as the classical Kantorovich method, the solution of minimizing a functional having the form of a double integral, $I(w) = \int_A L(w)dxdy$ is assumed to have a separable form of

$$w(x, y) = \sum f(x) \cdot g(y), \tag{B.19}$$

where either $f(x)$ or $g(y)$ is chosen *a priori*. The chosen function has to satisfy the geometric boundary conditions of the problem. After introducing the assumed function into the functional and performing the integration over its domain, the problem of minimizing the double integral is simplified to a problem of minimizing a simple integral. This method is different from the Rayleigh–Ritz or Galerkin methods by the fact that only a part of the approximate function is chosen *a priori*. The other part of the function is determined according to the remaining unknown in the integral. The double integration functional can be derived based on the Galerkin weighted residual method or on the minimum total potential energy.

An improvement to the above-described procedure was proposed by Kerr [7] and is known as the extended Kantorovich method, in which the iterative procedure is repeated until the solution converges to a desired degree of accuracy.

$$w_m(x, y) = \sum_{n=1}^{m} f_n(x) \cdot g_n(y), \tag{B.20}$$

where $f_n(x)$ are the functions determined in the previous step (the classical Kantorovich method) and $g_n(y)$ are the assumed unknowns. The g_n's are then determined from a set of m ordinary differential equations. This iterative process is continued by back substituting the determined $g_n(y)$ in Eq. (B.20), leading to the determination of the unknown f_n's, and so on till the convergence of the solutions. The iteration process was found to converge very rapidly [7–11] and the final form of the solution is independent of the initial solution choice. Moreover, it is claimed in [8, 11] that the initial functions are neither required to satisfy the geometric nor the natural boundary conditions; the iterative process will eventually force the solution to satisfy all boundary conditions of the posed problem.

Example: Determining the deflection of a laminated composite rectangular plate using the Kantorovich method (adapted from [8]).

The governing equation for the deflection of a laminated composite rectangular plate under a lateral distributed load $q(x, y)$ can be written as

$$D_{11}\frac{\partial^4 w(x, y)}{\partial x^4} + (2D_{12} + 4D_{66})\frac{\partial^4 w(x, y)}{\partial x^2 \partial y^2} + D_{22}\frac{\partial^4 w(x, y)}{\partial y^4} = q(x, y), \tag{B.21}$$

where D_{11}, D_{12}, D_{22} and D_{66} are the bending stiffness defined by Eq. (2.26).

To solve Eq. (B.21), the Galerkin method is applied leading to

$$\int_0^a \int_0^b \left[D_{11}\frac{\partial^4 \widetilde{w}(x, y)}{\partial x^4} + (2D_{12} + 4D_{66})\frac{\partial^4 \widetilde{w}(x, y)}{\partial x^2 \partial y^2} \right.$$
$$\left. + D_{22}\frac{\partial^4 \widetilde{w}(x, y)}{\partial y^4} - q(x, y) \right] \widetilde{w}(x, y) dx dy = 0. \tag{B.22}$$

Applying the Kantorovich method, the solution $\widetilde{w}(x,y)$ is assumed to be separable, namely

$$\widetilde{w}(x,y) = W(x)W(y). \tag{B.23}$$

Let us assume that $W(y)$ is known. Substituting in Eqs. (B.8) and (B.7) leads to the following double integral:

$$\int_0^a \left\{ \int_0^b \left[D_{11} \frac{d^4W(x)}{dx^4} W(y) + \delta \frac{d^2W(x)}{dx^2} \frac{d^2W(y)}{dy^2} \right.\right.$$
$$\left.\left. + D_{22} \frac{d^4W(y)}{dy^4} W(x) - q(x,y) \right] W(y)dy \right\} W(x)dx = 0,$$

where

$$\delta \equiv (2D_{12} + 4D_{66}). \tag{B.24}$$

To satisfy Eq. (B.24), the expression in the squared brackets must be zero. This leads to an ordinary differential equation for the determination of the $W(x)$ unknown function, namely

$$\alpha_1 \frac{d^4W(x)}{dx^4} + \alpha_2 \frac{d^2W(x)}{dx^2} + \alpha_3 W(x) = F(x), \text{ where}$$
$$\alpha_1 = D_{11} \int_0^b W^2(y)dy, \ \alpha_2 = \delta \int_0^b \frac{d^2W(y)}{dx^2} W(y)\, dy,$$
$$\alpha_1 = D_{22} \int_0^b \frac{d^4W(y)}{dx^4} W(y)\, dy, \ F(x) = \int_0^b q(x,y) W(y)\, dy. \tag{B.25}$$

The integrals in Eq. (B.25) are solvable using any commercial code available on the market, like the Maple code[n]. For a uniform pressure loading $q(x,y) = q_o$ the solution for $W(x)$ will have the following form:

$$W(x) = \sum_{i=1}^{4} X_i e^{\lambda_i x} + \frac{q_0}{\alpha_3}, \quad \text{where}$$
$$\lambda_i = \pm \frac{\sqrt[\sqrt{2}]{\alpha_1(\pm\alpha_2 + \sqrt{\alpha_2^2 - \alpha_1\alpha_3})}}{2\alpha_1}, \ i = 1,2,3,4. \tag{B.26}$$

Applying the extended Kantorovich, after the first Kantorovich solution was obtained (see Eq. (B.26)), the next step will be for $W(y)$ to be analytical solved while the solution $W(x)$ was obtained analytically solution in the previous step. This iterative procedure is repeated until the result converges to any desired degree of accuracy.

[n]Maple User Manual. 2014 Copyright©Maplesoft, a division of Waterloo Maple Inc.

References

[1] Pilkey, W. and Wunderlich, W., *Mechanics of Structures-Variational and Computational Methods*, 2nd Edition, 2016, CRC Press, Boca Raton, 874 p.

[2] Dym, C. and Shames, I., *Solid Mechanics — A Variational Approach*, McGraw-Hill, New York, 633 p.

[3] Allen, H.G. and Bulson, P.S., *Background to Buckling*, McGraw-Hill, London, 1980, 582 p.

[4] Galerkin, B.G., Rods and plates, series occurring in various questions concerning the elastic equilibrium of rods and plates, *Vestnik Inzhenerov Tekh., (Eng. Technol. Bull.)*, Vol. 19, 1915, pp. 897–908 (in Russian). (English Translation: 63-18925, Clearinghouse Fed. Sci. Tech. Info.1963).

[5] Kantorovich, L.V. and Krylov, V.I. (translated by Benster, C.D.), *Approximate Method of Higher Analysis*, Dover Publications, Inc. Mineola, New York, 2018, 671 p.

[6] Chen, P.C.T., Buckling analysis of a rectangular plate by the Kantorovich method, *Int. J. Mech. Sci.*, Vol. 14, 1972, pp. 15–24.

[7] Kerr, A.D., An extended Kantorovich method for the solution of eigenvalue problems, *Int. J. Solids Struct.*, Vol. 5, 1969, pp. 559–572.

[8] Edery-Azulay, L. and Abramovich, H., Piezolaminated plates-Highly accurate solutions based on the extended Kantorovich method, *Composite Struct.*, Vol. 84, 2008, pp. 241–247.

[9] Ungbhakorn, V. and Singhatanadgid, P., Buckling analysis of symmetrically laminated composite plates by the extended Kantorovich method, *Composite Struct.*, Vol. 73, 2006, pp. 120–128.

[10] Singhatanadgid, P. and Singhanart, T., The Kantorovich method applied to bending, buckling, vibration and 3D stress analyses of plates: A literature review, *Mech. Adv. Mater. Struct.*, 2017, doi: 10.1080/15376494.2017.1365984.

[11] Yuan, S. and Jin, Y., Computation of elastic buckling loads of rectangular thin plates using the extended Kantorovich method, *Computers Struct.*, Vol. 66(6), 1998, pp. 861–867.

Chapter 4

Vibrations of Thin-Walled Structures

4.1. Introduction

The present chapter is the complementary chapter to Chapter 3, which dealt with stability issues for isotropic and composite columns, plates and shells. It is aimed at presenting the vibration analysis of isotropic and composite columns, plates and shells based on the classical lamination theory (CLT) and the first-order shear deformation theory (FSDT) developed in Chapter 2. Some typical references dealing with one-dimensional behavior characteristically to beams are given in [1–17].

4.2. CLPT Approach

As described in the previous chapter, the classical lamination plate theory (CLPT) is an extension of the well-known Kirchhoff–Love classical plate theory to be applied to laminated composite plates. The equations of motion are as follows:

$$
\begin{aligned}
&\frac{\partial N_{xx}}{\partial x}+\frac{\partial N_{xy}}{\partial y}=I_0\frac{\partial^2 u_0}{\partial t^2}-I_1\frac{\partial^2}{\partial t^2}\left(\frac{\partial w_0}{\partial x}\right),\\
&\frac{\partial N_{xy}}{\partial x}+\frac{\partial N_{yy}}{\partial y}=I_0\frac{\partial^2 v_0}{\partial t^2}-I_1\frac{\partial^2}{\partial t^2}\left(\frac{\partial w_0}{\partial y}\right),\\
&\frac{\partial^2 M_{xx}}{\partial x^2}+2\frac{\partial^2 M_{xy}}{\partial x\partial y}+\frac{\partial^2 M_{yy}}{\partial y^2}+\frac{\partial}{\partial x}\left(N_{xx}\frac{\partial w_0}{\partial x}+N_{xy}\frac{\partial w_0}{\partial y}\right)\\
&\quad+\frac{\partial}{\partial y}\left(N_{xy}\frac{\partial w_0}{\partial x}+N_{yy}\frac{\partial w_0}{\partial y}\right)=q+I_0\frac{\partial^2 w_0}{\partial t^2}+I_1\frac{\partial^2}{\partial t^2}\left(\frac{\partial u_0}{\partial x}+\frac{\partial v_0}{\partial y}\right)\\
&\quad-I_2\frac{\partial^2}{\partial t^2}\left(\frac{\partial^2 w_0}{\partial x^2}+\frac{\partial^2 w_0}{\partial y^2}\right),
\end{aligned}
\tag{4.1}
$$

where q is the distributed pressure on the surface of the plate and the mass moments of inertia I_0, I_1 and I_2 are defined as

$$\begin{Bmatrix} I_0 \\ I_1 \\ I_2 \end{Bmatrix} = \int_{-h/2}^{+h/2} \begin{Bmatrix} 1 \\ z \\ z^2 \end{Bmatrix} \cdot \rho_0 \cdot dz, \tag{4.2}$$

where h is the total thickness of the plate and ρ_0 is the relevant density.

In Eq. (4.1), N_{xx}, N_{xy}, N_{yy} are the force resultants per unit length and M_{xx}, M_{xy}, M_{yy} are the moment resultants per unit length defined as:

$$\begin{Bmatrix} N_{xx} \\ N_{yy} \\ N_{xy} \end{Bmatrix} = \int_{-h/2}^{+h/2} \begin{Bmatrix} \sigma_{xx} \\ \sigma_{yy} \\ \tau_{xy} \end{Bmatrix} dz; \quad \begin{Bmatrix} M_{xx} \\ M_{yy} \\ M_{xy} \end{Bmatrix} = -\int_{-h/2}^{+h/2} \begin{Bmatrix} \sigma_{xx} \\ \sigma_{yy} \\ \tau_{xy} \end{Bmatrix} \cdot z \cdot dz. \tag{4.3}$$

In Eq. (4.3), z is a coordinate normal to the surface of the plate and σ_{xx} and σ_{yy} are the normal stresses in the x and y directions, respectively, while τ_{xy} is the shear stress.

The stress resultants N_{xx}, N_{xy}, N_{yy} and M_{xx}, M_{xy}, M_{yy} can be defined using the assumed displacements to yield:

$$\begin{Bmatrix} N_{xx} \\ N_{yy} \\ N_{xy} \end{Bmatrix} = \begin{bmatrix} A_{11} & A_{12} & A_{16} \\ A_{12} & A_{22} & A_{26} \\ A_{16} & A_{26} & A_{66} \end{bmatrix} \begin{Bmatrix} \frac{\partial u_0}{\partial x} + \frac{1}{2}\left(\frac{\partial w_0}{\partial x}\right)^2 \\ \frac{\partial v_0}{\partial y} + \frac{1}{2}\left(\frac{\partial w_0}{\partial y}\right)^2 \\ \frac{\partial u_0}{\partial y} + \frac{\partial v_0}{\partial x} + \frac{\partial^2 w_0}{\partial x \partial y} \end{Bmatrix} - \begin{bmatrix} B_{11} & B_{12} & B_{16} \\ B_{12} & B_{22} & B_{26} \\ B_{16} & B_{26} & B_{66} \end{bmatrix} \begin{Bmatrix} \frac{\partial^2 w_0}{\partial x^2} \\ \frac{\partial^2 w_0}{\partial y^2} \\ 2\frac{\partial^2 w_0}{\partial x \partial y} \end{Bmatrix}, \tag{4.4}$$

$$\begin{Bmatrix} M_{xx} \\ M_{yy} \\ M_{xy} \end{Bmatrix} = \begin{bmatrix} B_{11} & B_{12} & B_{16} \\ B_{12} & B_{22} & B_{26} \\ B_{16} & B_{26} & B_{66} \end{bmatrix} \begin{Bmatrix} \frac{\partial u_0}{\partial x} + \frac{1}{2}\left(\frac{\partial w_0}{\partial x}\right)^2 \\ \frac{\partial v_0}{\partial y} + \frac{1}{2}\left(\frac{\partial w_0}{\partial y}\right)^2 \\ \frac{\partial u_0}{\partial y} + \frac{\partial v_0}{\partial x} + \frac{\partial^2 w_0}{\partial x \partial y} \end{Bmatrix} - \begin{bmatrix} D_{11} & D_{12} & D_{16} \\ D_{12} & D_{22} & D_{26} \\ D_{16} & D_{26} & D_{66} \end{bmatrix} \begin{Bmatrix} \frac{\partial^2 w_0}{\partial x^2} \\ \frac{\partial^2 w_0}{\partial y^2} \\ 2\frac{\partial^2 w_0}{\partial x \partial y} \end{Bmatrix}, \tag{4.5}$$

where

$$A_{ij} = \sum_{k=1}^{N} \overline{Q}_{ij}^{(k)}(z_{k+1} - z_k), \quad B_{ij} = \frac{1}{2}\sum_{k=1}^{N} \overline{Q}_{ij}^{(k)}(z_{k+1}^2 - z_k^2),$$

$$D_{ij} = \frac{1}{3}\sum_{k=1}^{N} \overline{Q}_{ij}^{(k)}(z_{k+1}^3 - z_k^3) \tag{4.6}$$

with $\overline{Q}_{ij}^{(k)}$ being the lamina stiffness after transformation. Substituting Eqs. (4.4) and (4.5) into Eq. (4.1) leads to the equations of motion for a laminated composite plate expressed by the three assumed displacements (u_0, v_0 and w_0):

$$\begin{aligned}
&A_{11}\left(\frac{\partial^2 u_0}{\partial x^2} + \frac{\partial^3 w_0}{\partial x^3}\right) + A_{12}\left(\frac{\partial^2 v_0}{\partial x \partial y} + \frac{\partial^3 w_0}{\partial x \partial y^2}\right) \\
&\quad + A_{16}\left(2\frac{\partial^2 u_0}{\partial x \partial y} + 3\frac{\partial^3 w_0}{\partial x^2 \partial y} + \frac{\partial^2 v_0}{\partial x^2}\right) \\
&\quad + A_{26}\left(\frac{\partial^2 v_0}{\partial y^2} + \frac{\partial^3 w_0}{\partial y^3}\right) + A_{66}\left(\frac{\partial^2 u_0}{\partial y^2} + \frac{\partial^2 v_0}{\partial x \partial y} + 2\frac{\partial^3 w_0}{\partial x \partial y^2}\right) \\
&\quad - B_{11}\frac{\partial^3 w_0}{\partial x^3} - B_{12}\frac{\partial^3 w_0}{\partial x \partial y^2} - 3B_{16}\frac{\partial^3 w_0}{\partial x^2 \partial y} - B_{26}\frac{\partial^3 w_0}{\partial y^3} \\
&\quad - 2B_{66}\frac{\partial^3 w_0}{\partial x \partial y^2} = I_0\frac{\partial^2 u_0}{\partial t^2} - I_1\frac{\partial^3 w_0}{\partial x \partial t^2},
\end{aligned} \tag{4.7}$$

$$\begin{aligned}
&A_{22}\left(\frac{\partial^2 v_0}{\partial y^2} + \frac{\partial^3 w_0}{\partial y^3}\right) + A_{12}\left(\frac{\partial^2 u_0}{\partial x \partial y} + \frac{\partial^3 w_0}{\partial x^2 \partial y}\right) + A_{16}\left(\frac{\partial^2 u_0}{\partial x^2} + \frac{\partial^3 w_0}{\partial x^3}\right) \\
&\quad + A_{26}\left(\frac{\partial^2 u_0}{\partial y^2} + 2\frac{\partial^2 v_0}{\partial x \partial y} + 3\frac{\partial^3 w_0}{\partial x \partial y^2}\right) \\
&\quad + A_{66}\left(\frac{\partial^2 u_0}{\partial x \partial y} + \frac{\partial^2 v_0}{\partial x^2} + 2\frac{\partial^3 w_0}{\partial x^2 \partial y}\right) - B_{12}\frac{\partial^3 w_0}{\partial x^2 \partial y} \\
&\quad - B_{22}\frac{\partial^3 w_0}{\partial y^3} - B_{16}\frac{\partial^3 w_0}{\partial x^3} - 3B_{26}\frac{\partial^3 w_0}{\partial x \partial y^2} - 2B_{66}\frac{\partial^3 w_0}{\partial x^2 \partial y} \\
&\quad = I_0\frac{\partial^2 v_0}{\partial t^2} - I_1\frac{\partial^3 w_0}{\partial y \partial t^2},
\end{aligned} \tag{4.8}$$

$$\begin{aligned}
&B_{11}\left(\frac{\partial^3 u_0}{\partial x^3} + \frac{\partial^4 w_0}{\partial x^4}\right) + B_{12}\left(\frac{\partial^3 v_0}{\partial x^2 \partial y} + \frac{\partial^3 u_0}{\partial x \partial y^2} + 4\frac{\partial^3 w_0}{\partial x^2 \partial y^2}\right) \\
&\quad + B_{16}\left(3\frac{\partial^3 u_0}{\partial x^2 \partial y} + \frac{\partial^3 v_0}{\partial x^3} + 8\frac{\partial^4 w_0}{\partial x^3 \partial y}\right) + B_{22}\left(\frac{\partial^3 v_0}{\partial y^3} + 2\frac{\partial^4 w_0}{\partial y^4}\right)
\end{aligned}$$

$$
\begin{aligned}
&+ B_{26}\left(\frac{\partial^3 u_0}{\partial y^3} + 3\frac{\partial^3 v_0}{\partial x \partial y^2} + 8\frac{\partial^4 w_0}{\partial x \partial y^3}\right) \\
&+ 2B_{66}\left(\frac{\partial^3 u_0}{\partial x \partial y^2} + \frac{\partial^3 v_0}{\partial x^2 \partial y} + 4\frac{\partial^4 w_0}{\partial x^2 \partial y^2}\right) \\
&- D_{11}\frac{\partial^4 w_0}{\partial x^4} - 2D_{12}\frac{\partial^4 w_0}{\partial x^2 \partial y^2} - D_{22}\frac{\partial^4 w_0}{\partial y^4} - 4D_{16}\frac{\partial^4 w_0}{\partial x^3 \partial y} \\
&- 4D_{26}\frac{\partial^4 w_0}{\partial x \partial y^3} - 4D_{66}\frac{\partial^4 w_0}{\partial x^2 \partial y^2} + \mathrm{P}(w_0) \\
&= q + I_0\frac{\partial^2 w_0}{\partial t^2} - I_2\frac{\partial^2}{\partial t^2}\left(\frac{\partial^2 w_0}{\partial x^2} + \frac{\partial^2 w_0}{\partial y^2}\right) + I_1\frac{\partial^2}{\partial t^2}\left(\frac{\partial u_0}{\partial x} + \frac{\partial v_0}{\partial y}\right), \qquad (4.9)
\end{aligned}
$$

where

$$
\mathrm{P}(w_0) = \frac{\partial}{\partial x}\left(N_{xx}\frac{\partial w_0}{\partial x} + N_{xy}\frac{\partial w_0}{\partial y}\right) + \frac{\partial}{\partial y}\left(N_{xy}\frac{\partial w_0}{\partial x} + N_{yy}\frac{\partial w_0}{\partial y}\right). \qquad (4.10)
$$

Note that the derivation for isotropic columns can easily reached from the governing equations for laminated composite columns (see Chapter 3).

4.3. FSDPT Approach

The first-order shear deformation plate theory (FSDPT) described in details in Chapter 2 is an extension of the well-known Timoshenko beam theory and/or Mindlin–Reissner plate theory to be applied to laminated composite plates.

The equations of motion in terms of displacements for the FSDPT approach can be written as

$$
\begin{aligned}
&A_{11}\left(\frac{\partial^2 u_0}{\partial x^2} + \frac{\partial^3 w_0}{\partial x^3}\right) + A_{12}\left(\frac{\partial^2 v_0}{\partial x \partial y} + \frac{\partial^3 w_0}{\partial x \partial y^2}\right) \\
&\quad + A_{16}\left(2\frac{\partial^2 u_0}{\partial x \partial y} + 3\frac{\partial^3 w_0}{\partial x^2 \partial y} + \frac{\partial^2 v_0}{\partial x^2}\right) \\
&\quad + A_{26}\left(\frac{\partial^2 v_0}{\partial y^2} + \frac{\partial^3 w_0}{\partial y^3}\right) + A_{66}\left(\frac{\partial^2 u_0}{\partial y^2} + \frac{\partial^2 v_0}{\partial x \partial y} + 2\frac{\partial^3 w_0}{\partial x \partial y^2}\right) \\
&\quad + B_{11}\frac{\partial^2 \phi_x}{\partial x^2} + B_{12}\frac{\partial^2 \phi_y}{\partial x \partial y} + B_{16}\left(2\frac{\partial^2 \phi_x}{\partial x \partial y} + \frac{\partial^2 \phi_y}{\partial x^2}\right)
\end{aligned}
$$

$$+ B_{26}\frac{\partial^2 \phi_y}{\partial y^2} + B_{66}\left(\frac{\partial^2 \phi_x}{\partial y^2} + \frac{\partial^2 \phi_y}{\partial x \partial y}\right) = I_0\frac{\partial^2 u_0}{\partial t^2} + I_1\frac{\partial^2 \phi_x}{\partial t^2}, \tag{4.11}$$

$$A_{22}\left(\frac{\partial^2 v_0}{\partial y^2} + \frac{\partial^3 w_0}{\partial y^3}\right) + A_{12}\left(\frac{\partial^2 u_0}{\partial x \partial y} + \frac{\partial^3 w_0}{\partial x^2 \partial y}\right) + A_{16}\left(\frac{\partial^2 u_0}{\partial x^2} + \frac{\partial^3 w_0}{\partial x^3}\right)$$

$$+ A_{26}\left(\frac{\partial^2 u_0}{\partial y^2} + 2\frac{\partial^2 v_0}{\partial x \partial y} + 3\frac{\partial^3 w_0}{\partial x \partial y^2}\right) + A_{66}\left(\frac{\partial^2 u_0}{\partial x \partial y} + \frac{\partial^2 v_0}{\partial x^2} + 2\frac{\partial^3 w_0}{\partial x^2 \partial y}\right)$$

$$+ B_{12}\frac{\partial^2 \phi_x}{\partial x \partial y} + B_{22}\frac{\partial^2 \phi_y}{\partial y^2} + B_{16}\frac{\partial^2 \phi_x}{\partial x^2} + B_{26}\left(2\frac{\partial^2 \phi_y}{\partial x \partial y} + \frac{\partial^2 \phi_x}{\partial y^2}\right)$$

$$+ B_{66}\left(\frac{\partial^2 \phi_x}{\partial x \partial y} + \frac{\partial^2 \phi_y}{\partial x^2}\right) = I_0\frac{\partial^2 v_0}{\partial t^2} + I_1\frac{\partial^2 \phi_y}{\partial t^2}, \tag{4.12}$$

$$KA_{44}\left(\frac{\partial^2 w_0}{\partial y^2} + \frac{\partial \phi_y}{\partial y}\right) + KA_{45}\left(\frac{\partial^2 w_0}{\partial x \partial y} + \frac{\partial \phi_y}{\partial x}\right) + KA_{45}\left(\frac{\partial^2 w_0}{\partial x \partial y} + \frac{\partial \phi_x}{\partial y}\right)$$

$$+ KA_{55}\left(\frac{\partial^2 w_0}{\partial x^2} + \frac{\partial \phi_x}{\partial x}\right) + \mathrm{P}(w_0) = q + I_0\frac{\partial^2 w_0}{\partial t^2}, \tag{4.13}$$

$$B_{11}\left(\frac{\partial^2 u_0}{\partial x^2} + \frac{\partial^3 w_0}{\partial x^3}\right) + B_{12}\left(\frac{\partial^2 v_0}{\partial x \partial y} + \frac{\partial^3 w_0}{\partial x \partial y^2}\right)$$

$$+ B_{16}\left(2\frac{\partial^2 u_0}{\partial x \partial y} + \frac{\partial^2 v_0}{\partial x^2} + 3\frac{\partial^4 w_0}{\partial x^3 \partial y}\right) + B_{26}\left(\frac{\partial^2 v_0}{\partial y^2} + \frac{\partial^3 w_0}{\partial y^3}\right)$$

$$+ B_{66}\left(\frac{\partial^2 u_0}{\partial y^2} + \frac{\partial^2 v_0}{\partial x \partial y} + 2\frac{\partial^3 w_0}{\partial x \partial y^2}\right) + D_{11}\frac{\partial^2 \phi_x}{\partial x^2} + D_{12}\frac{\partial^2 \phi_y}{\partial x \partial y}$$

$$+ D_{16}\left(2\frac{\partial^2 \phi_x}{\partial x \partial y} + \frac{\partial^2 \phi_y}{\partial x^2}\right) + D_{26}\frac{\partial^2 \phi_y}{\partial y^2} + D_{66}\left(\frac{\partial^2 \phi_x}{\partial y^2} + \frac{\partial^2 \phi_y}{\partial x \partial y}\right)$$

$$- KA_{55}\left(\frac{\partial w_0}{\partial x} + \phi_x\right) - KA_{45}\left(\frac{\partial w_0}{\partial y} + \phi_y\right)$$

$$= I_2\frac{\partial^2 \phi_x}{\partial t^2} + I_1\frac{\partial^2 u_0}{\partial t^2}, \tag{4.14}$$

$$B_{12}\left(\frac{\partial^2 u_0}{\partial x \partial y} + \frac{\partial^3 w_0}{\partial x \partial y^2}\right) + B_{16}\left(\frac{\partial^2 u_0}{\partial x^2} + \frac{\partial^3 w_0}{\partial x^3}\right) + B_{22}\left(\frac{\partial^2 v_0}{\partial y^2} + \frac{\partial^3 w_0}{\partial y^3}\right)$$

$$+ B_{26}\left(\frac{\partial^2 u_0}{\partial y^2} + 2\frac{\partial^2 v_0}{\partial x \partial y} + 3\frac{\partial^3 w_0}{\partial x \partial y^2}\right) + B_{66}\left(\frac{\partial^2 v_0}{\partial x^2} + \frac{\partial^2 u_0}{\partial x \partial y} + 2\frac{\partial^3 w_0}{\partial x \partial y^2}\right)$$

$$+D_{12}\frac{\partial^2\phi_x}{\partial x\partial y}+D_{22}\frac{\partial^2\phi_y}{\partial y^2}+D_{16}\frac{\partial^2\phi_x}{\partial x^2}+D_{26}\left(\frac{\partial^2\phi_x}{\partial y^2}+2\frac{\partial^2\phi_y}{\partial x\partial y}\right)$$
$$+D_{66}\left(\frac{\partial^2\phi_y}{\partial x^2}+\frac{\partial^2\phi_x}{\partial x\partial y}\right)-KA_{45}\left(\frac{\partial w_0}{\partial x}+\phi_x\right)$$
$$-KA_{44}\left(\frac{\partial w_0}{\partial y}+\phi_y\right)=I_2\frac{\partial^2\phi_y}{\partial t^2}+I_1\frac{\partial^2 v_0}{\partial t^2}, \tag{4.15}$$

where $\mathrm{P}(w_0)$ is given by Eq. (4.10), u_0, v_0, w_0 are the displacements in the directions x, y and z, respectively, ϕ_x, ϕ_y are rotation about the x and y axes, respectively, and the various terms A_{ij}, B_{ij} and D_{ij} are defined in details in Chapter 2.

4.4. Vibrations of Columns — CLT Approach

Figure 4.1 presents schematically a laminated column under lateral and compressive loads. The equations of motion for calculating a general laminated column can be derived from Eq. (4.1) by assuming that the displacement of the model is only a function of the x coordinate, yielding the following equations (v_0 is assumed to be zero):

$$A_{11}\left(\frac{\partial^2 u_0}{\partial x^2}+\frac{\partial^3 w_0}{\partial x^3}\right)-B_{11}\frac{\partial^3 w_0}{\partial x^3}=I_0\frac{\partial^2 u_0}{\partial t^2}-I_1\frac{\partial^3 w_0}{\partial x\partial t^2}, \tag{4.16}$$

$$B_{11}\left(\frac{\partial^3 u_0}{\partial x^3}+\frac{\partial^4 w_0}{\partial x^4}\right)-D_{11}\frac{\partial^4 w_0}{\partial x^4}$$
$$=q+\overline{\mathrm{P}}(w_0)+I_0\frac{\partial^2 w_0}{\partial t^2}-I_2\frac{\partial^2}{\partial t^2}\left(\frac{\partial^2 w_0}{\partial x^2}\right)+I_1\frac{\partial^2}{\partial t^2}\left(\frac{\partial u_0}{\partial x}\right), \tag{4.17}$$

where

$$\overline{\mathrm{P}}(w_0)=\frac{\partial}{\partial x}\left(N_{xx}\frac{\partial w_0}{\partial x}\right), \tag{4.18}$$

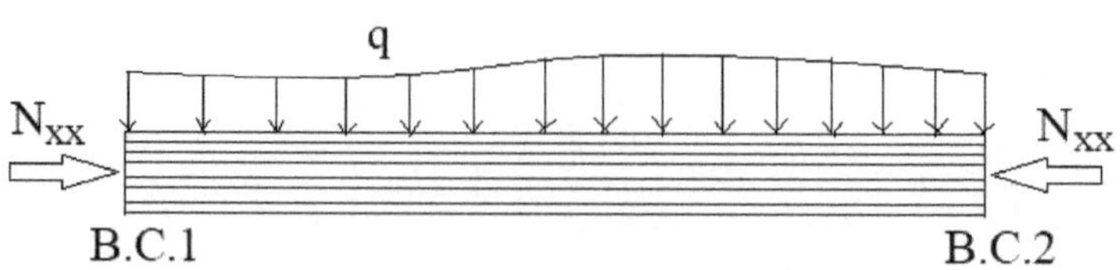

Fig. 4.1. A general laminated column — a schematic drawing.

$$\begin{Bmatrix} I_0 \\ I_1 \\ I_2 \end{Bmatrix} = \int_{-h/2}^{+h/2} \begin{Bmatrix} 1 \\ z \\ z^2 \end{Bmatrix} \cdot \rho_0 \cdot dz, \tag{4.19}$$

and h is the total thickness of the plate and ρ_0 is the relevant density.

To analyze the vibration problem of the columns, the lateral load q will be assumed to be zero while the axial compressive load, $\overline{\mathrm{P}}(w_0)$ will be assumed to be non-zero. This leads to the following coupled equations:

$$A_{11}\left(\frac{\partial^2 u_0}{\partial x^2} + \frac{\partial^3 w_0}{\partial x^3}\right) - B_{11}\frac{\partial^3 w_0}{\partial x^3} = I_0\frac{\partial^2 u_0}{\partial t^2} - I_1\frac{\partial^2}{\partial t^2}\left(\frac{\partial w_0}{\partial x}\right), \tag{4.20}$$

$$\begin{aligned} &B_{11}\left(\frac{\partial^3 u_0}{\partial x^3} + \frac{\partial^4 w_0}{\partial x^4}\right) - D_{11}\frac{\partial^4 w_0}{\partial x^4} \\ &\quad = \overline{\mathrm{P}}(w_0) + I_0\frac{\partial^2 w_0}{\partial t^2} - I_2\frac{\partial^2}{\partial t^2}\left(\frac{\partial^2 w_0}{\partial x^2}\right) + I_1\frac{\partial^2}{\partial t^2}\left(\frac{\partial u_0}{\partial x}\right). \end{aligned} \tag{4.21}$$

4.4.1. *Symmetric laminate* ($B_{11} = 0, I_1 = 0$)

For the case of symmetric laminate $B_{11} = 0$ and $I_1 = 0$, the following single equation is obtained, while Eq. (4.20) (assuming the term $\frac{\partial^3 w_0}{\partial x^3}$ is neglected) is solved independently,

$$-D_{11}\frac{\partial^4 w_0}{\partial x^4} - \overline{N}_{xx}\frac{\partial^2 w_0}{\partial x^2} = I_0\frac{\partial^2 w_0}{\partial t^2} - I_2\frac{\partial^2}{\partial t^2}\left(\frac{\partial^2 w_0}{\partial x^2}\right). \tag{4.22}$$

Assuming that $w_0(x,t) = W(x)e^{i\omega t}$, where ω is the circular natural frequency and substituting in Eq. (4.22) yields the following differential equation:

$$D_{11}\frac{d^4 W}{dx^4} + (\overline{N}_{xx} + \omega^2 I_2)\frac{d^2 W}{dx^2} - \omega^2 I_0 \cdot W = 0. \tag{4.23}$$

The solution of Eq. (4.23) has the following general form:

$$W(x) = A_1 \sinh(\alpha_1 x) + A_2 \cosh(\alpha_1 x) + A_3 \sin(\alpha_2 x) + A_4 \cos(\alpha_2 x), \tag{4.24}$$

where A_1, A_2, A_3 and A_4 are constants to be determined using boundary conditions and the terms α_1 and α_2 are defined as

$$\begin{aligned} \alpha_1 &= \sqrt{\frac{-(N_{xx} + \omega^2 I_2) + \sqrt{(\overline{N}_{xx} + \omega^2 I_2)^2 + 4D_{11}\omega^2 I_0}}{2D_{11}}}, \\ \alpha_2 &= \sqrt{\frac{(N_{xx} + \omega^2 I_2) + \sqrt{(\overline{N}_{xx} + \omega^2 I_2)^2 + 4D_{11}\omega^2 I_0}}{2D_{11}}}. \end{aligned} \tag{4.25}$$

To investigate the influence of the compression load, $\overline{N}_{xx}$ and the rotary inertia, I_2, on the natural frequency, the circular natural frequency squared (ω^2) is rewritten using Eq. (4.25) to yield

$$\begin{aligned} \omega^2 &= \frac{D_{11}\alpha_1^4 + \overline{N}_{xx}\alpha_1^2}{I_0 - I_2\alpha_1^2} = \alpha_1^4 \frac{D_{11}}{I_0} \frac{\left(1 + \frac{\overline{N}_{xx}}{D_{11}\alpha_1^2}\right)}{\left(1 - \frac{I_2}{I_0}\alpha_1^2\right)}, \\ \omega^2 &= \frac{D_{11}\alpha_2^4 - \overline{N}_{xx}\alpha_2^2}{I_0 + I_2\alpha_2^2} = \alpha_2^4 \frac{D_{11}}{I_0} \frac{\left(1 - \frac{\overline{N}_{xx}}{D_{11}\alpha_2^2}\right)}{\left(1 + \frac{I_2}{I_0}\alpha_2^2\right)}. \end{aligned} \tag{4.26}$$

The expressions in Eq. (4.26) are the same, and one of them can be used, once for instance α_2 is known. Also it is clear from Eq. (4.26) (the second expression) that the compressive load $\overline{N}_{xx}$ reduces the natural frequency and the same tendency is obtained by including the rotary inertia I_2. Assuming $\overline{N}_{xx} = I_2 = 0$ one obtains the known frequency equation used by Bernoulli–Euler theory, namely

$$\omega^2 = \alpha_2^4 \frac{D_{11}}{I_0}, \tag{4.27}$$

which holds true also for an isotropic column, by letting the bending stiffness, to be written as $D_{11} \equiv D = Eh^3/[12(1-\upsilon^2)]$, where E is the Young's modulus, υ is the Poisson's ratio and h is the thickness of the column.

Let us show the application of boundary conditions in Eq. (4.24) for the simply supported case. The boundary conditions for this case are as follows:

$$\begin{aligned} W(0) = 0; \quad M_{xx}(0) = 0 \Rightarrow \frac{d^2W(0)}{dx^2} = 0, \\ W(L) = 0; \quad M_{xx}(L) = 0 \Rightarrow \frac{d^2W(L)}{dx^2} = 0. \end{aligned} \tag{4.28}$$

Substituting the boundary conditions in Eq. (4.24) yields a set of four equations with four unknowns, having the following matrix form:

$$\begin{bmatrix} 0 & 1 & 0 & 1 \\ 0 & \alpha_1^2 & 0 & -\alpha_2^2 \\ \sinh(\alpha_1 L) & \cosh(\alpha_2 L) & \sin(\alpha_2 L) & \cos(\alpha_2 L) \\ \alpha_1^2 \sinh(\lambda L) & \alpha_1^2 \cosh(\lambda L) & -\alpha_2^2 \sin(\alpha_2 L) & -\alpha_2^2 \cos(\alpha_2 L) \end{bmatrix} \begin{Bmatrix} C_1 \\ C_2 \\ C_3 \\ C_4 \end{Bmatrix} = \begin{Bmatrix} 0 \\ 0 \\ 0 \\ 0 \end{Bmatrix}. \tag{4.29}$$

To obtain a unique solution, the determinant of the matrix appearing in Eq. (4.29) must vanish. This leads to the following characteristic equation:

$$\sin(\alpha_2 L) = 0 \Rightarrow \alpha_2 L = n\pi, \quad n = 1, 2, 3, 4, \ldots \Rightarrow \alpha_2^2 = \frac{n^2\pi^2}{L^2}. \tag{4.30}$$

Substituting the result of Eq. (4.30) into the second expression of Eq. (4.26) leads to the natural frequencies of a symmetric layered column, having a length L, under compressive load, including rotatory inertia on simply supported boundary conditions

$$\omega = \left(\frac{n\pi}{L}\right)^2 \sqrt{\frac{D_{11}}{I_0}} \sqrt{\frac{\left(1 - \frac{\overline{N}_{xx}}{D_{11}\left(\frac{n\pi}{L}\right)^2}\right)}{\left(1 + \frac{I_2}{I_0}\left(\frac{n\pi}{L}\right)^2\right)}}, \quad n = 1, 2, 3, 4, \ldots. \tag{4.31}$$

Neglecting the rotatory inertia leads to the following expression:

$$\omega = \left(\frac{n\pi}{L}\right)^2 \sqrt{\frac{D_{11}}{I_0}} \sqrt{\left(1 - \frac{\overline{N}_{xx}}{D_{11}\left(\frac{n\pi}{L}\right)^2}\right)}, \quad n = 1, 2, 3, 4, \ldots \tag{4.32}$$

showing that the axial compression load reduces the natural frequencies of a column, while for the case of no axial compression, but including the rotatory inertia the expression has the following form:

$$\omega = \left(\frac{n\pi}{L}\right)^2 \sqrt{\frac{D_{11}}{I_0}} \sqrt{\frac{1}{\left(1 + \frac{I_2}{I_0}\left(\frac{n\pi}{L}\right)^2\right)}}, \quad n = 1, 2, 3, 4, \ldots \tag{4.33}$$

displaying the same tendency as before, namely the rotatory inertia tends to reduce the natural frequencies of the column. For the simple case of a column, without axial compression and rotatory inertia, the expression for the natural frequencies is like for an isotropic case using the classical beam theory

$$\omega_n = \left(\frac{n\pi}{L}\right)^2 \sqrt{\frac{D_{11}}{I_0}}, \quad n = 1, 2, 3, 4, \ldots. \tag{4.34}$$

Tables 4.1 and 4.2 present some of the most encountered column cases having various boundary conditions (see similar tables in Chapter 3). The various expressions given in Table 4.1 are for the case without axial compression load and neglecting the rotary inertia term. For this case, Eq. (4.24) simplifies into the following equation:

$$W(x) = A_1 \sinh(\alpha_2 x) + A_2 \cosh(\alpha_2 x) + A_3 \sin(\alpha_2 x) + A_4 \cos(\alpha_2 x). \tag{4.35}$$

4.4.2. *Non-symmetric laminate* ($B_{11} \neq 0, I_1 \neq 0$)

To solve the case of a non-symmetric laminate, it is easier to present Eqs. (4.20) and (4.21) in a matrix form, where higher order terms are

Table 4.1. Characteristic equations and eigenvalues for natural vibrations of laminated composite columns using CLT approach.

No.	Name	Characteristic equation	Eigenvalues
1	SS–SS*	$\sin(\alpha_2 L) = 0$ $(\alpha_2 L)_n = n\pi, n = 1, 2, 3 \ldots$	$\omega_n = \left(\frac{n\pi}{L}\right)^2 \sqrt{\frac{D_{11}}{I_0}}$
2	C–C**	$\cos(\alpha_2 L)\cosh(\alpha_2 L) = 1$ $(\alpha_2 L)_n = 4.73004, 7.85321, 10.9956, \ldots, \frac{(2n+1)\pi}{2}$	$\omega_1 = \left(\frac{4.73004}{L}\right)^2 \sqrt{\frac{D_{11}}{I_0}}$
3	C–F***	$\cos(\alpha_2 L)\cosh(\alpha_2 L) = -1$ $(\alpha_2 L)_n = 1.87510, 4.69409, 7.85340, \ldots, \frac{(2n-1)\pi}{2}$	$\omega_1 = \left(\frac{1.87351}{L}\right)^2 \sqrt{\frac{D_{11}}{I_0}}$
4	F–F	$\cos(\alpha_2 L)\cosh(\alpha_2 L) = 1$ $(\alpha_2 L)_n = 4.73004, 7.85321, 10.9956, \ldots, \frac{(2n+1)\pi}{2}$	$\omega_1 = \left(\frac{4.73004}{L}\right)^2 \sqrt{\frac{D_{11}}{I_0}}$
5	SS–C	$\tan(\alpha_2 L) = \tanh(\alpha_2 L)$ $(\alpha_2 L)_n = 3.9266, 7.0686, 10.2102, \ldots, \frac{(4n+1)\pi}{4}$	$\omega_1 = \left(\frac{3.9266}{L}\right)^2 \sqrt{\frac{D_{11}}{I_0}}$
6	SS–F	$\tan(\alpha_2 L) = \tanh(\alpha_2 L)$ $(\alpha_2 L)_n = 3.9266, 7.0686, 10.2102, \ldots, \frac{(4n+1)\pi}{4}$	$\omega_1 = \left(\frac{3.9266}{L}\right)^2 \sqrt{\frac{D_{11}}{I_0}}$
7	G****–F	$\tan(\alpha_2 L) = -\tanh(\alpha_2 L)$ $(\alpha_2 L)_n = 2.3650, 5.4978, 8.6394, \ldots, \frac{(4n-1)\pi}{4}$	$\omega_1 = \left(\frac{2.3650}{L}\right)^2 \sqrt{\frac{D_{11}}{I_0}}$
8	G–SS	$\cos(\alpha_2 L) = 0$ $(\alpha_2 L)_n = (2n-1)\frac{\pi}{2}, n = 1, 2, 3, \ldots$	$\omega_n = \left[\frac{(2n-1)\pi}{2L}\right] \sqrt{\frac{D_{11}}{I_0}}$
9	G–G	$\sin(\alpha_2 L) = 0$ $(\alpha_2 L)_n = n\pi,\ n = 1, 2, 3, \ldots$	$\omega_n = \left(\frac{n\pi}{L}\right)^2 \sqrt{\frac{D_{11}}{I_0}}$
10	G–C	$\tan(\alpha_2 L) = -\tanh(\alpha_2 L)$ $(\alpha_2 L)_n = 2.3650, 5.4978, 8.6394, \ldots, \frac{(4n-1)\pi}{4}$	$\omega_1 = \left(\frac{2.3650}{L}\right)^2 \sqrt{\frac{D_{11}}{I_0}}$

*SS = Simply supported; **C = Clamped; ***F = Free; G**** = Guided.

neglected, leading to

$$\begin{bmatrix} A_{11}\frac{\partial^2}{\partial x^2} & -B_{11}\frac{\partial^3}{\partial x^3} \\ -B_{11}\frac{\partial^3}{\partial x^3} & D_{11}\frac{\partial^4}{\partial x^4} + \overline{N}_{xx}\frac{\partial^2}{\partial x^2} \end{bmatrix} \begin{Bmatrix} u_0(x,t) \\ w_0(x,t) \end{Bmatrix} + \begin{bmatrix} -I_0 & I_1\frac{\partial}{\partial x} \\ I_1\frac{\partial}{\partial x} & I_0 - I_2\frac{\partial^2}{\partial x^2} \end{bmatrix} \frac{\partial^2}{\partial t^2} \begin{Bmatrix} u_0(x,t) \\ w_0(x,t) \end{Bmatrix} = \begin{Bmatrix} 0 \\ 0 \end{Bmatrix}. \tag{4.36}$$

The matrix presentation in Eq. (4.36) includes axial compression $\overline{N}_{xx}$, rotary inertia I_2, and the coupling mass moment of inertia I_1 which couples both structurally and dynamically the two assumed deflections, u_0 and w_0. There is no general solution for Eq. (4.36). Only for simply supported boundary conditions we can present the natural frequencies of a non-symmetric beam in a closed-form solution. For this case, we assume that the deflections u_0

Table 4.2. Mode shapes and their relevant eigenvalues for natural vibrations of laminated composite columns using CLT approach.

No.	Name	Mode shape	Eigenvalues
1	SS–SS	$W_n(x) = \sin\left(\frac{n\pi x}{L}\right)$	$\omega_n = \left(\frac{n\pi}{L}\right)^2\sqrt{\frac{D_{11}}{I_0}}$
2†	C–C	$W_n(x) = \cosh\left[\frac{(\alpha_2 L)_n x}{L}\right] - \cos\left[\frac{(\alpha_2 L)_n x}{L}\right] - \frac{\cosh[(\alpha_2 L)_n] - \cos[(\alpha_2 L)_n]}{\sinh[(\alpha_2 L)_n] - \sin[(\alpha_2 L)_n]} \times \left\{\sinh\left[\frac{(\alpha_2 L)_n x}{L}\right] - \sin\left[\frac{(\alpha_2 L)_n x}{L}\right]\right\}$	$\omega_n = \left[\frac{(2n+1)\pi}{2L}\right]^2\sqrt{\frac{D_{11}}{I_0}}$
3†	C–F	$W_n(x) = \cosh\left[\frac{(\alpha_2 L)_n x}{L}\right] - \cos\left[\frac{(\alpha_2 L)_n x}{L}\right] - \frac{\cosh[(\alpha_2 L)_n] + \cos[(\alpha_2 L)_n]}{\sinh[(\alpha_2 L)_n] + \sin[(\alpha_2 L)_n]} \times \left\{\sinh\left[\frac{(\alpha_2 L)_n x}{L}\right] - \sin\left[\frac{(\alpha_2 L)_n x}{L}\right]\right\}$	$\omega_n = \left[\frac{(2n-1)\pi}{2L}\right]^2\sqrt{\frac{D_{11}}{I_0}}$
4†	F–F	$W_n(x) = \cosh\left[\frac{(\alpha_2 L)_n x}{L}\right] - \cos\left[\frac{(\alpha_2 L)_n x}{L}\right] - \frac{\cosh[(\alpha_2 L)_n] - \cos[(\alpha_2 L)_n]}{\sinh[(\alpha_2 L)_n] - \sin[(\alpha_2 L)_n]} \times \left\{\sinh\left[\frac{(\alpha_2 L)_n x}{L}\right] - \sin\left[\frac{(\alpha_2 L)_n x}{L}\right]\right\}$	$\omega_n = \left[\frac{(2n+1)\pi}{2L}\right]^2\sqrt{\frac{D_{11}}{I_0}}$
5†	SS–C	$W_n(x) = \cosh\left[\frac{(\alpha_2 L)_n x}{L}\right] - \cos\left[\frac{(\alpha_2 L)_n x}{L}\right] - \frac{\cosh[(\alpha_2 L)_n] - \cos[(\alpha_2 L)_n]}{\sinh[(\alpha_2 L)_n] - \sin[(\alpha_2 L)_n]} \times \left\{\sinh\left[\frac{(\alpha_2 L)_n x}{L}\right] - \sin\left[\frac{(\alpha_2 L)_n x}{L}\right]\right\}$	$\omega_n = \left[\frac{(4n+1)\pi}{4L}\right]^2\sqrt{\frac{D_{11}}{I_0}}$
6†	SS–F	$W_n(x) = \cosh\left[\frac{(\alpha_2 L)_n x}{L}\right] + \cos\left[\frac{(\alpha_2 L)_n x}{L}\right] - \frac{\cosh[(\alpha_2 L)_n] + \cos[(\alpha_2 L)_n]}{\sinh[(\alpha_2 L)_n] + \sin[(\alpha_2 L)_n]} \times \left\{\sinh\left[\frac{(\alpha_2 L)_n x}{L}\right] + \sin\left[\frac{(\alpha_2 L)_n x}{L}\right]\right\}$	$\omega_n = \left[\frac{(4n+1)\pi}{4L}\right]^2\sqrt{\frac{D_{11}}{I_0}}$
7†	G–F	$W_n(x) = \cosh\left[\frac{(\alpha_2 L)_n x}{L}\right] - \cos\left[\frac{(\alpha_2 L)_n x}{L}\right] - \frac{\cosh[(\alpha_2 L)_n] - \cos[(\alpha_2 L)_n]}{\sinh[(\alpha_2 L)_n] - \sin[(\alpha_2 L)_n]} \times \left\{\sinh\left[\frac{(\alpha_2 L)_n x}{L}\right] - \sin\left[\frac{(\alpha_2 L)_n x}{L}\right]\right\}$	$\omega_n = \left[\frac{(4n-1)\pi}{4L}\right]^2\sqrt{\frac{D_{11}}{I_0}}$
8	G–SS	$W_n(x) = \sin\left[\frac{(2n-1)\pi x}{2L}\right]$	$\omega_n = \left[\frac{(2n-1)\pi}{2L}\right]\sqrt{\frac{D_{11}}{I_0}}$

(*Continued*)

Table 4.2. (*Continued*)

No.	Name	Mode shape	Eigenvalues
9	G–G	$W_n(x) = \cos\left(\frac{n\pi x}{L}\right)$	$\omega_n = \left(\frac{n\pi}{L}\right)^2 \sqrt{\frac{D_{11}}{I_0}}$
10†	G–C	$W_n(x) = \cosh\left[\frac{(\alpha_2 L)_n x}{L}\right] - \cos\left[\frac{(\alpha_2 L)_n x}{L}\right] - \frac{\sinh[(\alpha_2 L)_n] + \sin[(\alpha_2 L)_n]}{\cosh[(\alpha_2 L)_n] - \cos[(\alpha_2 L)_n]} \times \left\{\sinh\left[\frac{(\alpha_2 L)_n x}{L}\right] - \sin\left[\frac{(\alpha_2 L)_n x}{L}\right]\right\}$	$\omega_n = \left[\frac{(4n-1)\pi}{4L}\right]^2 \sqrt{\frac{D_{11}}{I_0}}$

†The expression for the eigenvalues is given for $n \gg 1$. For the first modes, please refer to Table 4.1.

and w_0 have the following form:

$$\begin{Bmatrix} u_0(x,t) \\ w_0(x,t) \end{Bmatrix} = \begin{Bmatrix} \sum_{m=1}^{M} U_m \cos\left(\frac{m\pi x}{L}\right) e^{i\omega t} \\ \sum_{m=1}^{M} W_m \sin\left(\frac{m\pi x}{L}\right) e^{i\omega t} \end{Bmatrix} \equiv \begin{Bmatrix} \sum_{m=1}^{M} U_m \cos(\lambda x) e^{i\omega t} \\ \sum_{m=1}^{M} W_m \sin(\lambda x) e^{i\omega t} \end{Bmatrix}, \tag{4.37}$$

where $i = \sqrt{-1}$, ω^2 is the circular natural frequency squared and $\lambda = m\pi/L$. Substituting of Eq. (4.37) into Eq. (4.36) yields

$$\begin{bmatrix} A_{11}\lambda^2 & -B_{11}\lambda^3 \\ -B_{11}\lambda^3 & D_{11}\lambda^4 - \overline{N}_{xx}\lambda^2 \end{bmatrix} \begin{Bmatrix} U_m \\ W_m \end{Bmatrix} + \begin{bmatrix} -\omega^2 I_0 & \omega^2 I_1 \lambda \\ \omega^2 I_1 \lambda & -\omega^2 (I_0 - I_2\lambda^2) \end{bmatrix} \begin{Bmatrix} U_m \\ W_m \end{Bmatrix} = \begin{Bmatrix} 0 \\ 0 \end{Bmatrix}. \tag{4.38}$$

Then Eq. (4.38) can be casted in the following form:

$$\begin{bmatrix} A_{11}\lambda^2 - \omega^2 I_0 & -B_{11}\lambda^3 + \omega^2 I_1\lambda \\ -B_{11}\lambda^3 + \omega^2 I_1\lambda & D_{11}\lambda^4 - \overline{N}_{xx}\lambda^2 - \omega^2(I_0 - I_2\lambda^2) \end{bmatrix} \begin{Bmatrix} U_m \\ W_m \end{Bmatrix} = \begin{Bmatrix} 0 \\ 0 \end{Bmatrix}. \tag{4.39}$$

To obtain a unique solution, the determinant of the matrix in Eq. (4.39) must vanish, leading to the following characteristic equation:

$$A(\omega^2)^2 + B(\omega^2) + C = 0, \tag{4.40}$$

where

$$\begin{aligned} A &\equiv I_1^2\lambda^2 - I_0(I_0 - I_2\lambda^2), \\ B &\equiv A_{11}\lambda^2(I_0 - I_2\lambda^2) + I_0(D_{11}\lambda^2 - \overline{N}_{xx})\lambda^2 - 2I_1B_{11}\lambda^4, \\ C &\equiv [A_{11}\overline{N}_{xx} - (A_{11}D_{11} - B_{11}^2)\lambda^2]\lambda^4. \end{aligned} \tag{4.41}$$

Solution of Eq. (4.40) would provide the natural frequency for the simply supported non-symmetric case. For a symmetric case ($B_{11} = I_1 = 0$), Eq. (4.39) will not be coupled and the result will be the expression presented in Eq. (4.31).

A procedure to solve the case of a non-symmetric beam having other boundary conditions is presented in Appendix A.

4.5. Vibrations of Columns — FSDT Approach

As derived in Chapter 2, the equations of motion for a general laminate using the FSDT approach (see also [9–14]) can be written as (assuming $v_0 = \phi_y = 0$):

$$A_{11}\left(\frac{\partial^2 u_0}{\partial x^2} + \frac{\partial^3 w_0}{\partial x^3}\right) + B_{11}\frac{\partial^2 \phi_x}{\partial x^2} = I_0\frac{\partial^2 u_0}{\partial t^2} + I_1\frac{\partial^2 \phi_x}{\partial t^2}, \tag{4.42}$$

$$KA_{55}\left(\frac{\partial^2 w_0}{\partial x^2} + \frac{\partial \phi_x}{\partial x}\right) - \overline{N}_{xx}\frac{\partial^2 w_0}{\partial x^2} = q + I_0\frac{\partial^2 w_0}{\partial t^2}, \tag{4.43}$$

$$B_{11}\left(\frac{\partial^2 u_0}{\partial x^2} + \frac{\partial^3 w_0}{\partial x^3}\right) + D_{11}\frac{\partial^2 \phi_x}{\partial x^2} - KA_{55}\left(\frac{\partial w_0}{\partial x} + \phi_x\right)$$
$$= I_2\frac{\partial^2 \phi_x}{\partial t^2} + I_1\frac{\partial^2 u_0}{\partial t^2}. \tag{4.44}$$

One should note that Eqs. (4.42)–(4.44) are for the case of uniform properties along the beam. For the particular case of properties varying along the x coordinate of the beam, the reader is referred to [12–14]. To solve the vibration problem, the lateral load, q, is set to zero leading to the following three coupled equations of motion:

$$A_{11}\left(\frac{\partial^2 u_0}{\partial x^2} + \frac{\partial^3 w_0}{\partial x^3}\right) + B_{11}\frac{\partial^2 \phi_x}{\partial x^2} = I_0\frac{\partial^2 u_0}{\partial t^2} + I_1\frac{\partial^2 \phi_x}{\partial t^2}, \tag{4.45}$$

$$KA_{55}\left(\frac{\partial^2 w_0}{\partial x^2} + \frac{\partial \phi_x}{\partial x}\right) - \overline{N}_{xx}\frac{\partial^2 w_0}{\partial x^2} = I_0\frac{\partial^2 w_0}{\partial t^2}, \tag{4.46}$$

$$B_{11}\left(\frac{\partial^2 u_0}{\partial x^2} + \frac{\partial^3 w_0}{\partial x^3}\right) + D_{11}\frac{\partial^2 \phi_x}{\partial x^2} - KA_{55}\left(\frac{\partial w_0}{\partial x} + \phi_x\right)$$
$$= I_2\frac{\partial^2 \phi_x}{\partial t^2} + I_1\frac{\partial^2 u_0}{\partial t^2}. \tag{4.47}$$

Assuming harmonic vibrations with a circular frequency squared, ω^2 and neglecting higher order terms we can rewrite Eqs. (4.45)–(4.47) as

$$A_{11}\frac{d^2U}{dx^2} + B_{11}\frac{d^2\Phi}{dx^2} + \omega^2 I_0 U + \omega^2 I_1 \Phi = 0, \tag{4.48}$$

$$KA_{55}\left(\frac{d^2W}{dx^2} + \frac{d\Phi}{dx}\right) - \overline{N}_{xx}\frac{d^2W}{dx^2} + \omega^2 I_0 W = 0, \tag{4.49}$$

$$B_{11}\frac{d^2U}{dx^2} + D_{11}\frac{d^2\Phi}{dx^2} - KA_{55}\left(\frac{dW}{dx} + \Phi\right) + \omega^2 I_2 \Phi + \omega^2 I_1 U = 0. \tag{4.50}$$

For symmetric laminates, we can write the following expressions:

$$A_{11}\frac{d^2U}{dx^2} + \omega^2 I_0 U = 0, \tag{4.51}$$

$$KA_{55}\left(\frac{d^2W}{dx^2} + \frac{d\Phi}{dx}\right) - \overline{N}_{xx}\frac{d^2W}{dx^2} + \omega^2 I_0 W = 0, \tag{4.52}$$

$$D_{11}\frac{d^2\Phi}{dx^2} - KA_{55}\left(\frac{dW}{dx} + \Phi\right) + \omega^2 I_2 \Phi = 0. \tag{4.53}$$

One should note that for the symmetric case, Eq. (4.51) is not coupled with the Eqs. (4.52) and (4.53), which are coupled and have to be solved together.

4.5.1. *Symmetric laminate* ($B_{11} = 0, I_1 = 0$)

First, we shall present a general solution for the symmetric case presented by the two coupled equations (Eqs. (4.52) and (4.53)). The two equations are decoupled to yield uncoupled equations

$$\left[D_{11} - \frac{\overline{N}_{xx}}{KA_{55}}\right]\frac{d^4W}{dx^4} + \left\{\omega^2\left[\frac{D_{11}I_0}{KA_{55}} + \left(1 - \frac{\overline{N}_{xx}}{KA_{55}}\right)I_2 + \overline{N}_{xx}\right]\right\}$$
$$\times\frac{d^2W}{dx^2} + \omega^2 I_0\left[\frac{I_2\omega^2}{KA_{55}} - 1\right]W = 0, \tag{4.54}$$

$$\left[D_{11} - \frac{\overline{N}_{xx}}{KA_{55}}\right]\frac{d^4\Phi}{dx^4} + \left\{\omega^2\left[\frac{D_{11}I_0}{KA_{55}} + \left(1 - \frac{\overline{N}_{xx}}{KA_{55}}\right)I_2 + \overline{N}_{xx}\right]\right\}$$
$$\times\frac{d^2\Phi}{dx^2} + \omega^2 I_0\left[\frac{I_2\omega^2}{KA_{55}} - 1\right]\Phi = 0. \tag{4.55}$$

One should note that without axial compression, $\overline{N}_{xx} = 0$ Eqs. (4.54) and (4.55) will have the following form which is similar to what is given in [10]:

$$D_{11}\frac{d^4W}{dx^4} + \omega^2\left[\frac{D_{11}I_0}{KA_{55}} + I_2\right]\frac{d^2W}{dx^2} + \omega^2 I_0\left[\frac{I_2\omega^2}{KA_{55}} - 1\right]W = 0, \quad (4.56)$$

$$D_{11}\frac{d^4\Phi}{dx^4} + \omega^2\left[\frac{D_{11}I_0}{KA_{55}} + I_2\right]\frac{d^2\Phi}{dx^2} + \omega^2 I_0\left[\frac{I_2\omega^2}{KA_{55}} - 1\right]\Phi = 0. \quad (4.57)$$

The general solutions for Eqs. (4.54) and (4.55) (as well as for Eqs. (4.56) and (4.57)) have the following form:

$$W(x) = A_1\cosh(s_1x) + A_2\sinh(s_1x) + A_3\cos(s_2x) + A_4\sin(s_2x), \quad (4.58)$$

$$\Phi(x) = B_1\cosh(s_1x) + B_2\sinh(s_1x) + B_3\cos(s_2x) + B_4\sin(s_2x), \quad (4.59)$$

while the constants A_1, A_2, A_3, A_4 and B_1, B_2, B_3, B_4 are interconnected (by back substituting Eqs. (4.58) and (4.59) into the coupled equations, Eqs. (4.52) and (4.53)) and have the following form:

$$B_1 = \frac{KA_{55}\cdot s_1}{D_{11}\cdot s_1^2 - KA_{55} + \omega^2 I_2}A_2 \equiv \alpha A_2, \quad (4.60)$$

$$B_2 = \frac{KA_{55}\cdot s_1}{D_{11}\cdot s_1^2 - KA_{55} + \omega^2 I_2}A_1 \equiv \alpha A_1, \quad (4.61)$$

$$B_3 = \frac{KA_{55}\cdot s_2}{\omega^2 I_2 - D_{11}\cdot s_1^2 - KA_{55}}A_4 \equiv \beta A_4, \quad (4.62)$$

$$B_4 = -\frac{KA_{55}\cdot s_2}{\omega^2 I_2 - D_{11}\cdot s_1^2 - KA_{55}}A_3 \equiv -\beta A_3, \quad (4.63)$$

and

$$s_1 = \sqrt{-\frac{b}{2a} + \frac{\sqrt{(b^2 - 4ac)}}{2a}}, \quad (4.64)$$

$$s_2 = \sqrt{+\frac{b}{2a} + \frac{\sqrt{(b^2 - 4ac)}}{2a}}, \quad (4.65)$$

where

$$a \equiv D_{11} - \frac{\overline{N}_{xx}}{KA_{55}};\, b \equiv \omega^2\left[\frac{D_{11}I_0}{KA_{55}} + \left(1 - \frac{\overline{N}_{xx}}{KA_{55}}\right)I_2 + \overline{N}_{xx}\right];$$

$$c \equiv \omega^2 I_0\left[\frac{I_2\omega^2}{KA_{55}} - 1\right]. \quad (4.66)$$

The remaining constants, A_1, A_2, A_3 and A_4, will be found after imposing the relevant boundary conditions.

The general boundary conditions for a non-symmetric laminate are given in Table 4.3 and Eq. (4.67).

The in-plane boundary condition for the non-symmetric laminate is given in Eq. (4.67):

$$A_{11}\frac{dU}{dx} + B_{11}\frac{d\Phi}{dx} = -\overline{N}_{xx} \quad \text{or} \quad U = 0 \tag{4.67}$$

One should note that for a symmetric case without axial compression load the expressions for the boundary condition simplify, and are given in Table 4.4 and Eq. (4.68).

The in-plane boundary condition for the symmetric laminate axial non-compressed column is given in Eq. (4.68):

$$\frac{dU}{dx} = 0 \quad \text{or} \quad U = 0 \tag{4.68}$$

Table 4.5 presents the characteristic equations for various boundary conditions for symmetric laminate, axial non-compressed columns.

Table 4.3. Out-of-plane boundary conditions for a non-symmetric laminate column.

Name	Boundary conditions
Simply supported (or hinged) end	$W = 0$ and $B_{11}\frac{dU}{dx} + D_{11}\frac{d\Phi}{dx} = 0$
Clamped end	$W = 0$ and $\Phi = 0$
Free end	$B_{11}\frac{dU}{dx} + D_{11}\frac{d\Phi}{dx} = 0$ and $KA_{55}\left(\frac{dW}{dx} + \Phi\right) - \overline{N}_{xx}\frac{dW}{dx} = 0$
Guided end	$\Phi = 0$ and $KA_{55}\left(\frac{dW}{dx} + \Phi\right) - \overline{N}_{xx}\frac{dW}{dx} = 0$

Table 4.4. Out-of-plane boundary conditions for a symmetric laminate axial non-compressed column.

Name	Boundary conditions
Simply supported (or hinged) end	$W = 0$ and $\frac{d\Phi}{dx} = 0$
Clamped end	$W = 0$ and $\Phi = 0$
Free end	$\frac{d\Phi}{dx} = 0$ and $\frac{dW}{dx} + \Phi = 0$
Guided end	$\Phi = 0$ and $\frac{dW}{dx} + \Phi = 0$

Table 4.5. Characteristic equations for natural vibrations of laminated composite columns using FSDT approach.

No.	Name	Boundary conditions	Characteristic equation
1	SS–SS*	$W(0) = 0, \quad d\Phi(0)/dx = 0$ $W(L) = 0, \quad d\Phi(L)/dx = 0$	$\sin(s_2L) = 0$ $(s_2L)_n = n\pi, \ n = 1, 2, 3, \ldots$
2	C–C**	$W(0) = 0, \quad \Phi(0) = 0$ $W(L) = 0, \quad \Phi(L) = 0$	$2\alpha - 2\alpha\cos(s_2L)\cosh(s_1L)$ $+\left(\frac{\alpha^2-\beta^2}{\beta}\right)\sin(s_2L)\sinh(s_1L) = 0$
3	C–F***	$W(0) = 0, \quad \Phi(0) = 0$ $d\Phi(L)/dx = 0$ $dW(L)/dx + \Phi(L) = 0$	$\alpha[s_1(s_1-\alpha) - s_2(s_2+\beta)]$ $+\left[s_1s_2\left(\frac{\beta-\alpha^2}{\beta}\right) + \alpha(\beta s_2 - \alpha s_1)\right]$ $\cosh(s_1L)\cos(s_2L)$ $-\alpha(2s_1s_2 + s_1\beta + s_2\alpha)$ $\cosh(s_1L)\cos(s_2L) = 0$
4	F–F	$d\Phi(0)/dx = 0$ $dW(0)/dx + \Phi(0) = 0$ $d\Phi(L)/dx = 0$ $dW(L)/dx + \Phi(L) = 0$	$2\alpha s_1(s_1+\alpha) - 2\cos(s_2L)\cosh(s_1L)$ $+\frac{\beta^2s_2^2(s_1+\alpha)^2 - \alpha^2s_1^2(s_2+\beta)^2}{\beta s_2(s_2+\beta)}$ $\sin(s_2L)\sinh(s_1L) = 0$
5	SS–C	$W(0) = 0, \ d\Phi(0)/dx = 0$ $W(L) = 0, \ \Phi(L) = 0$	$\beta\tanh(s_1L) = \alpha\tan(s_2L)$
6	SS–F	$W(0) = 0, \ d\Phi(0)/dx = 0$ $d\Phi(L)/dx = 0$ $dW(L)/dx + \Phi(L) = 0$	$s_1\alpha(s_2+\beta)\tanh(s_1L)$ $= -s_2\beta(s_1+\alpha)\tan(s_2L)$
7	G****–F	$\Phi(0) = 0$ $dW(0)dx + \Phi(0) = 0$ $d\Phi(L)/dx = 0$ $dW(L)/dx + \Phi(L) = 0$	$s_1\alpha(s_2+\beta)\tan(s_2L)$ $= s_2\beta(s_1+\alpha)\tanh(s_1L)$
8	G–SS	$\Phi(0) = 0$ $dW(0)dx + \Phi(0) = 0$ $W(L) = 0, \ d\Phi(L)/dx = 0$	$\cos(s_2L) = 0$ $(s_2L)_n = (2n-1)\frac{\pi}{2}, \ n = 1, 2, 3, \ldots$
9	G–G	$\Phi(0) = 0$ $dW(0)dx + \Phi(0) = 0\Phi(L) = 0$ $dW(L)dx + \Phi(L) = 0$	$\sin(s_2L) = 0$ $(s_2L)_n = n\pi, \ n = 1, 2, 3, \ldots$
10	G–C	$\Phi(0) = 0$ $dW(0)dx + \Phi(0) = 0$ $W(L) = 0 \ \Phi(L) = 0,$	$\alpha\tanh(s_1L) = -\beta\tan(s_2L)$

*SS = Simply Supported; **C = Clamped; ***F = Free; G**** = Guided.

4.5.2. *Non-symmetric laminate* $(B_{11} \neq 0, I_1 \neq 0)$

Following the derivation performed by Abramovich and Livshits [17], the procedure to solve a non-symmetric laminate for the case of beams of columns is next presented.

The equations of motion, Eqs. (4.45)–(4.47), will be presented in a matrix formulation (for $\overline{N}_{xx} = 0$) namely,

$$\begin{bmatrix} A_{11}\frac{\partial^2}{\partial x^2} & 0 & B_{11}\frac{\partial^2}{\partial x^2} \\ 0 & KA_{55}\frac{\partial^2}{\partial x^2} & KA_{55}\frac{\partial}{\partial x} \\ B_{11}\frac{\partial^2}{\partial x^2} & -KA_{55}\frac{\partial}{\partial x} & \left(D_{11}\frac{\partial^2}{\partial x^2} - KA_{55}\right) \end{bmatrix} \begin{Bmatrix} u_0(x,t) \\ w_0(x,t) \\ \phi_x(x,t) \end{Bmatrix}$$
$$+ \begin{bmatrix} -I_0 & 0 & -I_1 \\ 0 & -I_0 & 0 \\ -I_1 & 0 & -I_2 \end{bmatrix} \frac{\partial^2}{\partial t^2} \begin{Bmatrix} u_0(x,t) \\ w_0(x,t) \\ \phi_x(x,t) \end{Bmatrix} = \begin{Bmatrix} 0 \\ 0 \\ 0 \end{Bmatrix}. \tag{4.69}$$

Equation (4.63) and Table 4.3 give the associated boundary conditions. Following [17], non-dimensional displacements and beam's length are defined as

$$\{\overline{q}\} = \{u_0/L, w_0/L, \phi_x\}^T, \quad \xi \equiv x/L \tag{4.70}$$

yielding

$$\{q\} = \text{diagonal}\{L, L, 1\}\{\overline{q}\} \tag{4.71}$$

To obtain expressions for free vibrations, we assume that the solution for Eq. (4.69) has the following form:

$$\{\overline{q}\} = \{U, W, \Phi\}^T e^{i\omega t} \equiv \{Q\}e^{i\omega t} \tag{4.72}$$

Substituting Eq. (4.72) into Eq. (4.69) leads to the following matrix equation:

$$\left[\begin{bmatrix} A_{11}\frac{\partial^2}{\partial x^2} & 0 & B_{11}\frac{\partial^2}{\partial x^2} \\ 0 & KA_{55}\frac{\partial^2}{\partial x^2} & KA_{55}\frac{\partial}{\partial x} \\ B_{11}\frac{\partial^2}{\partial x^2} & -KA_{55}\frac{\partial}{\partial x} & \left(D_{11}\frac{\partial^2}{\partial x^2} - KA_{55}\right) \end{bmatrix} + \omega^2 \begin{bmatrix} I_0 & 0 & I_1 \\ 0 & I_0 & 0 \\ I_1 & 0 & I_2 \end{bmatrix}\right]$$
$$\times \begin{Bmatrix} L \cdot U \\ L \cdot W \\ \Phi \end{Bmatrix} = \begin{Bmatrix} 0 \\ 0 \\ 0 \end{Bmatrix}. \tag{4.73}$$

Introducing non-dimensional parameters and after some algebraic transformation, Eq. (4.73) can be written as

$$\left[\begin{bmatrix} \zeta_1^2 & 0 & \zeta^2 \\ 0 & 1 & 0 \\ \zeta^2 b^2 & 0 & b^2 \end{bmatrix} \frac{\partial^2}{\partial \xi^2} + \begin{bmatrix} 0 & 0 & 0 \\ 0 & 0 & 1 \\ 0 & -1 & 0 \end{bmatrix} \frac{\partial}{\partial \xi} \right.$$

$$\left. + \begin{bmatrix} p^2 & 0 & \eta^2 p^2 \\ 0 & b^2 p^2 & 0 \\ \eta^2 b^2 p^2 & 0 & (r^2 b^2 p^2 - 1) \end{bmatrix}\right] \begin{Bmatrix} U \\ W \\ \Phi \end{Bmatrix} = 0, \tag{4.74}$$

where the various parameters are defined as:

$$p^2 \equiv \frac{\omega^2 I_0 L^4}{D_{11}}, \quad b^2 \equiv \frac{D_{11}}{K A_{55} L^2}, \quad \zeta^2 \equiv \frac{B_{11} L}{D_{11}},$$

$$\zeta_1^2 \equiv \frac{A_{11} L^2}{D_{11}}, \quad r^2 \equiv \frac{I_2}{I_0 L^2}, \quad \eta^2 \equiv \frac{I_1}{I_0 L}. \tag{4.75}$$

Assuming that the general solution for Eq. (4.75) has the following form:

$$\{Q\} = \{\overline{Q}\} e^{im\xi} \tag{4.76}$$

with m being the eigenvalue and $\{\overline{Q}\}$ the eigenvector and substituting into Eq. (4.75) lead to the following cubic algebraic equation:

$$A s^3 + B s^2 + C s + D = 0, \quad s \equiv \frac{m^2}{p^2} \tag{4.77}$$

with

$$A \equiv \zeta_1^2 - \zeta^4, \quad B \equiv 1 + (r^2 + b^2)\zeta_1^2 - \zeta^2(2\eta^2 + \zeta^2 b^2),$$

$$C \equiv \zeta_1^2 \left(r^2 b^2 - \frac{1}{p^2} \right) + (r^2 + b^2) - \eta^2(\eta^2 + 2\zeta^2 b^2),$$

$$D \equiv r^2 b^2 - \frac{1}{p^2} - b^2 \eta^4. \tag{4.78}$$

Solving Eq. (4.77), the general solution for Eq. (4.75) can be cast in the following form (see a detailed discussion regarding the format of the solution in [17]):

$$U = A_1 \gamma\mu \sinh(m_1\xi) + A_2 \gamma\mu \cosh(m_1\xi) + A_3 \lambda\delta \sin(m_2\xi)$$
$$- A_4 \lambda\delta \cos(m_2\xi) + A_5 \alpha\beta \sin(m_3\xi) - A_6 \alpha\beta \cos(m_3\xi),$$

$$\begin{aligned}W &= A_1\cosh(m_1\xi) + A_2\sinh(m_1\xi) + A_3\cos(m_2\xi)\\&\quad + A_4\sin(m_2\xi) + A_5\cos(m_3\xi) + A_6\sin(m_3\xi),\\\Phi &= -A_1\mu\sinh(m_1\xi) - A_2\mu\cosh(m_1\xi) - A_3\lambda\sin(m_2\xi)\\&\quad + A_4\lambda\cos(m_2\xi) - A_5\alpha\sin(m_3\xi) + A_6\alpha\cos(m_3\xi),\end{aligned} \tag{4.79}$$

where the various terms in Eq. (4.79) are defined as follows:

$$\begin{aligned}&m_1 \equiv \sqrt{s_1}p, \quad m_2 \equiv \sqrt{-s_2}p, \quad m_3 \equiv \sqrt{-s_3}p,\\&\gamma \equiv \frac{\zeta^2 s_1 + \eta^2}{\zeta_1^2 s_1 + 1}, \quad \delta \equiv \frac{\zeta^2 s_2 + \eta^2}{\zeta_1^2 s_2 + 1}, \quad \beta \equiv \frac{\zeta^2 s_3 + \eta^2}{\zeta_1^2 s_3 + 1},\\&\mu \equiv \frac{s_1 + b^2}{\sqrt{s_1}}p, \quad \lambda \equiv \frac{s_2 + b^2}{\sqrt{-s_2}}p, \quad \alpha \equiv \frac{s_3 + b^2}{\sqrt{-s_3}}p.\end{aligned} \tag{4.80}$$

Imposing the adequate boundary conditions would lead to the finding of both the eigenvalues (the natural frequencies) and the eigenvectors (the modes of vibration). For this problem six boundary conditions should be imposed as presented in Table 4.6.

One should note (see also [17]) that only for a column or a beam with two simply supported (hinged) movable ends an analytical solution exists. The characteristic equation, obtained after demanding that the determinant of the coefficients of Eq. (4.79) would be zero, has the following form:

$$\sin(m_3)\sin(m_2) = 0. \tag{4.81}$$

The solutions for Eq. (4.81) has two series: the first one having bending dominated vibrations

$$m_3 = k\pi, \quad k = 1, 2, 3, \ldots, n, \tag{4.82}$$

Table 4.6. Boundary conditions for the non-symmetrical laminated columns using the FSDT approach.

Name	Boundary conditions
Clamped immovable end	$U = W = \Phi = 0$
Clamped movable end	$\zeta_1^2\frac{dU}{d\xi} + \zeta^2\frac{d\Phi}{d\xi} = W = \Phi = 0$
Simply supported immovable end	$\zeta^2\frac{dU}{d\xi} + \frac{d\Phi}{d\xi} = W = U = 0$
Simply supported movable end	$\zeta^2\frac{dU}{d\xi} + \frac{d\Phi}{d\xi} = W = \zeta_1^2\frac{dU}{d\xi} + \zeta^2\frac{d\Phi}{d\xi} = 0$
Free end	$\zeta^2\frac{dU}{d\xi} + \frac{d\Phi}{d\xi} = \frac{1}{b^2}\left(\Phi + \frac{dW}{d\xi}\right) = \zeta_1^2\frac{dU}{d\xi} + \zeta^2\frac{d\Phi}{d\xi} = 0$

while the second series has longitudinal dominated vibrations

$$m_2 = k\pi, \quad k = 1, 2, 3, \ldots, n. \tag{4.83}$$

4.6. Vibrations of Plates — CLT Approach

4.6.1. *Simply supported special orthotropic plates*

Figure 4.2 presents the schematic view of a laminated rectangular plate.

The first case to be solved is sometimes called *special orthotropic plates* for which the bending-stretching coupling terms B_{ij} and the bending-twisting coefficients D_{16} and D_{26} are set to zero. Then taking Eq. (2.27) from Chapter 2 assuming symmetry ($B_{ij} = D_{16} = D_{26} = I_1 = 0$) and zeroing the in-plane and out-of-plane loads, one obtains

$$\begin{aligned} &D_{11}\frac{\partial^4 w_0}{\partial x^4} + 2(D_{12} + 2D_{66})\frac{\partial^4 w_0}{\partial x^2 \partial y^2} + D_{22}\frac{\partial^4 w_0}{\partial y^4} \\ &+ I_0\frac{\partial^2 w_0}{\partial t^2} - I_2\frac{\partial^2}{\partial t^2}\left(\frac{\partial^2 w_0}{\partial x^2} + \frac{\partial^2 w_0}{\partial y^2}\right) = 0, \end{aligned} \tag{4.84}$$

where I_0 and I_2 are defined in Eq. (4.2). Let us assume the following solution for the out-of-plane displacement w_0 (tacitly assuming harmonic vibrations with a circular frequency ω):

$$w_0(x, y, t) = W_{mn} \sin\left(\frac{m\pi x}{a}\right)\left(\frac{n\pi y}{b}\right) e^{i\omega t}, \tag{4.85}$$

where a is the length and b is the width of the plate having a total thickness of t, and substituting in Eq. (4.84) we get

$$\begin{aligned} &D_{11}\left(\frac{m\pi}{a}\right)^4 + 2(D_{12} + 2D_{66})\left(\frac{m\pi}{a}\right)^2\left(\frac{n\pi}{b}\right)^2 + D_{22}\left(\frac{n\pi}{b}\right)^4 \\ &-\omega^2\left\{I_0 + I_2\left[\left(\frac{m\pi}{a}\right)^2 + \left(\frac{n\pi}{b}\right)^2\right]\right\} = 0. \end{aligned} \tag{4.86}$$

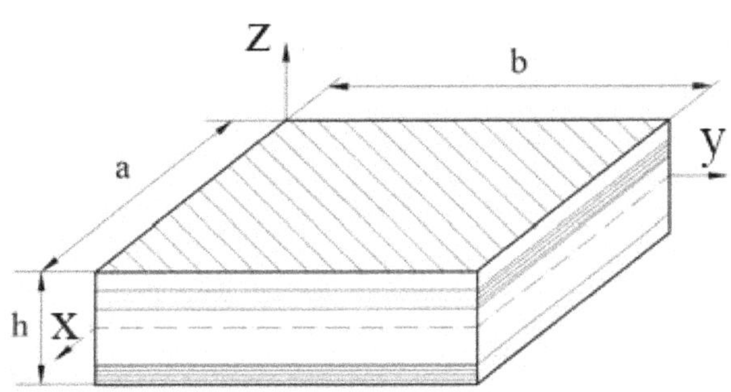

Fig. 4.2. A laminated rectangular plate — a schematic drawing.

The solution for Eq. (4.86) has the following form, which present the natural frequencies for a special orthotropic laminated plate

$$\omega_{mn}^2 = \left(\frac{\pi}{a}\right)^4 \frac{D_{11}m^4 + 2(D_{12} + 2D_{66})m^2n^2\left(\frac{a}{b}\right)^2 + D_{22}n^4\left(\frac{a}{b}\right)^4}{I_0 + I_2\left(\frac{\pi}{a}\right)^2\left[m^2 + n^2\left(\frac{a}{b}\right)^2\right]} \tag{4.87}$$

while the mode of vibration is given by

$$w_0(x, y) = W_{mn}\sin\left(\frac{m\pi x}{a}\right)\left(\frac{n\pi y}{b}\right). \tag{4.88}$$

One can see from Eq. (4.87) that inclusion of the rotary inertia I_2 tends to reduce the natural frequencies. For a square plate, $a = b$, when neglecting the rotary inertia, the general frequency would be

$$\omega_{mn}^2 = \frac{\pi^4}{I_0 a^4}[D_{11}m^4 + 2(D_{12} + 2D_{66})m^2n^2 + D_{22}n^4]. \tag{4.89}$$

The lowest frequency sometimes called also the fundamental frequency will occur at $m = n = 1$, namely for a rectangular plate we will have

$$\omega_{11}^2 = \frac{\pi^4}{I_0 a^4}\left[D_{11} + 2(D_{12} + 2D_{66})\left(\frac{a}{b}\right)^2 + D_{22}\left(\frac{a}{b}\right)^4\right]. \tag{4.90}$$

The matrix notation of the governing equations for a plate is presented in Appendix B.

4.6.2. *Simply supported on two opposite edges of special orthotropic plates*

Using the Lévy method,[a] one can solve the vibrations of rectangular plates, where two opposite edges are on simply-supported boundary conditions, while the other two sides can be either clamped, free or any other combination of boundaries. Therefore, the equilibrium equation, without in-plane loads, is given by Eq. (4.84). The out-of-plane deflections can be written as

$$w_0(x, y, t) = W_m(x)\left(\frac{n\pi y}{b}\right)e^{i\omega t}. \tag{4.91}$$

[a]M. Lévy, Memoire sur la theorie des plaques elastiques planes, *J. Math. Pures Appl.*, Vol. 3, 1899, p. 219.

Substituting Eq. (4.91) into (4.84) lead to the following differential equation:

$$D_{11}\frac{d^4W_m}{dx^4}+\left[\omega^2 I_2-2(D_{12}+2D_{66})\left(\frac{n\pi}{b}\right)^2\right]\frac{d^2W_m}{dx^2}$$
$$-\left\{\omega^2\left[I_0+I_2\left(\frac{n\pi}{b}\right)^2\right]-D_{22}\left(\frac{n\pi}{b}\right)^4\right\}W_m=0 \tag{4.92}$$

or

$$\widehat{a}\frac{d^4W_m}{dx^4}+\widehat{b}\frac{d^2W_m}{dx^2}-\widehat{c}W_m=0, \tag{4.93}$$

where

$$\widehat{a}\equiv D_{11},\quad b\,\widehat{\equiv}\,\omega^2 I_2-2(D_{12}+2D_{66})\left(\frac{n\pi}{b}\right)^2,$$
$$c\equiv\omega^2\left[I_0+I_2\left(\frac{n\pi}{b}\right)^2\right]-D_{22}\left(\frac{n\pi}{b}\right)^4. \tag{4.94}$$

The general solution of Eq. (4.92) has the following form:

$$W(x)=A_1\sinh(\alpha_1x)+A_2\cosh(\alpha_1x)+A_3\sin(\alpha_2x)+A_4\cos(\alpha_2x), \tag{4.95}$$

where

$$s_1=\sqrt{\frac{-\widehat{b}+\sqrt{\widehat{b}^2+4\widehat{a}\widehat{c}}}{2\widehat{a}}},\quad s_2=\sqrt{\frac{\widehat{b}+\sqrt{\widehat{b}^2+4\widehat{a}\widehat{c}}}{2\widehat{a}}}. \tag{4.96}$$

Application of the boundary conditions in the x direction will lead to the natural frequencies and their associated mode shapes in the x direction.

4.7. Vibration of Cylindrical Shells

The topic of shells was presented in detail in Chapter 3, which dealt with buckling of cylindrical shell under various loading. This chapter will present ways of calculating natural frequencies and their associated mode shapes. Typical studies on this topic can be found in [18–26].

4.7.1. *Isotropic shells*

Figure 4.3 presents the dimensions of an isotropic circular cylindrical shell, for which the natural frequencies can be calculated.

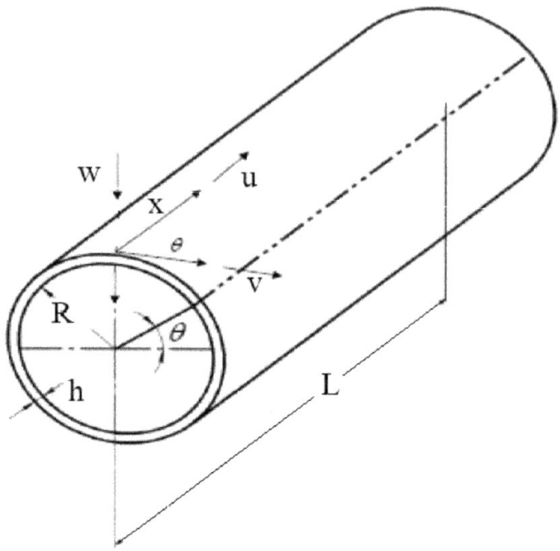

Fig. 4.3. A circular cylindrical shell — a schematic drawing.

As suggested in [25, 26], to study the free vibration of a circular cylindrical shell, the governing equations can be presented in a matrix form

$$\begin{bmatrix} L_{11} & L_{12} & L_{13} \\ L_{21} & L_{22} & L_{23} \\ L_{31} & L_{32} & L_{33} \end{bmatrix} \begin{Bmatrix} u(x,\theta,t) \\ v(x,\theta,t) \\ w(x,\theta,t) \end{Bmatrix} = \begin{Bmatrix} 0 \\ 0 \\ 0 \end{Bmatrix}, \quad \text{where}$$
$$L_{31} = -L_{13}, \quad L_{32} = -L_{23} \tag{4.97}$$

where $L_{i,j}$, $i, j = 1, 2, 3$ are differential operators with respect to the variables x, θ and t.

The operators depend on which shell theory is used to calculate the natural frequency of the free circular cylindrical shell (see Appendix C). According to [23, 25, 26] a first approximation for the axial, circumferential and radial displacements can be written as

$$\begin{aligned} u(x,\theta,t) &= \alpha e^{\lambda_n x} \sin(m\theta) e^{i\omega t}, \\ v(x,\theta,t) &= \beta e^{\lambda_n x} \cos(m\theta) e^{i\omega t}, \\ w(x,\theta,t) &= \gamma e^{\lambda_n x} \sin(m\theta) e^{i\omega t}, \end{aligned} \tag{4.98}$$

where $\lambda_n = n\pi/L$ and n and m are half wave number in the axial and circumferential directions, respectively, α, β, γ are constants to be determined , and ω is the circular frequency of the natural free vibration. Substituting Eq. (4.98) into Eq. (4.97) leads to the following set of homogenous equations, written in a matrix form:

$$\begin{bmatrix} A_{11} & A_{12} & A_{13} \\ A_{21} & A_{22} & A_{23} \\ A_{31} & A_{32} & A_{33} \end{bmatrix} \begin{Bmatrix} \alpha \\ \beta \\ \gamma \end{Bmatrix} = \begin{Bmatrix} 0 \\ 0 \\ 0 \end{Bmatrix}, \quad \text{where}$$

$$A_{21} = -A_{12}, \quad A_{31} = -A_{13}, \tag{4.99}$$

where the terms A_{ij}, $i, j = 1, 2, 3$, are functions of m, λ_n and what it is called a frequency parameter Φ, defined as

$$\Phi^2 = \frac{(1-\upsilon^2)}{E}\rho R^2\omega^2. \tag{4.100}$$

The matrix $[A_{ij}]$ is next presented for various shell theories.

Donnell–Mushtari theory [25]:

$$[A_{ij}] = \begin{bmatrix} \Phi^2 + \lambda_n^2 - \frac{1-\upsilon}{2}m^2 & -\frac{1+\upsilon}{2}m\lambda_n & \upsilon\lambda_n \\ \frac{1+\upsilon}{2}m\lambda_n & \Phi^2 + \frac{1-\upsilon}{2}\lambda_n^2 - m^2 & m \\ -\upsilon\lambda_n & m & \Phi^2 - [1 + k(\lambda_n^2 - m^2)^2] \end{bmatrix}, \tag{4.101}$$

where $k \equiv h^2/(12R^2)$ is the non-dimensional thickness parameter. Note that this parameter is very small for small h/R ratios.

Love–Timoshenko theory [25]:

$$[A_{ij}] = \begin{bmatrix} \Phi^2 + \lambda_n^2 - \frac{1-\upsilon}{2}m^2 & -\frac{1+\upsilon}{2}m\lambda_n & \upsilon\lambda_n \\ \frac{1+\upsilon}{2}m\lambda_n & \begin{matrix}\Phi^2 + (1+2k)\frac{1-\upsilon}{2}\lambda_n^2 \\ -(1+k)m^2\end{matrix} & m + mk(m^2 - \lambda_n^2) \\ -\upsilon\lambda_n & m + mk(m^2 - \lambda_n^2) & \Phi^2 - [1 + k(\lambda_n^2 - m^2)^2] \end{bmatrix}. \tag{4.102}$$

Arnold–Warburton theory [25]:

$$[A_{ij}] = \begin{bmatrix} \Phi^2 + \lambda_n^2 - \frac{1-\upsilon}{2}m^2 & -\frac{1+\upsilon}{2}m\lambda_n & \upsilon\lambda_n \\ \frac{1+\upsilon}{2}m\lambda_n & \begin{matrix}\Phi^2 + (1+4k)\frac{1-\upsilon}{2}\lambda_n^2 \\ -(1+k)m^2\end{matrix} & m + mk[m^2 - (2-\upsilon)\lambda_n^2] \\ -\upsilon\lambda_n & m + mk[m^2 - (2-\upsilon)\lambda_n^2] & \Phi^2 - [1 + k(\lambda_n^2 - m^2)^2] \end{bmatrix}. \tag{4.103}$$

Houghton–Johns theory [25]:

$$[A_{ij}]=\begin{bmatrix} \Phi^2+\lambda_n^2-\frac{1-\upsilon}{2}m^2 & -\frac{1+\upsilon}{2}m\lambda_n & \upsilon\lambda_n \\ \frac{1+\upsilon}{2}m\lambda_n & \Phi^2+\frac{1-\upsilon}{2}\lambda_n^2-m^2 & m+mk[m^2-(2-\upsilon)\lambda_n^2] \\ -\upsilon\lambda_n & m+mk[m^2-(2-\upsilon)\lambda_n^2] & \Phi^2-[1+k(\lambda_n^2-m^2)^2] \end{bmatrix}. \tag{4.104}$$

Flügge–Byrne–Lur'ye theory [25]:

$$[A_{ij}]=\begin{bmatrix} \begin{matrix}\Phi^2+\lambda_n^2\\ -(1+k)\frac{1-\upsilon}{2}m^2\end{matrix} & -\frac{1+\upsilon}{2}m\lambda_n & \upsilon\lambda_n-k\lambda_n[\lambda_n^2+(1-\upsilon)m^2] \\ \frac{1+\upsilon}{2}m\lambda_n & \begin{matrix}\Phi^2+(1+3k)\\ \frac{1-\upsilon}{2}\lambda_n^2-m^2\end{matrix} & m\left[1-\frac{3-\upsilon}{2}k\lambda_n^2\right] \\ \begin{matrix}-\upsilon\lambda_n+k\lambda_n\\ [\lambda_n^2+(1-\upsilon)m^2]\end{matrix} & m\left[1-\frac{3-\upsilon}{2}k\lambda_n^2\right] & \begin{matrix}\Phi^2-k[(\lambda_n^2-m^2)^2-2m^2]\\ -(1+k)\end{matrix} \end{bmatrix}. \tag{4.105}$$

Reissner–Naghdi–Berry theory [25]:

$$[A_{ij}]=\begin{bmatrix} \Phi^2+\lambda_n^2-\frac{1-\upsilon}{2}m^2 & -\frac{1+\upsilon}{2}m\lambda_n & \upsilon\lambda_n \\ \frac{1+\upsilon}{2}m\lambda_n & \Phi^2+(1+k)\left[\frac{1-\upsilon}{2}\lambda_n^2-m^2\right] & m[1+k(m^2-\lambda_n^2)] \\ -\upsilon\lambda_n & m[1+k(m^2-\lambda_n^2)] & \Phi^2-[1+k(\lambda_n^2-m^2)^2] \end{bmatrix}. \tag{4.106}$$

Sanders theory [25]:

$$[A_{ij}]=\begin{bmatrix} \begin{matrix}\Phi^2+\lambda_n^2\\ -\left(1+\frac{k}{4}\right)\frac{1-\upsilon}{2}m^2\end{matrix} & -m\lambda_n\left[\frac{1+\upsilon}{2}-\frac{3k(1-\upsilon)}{8}\right] & \lambda_n\left(\upsilon-\frac{1-\upsilon}{2}km^2\right) \\ m\lambda_n\left[\frac{1+\upsilon}{2}-\frac{3k(1-\upsilon)}{8}\right] & \begin{matrix}\Phi^2+(1+k)m^2\\ +\left(1+\frac{9k}{4}\right)\frac{1-\upsilon}{2}\lambda_n^2\end{matrix} & m\left[1+k\left(m^2-\frac{3-\upsilon}{2}\lambda_n^2\right)\right] \\ \lambda_n\left(\frac{1-\upsilon}{2}km^2-\upsilon\right) & m\left[1+k\left(m^2-\frac{3-\upsilon}{2}\lambda_n^2\right)\right] & \Phi^2-[1+k(\lambda_n^2-m^2)^2] \end{bmatrix}. \tag{4.107}$$

Vlasov theory [25]:

$$[A_{ij}]=\begin{bmatrix} \Phi^2+\lambda_n^2-\frac{1-\upsilon}{2}m^2 & -\frac{1+\upsilon}{2}m\lambda_n & \lambda_n\left[\upsilon-k\left(\frac{1-\upsilon}{2}m^2+\lambda_n\right)\right] \\ \frac{1+\upsilon}{2}m\lambda_n & \Phi^2-m^2+\frac{1-\upsilon}{2}\lambda_n^2 & m\left(1-\frac{3-\upsilon}{2}k\lambda_n^2\right) \\ \lambda_n\left[k\left(\frac{1-\upsilon}{2}m^2+\lambda_n\right)-\upsilon\right] & m\left(1-\frac{3-\upsilon}{2}k\lambda_n^2\right) & \begin{matrix}\Phi^2-(1+k)\\ -k[(\lambda_n^2-m^2)^2-2m^2]\end{matrix} \end{bmatrix}. \tag{4.108}$$

Kennard simplified theory [25]:

$$[A_{ij}] = \begin{bmatrix} \Phi^2 + \lambda_n^2 - \frac{1-\upsilon}{2}m^2 & -\frac{1+\upsilon}{2}m\lambda_n & \upsilon\lambda_n \\ \frac{1+\upsilon}{2}m\lambda_n & \Phi^2 - m^2 + \frac{1-\upsilon}{2}\lambda_n^2 & m\left[1 + \frac{3k\upsilon}{2(1-\upsilon)}(1-m^2)\right] \\ -\upsilon\lambda_n & 0 & \begin{matrix} \Phi^2 - \left[1 + \frac{2+\upsilon}{2(1-\upsilon)}\right] \\ -k\left[(\lambda_n^2 - m^2)^2 - \frac{4-\upsilon}{2(1-\upsilon)}m^2\right] \end{matrix} \end{bmatrix}. \tag{4.109}$$

To obtain the natural frequencies, the determinant of the matrix presented by Eq. (4.99) should vanish. Solving the determinant leads to a cubic equations in terms of the frequency parameter, Φ^2. For a fixed value of m and λ_n, ones obtains three positive and three negative roots, leading to determination of Φ^2.

As claimed in [23] the results based on Soedel (see next chapter) and Kennard simplified theories were the best ones, in comparison with the other methods. Those results were obtained using the beam function approach as an approximation for boundary condition.

4.7.2. *Soedel theory*

Soedel [26] presents a solution for a simply supported, isotropic circular cylindrical shell, having a length L and radius R. The boundary conditions can be written as

$$\begin{aligned} &w(0,\theta,t) = 0, && w(L,\theta,t) = 0, \\ &v(0,\theta,t) = 0, && v(L,\theta,t) = 0, \\ &M_{xx}(0,\theta,t) = 0, && M_{xx}(L,\theta,t) = 0, \\ &N_{xx}(0,\theta,t) = 0, && N_{xx}(L,\theta,t) = 0. \end{aligned} \tag{4.110}$$

One should note that for circular cylindrical shell, there are four types of simply supported boundary conditions

$$\begin{aligned} &\text{SS1}: && w = 0, && M_{xx} = w_{,xx} = 0, && N_{xx} = 0, && N_{x\theta} = 0, \\ &\text{SS2}: && w = 0, && M_{xx} = w_{,xx} = 0, && u = 0, && N_{x\theta} = 0, \\ &\text{SS3}: && w = 0, && M_{xx} = w_{,xx} = 0, && N_{xx} = 0, && v = 0, \\ &\text{SS4}: && w = 0, && M_{xx} = w_{,xx} = 0, && u = 0, && v = 0. \end{aligned} \tag{4.111}$$

The SS3 is known in the literature as the classical simply supported boundary conditions for circular cylindrical shells. The same is for the clamped boundary conditions

$$\begin{aligned}
&\text{C1}: \quad w=0, \quad w_{,x}=0, \quad N_{xx}=0, \quad N_{x\theta}=0,\\
&\text{C2}: \quad w=0, \quad w_{,x}=0, \quad u=0, \quad N_{x\theta}=0,\\
&\text{C3}: \quad w=0, \quad w_{,x}=0, \quad N_{xx}=0, \quad v=0,\\
&\text{C4}: \quad w=0, \quad w_{,x}=0, \quad u=0, \quad v=0.
\end{aligned} \tag{4.112}$$

The equations of motions are [26]

$$\begin{aligned}
&\frac{\partial N_{xx}}{\partial x}+\frac{1}{R}\frac{\partial N_{x\theta}}{\partial \theta}-\rho h\frac{\partial^2 u}{\partial t^2}=0,\\
&\frac{\partial N_{x\theta}}{\partial x}+\frac{1}{R}\frac{\partial N_{\theta\theta}}{\partial \theta}+\frac{Q_{\theta z}}{R}-\rho h\frac{\partial^2 v}{\partial t^2}=0,\\
&\frac{\partial Q_{xz}}{\partial x}+\frac{1}{R}\frac{\partial Q_{\theta z}}{\partial \theta}+\frac{N_{\theta\theta}}{R}-\rho h\frac{\partial^2 w}{\partial t^2}=0,
\end{aligned} \tag{4.113}$$

where ρ is the mass density and h is the shell thickness.

The assumed displacements have the following form:

$$\begin{aligned}
u(x,\theta,t)&=\alpha\cos\frac{n\pi x}{L}\cos m(\theta-\phi)e^{i\omega t},\\
v(x,\theta,t)&=\beta\sin\frac{n\pi x}{L}\sin m(\theta-\phi)e^{i\omega t},\\
w(x,\theta,t)&=\gamma\sin\frac{n\pi x}{L}\cos m(\theta-\phi)e^{i\omega t}.
\end{aligned} \tag{4.114}$$

The assumption for the u displacement is that a longitudinal node line will not experience deflections in the x direction (the term $\cos m(\theta-\phi)$) while the term $\cos(n\pi x/L)$ was chosen to satisfy zero axial force at both ends.

Substituting the expression for the displacement, Eq. (4.114) into Eq. (4.113), using the relations between the resultants and the displacement (see [26]) leads to

$$\begin{bmatrix} \rho h\omega^2-f_{11} & f_{12} & f_{13}\\ f_{21} & \rho h\omega^2-f_{22} & f_{23}\\ f_{31} & f_{32} & \rho h\omega^2-f_{33}\end{bmatrix}\begin{Bmatrix}\alpha\\ \beta\\ \gamma\end{Bmatrix}=\begin{Bmatrix}0\\0\\0\end{Bmatrix}, \quad \text{where}$$

$$f_{11}=EA\left[\left(\frac{n\pi}{L}\right)^2+\frac{1-\upsilon}{2}\left(\frac{m}{R}\right)^2\right],$$

$$f_{22}=\left(EA+\frac{D}{R^2}\right)\left[\frac{1-\upsilon}{2}\left(\frac{n\pi}{L}\right)^2+\left(\frac{m}{R}\right)^2\right],$$

$$f_{12} = f_{21} = EA\frac{1+\upsilon}{2}\frac{n\pi}{L}\frac{m}{R},$$

$$f_{23} = f_{23} = -\frac{m}{R^2}(EA - D)\left[\left(\frac{n\pi}{L}\right)^2 + \left(\frac{m}{R}\right)^2\right],$$

$$f_{13} = f_{31} = \frac{\upsilon EA}{R}\frac{n\pi}{L}, \quad f_{33} = D\left[\left(\frac{n\pi}{L}\right)^2 + \left(\frac{m}{R}\right)^2\right] + \frac{EA}{R^2}. \tag{4.115}$$

To find a non-trivial solution, the n of the matrix in Eq. (4.115) should vanish. This leads to the following cubic equation for ω^2

$$\begin{aligned}
&\omega^6 + c_1\omega^4 + c_2\omega^2 + c_3 = 0, \quad \text{where} \\
&c_1 = -\frac{1}{\rho h}(f_{11} + f_{22} + f_{33}), \\
&c_2 = \frac{1}{(\rho h)^2}(f_{11}f_{33} + f_{22}f_{33} + f_{11}f_{22} - f_{23}^2 - f_{12}^2 - f_{13}^2), \\
&c_3 = \frac{1}{(\rho h)^3}(f_{11}f_{23}^2 + f_{22}f_{13}^2 + f_{33}f_{12}^2 + 2f_{12}f_{23}f_{13} - f_{11}f_{22}f_{33}).
\end{aligned} \tag{4.116}$$

The solutions for the cubic equations have the following form [26]:

$$\begin{aligned}
\omega_{1mn}^2 &= -\frac{2}{3}\sqrt{c_1^2 - 3c_2}\cos\frac{\kappa}{3} - \frac{c_1}{3}, \\
\omega_{2mn}^2 &= -\frac{2}{3}\sqrt{c_1^2 - 3c_2}\cos\frac{\kappa + 2\pi}{3} - \frac{c_1}{3}, \\
\omega_{3mn}^2 &= -\frac{2}{3}\sqrt{c_1^2 - 3c_2}\cos\frac{\kappa + 4\pi}{3} - \frac{c_1}{3}, \quad \text{where} \\
\kappa &= \cos^{-1}\frac{27c_3 + 2c_1^3 - 9c_1c_2}{2\sqrt{(c_1^2 - 3c_2)^3}}.
\end{aligned} \tag{4.117}$$

Three frequencies are obtained for each combination of m, n. The lowest one is normally associated with mode in the z direction. To obtain the modes, we shall define α, β in terms of γ leading to

$$\begin{bmatrix} \rho h\omega^2 - f_{11} & f_{12} \\ f_{21} & \rho h\omega^2 - f_{22} \end{bmatrix}\begin{Bmatrix} \alpha_i \\ \beta_i \end{Bmatrix} = -\gamma_i\begin{Bmatrix} f_{13} \\ f_{23} \end{Bmatrix}, \quad i = 1, 2, 3. \tag{4.118}$$

The ratio for the mode shape is calculated from

$$\frac{\alpha_i}{\gamma_i} = -\frac{\begin{vmatrix} f_{13} & f_{12} \\ f_{23} & \rho h\omega_{imn}^2 - f_{22} \end{vmatrix}}{\begin{vmatrix} \rho h\omega_{imn}^2 - f_{11} & f_{12} \\ f_{21} & \rho h\omega_{imn}^2 - f_{22} \end{vmatrix}}$$

$$= -\frac{f_{13}(\rho h\omega_{imn}^2 - f_{22}) - f_{12}f_{23}}{(\rho h\omega_{imn}^2 - f_{11})(\rho h\omega_{imn}^2 - f_{22}) - f_{12}^2}, \tag{4.119a}$$

$$\frac{\beta_i}{\gamma_i} = -\frac{\begin{vmatrix} \rho h\omega_{imn}^2 - f_{11} & f_{13} \\ f_{21} & f_{23} \end{vmatrix}}{\begin{vmatrix} \rho h\omega_{imn}^2 - f_{11} & f_{12} \\ f_{21} & \rho h\omega_{imn}^2 - f_{22} \end{vmatrix}}$$

$$= -\frac{f_{23}(\rho h\omega_{imn}^2 - f_{11}) - f_{21}f_{13}}{(\rho h\omega_{imn}^2 - f_{11})(\rho h\omega_{imn}^2 - f_{22}) - f_{12}^2}. \tag{4.119b}$$

To sum-up, the mode of the shell will have the following form:

$$\frac{\alpha_i}{\gamma_i} = -\frac{\begin{vmatrix} f_{13} & f_{12} \\ f_{23} & \rho h\omega_{imn}^2 - f_{22} \end{vmatrix}}{\begin{vmatrix} \rho h\omega_{imn}^2 - f_{11} & f_{12} \\ f_{21} & \rho h\omega_{imn}^2 - f_{22} \end{vmatrix}}$$

$$= -\frac{f_{13}(\rho h\omega_{imn}^2 - f_{22}) - f_{12}f_{23}}{(\rho h\omega_{imn}^2 - f_{11})(\rho h\omega_{imn}^2 - f_{22}) - f_{12}^2},$$

$$\frac{\beta_i}{\gamma_i} = -\frac{\begin{vmatrix} \rho h\omega_{imn}^2 - f_{11} & f_{13} \\ f_{21} & f_{23} \end{vmatrix}}{\begin{vmatrix} \rho h\omega_{imn}^2 - f_{11} & f_{12} \\ f_{21} & \rho h\omega_{imn}^2 - f_{22} \end{vmatrix}}$$

$$= -\frac{f_{23}(\rho h\omega_{imn}^2 - f_{11}) - f_{21}f_{13}}{(\rho h\omega_{imn}^2 - f_{11})(\rho h\omega_{imn}^2 - f_{22}) - f_{12}^2}, \tag{4.119c}$$

$$\begin{Bmatrix} u \\ v \\ w \end{Bmatrix} = \gamma_i \begin{Bmatrix} \dfrac{\alpha_i}{\gamma_i}\cos\dfrac{n\pi x}{L}\cos m(\theta - \phi) \\ \dfrac{\beta_i}{\gamma_i}\sin\dfrac{n\pi x}{L}\sin m(\theta - \phi) \\ \sin\dfrac{n\pi x}{L}\cos m(\theta - \phi) \end{Bmatrix}. \tag{4.120}$$

4.8. Laminated Composite Shells — CLT Approach

Leissa [25] presented the Donnell–Mushtari governing equations for an orthotropic circular cylindrical shell having the following formulation:

$$\begin{aligned}
&\frac{\partial^2 u}{\partial s^2} + \frac{G_{x\theta}(1-\upsilon_x\upsilon_\theta)}{E_x}\frac{\partial^2 u}{\partial \theta^2} + \frac{G_{x\theta}(1-\upsilon_x\upsilon_\theta)+\upsilon_x E_\theta}{E_x}\frac{\partial^2 v}{\partial s\partial\theta} + \frac{\upsilon_x E_\theta}{E_x}\frac{\partial w}{\partial s}\\
&\quad = \frac{\rho R^2(1-\upsilon_x\upsilon_\theta)}{E_x}\frac{\partial^2 u}{\partial t^2}\\
&\frac{G_{x\theta}(1-\upsilon_x\upsilon_\theta)+\upsilon_x E_\theta}{E_x}\frac{\partial^2 u}{\partial s\partial\theta} + \frac{G_{x\theta}(1-\upsilon_x\upsilon_\theta)}{E_x}\frac{\partial^2 v}{\partial s^2} + \frac{E_\theta}{E_x}\frac{\partial^2 v}{\partial \theta^2}\\
&\quad + \frac{E_\theta}{E_x}\frac{\partial w}{\partial \theta} = \frac{\rho R^2(1-\upsilon_x\upsilon_\theta)}{E_x}\frac{\partial^2 v}{\partial t^2}\\
&\frac{\upsilon_x E_\theta}{E_x}\frac{\partial u}{\partial s} + \frac{E_\theta}{E_x}\frac{\partial v}{\partial \theta} + \frac{E_\theta}{E_x}w\\
&\quad + k\left[\frac{\partial^4 w}{\partial s^4} + 2\frac{\upsilon_x E_\theta + 2G_{x\theta}(1-\upsilon_x\upsilon_\theta)}{E_x}\frac{\partial^4 w}{\partial s^2\partial\theta^2} + \frac{E_\theta}{E_x}\frac{\partial^4 w}{\partial \theta^4}\right]\\
&\quad = -\frac{\rho R^2(1-\upsilon_x\upsilon_\theta)}{E_x}\frac{\partial^2 w}{\partial t^2}, \quad \text{where } s = x/R, k = h^2/(12R^2), \qquad (4.121)
\end{aligned}$$

and E_x, υ_x, E_θ, υ_θ are the longitudinal and circumferential Young's modulus, Poisson's ratio, respectively. More shell theories can be found in Appendix D.

References

[1] Murty, A.V.K. and Shimpi, R.P., Vibrations of laminated beams, *J. Sound Vibration*, Vol. 36(2), 1974, pp. 273–284.

[2] Teoh, L.S. and Huang, C.C., The vibration of beams of fiber-reinforced material, *J. Sound Vibration*, Vol. 51(4), 1977, pp. 467–473.

[3] Chandrashekhara, K., Krishnamurthy, K. and Roy, S., Free vibration of composite beams including rotary inertia and shear deformation, *Composite Struct.*, Vol. 14, 1990, pp. 269–279.

[4] Singh, G., Rao, G.V. and Iyengar, N.G.R., Analysis of the nonlinear vibrations of unsymmetrically laminated composite beams, *AIAA J.*, Vol. 29(10), 1991, pp. 1727–1735.

[5] Krishnaswamy, S., Chandrashekhara, K. and Wu, W.Z.B., Analytical solutions to vibration of generally layered composite beams, *J. Sound Vibration*, Vol. 159(1), 1992, pp. 85–99.

[6] Chandrashekhara, K. and Bangera, K.M., Free vibration of composite beams using a refined shear flexible beam element, *Computers Struct.*, Vol. 43(4), 1992, pp. 719–727.

[7] Kant, T. and Swaminathan, K., Analytical solutions for free vibration of laminated composite and sandwich plates based on a higher-order refined theory, *Composite Struct.*, Vol. 53, 2001, pp. 73–85.

[8] Aagaah, M.R., Mahinfalah, M. and Jazar, G.N., Natural frequencies of laminated composite plates using third order shear deformation theory, *Composite Struct.*, Vol. 72, 2006, pp. 273–279.

[9] Abramovich, H. and Hamburger, O., Vibration of a uniform cantilever Timoshenko beam with translational and rotational springs and with a tip mass, *J. Sound Vibration*, Vol. 154(1), 1992, pp. 67–80.

[10] Abramovich, H., Shear deformation and rotary inertia effects of vibrating composite beams, *Composite Struct.*, Vol. 20, 1992, pp. 165–173.

[11] Abramovich, H., A note on experimental investigation on a vibrating Timoshenko cantilever beam, *J. Sound Vibration*, Vol. 160(1), 1993, pp. 167–171.

[12] Abramovich, H., Thermal buckling of cross-ply composite laminates using a first-order shear deformation theory, *Composites Struct.*, Vol. 28, 1994, pp. 201–213.

[13] Abramovich, H., Deflection control of laminated composite beams with piezoceramic layers-closed form solutions, *Composite Struct.*, Vol. 43, 1998, pp. 217–231.

[14] Abramovich, H., Eisenberger, M. and Shulepov, O., Vibrations and buckling of cross-ply non-symmetric laminated composite beams, *AIAA J.*, Vol. 34(5), 1996, pp. 1064–1069.

[15] Abramovich, H. and Livshits, A., Free vibrations of non-symmetric cross-ply laminated composite beams, *J. Sound Vibration*, Vol. 176(5), 1994, pp. 597–612.

[16] Abramovich, H. and Livshits, A., Dynamic behavior of cross-ply laminated beams with piezoelectric layers, *Composite Struct.*, Vol. 25(1–4), 1993, pp. 371–379.

[17] Abramovich, H. and Livshits, A., Free vibrations of non-symmetric cross-ply laminated composite beams, *J. Sound Vibration*, Vol. 176(5), 1994, pp. 597–612.

[18] Bert, C.W. and Kumar M., Vibration of cylindrical shells of bimodulus composite materials, *J. Sound Vibrations*, Vol. 81(1), 1982, pp. 107–121.

[19] Soedel, W., Simplified equations and solutions for the vibration of orthotropic cylindrical shells, *J. Sound Vibration*, Vol. 87(4), 1983, pp. 555–566.

[20] Soldatos, K.P., A comparison of some shell theories used for the dynamic analysis of cross-ply laminated circular cylindrical panels, *J. Sound Vibrations*, Vol. 97(2), 1984, pp. 305–319.

[21] Jones, R.M. and Morgan, H.S., Buckling and vibration of cross-ply laminated circular cylindrical shells, *AIAA J.*, Vol. 13(5), 1975, pp. 664–671.

[22] Lim, C.W., Kitipornchai, S. and Liew, K.M., Comparative accuracy of shallow and deep shell theories for vibrations of cylindrical shells, *J. Vibration Control*, Vol. 3, 1997, pp. 119–143.

[23] Farshidianfar, A. and Oliazadeh, P., Free vibration analysis of circular cylindrical shell: comparison of different shell theories, *Int. J. Mech. Appl.*, Vol. 2(5), 2012, pp. 74–80.

[24] Singer, J. and Abramovich, H., Vibration techniques for definition of practical boundary conditions in stiffened shells, *AIAA J.*, Vol. 17(7), 1979, pp. 762–769.

[25] Leissa, W., Vibration of shells, NASA SP-288, US Government Printing Office, Washington, DC, 1973.

[26] Soedel, W., *Vibrations of Shells and Plates*, 3rd Edition, Marcel Dekker, Inc., 2004, 553 p.

[27] DiGiovanni, P.R. and Dugundji, J., Vibrations of freely-supported orthotropic cylindrical shells under internal pressure, AFOSR Scientific Report, AFOSR 65-0640, ASRL TR 112-4 (AD 617 269), Feb. 1965.

Appendix A: General Solution for a Nonsymmetrical Beam Resting on any Boundary Conditions

Using Eq. (4.36), and assuming no axial compression is applied ($\overline{N}_{xx} = 0$), while $I_1 = 0$ (by placing the beam's coordinate system in the middle plane of the beam) and the rotary moment of inertia, assumed to be being negligible ($I_2 = 0$) we have the following matrix notation:

$$\begin{bmatrix} A_{11}\frac{\partial^2}{\partial x^2} & -B_{11}\frac{\partial^3}{\partial x^3} \\ -B_{11}\frac{\partial^3}{\partial x^3} & D_{11}\frac{\partial^4}{\partial x^4} \end{bmatrix} \begin{Bmatrix} u_0(x,t) \\ w_0(x,t) \end{Bmatrix} + \begin{bmatrix} -I_0 & 0 \\ 0 & I_0 \end{bmatrix} \frac{\partial^2}{\partial t^2} \begin{Bmatrix} u_0(x,t) \\ w_0(x,t) \end{Bmatrix} = \begin{Bmatrix} 0 \\ 0 \end{Bmatrix}. \tag{A.1}$$

Assuming that the non-dimensional displacements have the following form:

$$\begin{Bmatrix} \frac{u_0(x,t)}{L} \\ \frac{w_0(x,t)}{L} \end{Bmatrix} = \begin{Bmatrix} U(x)e^{i\omega t} \\ W(x)e^{i\omega t} \end{Bmatrix}, \tag{A.2}$$

while the axial non-dimensional axis being $\xi = x/L$ (L is the length of the beam) and substituting the non-dimensional expressions of the two beam's displacements into Eq. (A.1) we get

$$\begin{bmatrix} \zeta_1^2\frac{d^2}{d\xi^2} + p^2 & -\zeta^2\frac{d^3}{d\xi^3} \\ -\zeta^2\frac{d^3}{d\xi^3} & \frac{d^4}{d\xi^4} - p^2 \end{bmatrix} \begin{Bmatrix} U(\xi) \\ W(\xi) \end{Bmatrix} = \begin{Bmatrix} 0 \\ 0 \end{Bmatrix}, \tag{A.3}$$

where

$$p^2 \equiv \frac{\omega^2 I_0 L^4}{D_{11}}, \quad b^2 \equiv \frac{D_{11}}{KA_{55}L^2}, \quad \zeta^2 \equiv \frac{B_{11}L}{D_{11}}, \quad \zeta_1^2 \equiv \frac{A_{11}L^2}{D_{11}}. \tag{A.4}$$

The characteristic equation of Eq. (A.3) has the following form:

$$(\zeta_1^2 - \zeta^4)s^3 + s^2 - \frac{\zeta_1^2}{p^2}s - \frac{1}{p^2} = 0, \quad s \equiv \frac{m^2}{p^2} \tag{A.5}$$

Solving Eq. (A.5) we can write the general solution for (A.3) in the following form (see a detailed discussion regarding the format of the solution in [17]):

$$\begin{aligned} U &= A_1\gamma\mu\sinh(m_1\xi) + A_2\gamma\mu\cosh(m_1\xi) + A_3\lambda\delta\sin(m_2\xi) \\ &\quad - A_4\lambda\delta\cos(m_2\xi) + A_5\alpha\beta\sin(m_3\xi) - A_6\alpha\beta\cos(m_3\xi), \\ W &= A_1\cosh(m_1\xi) + A_2\sinh(m_1\xi) + A_3\cos(m_2\xi) \\ &\quad + A_4\sin(m_2\xi) + A_5\cos(m_3\xi) + A_6\sin(m_3\xi), \end{aligned} \tag{A.6}$$

Table A.1. Boundary conditions for a non-symmetric laminate beam (CLT approach).

Name	Boundary conditions
Clamped immovable end	$U = W = \frac{dW}{d\xi} = 0$
Clamped movable end	$\zeta_1^2 \frac{dU}{d\xi} - \zeta^2 \frac{d^2W}{d\xi^2} = W = \frac{dW}{d\xi} = 0$
Simply supported immovable end	$U = W = \zeta^2 \frac{dU}{d\xi} - \frac{d^2W}{d\xi^2} = 0$
Simply supported movable end	$\zeta^2 \frac{dU}{d\xi} - \frac{d^2W}{d\xi^2} = W = \zeta_1^2 \frac{dU}{d\xi} - \zeta^2 \frac{d^2W}{d\xi^2} = 0$
Free end	$\zeta^2 \frac{dU}{d\xi} - \frac{d^2W}{d\xi^2} = \zeta_1^2 \frac{d^2U}{d\xi^2} - \zeta^2 \frac{d^3W}{d\xi^3} = \zeta_1^2 \frac{dU}{d\xi} - \zeta^2 \frac{d^2W}{d\xi^2} = 0$

where the various terms in Eq. (A.6) are defined as follows:

$$\begin{aligned}
&m_1 \equiv \sqrt{s_1}p, \quad m_2 \equiv \sqrt{-s_2}p, \quad m_3 \equiv \sqrt{-s_3}p, \\
&\gamma \equiv \frac{\zeta^2 s_1}{\zeta_1^2 s_1 + 1}, \quad \delta \equiv \frac{\zeta^2 s_2}{\zeta_1^2 s_2 + 1}, \quad \beta \equiv \frac{\zeta^2 s_3}{\zeta_1^2 s_3 + 1}, \\
&\mu \equiv \sqrt{s_1}p, \quad \lambda \equiv \sqrt{-s_2}p, \quad \alpha \equiv \sqrt{-s_3}p.
\end{aligned} \tag{A.7}$$

Imposing the adequate boundary conditions would lead to the finding of both the eigenvalues (the natural frequencies) and the eigenvectors (the modes of vibration). For this problem, six boundary conditions should be imposed at each end as presented in Table A.1.

Appendix B: Matrix Notation for the Equilibrium Equations Using CLT Approach

A convenient way of presenting the equilibrium equations at buckling and/or vibration is the following matrix form:

$$\begin{bmatrix} \alpha_{11} & \alpha_{12} & \alpha_{13} \\ \alpha_{21} & \alpha_{22} & \alpha_{23} \\ \alpha_{31} & \alpha_{32} & [\alpha_{33} - N] \end{bmatrix} \begin{Bmatrix} u(x,y,t) \\ v(x,y,t) \\ w_0(x,y,t) \end{Bmatrix}$$

$$+ \frac{\partial^2}{\partial t^2} \begin{bmatrix} m_{11} & 0 & m_{13} \\ 0 & m_{22} & m_{23} \\ m_{13} & m_{23} & m_{33} \end{bmatrix} \begin{Bmatrix} u(x,y,t) \\ v(x,y,t) \\ w_0(x,y,t) \end{Bmatrix} = \begin{Bmatrix} 0 \\ 0 \\ q \end{Bmatrix}, \tag{B.1}$$

where the various operators are given by

$$\alpha_{11} \equiv A_{11} \frac{\partial^2}{\partial x^2} + 2A_{16} \frac{\partial^2}{\partial x \partial y} + A_{66} \frac{\partial^2}{\partial y^2},$$

$$
\begin{aligned}
\alpha_{22} &\equiv A_{22}\frac{\partial^2}{\partial y^2} + 2A_{26}\frac{\partial^2}{\partial x\partial y} + A_{66}\frac{\partial^2}{\partial x^2},\\
\alpha_{33} &\equiv D_{11}\frac{\partial^4}{\partial x^4} + 4D_{16}\frac{\partial^4}{\partial x^3\partial y} + 2(D_{12}+2D_{66})\frac{\partial^4}{\partial x^2\partial y^2}\\
&\quad + 4D_{26}\frac{\partial^4}{\partial x\partial y^3} + D_{22}\frac{\partial^4}{\partial y^4},\\
\alpha_{12} &= \alpha_{21} \equiv A_{16}\frac{\partial^2}{\partial x^2} + (A_{12}+A_{66})\frac{\partial^2}{\partial x\partial y} + A_{26}\frac{\partial^2}{\partial y^2},\\
\alpha_{13} &= \alpha_{31} \equiv -B_{11}\frac{\partial^3}{\partial x^3} - 3B_{16}\frac{\partial^3}{\partial x^2\partial y} - (B_{12}+2B_{66})\frac{\partial^3}{\partial x\partial y^2} - B_{26}\frac{\partial^3}{\partial y^3},\\
\alpha_{23} &= \alpha_{32} \equiv -B_{16}\frac{\partial^3}{\partial x^3} - (B_{12}+2B_{66})\frac{\partial^3}{\partial x^2\partial y} - 3B_{26}\frac{\partial^3}{\partial x\partial y^2} - B_{22}\frac{\partial^3}{\partial y^3},\\
N &\equiv \overline{N}_{xx}\frac{\partial^2}{\partial x^2} + 2\overline{N}_{xy}\frac{\partial^2}{\partial x\partial y} + \overline{N}_{yy}\frac{\partial^2}{\partial y^2},
\end{aligned}
\tag{B.2}
$$

and

$$
\begin{gathered}
m_{11} \equiv -I_0, \quad m_{22} \equiv -\,I_0, \quad m_{33} \equiv I_0 - I_2\left(\frac{\partial^2}{\partial x^2} + \frac{\partial^2}{\partial y^2}\right)\\
m_{13} \equiv I_1\frac{\partial}{\partial x}, \quad m_{23} \equiv I_1\frac{\partial}{\partial y}.
\end{gathered}
\tag{B.3}
$$

Appendix C: Governing Equations According to Various Shell Theories

Donnell–Mushtari theory [25]:

$$
\begin{aligned}
\frac{\partial^2 u}{\partial s^2} + \frac{(1-\upsilon)}{2}\frac{\partial^2 u}{\partial\theta^2} + \frac{(1+\upsilon)}{2}\frac{\partial^2 v}{\partial s\partial\theta} - \upsilon\frac{\partial w}{\partial s} &= \rho\frac{(1-\upsilon^2)}{E}R^2\frac{\partial^2 u}{\partial t^2},\\
\frac{(1+\upsilon)}{2}\frac{\partial^2 u}{\partial s\partial\theta} + \frac{(1-\upsilon)}{2}\frac{\partial^2 v}{\partial s^2} + \frac{\partial^2 v}{\partial\theta^2} + \frac{\partial w}{\partial\theta} &= \rho\frac{(1-\upsilon^2)}{E}R^2\frac{\partial^2 v}{\partial t^2},\\
\upsilon\frac{\partial u}{\partial s} + \frac{\partial v}{\partial\theta} + w + k\left[\frac{\partial^4 w}{\partial s^4} + 2\frac{\partial^4 w}{\partial s^2\partial\theta^2} + \frac{\partial^4 w}{\partial\theta^4}\right] &= -\rho\frac{(1-\upsilon^2)}{E}R^2\frac{\partial^2 w}{\partial t^2},
\end{aligned}
\tag{C.1}
$$

where $s = x/R$ and $k = h^2/(12R^2)$.

To obtain the equations of motion using other shell theories, additional terms are added to Donnell–Mushtari governing equations. These terms will be underlined.

Love-Timoshenko theory [25]:

$$\begin{aligned}
&\frac{\partial^2 u}{\partial s^2} + \frac{(1-\upsilon)}{2}\frac{\partial^2 u}{\partial \theta^2} + \frac{(1+\upsilon)}{2}\frac{\partial^2 v}{\partial s \partial \theta} - \upsilon\frac{\partial w}{\partial s} = \rho\frac{(1-\upsilon^2)}{E}R^2\frac{\partial^2 u}{\partial t^2},\\
&\frac{(1+\upsilon)}{2}\frac{\partial^2 u}{\partial s \partial \theta} + \frac{(1-\upsilon)}{2}\frac{\partial^2 v}{\partial s^2} + \frac{\partial^2 v}{\partial \theta^2} + \underline{k\left[(1-\upsilon)\frac{\partial^2 v}{\partial s^2} + \frac{\partial^2 v}{\partial \theta^2}\right]}\\
&\quad + \frac{\partial w}{\partial \theta} - \underline{k\left[\frac{\partial^3 w}{\partial s^2 \partial \theta} + \frac{\partial^3 w}{\partial \theta^3}\right]} = \rho\frac{(1-\upsilon^2)}{E}R^2\frac{\partial^2 v}{\partial t^2},\\
&\upsilon\frac{\partial u}{\partial s} + \frac{\partial v}{\partial \theta} - \underline{k\left[(2-\upsilon)\frac{\partial^3 v}{\partial s^2 \partial \theta} - \frac{\partial^3 v}{\partial \theta^3}\right]} + w + k\left[\frac{\partial^4 w}{\partial s^4} + 2\frac{\partial^4 w}{\partial s^2 \partial \theta^2} + \frac{\partial^4 w}{\partial \theta^4}\right]\\
&\quad = -\rho\frac{(1-\upsilon^2)}{E}R^2\frac{\partial^2 w}{\partial t^2}. \qquad \text{(C.2)}
\end{aligned}$$

Goldenveizer–Novozhilov (also known as Arnold–Warburton) theory [25]:

$$\begin{aligned}
&\frac{\partial^2 u}{\partial s^2} + \frac{(1-\upsilon)}{2}\frac{\partial^2 u}{\partial \theta^2} + \frac{(1+\upsilon)}{2}\frac{\partial^2 v}{\partial s \partial \theta} - \upsilon\frac{\partial w}{\partial s} = \rho\frac{(1-\upsilon^2)}{E}R^2\frac{\partial^2 u}{\partial t^2},\\
&\frac{(1+\upsilon)}{2}\frac{\partial^2 u}{\partial s \partial \theta} + \frac{(1-\upsilon)}{2}\frac{\partial^2 v}{\partial s^2} + \frac{\partial^2 v}{\partial \theta^2} + \underline{k\left[2(1-\upsilon)\frac{\partial^2 v}{\partial s^2} + \frac{\partial^2 v}{\partial \theta^2}\right]}\\
&\quad + \frac{\partial w}{\partial \theta} - \underline{k\left[(2-\upsilon)\frac{\partial^3 w}{\partial s^2 \partial \theta} + \frac{\partial^3 w}{\partial s^3}\right]} = \rho\frac{(1-\upsilon^2)}{E}R^2\frac{\partial^2 v}{\partial t^2},\\
&\upsilon\frac{\partial u}{\partial s} + \frac{\partial v}{\partial \theta} - \underline{k\left[(2-\upsilon)\frac{\partial^3 v}{\partial s^2 \partial \theta} + \frac{\partial^3 v}{\partial s^3}\right]}\\
&\quad + w + k\left[\frac{\partial^4 w}{\partial s^4} + 2\frac{\partial^4 w}{\partial s^2 \partial \theta^2} + \frac{\partial^4 w}{\partial \theta^4}\right] = -\rho\frac{(1-\upsilon^2)}{E}R^2\frac{\partial^2 w}{\partial t^2}. \qquad \text{(C.3)}
\end{aligned}$$

Houghton–Johns (simplified Goldenveizer–Novozhilov) theory [25]:

$$\begin{aligned}
&\frac{\partial^2 u}{\partial s^2} + \frac{(1-\upsilon)}{2}\frac{\partial^2 u}{\partial \theta^2} + \frac{(1+\upsilon)}{2}\frac{\partial^2 v}{\partial s \partial \theta} - \upsilon\frac{\partial w}{\partial s} = \rho\frac{(1-\upsilon^2)}{E}R^2\frac{\partial^2 u}{\partial t^2},\\
&\frac{(1+\upsilon)}{2}\frac{\partial^2 u}{\partial s \partial \theta} + \frac{(1-\upsilon)}{2}\frac{\partial^2 v}{\partial s^2} + \frac{\partial^2 v}{\partial \theta^2} + \frac{\partial w}{\partial \theta} - \underline{k\left[(2-\upsilon)\frac{\partial^3 w}{\partial s^2 \partial \theta} + \frac{\partial^3 w}{\partial s^3}\right]}\\
&\quad = \rho\frac{(1-\upsilon^2)}{E}R^2\frac{\partial^2 v}{\partial t^2},
\end{aligned}$$

$$\upsilon\frac{\partial u}{\partial s}+\frac{\partial v}{\partial\theta}-\underline{k\left[(2-\upsilon)\frac{\partial^3 v}{\partial s^2\partial\theta}+\frac{\partial^3 v}{\partial s^3}\right]}$$
$$+w+k\left[\frac{\partial^4 w}{\partial s^4}+2\frac{\partial^4 w}{\partial s^2\partial\theta^2}+\frac{\partial^4 w}{\partial\theta^4}\right]=-\rho\frac{(1-\upsilon^2)}{E}R^2\frac{\partial^2 w}{\partial t^2}. \tag{C.4}$$

Flügge–Byrne-Lur'ye (also known as Biezeno–Grammel) theory [25]:

$$\frac{\partial^2 u}{\partial s^2}+\frac{(1-\upsilon)}{2}\frac{\partial^2 u}{\partial\theta^2}+\underline{k\frac{(1-\upsilon)}{2}\frac{\partial^2 u}{\partial\theta^2}}+\frac{(1+\upsilon)}{2}\frac{\partial^2 v}{\partial s\partial\theta}-\upsilon\frac{\partial w}{\partial s}$$
$$\underline{+k\left[-\frac{\partial^3 w}{\partial s^3}+\frac{(1-\upsilon)}{2}\frac{\partial^3 w}{\partial s\partial\theta^2}\right]}=\rho\frac{(1-\upsilon^2)}{E}R^2\frac{\partial^2 u}{\partial t^2},$$
$$\frac{(1+\upsilon)}{2}\frac{\partial^2 u}{\partial s\partial\theta}+\frac{(1-\upsilon)}{2}\frac{\partial^2 v}{\partial s^2}+\frac{\partial^2 v}{\partial\theta^2}+\underline{k\frac{3(1-\upsilon)}{2}\frac{\partial^2 v}{\partial s^2}}$$
$$+\frac{\partial w}{\partial\theta}\underline{-k\frac{(3-\upsilon)}{2}\frac{\partial^3 w}{\partial s^2\partial\theta}}=\rho\frac{(1-\upsilon^2)}{E}R^2\frac{\partial^2 v}{\partial t^2},$$
$$\upsilon\frac{\partial u}{\partial s}+\underline{k\left[-\frac{\partial^3 u}{\partial s^3}+\frac{(1-\upsilon)}{2}\frac{\partial^3 u}{\partial s\partial\theta^2}\right]}+\frac{\partial v}{\partial\theta}-k\frac{(3-\upsilon)}{2}\frac{\partial^3 v}{\partial s^2\partial\theta}$$
$$+w+k\left[\frac{\partial^4 w}{\partial s^4}+2\frac{\partial^4 w}{\partial s^2\partial\theta^2}+\frac{\partial^4 w}{\partial\theta^4}\right]+k\left(w+2\frac{\partial w^2}{\partial\theta^2}\right)$$
$$=-\rho\frac{(1-\upsilon^2)}{E}R^2\frac{\partial^2 w}{\partial t^2}. \tag{C.5}$$

Reissner–Naghdi–Berry theory [25]:

$$\frac{\partial^2 u}{\partial s^2}+\frac{(1-\upsilon)}{2}\frac{\partial^2 u}{\partial\theta^2}+\frac{(1+\upsilon)}{2}\frac{\partial^2 v}{\partial s\partial\theta}-\upsilon\frac{\partial w}{\partial s}=\rho\frac{(1-\upsilon^2)}{E}R^2\frac{\partial^2 u}{\partial t^2},$$
$$\frac{(1+\upsilon)}{2}\frac{\partial^2 u}{\partial s\partial\theta}+\frac{(1-\upsilon)}{2}\frac{\partial^2 v}{\partial s^2}+\frac{\partial^2 v}{\partial\theta^2}+\underline{k\left[\frac{(1-\upsilon)}{2}\frac{\partial^2 v}{\partial s^2}+\frac{\partial^2 v}{\partial\theta^2}\right]}$$
$$+\frac{\partial w}{\partial\theta}-\underline{k\left[\frac{\partial^3 w}{\partial s^2\partial\theta}+\frac{\partial^3 w}{\partial\theta^3}\right]}=\rho\frac{(1-\upsilon^2)}{E}R^2\frac{\partial^2 v}{\partial t^2},$$
$$\upsilon\frac{\partial u}{\partial s}+\frac{\partial v}{\partial\theta}--\underline{k\left[\frac{\partial^3 v}{\partial s^2\partial\theta}+\frac{\partial^3 v}{\partial\theta^3}\right]}$$
$$+w+k\left[\frac{\partial^4 w}{\partial s^4}+2\frac{\partial^4 w}{\partial s^2\partial\theta^2}+\frac{\partial^4 w}{\partial\theta^4}\right]=-\rho\frac{(1-\upsilon^2)}{E}R^2\frac{\partial^2 w}{\partial t^2}. \tag{C.6}$$

Sanders theory [25]:

$$\frac{\partial^2 u}{\partial s^2}+\frac{(1-\upsilon)}{2}\frac{\partial^2 u}{\partial \theta^2}+\underline{k\frac{(1-\upsilon)}{2}\frac{\partial^2 u}{\partial \theta^2}}+\frac{(1+\upsilon)}{2}\frac{\partial^2 v}{\partial s\partial \theta}-\underline{k\frac{3(1-\upsilon)}{8}\frac{\partial^2 v}{\partial s\partial \theta}}$$

$$-\upsilon\frac{\partial w}{\partial s}+\underline{k\frac{(1-\upsilon)}{2}\frac{\partial^3 w}{\partial s\partial \theta^2}}=\rho\frac{(1-\upsilon^2)}{E}R^2\frac{\partial^2 u}{\partial t^2},$$

$$\frac{(1+\upsilon)}{2}\frac{\partial^2 u}{\partial s\partial \theta}-\underline{k\frac{3(1-\upsilon)}{8}\frac{\partial^2 u}{\partial s\partial \theta}}+\frac{(1-\upsilon)}{2}\frac{\partial^2 v}{\partial s^2}+\frac{\partial^2 v}{\partial \theta^2}$$

$$\underline{+k\left[\frac{9(1-\upsilon)}{8}\frac{\partial^2 v}{\partial s^2}+\frac{\partial^2 v}{\partial \theta^2}\right]}$$

$$+\frac{\partial w}{\partial \theta}-\underline{k\left[\frac{(3-\upsilon)}{2}\frac{\partial^3 w}{\partial s^2\partial \theta}+\frac{\partial^3 w}{\partial \theta^3}\right]}=\rho\frac{(1-\upsilon^2)}{E}R^2\frac{\partial^2 v}{\partial t^2},$$

$$\upsilon\frac{\partial u}{\partial s}+\underline{k\frac{(1-\upsilon)}{2}\frac{\partial^3 u}{\partial s\partial \theta^2}}+\frac{\partial v}{\partial \theta}-\underline{k\left[\frac{(3-\upsilon)}{2}\frac{\partial^3 v}{\partial s^2\partial \theta}+\frac{\partial^3 v}{\partial \theta^3}\right]},$$

$$+w+k\left[\frac{\partial^4 w}{\partial s^4}+2\frac{\partial^4 w}{\partial s^2\partial \theta^2}+\frac{\partial^4 w}{\partial \theta^4}\right]=-\rho\frac{(1-\upsilon^2)}{E}R^2\frac{\partial^2 w}{\partial t^2}. \tag{C.7}$$

Vlasov theory [25]:

$$\frac{\partial^2 u}{\partial s^2}+\frac{(1-\upsilon)}{2}\frac{\partial^2 u}{\partial \theta^2}+\frac{(1+\upsilon)}{2}\frac{\partial^2 v}{\partial s\partial \theta}-\upsilon\frac{\partial w}{\partial s}$$

$$\underline{+k\left[-\frac{\partial^3 w}{\partial s^3}+\frac{(1-\upsilon)}{2}\frac{\partial^3 w}{\partial s\partial \theta^2}\right]}=\rho\frac{(1-\upsilon^2)}{E}R^2\frac{\partial^2 u}{\partial t^2},$$

$$\frac{(1+\upsilon)}{2}\frac{\partial^2 u}{\partial s\partial \theta}+\frac{(1-\upsilon)}{2}\frac{\partial^2 v}{\partial s^2}+\frac{\partial^2 v}{\partial \theta^2}$$

$$+\frac{\partial w}{\partial \theta}-\underline{k\left[\frac{(3-\upsilon)}{2}\frac{\partial^3 w}{\partial s^2\partial \theta}\right]}=\rho\frac{(1-\upsilon^2)}{E}R^2\frac{\partial^2 v}{\partial t^2},$$

$$\upsilon\frac{\partial u}{\partial s}+\underline{k\left[-\frac{\partial^3 w}{\partial s^3}+\frac{(1-\upsilon)}{2}\frac{\partial^3 w}{\partial s\partial \theta^2}\right]}\frac{\partial v}{\partial \theta}-\underline{k\left[\frac{(3-\upsilon)}{2}\frac{\partial^3 v}{\partial s^2\partial \theta}\right]}$$

$$+w+k\left[\frac{\partial^4 w}{\partial s^4}+2\frac{\partial^4 w}{\partial s^2\partial \theta^2}+\frac{\partial^4 w}{\partial \theta^4}\right]$$

$$\underline{+k\left[w+2\frac{\partial^2 w}{\partial \theta^2}\right]}=-\rho\frac{(1-\upsilon^2)}{E}R^2\frac{\partial^2 w}{\partial t^2}. \tag{C.8}$$

Kennard simplified theory [25]:

$$\frac{\partial^2 u}{\partial s^2} + \frac{(1-\upsilon)}{2}\frac{\partial^2 u}{\partial \theta^2} + \frac{(1+\upsilon)}{2}\frac{\partial^2 v}{\partial s \partial \theta} - \upsilon \frac{\partial w}{\partial s} = \rho \frac{(1-\upsilon^2)}{E} R^2 \frac{\partial^2 u}{\partial t^2},$$

$$\frac{(1+\upsilon)}{2}\frac{\partial^2 u}{\partial s \partial \theta} + \frac{(1-\upsilon)}{2}\frac{\partial^2 v}{\partial s^2} + \frac{\partial^2 v}{\partial \theta^2} + \frac{\partial w}{\partial \theta}$$
$$\underline{+k\left[\frac{3\upsilon}{2(1-\upsilon)}\frac{\partial w}{\partial \theta} + \frac{3\upsilon}{2(1-\upsilon)}\frac{\partial^3 w}{\partial \theta^3}\right]} = \rho \frac{(1-\upsilon^2)}{E} R^2 \frac{\partial^2 v}{\partial t^2},$$

$$\upsilon \frac{\partial u}{\partial s} + \frac{\partial v}{\partial \theta} + w + k\left[\frac{\partial^4 w}{\partial s^4} + 2\frac{\partial^4 w}{\partial s^2 \partial \theta^2} + \frac{\partial^4 w}{\partial \theta^4}\right]$$
$$\underline{+k\left[\frac{(2+\upsilon)}{2(1-\upsilon)} w + \frac{(4-\upsilon)}{2(1-\upsilon)}\frac{\partial^2 w}{\partial \theta^2}\right]} = -\rho \frac{(1-\upsilon^2)}{E} R^2 \frac{\partial^2 w}{\partial t^2}. \quad \text{(C.9)}$$

Appendix D: Governing Equations for Orthotropic Circular Cylindrical Shells

As developed by DiGiovanni @Dugundji [27] and published also in [25], additional equations of motion can be presented based on the Donnell–Mushtari equations with additional terms being added in accordance to the theory presented.

Donnell–Mushtari equations are again written in a compact form using the following notations:

$$A_{11} = \frac{E_x h}{1-\upsilon_x \upsilon_\theta}, \quad A_{22} = \frac{E_\theta h}{1-\upsilon_x \upsilon_\theta}, \quad A_{12} = A_{21} = \frac{\upsilon_\theta E_x h}{1-\upsilon_x \upsilon_\theta} = \frac{\upsilon_x E_x h}{1-\upsilon_x \upsilon_\theta},$$
$$A_{66} = Gh, \quad \text{(D.1)}$$

$$D_{11} = \frac{E_x h^3}{12(1-\upsilon_x \upsilon_\theta)}, \quad D_{22} = \frac{E_\theta h^3}{12(1-\upsilon_x \upsilon_\theta)},$$

$$D_{12} = D_{21} = \frac{\upsilon_\theta E_x h^3}{12(1-\upsilon_x \upsilon_\theta)} = \frac{\upsilon_x E_x h^3}{12(1-\upsilon_x \upsilon_\theta)}, \quad D_{66} = \frac{G_{x\theta} h^3}{12}. \quad \text{(D.2)}$$

The coupling terms A_{16}, A_{26} and D_{16}, D_{26} are function of the angle θ and are given as

$$A_{16} = \left[(Q_{11} - Q_{12} - 2Q_{66})\sin\alpha\cos^3\alpha - (Q_{22} - Q_{12} - 2Q_{66})\sin^3\alpha\cos\alpha\right] h,$$
$$A_{26} = \left[(Q_{11} - Q_{12} - 2Q_{66})\sin^3\alpha\cos\alpha - (Q_{22} - Q_{12} - 2Q_{66})\sin\alpha\cos^3\alpha\right] h,$$

$$D_{16} = \left[(Q_{11} - Q_{12} - 2Q_{66})\sin\alpha\cos^3\alpha - (Q_{22} - Q_{12} - 2Q_{66})\sin^3\alpha\cos\alpha\right]\frac{h^3}{12},$$
$$D_{26} = \left[(Q_{11} - Q_{12} - 2Q_{66})\sin^3\alpha\cos\alpha - (Q_{22} - Q_{12} - 2Q_{66})\sin\alpha\cos^3\alpha\right]\frac{h^3}{12}, \quad \text{(D.3)}$$

where

$$Q_{11} = \frac{E_x}{1 - \upsilon_x\upsilon_\theta},\ Q_{22} = \frac{E_\theta}{1 - \upsilon_x\upsilon_\theta},$$
$$Q_{12} = Q_{12} = \frac{\upsilon_\theta E_x}{1 - \upsilon_x\upsilon_0} = \frac{\upsilon_x E_\theta}{1 - \upsilon_x\upsilon_\theta},\ Q_{66} = G_{x\theta}.$$

Then, the Donnell–Mushtari equations will look like

$$\frac{A_{11}}{A_{22}}\frac{\partial^2 u}{\partial s^2} + 2\frac{A_{16}}{A_{22}}\frac{\partial^2 u}{\partial s\partial\theta} + \frac{A_{66}}{A_{22}}\frac{\partial^2 u}{\partial\theta^2} + \frac{A_{16}}{A_{22}}\frac{\partial^2 v}{\partial s^2} + \left(\frac{A_{12} + A_{66}}{A_{22}}\right)\frac{\partial^2 v}{\partial s\partial\theta} + \frac{A_{26}}{A_{22}}\frac{\partial^2 v}{\partial\theta^2} + \frac{A_{12}}{A_{22}}\frac{\partial w}{\partial s} + \frac{A_{26}}{A_{22}}\frac{\partial w}{\partial\theta} = \frac{\rho h R^2}{A_{22}}\frac{\partial^2 u}{\partial t^2},$$

$$\frac{A_{16}}{A_{22}}\frac{\partial^2 u}{\partial s^2} + (\frac{A_{12} + A_{66}}{A_{22}})\frac{\partial^2 u}{\partial s\partial\theta} + \frac{A_{26}}{A_{22}}\frac{\partial^2 u}{\partial\theta^2} + \frac{A_{66}}{A_{22}}\frac{\partial^2 v}{\partial s^2} + 2\frac{A_{26}}{A_{22}}\frac{\partial^2 v}{\partial s\partial\theta} + \frac{\partial^2 v}{\partial\theta^2} + \frac{A_{26}}{A_{22}}\frac{\partial w}{\partial s} + \frac{\partial w}{\partial\theta} = \frac{\rho h R^2}{A_{22}}\frac{\partial^2 v}{\partial t^2},$$

$$\frac{A_{12}}{A_{22}}\frac{\partial u}{\partial s} + \frac{A_{26}}{A_{22}}\frac{\partial u}{\partial\theta} + \frac{A_{26}}{A_{22}}\frac{\partial v}{\partial s} + \frac{\partial v}{\partial\theta} + w + k\left[\frac{D_{11}}{D_{22}}\frac{\partial^4 w}{\partial s^4} + 2\left(\frac{D_{12} + 2D_{66}}{D_{22}}\right)\frac{\partial^4 w}{\partial s^2\partial\theta^2} + \frac{\partial^4 w}{\partial\theta^4}\right] = -\frac{\rho h R^2}{A_{22}}\frac{\partial^2 w}{\partial t^2}. \quad \text{(D.4)}$$

The governing equations for the Love–Timoshenko theory [25] are

$$\frac{A_{11}}{A_{22}}\frac{\partial^2 u}{\partial s^2} + 2\frac{A_{16}}{A_{22}}\frac{\partial^2 u}{\partial s\partial\theta} + \frac{A_{66}}{A_{22}}\frac{\partial^2 u}{\partial\theta^2} + \frac{A_{16}}{A_{22}}\frac{\partial^2 v}{\partial s^2} + \left(\frac{A_{12} + A_{66}}{A_{22}}\right)\frac{\partial^2 v}{\partial s\partial\theta} + \frac{A_{26}}{A_{22}}\frac{\partial^2 v}{\partial\theta^2} + \frac{A_{12}}{A_{22}}\frac{\partial w}{\partial s} + \frac{A_{26}}{A_{22}}\frac{\partial w}{\partial\theta} = \frac{\rho h R^2}{A_{22}}\frac{\partial^2 u}{\partial t^2},$$

$$\begin{aligned}
&\frac{A_{16}}{A_{22}}\frac{\partial^2 u}{\partial s^2} + \left(\frac{A_{12}+A_{66}}{A_{22}}\right)\frac{\partial^2 u}{\partial s\partial\theta} + \frac{A_{26}}{A_{22}}\frac{\partial^2 u}{\partial\theta^2} + \frac{A_{66}}{A_{22}}\frac{\partial^2 v}{\partial s^2} + 2\frac{A_{26}}{A_{22}}\frac{\partial^2 v}{\partial s\partial\theta} + \frac{\partial^2 v}{\partial\theta^2}\\
&\quad + \frac{A_{26}}{A_{22}}\frac{\partial w}{\partial s} + \frac{\partial w}{\partial\theta} + \underline{2\frac{D_{66}}{D_{22}}\frac{\partial^2 v}{\partial s^2}} + \underline{3\frac{D_{26}}{D_{22}}\frac{\partial^2 v}{\partial s\partial\theta}} + \underline{\frac{\partial^2 v}{\partial\theta^2}}\\
&\quad - \underline{\frac{D_{16}}{D_{22}}\frac{\partial^3 w}{\partial s^3}} - \underline{\left(\frac{D_{12}+2D_{66}}{D_{22}}\right)\frac{\partial^3 w}{\partial s^2\partial\theta}} - \underline{3\frac{D_{26}}{D_{22}}\frac{\partial^3 w}{\partial s\partial\theta^2}} = \frac{\rho h R^2}{A_{22}}\frac{\partial^2 v}{\partial t^2},\\
&\frac{A_{12}}{A_{22}}\frac{\partial u}{\partial s} + \frac{A_{26}}{A_{22}}\frac{\partial u}{\partial\theta} + \frac{A_{26}}{A_{22}}\frac{\partial v}{\partial s} + \frac{\partial v}{\partial\theta} - \underline{2\frac{D_{16}}{D_{22}}\frac{\partial^3 v}{\partial s^3}}\\
&\quad - \underline{\left(\frac{D_{12}+4D_{66}}{D_{22}}\right)\frac{\partial^3 v}{\partial s^2\partial\theta}} - \underline{4\frac{D_{26}}{D_{22}}\frac{\partial^3 v}{\partial s\partial\theta^2}}\\
&\quad - \underline{\frac{\partial^3 v}{\partial\theta^3}} + w + k\left[\frac{D_{11}}{D_{22}}\frac{\partial^4 w}{\partial s^4} + 2\left(\frac{D_{12}+2D_{66}}{D_{22}}\right)\frac{\partial^4 w}{\partial s^2\partial\theta^2} + \frac{\partial^4 w}{\partial\theta^4}\right]\\
&\quad + \underline{4\frac{D_{16}}{D_{22}}\frac{\partial^4 w}{\partial s^3\partial\theta} + \underline{4\frac{D_{26}}{D_{22}}\frac{\partial^4 w}{\partial s\partial\theta^3}}} = -\frac{\rho h R^2}{A_{22}}\frac{\partial^2 w}{\partial t^2} \qquad \text{(D.5)}
\end{aligned}$$

where the underlined terms are additions due to Love–Timoshenko theory to the Donnell–Mushtari equations.

The governing equations for the Goldenveizer–Novozhilov theory [25] are

$$\begin{aligned}
&\frac{A_{11}}{A_{22}}\frac{\partial^2 u}{\partial s^2} + 2\frac{A_{16}}{A_{22}}\frac{\partial^2 u}{\partial s\partial\theta} + \frac{A_{66}}{A_{22}}\frac{\partial^2 u}{\partial\theta^2} + \frac{A_{16}}{A_{22}}\frac{\partial^2 v}{\partial s^2}\\
&\quad + \left(\frac{A_{12}+A_{66}}{A_{22}}\right)\frac{\partial^2 v}{\partial s\partial\theta} + \frac{A_{26}}{A_{22}}\frac{\partial^2 v}{\partial\theta^2} + \frac{A_{12}}{A_{22}}\frac{\partial w}{\partial s} + \frac{A_{26}}{A_{22}}\frac{\partial w}{\partial\theta} = \frac{\rho h R^2}{A_{22}}\frac{\partial^2 u}{\partial t^2},\\
&\frac{A_{16}}{A_{22}}\frac{\partial^2 u}{\partial s^2} + \left(\frac{A_{12}+A_{66}}{A_{22}}\right)\frac{\partial^2 u}{\partial s\partial\theta} + \frac{A_{26}}{A_{22}}\frac{\partial^2 u}{\partial\theta^2} + \frac{A_{66}}{A_{22}}\frac{\partial^2 v}{\partial s^2} + 2\frac{A_{26}}{A_{22}}\frac{\partial^2 v}{\partial s\partial\theta} + \frac{\partial^2 v}{\partial\theta^2}\\
&\quad + \underline{4\frac{D_{66}}{D_{22}}\frac{\partial^2 v}{\partial s^2}} + \underline{4\frac{D_{26}}{D_{22}}\frac{\partial^2 v}{\partial s\partial\theta}} + \underline{\frac{\partial^2 v}{\partial\theta^2}}\\
&\quad - \underline{2\frac{D_{16}}{D_{22}}\frac{\partial^3 w}{\partial s^3}} - \underline{\left(\frac{D_{12}+4D_{66}}{D_{22}}\right)\frac{\partial^3 w}{\partial s^2\partial\theta}} - \underline{4\frac{D_{26}}{D_{22}}\frac{\partial^3 w}{\partial s\partial\theta^2}}\\
&\quad - \underline{\frac{\partial^3 w}{\partial\theta^3}} + \frac{A_{26}}{A_{22}}\frac{\partial w}{\partial s} + \frac{\partial w}{\partial\theta} = \frac{\rho h R^2}{A_{22}}\frac{\partial^2 v}{\partial t^2},\\
&\frac{A_{12}}{A_{22}}\frac{\partial u}{\partial s} + \frac{A_{26}}{A_{22}}\frac{\partial u}{\partial\theta} + \frac{A_{26}}{A_{22}}\frac{\partial v}{\partial s} + \frac{\partial v}{\partial\theta} - \underline{2\frac{D_{16}}{D_{22}}\frac{\partial^3 v}{\partial s^3}} - \underline{\left(\frac{D_{12}+4D_{66}}{D_{22}}\right)\frac{\partial^3 v}{\partial s^2\partial\theta}}
\end{aligned}$$

$$
\begin{aligned}
&\underline{-4\frac{D_{26}}{D_{22}}\frac{\partial^3 v}{\partial s\partial\theta^2}} - \underline{\frac{\partial^3 v}{\partial\theta^3}} + \underline{4\frac{D_{16}}{D_{22}}\frac{\partial^4 w}{\partial s^3\partial\theta}} + 4\frac{D_{26}}{D_{22}}\frac{\partial^4 w}{\partial s\partial\theta^3}\\
&+ w + k\left[\frac{D_{11}}{D_{22}}\frac{\partial^4 w}{\partial s^4} + 2\left(\frac{D_{12}+2D_{66}}{D_{22}}\right)\frac{\partial^4 w}{\partial s^2\partial\theta^2} + \frac{\partial^4 w}{\partial\theta^4}\right]\\
&= -\frac{\rho h R^2}{A_{22}}\frac{\partial^2 w}{\partial t^2}
\end{aligned}
\tag{D.6}
$$

where, as before, the underlined terms are additions due to Goldenveizer–Novozhilov theory to the Donnell–Mushtari equations.

The governing equations for the Flügge–Byrne–Lur'ye theory [25] are

$$
\begin{aligned}
&\frac{A_{11}}{A_{22}}\frac{\partial^2 u}{\partial s^2} + 2\frac{A_{16}}{A_{22}}\frac{\partial^2 u}{\partial s\partial\theta} + \frac{A_{66}}{A_{22}}\frac{\partial^2 u}{\partial\theta^2} + \underline{\frac{D_{66}}{D_{22}}\frac{\partial^2 u}{\partial\theta^2}} + \frac{A_{16}}{A_{22}}\frac{\partial^2 v}{\partial s^2}\\
&\quad + \left(\frac{A_{12}+A_{66}}{A_{22}}\right)\frac{\partial^2 v}{\partial s\partial\theta} + \frac{A_{26}}{A_{22}}\frac{\partial^2 v}{\partial\theta^2},\\
&\underline{\frac{D_{16}}{D_{22}}\frac{\partial^2 v}{\partial s^2}} + \frac{A_{12}}{A_{22}}\frac{\partial w}{\partial s} + \frac{A_{26}}{A_{22}}\frac{\partial w}{\partial\theta} - \underline{\frac{D_{11}}{D_{22}}\frac{\partial^3 w}{\partial s^3}} - \underline{\frac{D_{16}}{D_{22}}\frac{\partial^3 w}{\partial s^2\partial\theta}}\\
&\quad + \underline{\frac{D_{66}}{D_{22}}\frac{\partial^3 w}{\partial s\partial\theta^2}} + \underline{\frac{D_{26}}{D_{22}}\frac{\partial^3 w}{\partial\theta^3}} + \underline{\frac{D_{26}}{D_{22}}\frac{\partial w}{\partial\theta}} = \frac{\rho h R^2}{A_{22}}\frac{\partial^2 u}{\partial t^2},\\
&\frac{A_{16}}{A_{22}}\frac{\partial^2 u}{\partial s^2} + \left(\frac{A_{12}+A_{66}}{A_{22}}\right)\frac{\partial^2 u}{\partial s\partial\theta} + \frac{A_{26}}{A_{22}}\frac{\partial^2 u}{\partial\theta^2} + \underline{\frac{D_{16}}{D_{22}}\frac{\partial^2 u}{\partial s^2}}\\
&\quad + \frac{A_{66}}{A_{22}}\frac{\partial^2 v}{\partial s^2} + 2\frac{A_{26}}{A_{22}}\frac{\partial^2 v}{\partial s\partial\theta} + \frac{\partial^2 v}{\partial\theta^2}\\
&\quad + \underline{3\frac{D_{66}}{D_{22}}\frac{\partial^2 v}{\partial s^2}} + \underline{2\frac{D_{26}}{D_{22}}\frac{\partial^2 v}{\partial s\partial\theta}} + \frac{A_{26}}{A_{22}}\frac{\partial w}{\partial s} + \frac{\partial w}{\partial\theta} - \underline{2\frac{D_{16}}{D_{22}}\frac{\partial^3 w}{\partial s^3}}\\
&\quad - \left(\frac{D_{12}+3D_{66}}{D_{22}}\right)\frac{\partial^3 w}{\partial s^2\partial\theta}\\
&\quad - \underline{2\frac{D_{26}}{D_{22}}\frac{\partial^3 w}{\partial s\partial\theta^2}} = \frac{\rho h R^2}{A_{22}}\frac{\partial^2 v}{\partial t^2},\\
&\frac{A_{12}}{A_{22}}\frac{\partial u}{\partial s} + \frac{A_{26}}{A_{22}}\frac{\partial u}{\partial\theta} - \underline{\frac{D_{11}}{D_{22}}\frac{\partial^3 u}{\partial s^3}} - \underline{\frac{D_{16}}{D_{22}}\frac{\partial^3 u}{\partial s^2\partial\theta}} + \underline{\frac{D_{66}}{D_{22}}\frac{\partial^3 u}{\partial s\partial\theta^2}} + \underline{\frac{D_{26}}{D_{22}}\frac{\partial u}{\partial\theta}}\\
&\frac{A_{26}}{A_{22}}\frac{\partial v}{\partial s} + \frac{\partial v}{\partial\theta} - \underline{2\frac{D_{16}}{D_{22}}\frac{\partial^3 v}{\partial s^3}} - \underline{\left(\frac{D_{12}+3D_{66}}{D_{22}}\right)\frac{\partial^3 v}{\partial s^2\partial\theta}} - \underline{2\frac{D_{26}}{D_{22}}\frac{\partial^3 v}{\partial s\partial\theta^2}}
\end{aligned}
$$

$$
\begin{aligned}
&+w+k\left[\frac{D_{11}}{D_{22}}\frac{\partial^4 w}{\partial s^4}+2\left(\frac{D_{12}+2D_{66}}{D_{22}}\right)\frac{\partial^4 w}{\partial s^2\partial\theta^2}+\frac{\partial^4 w}{\partial\theta^4}\right]\\
&+\underline{4\frac{D_{16}}{D_{22}}\frac{\partial^4 w}{\partial s^3\partial\theta}}+\underline{4\frac{D_{26}}{D_{22}}\frac{\partial^4 w}{\partial s\partial\theta^3}}+\underline{2\frac{D_{26}}{D_{22}}\frac{\partial^2 w}{\partial s\partial\theta}}\\
&+\underline{2\frac{\partial^2 w}{\partial\theta^2}}+\underline{w}=-\frac{\rho hR^2}{A_{22}}\frac{\partial^2 w}{\partial t^2}
\end{aligned}
\tag{D.7}
$$

where the underlined terms are additions due to Flügge–Byrne–Lur'ye theory to the Donnell–Mushtari equations.

The governing equations for the Sanders theory [25] are

$$
\begin{aligned}
&\frac{A_{11}}{A_{22}}\frac{\partial^2 u}{\partial s^2}+2\frac{A_{16}}{A_{22}}\frac{\partial^2 u}{\partial s\partial\theta}+\frac{A_{66}}{A_{22}}\frac{\partial^2 u}{\partial\theta^2}+\underline{\frac{1}{4}\frac{D_{66}}{D_{22}}\frac{\partial^2 u}{\partial\theta^2}}\\
&\quad+\frac{A_{16}}{A_{22}}\frac{\partial^2 v}{\partial s^2}+\left(\frac{A_{12}+A_{66}}{A_{22}}\right)\frac{\partial^2 v}{\partial s\partial\theta}\\
&\quad+\frac{A_{26}}{A_{22}}\frac{\partial^2 v}{\partial\theta^2}-\underline{\frac{3}{4}\frac{D_{66}}{D_{22}}\frac{\partial^2 v}{\partial s\partial\theta}}-\underline{\frac{1}{2}\frac{D_{26}}{D_{22}}\frac{\partial^2 v}{\partial\theta^2}}+\frac{A_{12}}{A_{22}}\frac{\partial w}{\partial s}+\frac{A_{26}}{A_{22}}\frac{\partial w}{\partial\theta}\\
&\quad+\underline{\frac{1}{2}\frac{D_{16}}{D_{22}}\frac{\partial^3 w}{\partial s^2\partial\theta}}+\underline{\frac{D_{66}}{D_{22}}\frac{\partial^3 w}{\partial s\partial\theta^2}}+\underline{\frac{1}{2}\frac{D_{26}}{D_{22}}\frac{\partial^3 w}{\partial\theta^3}}=\frac{\rho hR^2}{A_{22}}\frac{\partial^2 u}{\partial t^2},\\
&\frac{A_{16}}{A_{22}}\frac{\partial^2 u}{\partial s^2}+\left(\frac{A_{12}+A_{66}}{A_{22}}\right)\frac{\partial^2 u}{\partial s\partial\theta}+\frac{A_{26}}{A_{22}}\frac{\partial^2 u}{\partial\theta^2}-\underline{\frac{3}{4}\frac{D_{66}}{D_{22}}\frac{\partial^2 u}{\partial s\partial\theta}}-\underline{\frac{1}{2}\frac{D_{26}}{D_{22}}\frac{\partial^2 u}{\partial\theta^2}}\\
&\quad+\frac{A_{66}}{A_{22}}\frac{\partial^2 v}{\partial s^2}+2\frac{A_{26}}{A_{22}}\frac{\partial^2 v}{\partial s\partial\theta}+\frac{\partial^2 v}{\partial\theta^2}+\underline{\frac{9}{4}\frac{D_{66}}{D_{22}}\frac{\partial^2 v}{\partial s^2}}+\underline{3\frac{D_{26}}{D_{22}}v}+\underline{\frac{\partial^2 v}{\partial\theta^2}}\\
&\quad+\frac{A_{26}}{A_{22}}\frac{\partial w}{\partial s}+\frac{\partial w}{\partial\theta}-\underline{\frac{3}{2}\frac{D_{16}}{D_{22}}\frac{\partial^3 w}{\partial s^3}}-\underline{\left(\frac{D_{12}+3D_{66}}{A_{22}}\right)\frac{\partial^3 w}{\partial s^2\partial\theta}}\\
&\quad-\underline{\frac{7}{2}\frac{D_{26}}{D_{22}}\frac{\partial^3 w}{\partial s\partial\theta^2}}-\underline{\frac{\partial^3 w}{\partial\theta^3}}=\frac{\rho hR^2}{A_{22}}\frac{\partial^2 v}{\partial t^2},\\
&\frac{A_{12}}{A_{22}}\frac{\partial u}{\partial s}+\frac{A_{26}}{A_{22}}\frac{\partial u}{\partial\theta}+\underline{\frac{1}{2}\frac{D_{16}}{D_{22}}\frac{\partial^3 u}{\partial s^2\partial\theta}}+\underline{\frac{D_{66}}{D_{22}}\frac{\partial^3 u}{\partial s\partial\theta^2}}+\underline{\frac{1}{2}\frac{D_{26}}{D_{22}}\frac{\partial^3 u}{\partial\theta^3}},\\
&\frac{A_{26}}{A_{22}}\frac{\partial v}{\partial s}+\frac{\partial v}{\partial\theta}-\underline{\frac{3}{2}\frac{D_{16}}{D_{22}}\frac{\partial^3 v}{\partial s^3}}-\underline{\left(\frac{D_{12}+3D_{66}}{A_{22}}\right)\frac{\partial^3 v}{\partial s^2\partial\theta}}-\underline{\frac{7}{2}\frac{D_{26}}{D_{22}}\frac{\partial^3 v}{\partial s\partial\theta^2}}-\underline{\frac{\partial^3 v}{\partial\theta^3}}
\end{aligned}
$$

$$+w+k\left[\frac{D_{11}}{D_{22}}\frac{\partial^4 w}{\partial s^4}+2\left(\frac{D_{12}+2D_{66}}{D_{22}}\right)\frac{\partial^4 w}{\partial s^2\partial\theta^2}+\frac{\partial^4 w}{\partial\theta^4}\right]$$

$$+\underline{4\frac{D_{16}}{D_{22}}\frac{\partial^4 w}{\partial s^3\partial\theta}}+\underline{4\frac{D_{26}}{D_{22}}\frac{\partial^4 w}{\partial s\partial\theta^3}}=-\frac{\rho h R^2}{A_{22}}\frac{\partial^2 w}{\partial t^2} \tag{D.8}$$

where the underlined terms are additions due to Sanders theory to the Donnell–Mushtari equations.

Chapter 5

The Vibration Correlation Technique

5.1. Introduction

The Vibration Correlation Technique (VCT) is a non-destructive *in-situ* method to assess buckling loads of thin walled-structures.

The term was proposed by the late Prof. Joseph Singer [1] and consists of measuring the natural frequencies of a loaded structure, and monitoring their change, while increasing the applied load.

Assuming that the vibrational modes are similar to the buckling ones, one can draw a curve, displaying the natural frequencies squared vs. the applied load, and extrapolating the curve to zero frequency would yield the predicted buckling load of the tested structure.

The VCT is usually classified in two main groups according to their approach:

(1) Determination of *in-situ* boundary conditions (Fig. 5.1).
(2) Direct prediction of buckling loads (Fig. 5.3).

Figure 5.1 displays the way the boundary conditions are determined for a column under compressive load. A column having boundary conditions between hinged–hinged and clamped–clamped is tested under compression, with the first bending frequency being recorded for increasing axial loads. To ensure the non-destructiveness of the procedure the acquisition of the natural frequencies is stopped at 50% of the critical buckling load, P_{cr} (see Fig. 5.1(a)). Then the buckling load is extrapolated, to yield the predicted buckling load, P_{extrap} (see Fig. 5.1(a)). One should note that the experimental determination of the boundary conditions for more advanced structures, like shells, follows the above described method and will be discussed below, in the section devoted to shells.

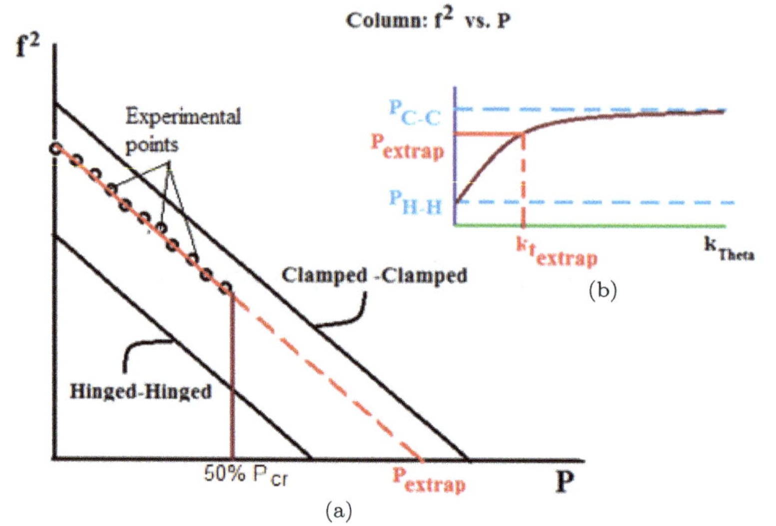

Fig. 5.1. Determination of *in-situ* boundary conditions of a thin-walled column: (a) extrapolating the buckling load; (b) determining the *in-situ* torsional spring constant.

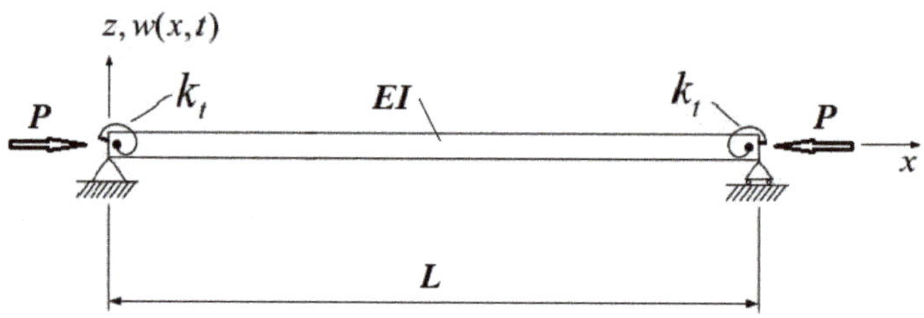

Fig. 5.2. A compressed column with torsional springs at its both ends.

To be able to define the *in-situ* boundary conditions of the tested column, a model is constructed as presented in Fig. 5.2. It consists of a column, having a stiffness of EI, and a length L. The column has two torsional springs, k_t, and zero deflection at its both ends.

When $k_t \to 0$ the boundary condition will be hinged–hinged while for $k_t \to \infty$ the boundary conditions will become clamped–clamped. A general solution for the buckling load of a column (see Fig. 5.2) is then obtained and the influence of the torsional spring constant is calculated, as schematically plotted in Fig. 5.1(b). Entering the value of the extrapolated buckling load, P_{extrap}, obtained before, yields the k_{textrap} as depicted in Fig. 5.1(b). The direct prediction of the buckling load is presented in Fig. 5.3. The first natural frequencies squared of an axially loaded column are plotted as a function of the compressive force, P. A straight line is then drawn (using least mean

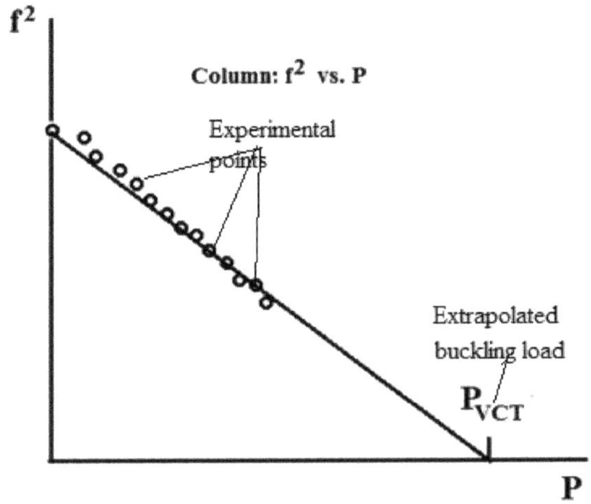

Fig. 5.3. Direct prediction of buckling load for an *in-situ* thin-walled column.

squares) and extrapolating for $f^2 = 0$, yields the experimentally predicted buckling load P_{VCT} (see Fig. 5.3). Note that a reliable buckling prediction would be for values of $P_{\mathrm{VCT}} \leq P_{\mathrm{exp.}}$, where $P_{\mathrm{exp.}}$ denotes the experimental buckling load of the tested column.

5.2. Columns, Beams and Rods

In his seminal works from 1950–1952, Lurie [2–4] claims that mathematically the equations of motion for both the stability and natural frequencies for columns and plates are similar and some kind of correlation between the compression load and the natural frequencies of the same thin-walled structure should exist. To show this correlation, the equation of motion for an uniform isotropic column with a length L, axially compressed by a load P (compression), with a bending stiffness of EI and having a mass per unit length (ρA), where A and E are the cross-section of the column and its Young's modulus, respectively, is next presented

$$EI\frac{\partial^4 w(x,t)}{\partial x^4} + P\frac{\partial^2 w(x,t)}{\partial x^2} + \rho A\frac{\partial^2 w(x,t)}{\partial t^2} = 0, \tag{5.1}$$

where $w(x,t)$ is the out-of-plane displacement. Assuming harmonic vibrations, the solution of Eq. (5.1) has the following form:

$$w(x,t) = W(x)e^{i\omega t}, \tag{5.2}$$

where ω is the angular (or circular) frequency of the column and $W(x)$ is a function to be next determined. Substituting Eq. (5.2) into Eq. (5.1) provides the following expression:

$$EIW_{,xxxx} + PW_{,xx} - \rho A\omega^2 W = 0, \tag{5.3}$$

where $[\ ]_{,xxxx}$ and $[\ ]_{,xx}$ are four and two times differentiation with respect with x, respectively. Division of Eq. (5.3) by EI and defining the following terms as

$$\lambda^2 = \frac{P}{EI}, \quad c^2 = \frac{\rho A}{EI}, \tag{5.4}$$

we get the following differential equation:

$$W_{,xxxx} + \lambda^2 W_{,xx} - c^2\omega^2 W = 0. \tag{5.5}$$

The differential equation 5.5 can be transformed in a characteristic algebraic equation by assuming $W(x) = \widetilde{W}(x)e^{\alpha x}$, yielding,

$$\alpha^4 + \lambda^2\alpha^2 - c^2\omega^2 = 0 \tag{5.6}$$

The solution for the characteristic equation presented by Eq. (5.6) is

$$\alpha_{1,2}^2 = -\frac{\lambda^2}{2} \pm \sqrt{\left(\frac{\lambda^2}{2}\right)^2 + c^2\omega^2}. \tag{5.7}$$

Therefore, the general solution for Eq. (5.5) has the following form:

$$W(x) = C_1\cosh(\alpha_1 x) + C_2\sinh(\alpha_1 x) + C_3\cos(\alpha_2 x) + C_4\sin(\alpha_2 x), \tag{5.8}$$

where

$$\alpha_1 = \sqrt{-\frac{\lambda^2}{2} + \sqrt{\left(\frac{\lambda^2}{2}\right)^2 + c^2\omega^2}}, \quad \alpha_2 = \sqrt{\frac{\lambda^2}{2} + \sqrt{\left(\frac{\lambda^2}{2}\right)^2 + c^2\omega^2}}. \tag{5.9}$$

The constants C_1–C_4 in Eq. (5.8) are to be found by imposing the adequate boundary conditions. For a column with simply supported boundary conditions which can be written as

$$\begin{aligned} @x = 0: \quad & W(0) = W_{,xx}(0) = 0, \\ @x = L: \quad & W(L) = W_{,xx}(L) = 0, \end{aligned} \tag{5.10}$$

we get the following eigenvalue problem written in matrix notation:

$$\begin{bmatrix} 1 & 0 & 1 & 0 \\ \alpha_1^2 & 0 & -\alpha_2^2 & 0 \\ \cosh(\alpha_1 L) & \sinh(\alpha_1 L) & \cos(\alpha_2 L) & \sin(\alpha_2 L) \\ \alpha_1^2\cosh(\alpha_1 L) & \alpha_1^2\sinh(\alpha_1 L) & -\alpha_2^2\cos(\alpha_2 L) & -\alpha_2^2\sin(\alpha_2 L) \end{bmatrix} \begin{Bmatrix} C_1 \\ C_2 \\ C_3 \\ C_4 \end{Bmatrix}$$

$$= \begin{Bmatrix} 0 \\ 0 \\ 0 \\ 0 \end{Bmatrix}. \tag{5.11}$$

To obtain a unique solution, the determinant of the matrix in Eq. (5.11) must vanish, leading to the following expression:

$$C_4(\alpha_1^2+\alpha_2^2)\sin(\alpha_2 L)=0$$
$$\Rightarrow \sin(\alpha_2 L)=0 \quad \text{or} \quad \alpha_2 L = n\pi, \quad n=1,2,3,\ldots. \tag{5.12}$$

Substituting the expression for α_2 and defining the following terms:

$$P_{cr}=\frac{n^2\pi^2 EI}{L^2}, \quad \omega_n^2=\frac{n^2\pi^4 EI}{\rho A L^4}, \tag{5.13}$$

we get the following relation

$$\frac{\lambda^2}{2}+\sqrt{\left(\frac{\lambda^2}{2}\right)^2+c^2\omega^2}=\frac{n^2\pi^2}{L^2}\Rightarrow \frac{P}{2P_{cr}}+\sqrt{\left(\frac{P}{2P_{cr}}\right)^2+\frac{\omega^2}{\omega_n^2}}=1,$$
$$1-\frac{P}{P_{cr}}=\left(\frac{\omega}{\omega_n}\right)^2. \tag{5.14}$$

Equation (5.14) shows that for a column on simply supported boundary conditions, the square of the natural frequency is inverse linear with the applied axial compression load (see Fig. 5.4).

For a clamped-clamped column, we will get the following expressions for the boundary conditions

$$\begin{aligned} @x=0: &\quad W(0)=W_{,x}(0)=0,\\ @x=L: &\quad W(L)=W_{,x}(L)=0. \end{aligned} \tag{5.15}$$

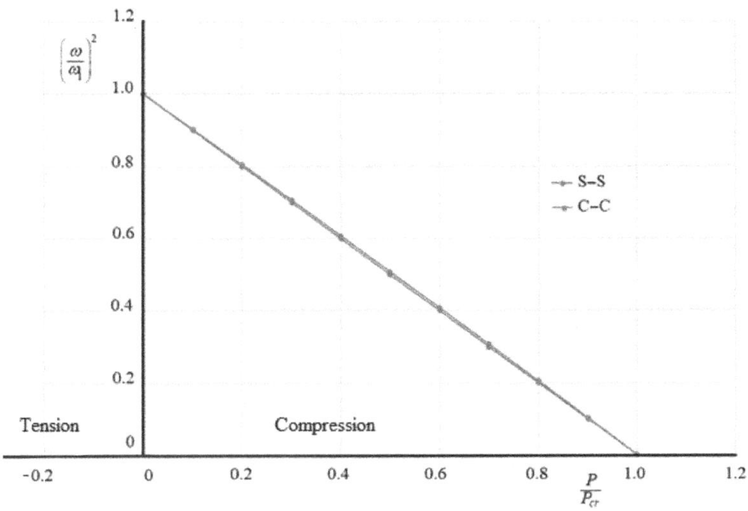

Fig. 5.4. A thin-walled column: non-dimensional circular frequency squared vs. non-dimensional axial compression.

Table 5.1. Buckling loads and natural frequencies of an axially compressed column.

Boundary condition	P_{cr}	ω_n^2 $n = 1, 2, 3$
Clamped–clamped	$\frac{4\pi^2 EI}{L^2}$	$\frac{(22.4)^2\text{EI}}{\rho\text{AL}^4}; \frac{(61.7)^2\text{EI}}{\rho\text{AL}^4}; \frac{(121.0)^2\text{EI}}{\rho\text{AL}^4}$
Simply supported – Simply supported	$\frac{\pi^2 EI}{L^2}$	$\frac{\pi^4\text{EI}}{\rho\text{AL}^4}; \frac{4\pi^4\text{EI}}{\rho\text{AL}^4}; \frac{9\pi^4\text{EI}}{\rho\text{AL}^4}$

with the following eigenvalue problem written in matrix notation:

$$\begin{bmatrix} 1 & 0 & 1 & 0 \\ 0 & \alpha_1 & 0 & \alpha_2 \\ \cosh(\alpha_1 L) & \sinh(\alpha_1 L) & \cos(\alpha_2 L) & \sin(\alpha_2 L) \\ \alpha_1 \sinh(\alpha_1 L) & \alpha_1 \cosh(\alpha_1 L) & -\alpha_2 \sin(\alpha_2 L) & \alpha_2 \cos(\alpha_2 L) \end{bmatrix} \begin{Bmatrix} C_1 \\ C_2 \\ C_3 \\ C_4 \end{Bmatrix} = \begin{Bmatrix} 0 \\ 0 \\ 0 \\ 0 \end{Bmatrix}. \tag{5.16}$$

Demanding the vanishing of the determinant of the matrix in Eq. (5.16) yields the following characteristic equation connecting the axial compression load and the natural frequency of the column

$$2c\omega[1 - \cosh(\alpha_1 L)\cos(\alpha_2 L)] - \lambda^2 \sinh(\alpha_1 L)\sin(\alpha_2 L) = 0, \tag{5.17}$$

where c and λ were defined in Eq. (5.4). Note that Eq. (5.17) is an implicit function, with no closed–form expression for P/P_{cr} vs. $(\omega/\omega_n)^2$ as for the case of a column on simply supported boundary conditions (Eq. (5.12)). An iterative procedure is applied to draw the graph presented in Fig. 5.4 (see Appendix A) for the clamped–clamped case (see also Appendix B for an approximated method). To complete this part, the buckling loads and the natural frequencies for clamped–clamped and simply supported at both ends of a column are presented in Table 5.1.

As can be seen in Fig. 5.4, the dependency of the of the circular frequency squared is linear with the applied axial compression load for a column on simply supported boundaries at both ends, while for a clamped–clamped column the dependency is almost linear (see Table A in Appendix A for the relevant values), with the largest deviation from the linear line occurring at $P/P_{cr} = 0.5$. Due to the very small deviations from linearity for the clamped–clamped case, one can assume also for this type of boundary conditions (and for other boundary conditions, see [1]) a linear relationship between the square of the frequency and the axial compression load. It is interested to note that also for tensile loads the square of the natural frequency is linear with the applied load, with the following relation:

$$1 + \frac{P}{P_{cr}} = \left(\frac{\omega}{\omega_n}\right)^2. \tag{5.18}$$

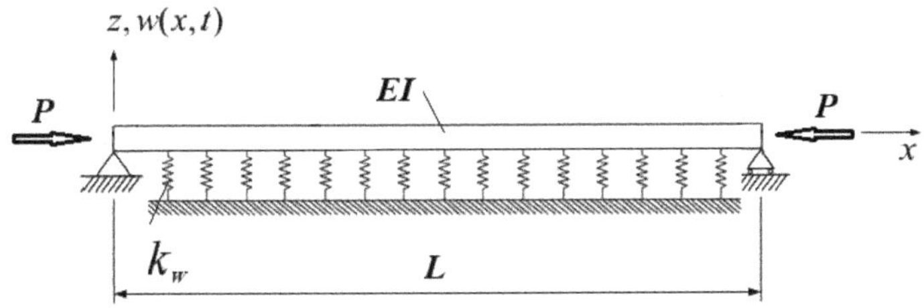

Fig. 5.5. A compressed column on Winkler-type foundation.

This means that for tension loads, the stiffness is increasing, due to the axial load, leading to an increase in the bending natural frequencies (see also [5]), while for compression forces, the stiffness is reduced, leading to lower bending natural frequencies (see also [[6]–[8]]).

An alternative approach is presented in Appendix B, where Rayleigh principle is pursued yielding a similar graph to the one presented in Fig. 5.4.

To exhibit the versatility and the applicability of the VCT to beams (or columns), three additional examples will be next presented. First, a column resting on a Winkler type foundation axially compressed as presented in Fig. 5.5.

The equation of motion for an uniform isotropic column resting on a Winkler-type foundation (k_W its spring coefficient) having a length L, axially compressed by a load P (compression), with a bending stiffness of EI and a mass per unit length (ρA), where A and E are the cross-section of the column and its Young's modulus, respectively, is

$$EI\frac{\partial^4 w(x,t)}{\partial x^4} + P\frac{\partial^2 w(x,t)}{\partial x^2} + k_W \cdot w(x,t) + \rho A\frac{\partial^2 w(x,t)}{\partial t^2} = 0. \qquad (5.19)$$

Following the steps presented above for a simply supported compressed column (Eqs. (5.1)–(5.14)) we assume the following solution for Eq. (5.19):

$$w(x,t) = W(x)e^{i\omega t} \qquad (5.20)$$

where ω is the circular (or angular) frequency of the column and $W(x)$ is a function to be next determined. Substituting Eq. (5.20) into Eq. (5.19) yields

$$EIW_{,xxxx} + PW_{,xx} + k_W \cdot W - \rho A\omega^2 W = 0, \qquad (5.21)$$

where, as before, $[\,]_{,xxxx}$ and $[\,]_{,xx}$ are four and two times differentiation with respect with x, respectively. Division of Eq. (5.21) by EI and defining the following terms as

$$\lambda^2 = \frac{P}{EI}, \quad c^2 = \frac{\rho A}{EI}, \quad b^2 = \frac{k_W}{EI}, \qquad (5.22)$$

we obtain following differential equation:

$$W_{,xxxx} + \lambda^2 W_{,xx} + b^2 W - c^2\omega^2 W = 0. \tag{5.23}$$

Assuming $W(x) = \widetilde{W}(x)e^{\alpha x}$, the differential equation (5.23) transforms in a characteristic algebraic equation,

$$\alpha^4 + \lambda^2\alpha^2 + b^2 - c^2\omega^2 = 0. \tag{5.24}$$

The solution for the characteristic equation presented by Eq. (5.24) is

$$\alpha_{1,2}^2 = -\frac{\lambda^2}{2} \pm \sqrt{\left(\frac{\lambda^2}{2}\right)^2 - (b^2 - c^2\omega^2)}. \tag{5.25}$$

Therefore, the general solution for Eq. (5.24) has the following form:

$$W(x) = B_1\cosh(\alpha_1 x) + B_2\sinh(\alpha_1 x) + B_3\cos(\alpha_2 x) + B_4\sin(\alpha_2 x), \tag{5.26}$$

where

$$\alpha_1 = \sqrt{-\frac{\lambda^2}{2} + \sqrt{\left(\frac{\lambda^2}{2}\right)^2 + c^2\omega^2 - b^2}},\ \alpha_2 = \sqrt{\frac{\lambda^2}{2} + \sqrt{\left(\frac{\lambda^2}{2}\right)^2 + c^2\omega^2 - b^2}}. \tag{5.27}$$

The constants B_1–B_4 in Eq. (5.26) are determined by imposing the adequate boundary conditions. For a column with simply supported boundary conditions (see Eq. (5.10), above), we get the following eigenvalue problem written in matrix notation, identical to the expression presented in Eq. (5.11), above, namely

$$\begin{bmatrix} 1 & 0 & 1 & 0 \\ \alpha_1^2 & 0 & -\alpha_2^2 & 0 \\ \cosh(\alpha_1 L) & \sinh(\alpha_1 L) & \cos(\alpha_2 L) & \sin(\alpha_2 L) \\ \alpha_1^2\cosh(\alpha_1 L) & \alpha_1^2\sinh(\alpha_1 L) & -\alpha_2^2\cos(\alpha_2 L) & -\alpha_2^2\sin(\alpha_2 L) \end{bmatrix} \begin{Bmatrix} B_1 \\ B_2 \\ B_3 \\ B_4 \end{Bmatrix} \begin{Bmatrix} 0 \\ 0 \\ 0 \\ 0 \end{Bmatrix}. \tag{5.28}$$

To obtain a unique solution, the determinant of the matrix in Eq. (5.28) must vanish, leading to the following expression:

$$\begin{aligned} &(\alpha_1^2 + \alpha_2^2)\sin(\alpha_2 L) = 0 \\ &\Rightarrow \sin(\alpha_2 L) = 0 \quad \text{or} \quad \alpha_2 L = n\pi, \quad n = 1, 2, 3, \ldots \end{aligned} \tag{5.29}$$

Substituting the expression for α_2 and defining the following terms:

$$P_{cr} = EI\left[\frac{n^2\pi^2}{L^2} + \frac{k_W \cdot L^2}{n^2\pi^2}\right] = EI\frac{n^2\pi^2}{L^2}\left[1 + \frac{k_W \cdot L^4}{n^4\pi^4 EI}\right] \equiv \mathrm{K} \cdot EI\frac{n^2\pi^2}{L^2},$$

$$\omega_n^2 = \frac{n^4\pi^4 EI}{\rho A L^4} + \frac{k_W}{\rho A} = \frac{n^4\pi^4 EI}{\rho A L^4}\left[1 + \frac{k_W \cdot L^4}{n^4\pi^4 EI}\right] \equiv \mathrm{K} \cdot \frac{n^4\pi^4 EI}{\rho A L^4}, 9,$$

where

$$\mathrm{K} \equiv \left[1 + \frac{k_W \cdot L^4}{n^4\pi^4 EI}\right], \tag{5.30}$$

we get the following relation:

$$\frac{\lambda^2}{2} + \sqrt{\left(\frac{\lambda^2}{2}\right)^2 + c^2\omega^2} = \frac{n^2\pi^2}{L^2} \Rightarrow \mathrm{K} = \frac{\omega^2}{\left(\frac{EI \cdot n^4\pi^4}{\rho A \cdot L^4}\right)} + \frac{P}{\left(\frac{EI \cdot n^2\pi^2}{L^2}\right)}$$

$$1 - \frac{P}{P_{cr}} = \left(\frac{\omega}{\omega_n}\right)^2. \tag{5.31}$$

Equation (5.31) shows that also for a compressed column resting on a Winkler-type foundation, the square of the natural frequency is linear with the compression load, similar to what had been shown above for a simply supported compressed column.

Another example is a symmetric-laminated composite column under compression. The first solution will be presented using CLT (Classical Lamination Theory) followed by a model based on FOSDT (First-Order Shear Deformation Theory).

As presented in [9], the equation of motion for a compressed symmetric composite column is given by the following expression:

$$D_{11}\frac{\partial^4 w(x,t)}{\partial x^4} + \bar{N}_{xx}\frac{\partial^2 w(x,t)}{\partial x^2} + I_2\frac{\partial^4 w(x,t)}{\partial t^2 \partial x^2} - I_0\frac{\partial^2 w(x,t)}{\partial t^2} = 0, \tag{5.32}$$

where D_{11} is the bending stiffness and $\bar{N}_{xx}$ is the compression load, both per unit width, s, while $I_0 = \rho(h^* s)/s = \rho A/s$ (mass per unit length per unit width), $I_2 = \rho h^3/12$ (rotary inertia mass per unit width). Assuming harmonic vibration (see Eqs. (5.2) and (5.20)), we get

$$D_{11}W_{,xxxx} + (\bar{N}_{xx} + \omega^2 I_2)W_{,xx} - \omega^2 I_0 \cdot W = 0, \tag{5.33a}$$

Note that if we multiply Eq. (5.33), by the width of the column, s, we get

$$\begin{aligned}(D_{11} \cdot s) \cdot W_{,xxxx} &+ [(\bar{N}_{xx} \cdot s) + \omega^2(I_2 \cdot s)]W_{,xx} - \omega^2(I_0 \cdot s) \cdot W \\ &= (EI) \cdot W_{,xxxx} + [P + \omega^2(\rho I)]W_{,xx} - \omega^2(\rho A) \cdot W = 0\end{aligned} \tag{5.33b}$$

The expression in Eq. (5.33b) is identical to Eq. (5.3) if the rotatory inertia I_2 is neglected.

Assuming a simply supported case, we can follow the derivation presented by Eqs. (5.3)–(5.14), while defining the following relationships:

$$P_{cr} = EI\frac{n^2\pi^2}{L^2} \Rightarrow P_{cr(n=1)} = EI\frac{\pi^2}{L^2} \text{ or } (\bar{N}_{xx})_{cr(n=1)} = D_{11}\frac{\pi^2}{L^2}$$

$$\omega_n^2 = \frac{n^4\pi^4 EI}{L^4\rho A}\cdot\frac{1}{\left[1+\frac{\rho \mathrm{I}}{\rho \mathrm{A}}\left(\frac{\mathrm{n}\pi}{\mathrm{L}}\right)^2\right]} \Rightarrow \omega_1^2 = \frac{\pi^4 EI}{L^4\rho A}\cdot\frac{1}{\left[1+\frac{\rho \mathrm{I}}{\rho \mathrm{A}}\left(\frac{\pi}{\mathrm{L}}\right)^2\right]}$$

or

$$\omega_n^2 = \frac{n^4\pi^4 D_{11}}{L^4 I_0}\cdot\frac{1}{\left[1+\frac{\mathrm{I}_2}{\mathrm{I}_0}\left(\frac{\mathrm{n}\pi}{\mathrm{L}}\right)^2\right]} \Rightarrow \omega_1^2 = \frac{\pi^4 D_{11}}{L^4 I_0}\cdot\frac{1}{\left[1+\frac{\mathrm{I}_2}{\mathrm{I}_0}\left(\frac{\pi}{\mathrm{L}}\right)^2\right]}. \quad (5.34)$$

Then we obtain

$$\frac{\omega^2}{\frac{\pi^4 \mathrm{EI}}{\mathrm{L}^4\rho \mathrm{A}}\cdot\frac{1}{\left[1+\frac{\rho \mathrm{I}}{\rho \mathrm{A}}\left(\frac{\pi}{\mathrm{L}}\right)^2\right]}} = 1 - \frac{\bar{N}_{xx}}{D_{11}\left(\frac{\pi}{\mathrm{L}}\right)^2}$$

$$\text{or} \quad \left(\frac{\omega}{\omega_1}\right)^2 = 1 - \frac{P}{P_{cr(n=1)}}. \quad (5.35)$$

Equation (5.35) shows again the linear relationship between the compression load and the first natural frequency squared, as for the previous presented cases.

To include non-symmetric laminates, as a first approximation, the bending stiffness D_{11}, appearing in Eq. (5.35), should be replaced by $(D_{11} - B_{11}^2/A_{11})$, where B_{11} is the coupling stiffness and A_{11} is the axial stiffness, thus reflecting the reduction of the bending stiffness due to unsymmetry of the laminate (see [9]).

Taking into account the shear deformation, namely using a model based on FOSDT (First-Order Shear Deformation Theory), the equations of motion for a column with constant properties (see [9], [10]) would be

$$A_{11}\left(\frac{\partial^2 u}{\partial x^2} + \frac{\partial^3 w}{\partial x^3}\right) + B_{11}\frac{\partial^2 \phi}{\partial x^2} = I_0\frac{\partial^2 u}{\partial t^2} + I_1\frac{\partial^2 \phi}{\partial t^2}, \quad (5.36)$$

$$KA_{55}\left(\frac{\partial^2 w}{\partial x^2} + \frac{\partial \phi}{\partial x}\right) - \overline{N}_{xx}\frac{\partial^2 w}{\partial x^2} = q + I_0\frac{\partial^2 w}{\partial t^2}, \quad (5.37)$$

$$B_{11}\left(\frac{\partial^2 u}{\partial x^2} + \frac{\partial^3 w}{\partial x^3}\right) + D_{11}\frac{\partial^2 \phi}{\partial x^2} - KA_{55}\left(\frac{\partial w}{\partial x} + \phi\right)$$

$$= I_2\frac{\partial^2 \phi}{\partial t^2} + I_1\frac{\partial^2 u}{\partial t^2}, \quad (5.38)$$

where u, w being the longitudinal and lateral displacements, respectively and ϕ is the rotation due to shear and the various expression for the stiffness are given by

$$A_{ij} = \sum_{k=1}^{N} \overline{Q}_{ij}^{(k)}(z_{k+1} - z_k); \quad B_{ij} = \frac{1}{2}\sum_{k=1}^{N} \overline{Q}_{ij}^{(k)}(z_{k+1}^2 - z_k^2);$$

$$D_{ij} = \frac{1}{3}\sum_{k=1}^{N} \overline{Q}_{ij}^{(k)}(z_{k+1}^3 - z_k^3) \tag{5.39}$$

with $\overline{Q}_{ij}^{(k)}$ being the lamina stiffness after transformation, q is the lateral loading per unit length, and K is the shear correction coefficient, being defined as the ratio between the shear strain energy calculated by the actual shear distribution and the constant distribution assumed in the FOSDT theory. The value of K is taken as 5/6 for a rectangular cross-section[a].

The mass moments of inertia I_0, I_1 and I_2, appearing in Eqs. (5.36)–(5.38) are defined as

$$\begin{Bmatrix} I_0 \\ I_1 \\ I_2 \end{Bmatrix} = \int_{-h/2}^{+h/2} \begin{Bmatrix} 1 \\ z \\ z^2 \end{Bmatrix} \cdot \rho \cdot dz, \tag{5.40}$$

while ρ is the relevant density.

Assuming harmonic vibrations (ω being the circular frequency of the column), and setting $q = 0$, while neglecting high order terms, we get the following equations:

$$A_{11}\frac{d^2U}{dx^2} + B_{11}\frac{d^2\Phi}{dx^2} + \omega^2 I_0 U + \omega^2 I_1 \Phi = 0, \tag{5.41}$$

$$KA_{55}\left(\frac{d^2W}{dx^2} + \frac{d\Phi}{dx}\right) - \overline{N}_{xx}\frac{d^2W}{dx^2} + \omega^2 I_0 W = 0, \tag{5.42}$$

$$B_{11}\frac{d^2U}{dx^2} + D_{11}\frac{d^2\Phi}{dx^2} - KA_{55}\left(\frac{dW}{dx} + \Phi\right) + \omega^2 I_2 \Phi + \omega^2 I_1 U = 0. \tag{5.43}$$

To present the relation between the circular natural frequency, ω and the axial load, $\overline{N}_{xx}$, we shall treat only the symmetrical case, namely

---

[a]The accurate value of the shear coefficient is $\kappa = \frac{10(1+\upsilon)}{12+11\upsilon}$ for a rectangular cross-section and $\kappa = \frac{6(1+\upsilon)}{7+6\upsilon}$ for a solid circular cross-section, where υ is the Poisson's ratio.

$B_{11} = I_1 = 0$, leading to

$$A_{11}\frac{d^2U}{dx^2} + \omega^2 I_0 U = 0, \tag{5.44}$$

$$KA_{55}\left(\frac{d^2W}{dx^2} + \frac{d\Phi}{dx}\right) - \overline{N}_{xx}\frac{d^2W}{dx^2} + \omega^2 I_0 W = 0, \tag{5.45}$$

$$D_{11}\frac{d^2\Phi}{dx^2} - KA_{55}\left(\frac{dW}{dx} + \Phi\right) + \omega^2 I_2 \Phi = 0. \tag{5.46}$$

While Eq. (5.44) is uncoupled, Eqs. (5.45) and (5.46) are coupled and need to be uncoupled to enable a solution for the problem. The uncoupled form has the following two equations:

$$\left[D_{11} - \frac{\overline{N}_{xx}}{KA_{55}}\right]\frac{d^4W}{dx^4} + \left\{\omega^2\left[\frac{D_{11}I_0}{KA_{55}} + \left(1 - \frac{\overline{N}_{xx}}{KA_{55}}\right)I_2 + \overline{N}_{xx}\right]\right\}\frac{d^2W}{dx^2}$$

$$+\omega^2 I_0\left[\frac{I_2\omega^2}{KA_{55}} - 1\right]W = 0, \tag{5.47}$$

$$\left[D_{11} - \frac{\overline{N}_{xx}}{KA_{55}}\right]\frac{d^4\Phi}{dx^4} + \left\{\omega^2\left[\frac{D_{11}I_0}{KA_{55}} + \left(1 - \frac{\overline{N}_{xx}}{KA_{55}}\right)I_2 + \overline{N}_{xx}\right]\right\}\frac{d^2\Phi}{dx^2}$$

$$+\omega^2 I_0\left[\frac{I_2\omega^2}{KA_{55}} - 1\right]\Phi = 0. \tag{5.48}$$

The solutions for Eqs. (5.47) and (5.48) are given as

$$W(x) = A_1\cosh(s_1x) + A_2\sinh(s_1x) + A_3\cos(s_2x) + A_4\sin(s_2x), \tag{5.49}$$

$$\Phi(x) = B_1\cosh(s_1x) + B_2\sinh(s_1x) + B_3\cos(s_2x) + B_4\sin(s_2x), \tag{5.50}$$

while the constants A_1, A_2, A_3, A_4 and B_1, B_2, B_3, B_4 are interconnected (by back substituting Eqs. (5.49) and (5.50) into the coupled equations, Eqs. (5.45) and (5.46)) and have the following form:

$$B_1 = \frac{KA_{55}\cdot s_1}{D_{11}\cdot s_1^2 - KA_{55} + \omega^2 I_2}A_2 \equiv \alpha A_2, \tag{5.51}$$

$$B_2 = \frac{KA_{55}\cdot s_1}{D_{11}\cdot s_1^2 - KA_{55} + \omega^2 I_2}A_1 \equiv \alpha A_1, \tag{5.52}$$

$$B_3 = \frac{KA_{55}\cdot s_2}{\omega^2 I_2 - D_{11}\cdot s_2^2 - KA_{55}}A_4 \equiv \beta A_4, \tag{5.53}$$

$$B_4 = -\frac{KA_{55}\cdot s_2}{\omega^2 I_2 - D_{11}\cdot s_2^2 - KA_{55}}A_3 \equiv -\beta A_3 \tag{5.54}$$

and

$$s_1 = \sqrt{-\frac{b}{2a} + \frac{\sqrt{(b^2 - 4ac)}}{2a}}, \tag{5.55}$$

$$s_2 = \sqrt{+\frac{b}{2a} + \frac{\sqrt{(b^2 - 4ac)}}{2a}}, \tag{5.56}$$

where

$$a \equiv D_{11} - \frac{\overline{N}_{xx}}{KA_{55}}; \quad b \equiv \omega^2 \left[\frac{D_{11} I_0}{KA_{55}} + \left(1 - \frac{\overline{N}_{xx}}{KA_{55}}\right) I_2 + \overline{N}_{xx}\right];$$
$$c \equiv \omega^2 I_0 \left[\frac{I_2 \omega^2}{KA_{55}} - 1\right]. \tag{5.57}$$

The remaining constants, A_1, A_2, A_3 and A_4, will be found after imposing the relevant boundary conditions. For a simply supported column, the following boundary conditions are imposed at each side of the column ($x = 0$ and $X = L$):

$$W = 0 \quad \text{and} \quad \frac{d\Phi}{dx} = 0. \tag{5.58}$$

This leads to the following relation:

$$\left(\frac{\omega}{\omega_1}\right)^2 = 1 - \frac{\bar{N}_{xx}}{\bar{N}_{xx,cr(n=1)}}, \tag{5.59}$$

which is identical to Eq. (5.35).

Another topic is the case in which the compressive force is due to thermal heating, as presented in Fig. 5.6, for a simply supported column, which cannot move axially.

The stretching due to temperature increase ΔT (where α is the linear coefficient of thermal expansion) is then equated to the axial displacement of a column to yield

$$\alpha \cdot \Delta T \cdot L = \frac{P \cdot L}{A \cdot E} \Rightarrow P = A \cdot E \cdot \alpha \cdot \Delta T, \tag{5.60}$$

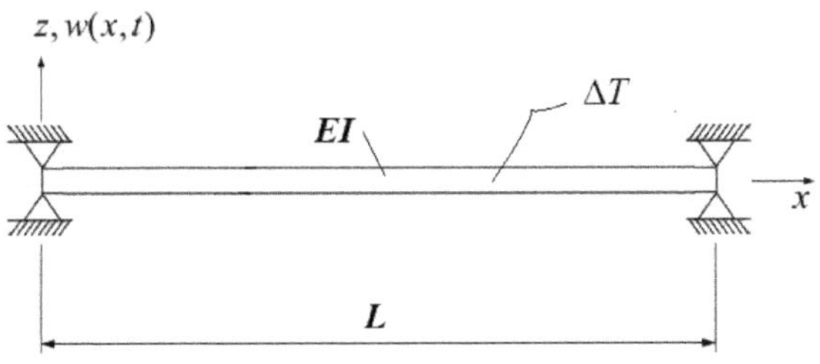

Fig. 5.6. An unmovable column under ΔT heating.

where A is the cross-section of the column and E stands for its Young's modulus. Substituting the expression for P, presented by Eq. (5.60) into Eq. (5.1), provides the following governing equation:

$$EI\frac{\partial^4 w(x,t)}{\partial x^4} + (A \cdot E \cdot \alpha \cdot \Delta T)\frac{\partial^2 w(x,t)}{\partial x^2} + \rho A\frac{\partial^2 w(x,t)}{\partial t^2} = 0. \tag{5.61}$$

Assuming harmonic vibrations, and division of Eq. (5.61) by EI, leads to the following expression:

$$\frac{d^4 W}{dx^4} + \lambda^2\frac{d^2 W}{dx^2} - c^2\omega^2 W = 0, \tag{5.62}$$

while

$$\lambda^2 = \frac{AE\alpha\Delta T}{EI} = \frac{\alpha\Delta T}{r^2}, \quad c^2 = \frac{\rho A}{EI}$$
$$\text{where} \quad r^2 = \frac{I}{A}. \tag{5.63}$$

Then, following the procedure presented in Eqs. (5.5)–(5.14) we get an identical expression to Eq. (5.14), namely

$$(\Delta T)_{cr} = \frac{n^2\pi^2 r^2}{L^2\alpha}, \quad \omega_n^2 = \frac{n^2\pi^4 r^2 E}{\rho L^4}, \quad n = 1, 2, 3, \ldots$$
$$\rightarrow (\Delta T)_{cr_{n=1}} = \frac{\pi^2 r^2}{L^2\alpha}, \quad \omega_{n=1}^2 = \frac{\pi^4 r^2 E}{\rho L^4} \tag{5.64}$$

and

$$\frac{\lambda^2}{2} + \sqrt{\left(\frac{\lambda^2}{2}\right)^2 + c^2\omega^2} = \frac{n^2\pi^2}{L^2} \Rightarrow \frac{\Delta T}{2P_{cr}} + \sqrt{\left(\frac{\Delta T}{2\left(\Delta T_{cr_{n=1}}\right)}\right)^2 + \frac{\omega^2}{\omega_{n=1}^2}} = 1$$
$$1 - \frac{\Delta T}{\Delta T_{cr_{n=1}}} = \left(\frac{\omega}{\omega_{n=1}}\right)^2. \tag{5.65}$$

When using axially constrained laminated composite columns under a thermal loading ΔT, the governing equation using the classical lamination theory would be Eq. (5.32), where the axial compression load $\bar{N}_{xx}$ would be replaced by the following expression:

$$\bar{N}_{xx} = \Delta T \cdot \sum (\bar{Q}_{11}\alpha_x + \bar{Q}_{12}\alpha_y)(t_{n+1} - t_n) \equiv \Delta T \cdot R, \tag{5.66}$$

while α_x and α_y are the thermal coefficients in the x and y directions, respectively, and $\bar{Q}_{11}, \bar{Q}_{12}$ are the stiffness after transformation (see [11]), thus yielding the following equation:

$$D_{11}\frac{\partial^4 w(x,t)}{\partial x^4} + \Delta T \cdot R\frac{\partial^2 w(x,t)}{\partial x^2} + I_2\frac{\partial^4 w(x,t)}{\partial t^2 \partial x^2} - I_0\frac{\partial^2 w(x,t)}{\partial t^2} = 0. \tag{5.67}$$

Equation (5.67) is used to perform all the manipulations presented by Eqs. (5.33)–(5.35) to yield identical expressions as presented in Eq. (5.34) and (5.35), namely

$$(\Delta T)_{cr} = D_{11}\frac{n^2\pi^2}{L^2 R} \quad \text{or} \quad (\Delta T)_{cr(n=1)} = D_{11}\frac{\pi^2}{L^2 R},$$
$$\omega_n^2 = \frac{n^4\pi^4 D_{11}}{L^4 I_0} \cdot \frac{1}{\left[1 + \frac{I_2}{I_0}\left(\frac{n\pi}{L}\right)^2\right]} \Rightarrow \omega_1^2 = \frac{\pi^4 D_{11}}{L^4 I_0} \cdot \frac{1}{\left[1 + \frac{I_2}{I_0}\left(\frac{\pi}{L}\right)^2\right]}. \tag{5.68}$$

This leads to the following relationship (identical to Eq. (5.65)), namely

$$\frac{\omega^2}{\frac{\pi^4 EI}{L^4 I_0} \cdot \frac{1}{\left[1+\frac{\mathrm{I}_2}{\mathrm{I}_0}\left(\frac{\pi}{\mathrm{L}}\right)^2\right]}} = 1 - \frac{\Delta T \cdot R}{D_{11}\left(\frac{\pi}{\mathrm{L}}\right)^2}$$
$$\text{or} \quad \left(\frac{\omega}{\omega_1}\right)^2 = 1 - \frac{\Delta T}{\Delta T_{cr(n=1)}}. \tag{5.69}$$

Application of a first-order shear deformation theory the thermal loading ΔT problem presented above (see Eqs. (5.36)–(5.59) would again require the replacement of the axial compression load $\bar{N}_{xx}$ by the expression presented in Eq. (5.66). Then, following the calculations presented by Eqs. (5.36)–(5.58), one obtains

$$\left(\frac{\omega}{\omega_1}\right)^2 = 1 - \frac{\Delta T}{\Delta T_{cr(n=1)}}, \tag{5.70}$$

which is identical to Eq. (5.69).

An interesting study is presented in [12, 13]. A simple example of a cantilevered beam, where the lateral vibrations are altered significantly by applying a constant voltage to collocated surface bonded piezoelectric patches is presented in Fig. 5.7. From the physical point of view, the heating temperatures ΔT can represent the applied voltages V to the piezoelectric patches, and Eqs. (5.60)–(5.65) represent the theoretical part of this problem. To prevent the movement of the free part of the cantilevered column, a wire is added, which connects the two parts of the column. Two axial forces act: a resultant compressive force (with a line of action going through the

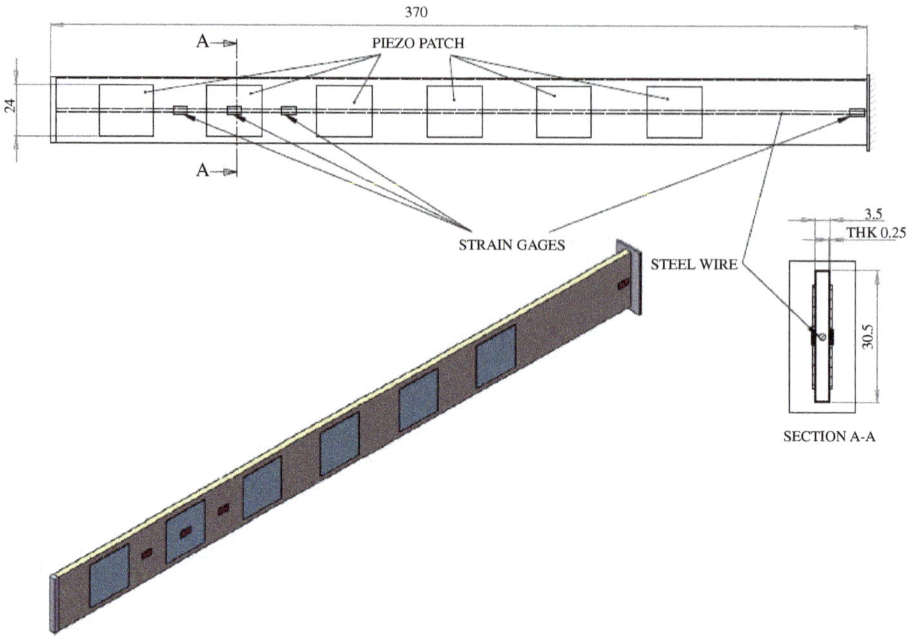

Fig. 5.7. The schematic experimental model-schematic (dimensions in mm) from [12].

cross-section center) and an equal tension force along the line. The resultant force is zero. When the column has zero lateral deflections, the lines of action of both forces coincide. However, in the presence of lateral deflections (like for the vibration case) the two forces are no more collinear, yielding a bending couple that affects the stiffness of the column. When this couple reach a critical value, it causes the buckling of the column.

The variation of the natural frequency squared with increasing the applied voltage is presented in Fig. 5.8. As expected, the variation is linear, with the buckling voltage is reported to be 1168 V in very good agreement with the calculated value of 1135 V (see [12, 13]).

The VCT as applied to beams (or columns) is widely presented in the literature. Both numerical and experimental studies were published. Since the early works of Lurie [2–4], many other articles tackled the topic of vibrations vs. compressive loads, and their application to nondestructively define the buckling load of beams (or columns) (see [14–30] as typical examples). The topic was summarized in a dedicated book authored by Virgin [24] and other books like [1], [9] and [31]. It is interesting to note that Jubb *et al.* [17] proposed the following relationship for structural unstable frames:

$$\left(\frac{f}{f_0}\right)^4 = 1 - \frac{P}{P_{cr}} \rightarrow \quad \left(\frac{\omega}{\omega_1}\right)^4 = 1 - \frac{P}{P_{cr(n=1)}}, \tag{5.71}$$

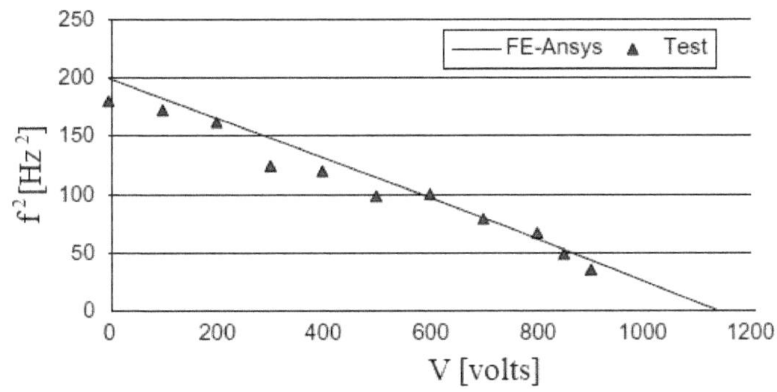

Fig. 5.8. Frequency squared vs. applied voltage using six piezoelectric patches-experimental and numerical results, from [12].

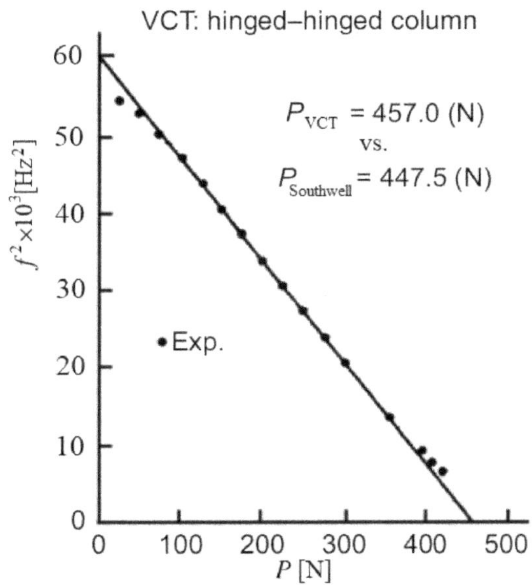

Fig. 5.9. Frequency squared vs. applied axial compression-experimental results, from [31].

and their comparison of the test results presented by Lurie [4] with their numerical results based on Eq. (5.71) displayed a very good correlation.

The experimental results presented in the literature, show, as expected, a linear relationship between the axial compression load and the squared of the first natural frequency of the column, as presented for typical results obtained in [25] and shown in Fig. 5.9. A very good comparison is reported, with a buckling load of 457 N being determined using VCT vs. a buckling load of 447.5 N being calculated using the Southwell plot[b] (see a detailed derivation of the method in Appendix C) for laminated composite columns.

5.3. Plates and Panels

The next basic structure, a plate or a panel, had been also extensively investigated in the literature, yielding natural frequencies and buckling loads for various types of plates, boundary conditions, and shapes (typical results can be found in [32–34] and recently in [31]). A theoretical relationship between the natural frequencies and the applied in-plane load can be obtained also for plates or panels.

For perfect isotropic plates (without initial geometric imperfection) with a length a and width b, the general relationship connecting the in-plane loads with the lateral vibration can be written as (see Fig. 5.10)

$$\begin{aligned} &D[W_{,xxxx}(x,y,t) + 2W_{,xxyy}(x,y,t) + W_{,yyyy}(x,y,t)] + \rho h W_{,tt}(x,y,t) \\ &\quad = N_x W_{,xx}(x,y,t) + N_y W_{,yy}(x,y,t) + 2N_{xy} W_{,xy}(x,y,t), \end{aligned} \tag{5.72}$$

where N_x, and N_y are tensile loads per unit width in the x and y directions, N_{xy} is the shear load per unit length, h is the thickness of the plate, ρ is its density, $W(x,y,t)$ is the lateral deflection and the flexural rigidity per unit width of the plate, D, is defined as

$$D = \frac{Eh^3}{12(1-v^2)}. \tag{5.73}$$

Now we shall demonstrate the derivation of the relationship between the square of the natural frequency and the applied in-plane compressive loads for a perfect (no initial geometric imperfections) isotropic plate being simply supported along its four edges. At this point, the shear load, N_{xy} is assumed zero.

Assuming the following solution for the out-of-plane deflection, $W(x,y,t)$, which satisfies the simply supported boundary conditions of the plate,

$$W(x,y,t) = \left(\sum_{m=1}^{\infty}\sum_{n=1}^{\infty} \bar{W}_{m,n} \sin\frac{m\pi x}{a} \sin\frac{n\pi y}{b}\right) e^{i\omega t}. \tag{5.74}$$

---

[b] Southwell plot is a graphical method to experimental determine the critical buckling load of a column. This method is considered a non-destructive approach to predict the buckling load, by plotting the lateral deflection, w, as a function of the ratio w/P (P the axial compressive load). The slope of the linear curve would be the buckling load, P_{cr}.

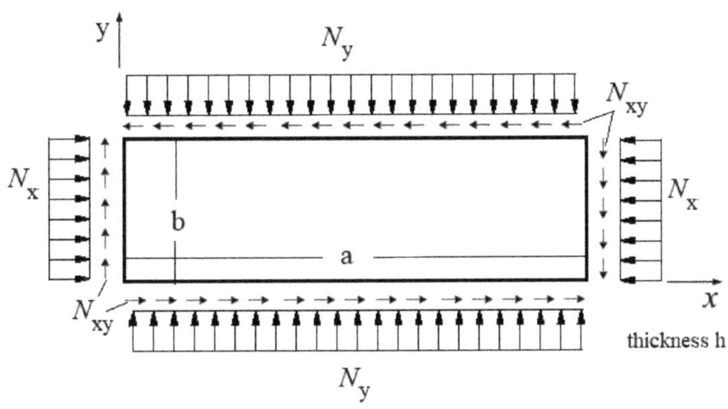

Fig. 5.10. A schematic drawing of a plate loaded in its plane.

and substituting back in Eq. (5.72), yields

$$D\left[\left(\frac{m\pi}{a}\right)^4 + 2\left(\frac{m\pi}{a}\right)^2\left(\frac{n\pi}{b}\right)^2 + \left(\frac{n\pi}{b}\right)^4\right] - \rho h\omega^2$$
$$= -N_x\left(\frac{m\pi}{a}\right)^2 - N_y\left(\frac{n\pi}{b}\right)^2. \tag{5.75}$$

Substituting $N_x = N_y = 0$ in Eq. (5.75), provides the natural frequencies for an unloaded plate, namely

$$\hat{\omega}^2_{mn} = D\frac{\pi^4}{\rho h}\left[\left(\frac{m}{a}\right)^2 + \left(\frac{n}{b}\right)^2\right]^2. \tag{5.76}$$

The critical buckling loads, in the x and y directions, would be calculating by setting the circular natural frequency to zero ($\omega^2 = 0$) in Eq. (5.76) and calculating the N_{xcr} (with $N_y = 0$) and then N_{ycr} (while $N_x = 0$). The calculated expressions for N_{xcr} and N_{ycr} are

$$N^{mn}_{xcr} = -D\pi^2\frac{a^2}{m^2}\left[\left(\frac{m}{a}\right)^2 + \left(\frac{n}{b}\right)^2\right]^2,$$
$$N^{mn}_{ycr} = -D\pi^2\frac{b^2}{n^2}\left[\left(\frac{m}{a}\right)^2 + \left(\frac{n}{b}\right)^2\right]^2. \tag{5.77}$$

Equation (5.75) can be rewritten as

$$1 - \frac{\rho h\omega^2}{D\pi^4\left[\left(\frac{m}{a}\right)^2 + \left(\frac{n}{b}\right)^2\right]^2} = +\frac{N_x\pi^2\left(\frac{m}{a}\right)^2}{D\pi^4\left[\left(\frac{m}{a}\right)^2 + \left(\frac{n}{b}\right)^2\right]^2} + \frac{N_y\pi^2\left(\frac{n}{b}\right)^2}{D\pi^4\left[\left(\frac{m}{a}\right)^2 + \left(\frac{n}{b}\right)^2\right]^2}. \tag{5.78}$$

Having the expressions for the natural frequencies at zero load, and the critical buckling loads in the x and y directions (Eqs. (5.76)–(5.77)), Eq. (5.78) yields

$$1-\left(\frac{\omega}{\hat{\omega}_{mn}}\right)^2=\frac{N}{N_{xcr}^{mn}}+\frac{N}{N_{ycr}^{mn}} \tag{5.79}$$

Equation (5.79) is identical to Eqs. (5.14), (5.31), (5.35) and (5.59), developed earlier for various types of columns, if we let the compressive load in Eq. (5.79) to act only in one direction. Note that the relationship presented by Eq. (5.79) is identical to the result presented in [33].

Presenting the governing equation for a laminated composite perfect shell, using the Classical Lamination Theory (CLT) yields the following formulation [9, 31]:

$$\begin{aligned}
&D_{11}W_{,xxxx}(x,y,t)+2(D_{12}+2D_{66})W_{,xxyy}(x,y,t)+D_{22}W_{,yyyy}(x,y,t)\\
&\quad+N_xW_{,xx}(x,y,t)+2N_{xy}W_{,xy}(x,y,t)+N_yW_{,yy}(x,y,t)\\
&\quad+I_0W_{,tt}(x,y,t)-I_2[W_{,tt_{xx}}(x,y,t)+W_{,tt_{yy}}(x,y,t)]=0,
\end{aligned} \tag{5.80}$$

where D_{11}, D_{12}, D_{66}, D_{22}, I_0 and I_2 are defined by Eqs. (5.38) and (5.40).

Assuming a solution for the out-of-plane deflection, $W(x,y,t)$, which satisfies the simply supported boundary conditions of the plate, in the form presented in Eq. (5.74), and substituting back in Eq. (5.80) we get (for the case $N_{xy}=0$)

$$\begin{aligned}
&D_{11}\left(\frac{m\pi}{a}\right)^4+2\left(D_{12}+2D_{66}\right)\left(\frac{m\pi}{a}\right)^2\left(\frac{n\pi}{b}\right)^2+D_{22}\left(\frac{n\pi}{b}\right)^4\\
&N_x\left(\frac{m\pi}{a}\right)^2+N_y\left(\frac{n\pi}{b}\right)^2+\omega^2\left\{I_0+I_2\left[\left(\frac{m\pi}{a}\right)^2+\left(\frac{n\pi}{b}\right)^2\right]\right\}=0.
\end{aligned} \tag{5.81}$$

Defining the buckling loads in the x and y directions, as well as the circular natural frequencies at zero in-plane loads as

$$\begin{aligned}
N_{xcr}^{mn}&=\frac{D_{11}\left(\frac{m\pi}{a}\right)^4+2\left(D_{12}+2D_{66}\right)\left(\frac{m\pi}{a}\right)^2\left(\frac{n\pi}{b}\right)^2+D_{22}\left(\frac{n\pi}{b}\right)^4}{\left(\frac{m\pi}{a}\right)^2},\\
N_{ycr}^{mn}&=\frac{D_{11}\left(\frac{m\pi}{a}\right)^4+2\left(D_{12}+2D_{66}\right)\left(\frac{m\pi}{a}\right)^2\left(\frac{n\pi}{b}\right)^2+D_{22}\left(\frac{n\pi}{b}\right)^4}{\left(\frac{n\pi}{b}\right)^2},\\
\hat{\omega}_{mn}^2&=\frac{D_{11}\left(\frac{m\pi}{a}\right)^4+2\left(D_{12}+2D_{66}\right)\left(\frac{m\pi}{a}\right)^2\left(\frac{n\pi}{b}\right)^2+D_{22}\left(\frac{n\pi}{b}\right)^4}{I_0+I_2\left[\left(\frac{m\pi}{a}\right)^2+\left(\frac{n\pi}{b}\right)^2\right]}.
\end{aligned} \tag{5.82}$$

One can rearrange Eq. (5.81) to the following form:

$$\frac{N_x\left(\frac{m\pi}{a}\right)^2+N_y\left(\frac{n\pi}{b}\right)^2+\omega^2\left\{I_0+I_2\left[\left(\frac{m\pi}{a}\right)^2+\left(\frac{n\pi}{b}\right)^2\right]\right\}}{D_{11}\left(\frac{m\pi}{a}\right)^4+2\left(D_{12}+2D_{66}\right)\left(\frac{m\pi}{a}\right)^2\left(\frac{n\pi}{b}\right)^2+D_{22}\left(\frac{n\pi}{b}\right)^4}=1. \tag{5.83}$$

Using the definitions presented by Eq. (5.82), we obtain the linear relation between the square of the natural frequencies and the axial in-plane load for a composite laminated plate

$$1-\frac{N}{N_{xcr}^{mn}}-\frac{N}{N_{ycr}^{mn}}=\left(\frac{\omega}{\hat{\omega}_{mn}}\right)^2, \tag{5.84}$$

which is identical to Eq. (5.79).

In the case, the in-plane load is pure shear Eqs. (5.72) and (5.80) will change to

$$\begin{aligned}&D[W_{,xxxx}(x,y)+2W_{,xxyy}(x,y)+W_{,yyyy}(x,y)]-\rho h\omega^2W(x,y)\\&\quad=2N_{xy}W_{,xy}(x,y),\end{aligned} \tag{5.85}$$

for an isotropic plate, while for a laminated composite plate, the governing equation will be presented by

$$\begin{aligned}&D_{11}W_{,xxxx}(x,y)+2(D_{12}+2D_{66}))W_{,xxyy}(x,y)+D_{22}W_{,yyyy}(x,y)\\&\quad+2N_{xy}W_{,xy}(x,y)-I_0\omega^2W(x,y,t)+I_2\omega^2[W_{,xx}(x,y)+W_{,yy}(x,y)]=0.\end{aligned} \tag{5.86}$$

No closed-form solutions for an all-around simply supported plate (or any other boundary conditions) under shear loads are available [35]. Moreover, as it is known and also claimed in [35], the buckling mode shape of the plate under shear would be skewed while the mode shape of the natural frequency under zero load would be sinusoidal, and transformed into a skewed one with the increasing of the shear load [35]. No formula can be derived for this case, and in [35] a non-linear relation between the shear load and the natural frequency squared was numerically calculated.

When trying to confirm experimentally, Eqs. (5.79) and/or (5.84), it turns out that only perfect plates, without initial geometric imperfections, obey the relation. This finding was highlighted by the calculations performed already in 1948 by Massonnet [36] on circular uniform plates under uniform compression and presented in Fig. 5.11. The term $f_0/\gamma h$ appearing in Fig. 5.8 is defined as

$$\frac{f_0}{\gamma h}=\frac{f_0}{\sqrt{12(1-\upsilon^2)}h}, \tag{5.87}$$

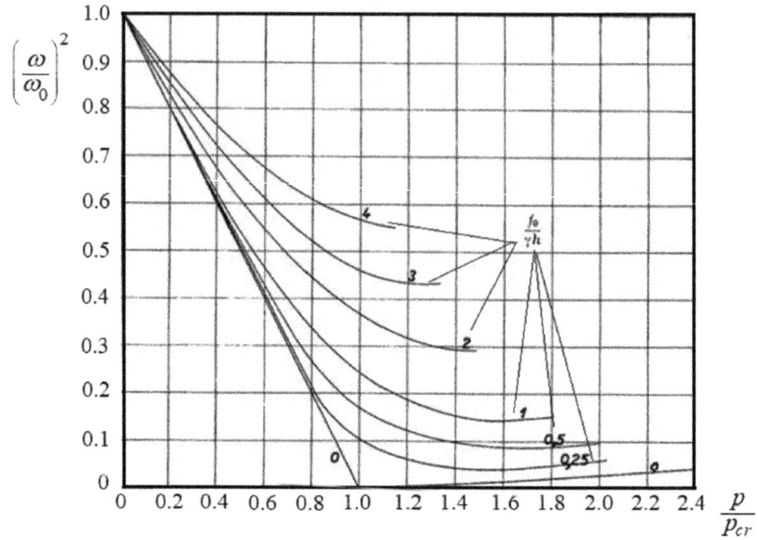

Fig. 5.11. Non-dimensional circular frequency squared vs. non-dimensional axial compression — a circular uniform plate compressed on its perimeter (adapted from [31]).

where v is the Poisson's ratio and f_0 is the initial deflection of the plate, or the initial amplitude of the geometric imperfection. It can be seen that a linear relationship would exist if the plate is perfectly flat and even small initial geometric imperfection would change the behavior. Therefore it is very important to measure the initial geometric imperfections of a plate prior of testing it, to obtain correctly the relation between the compressive (or tensile) load to the square of the natural frequencies of the plate. Note that also Lurie [2] claimed that the initial geometric imperfections are the main factor to change the linear relation load vs. natural frequency squared, as presented by Massonnet [36].

Similar results are reported in [37, 38], for circular aluminum plates, as presented in Fig. 5.12. Defining the load at which the squared nondimensional frequency changes its path from descent to rising as the predicted buckling load by VCT, one gets P_{cr} (VCT) = 275.6 kN as compared to P_{cr} (Exp.) = 275.8 kN, which is very good nondestructive buckling prediction. Due to its stable postbuckling behavior of a plate, one can record its natural frequency up to the buckling and beyond, without destroying the tested specimen and thus correctly define the postbuckling of a plate having initial geometric imperfections. More details are given in Chapter 6.

Additional experimental results are presented in Fig. 5.13, for a simply supported rectangular plate, compressed in a single direction. The linear

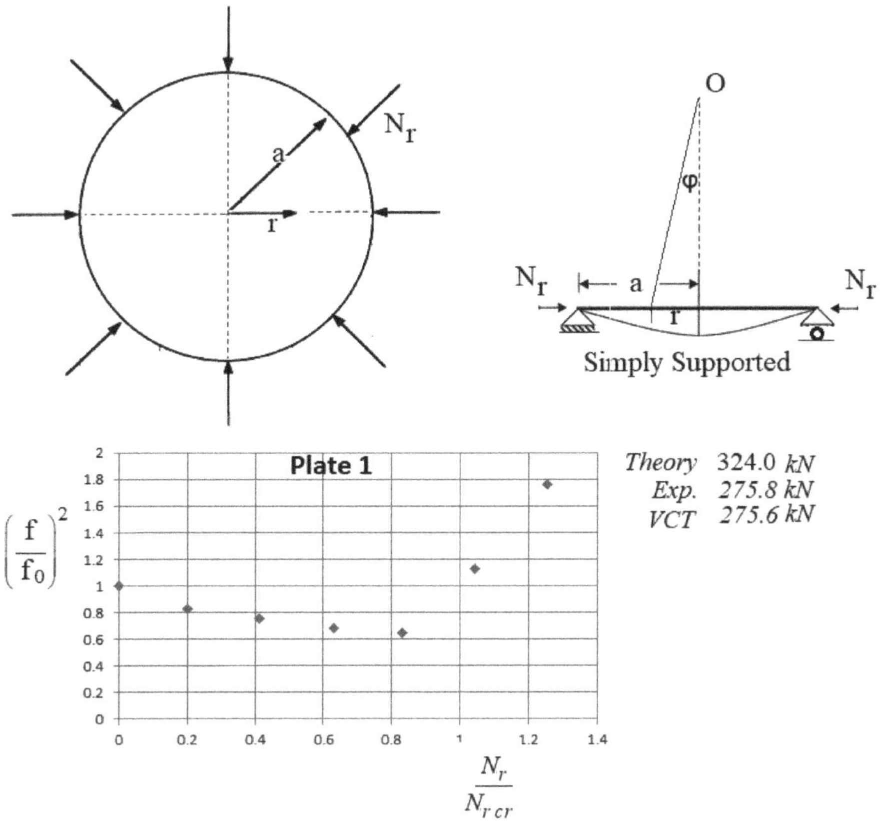

Fig. 5.12. Non-dimensional circular frequency squared vs. non-dimensional axial compression — a circular simply supported uniform plate compressed on its perimeter (adapted from [37, 38]).

relationship between the frequency squared and applied load is clearly displayed, although at about 860 N the curve starts to curl up, as stated by Massonnet [36]. Extrapolating the linear curve it yields a predicted buckling load of $P_{\text{VCT}} = 1175$ N which perfectly matches the buckling of the plate using the Southwell method $P_{\text{Southwell}} = 1170.0$ N (for more results see [25, 31]).

Figures 5.14 and 5.15 present experimental results of the first four natural frequencies of curved panels obtained by modal hammer (see detailed description of the method in Appendix D) and accelerometers bonded on the specimens [39]. As can been in Fig. 5.14, the natural frequencies for panel D_{com1} (a laminated curved panel with two aluminum longerons) are used to predict its buckling load using VCT, yielding $P_{\text{VCT}} = 16.4$ kN in comparison with the real buckling load $P_{\text{exp.}} = 16.19$ kN. Note the sharp bend in the

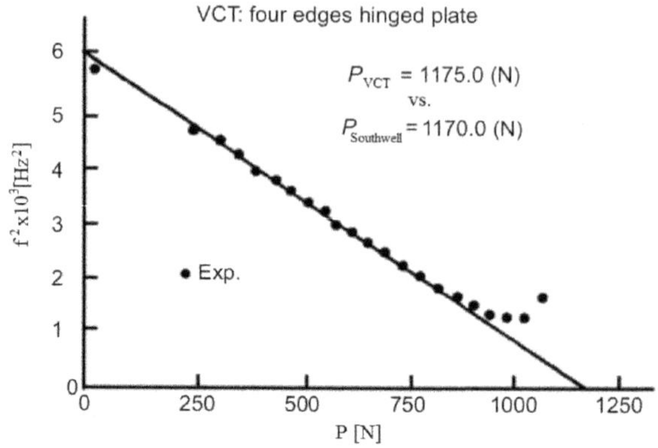

Fig. 5.13. Frequency squared vs. applied axial compression — experimental results for rectangular laminated composite plates from [31].

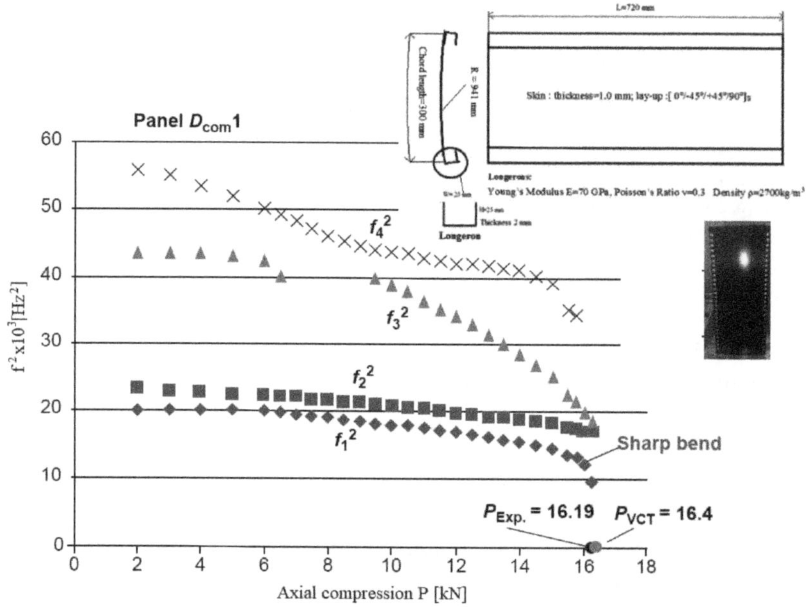

Fig. 5.14. Panel D_{com1} — Frequency squared vs. applied axial compression — experimental results to determine the first buckling load: $P_{VCT} = 16.4$ kN compared to $P_{exp.} = 16.19$ kN (adapted from [39]).

first frequency squared vs. the axial load in the vicinity of the experimental buckling load. A similar behavior is shown in Fig. 5.15 for the stringer stiffened (two stringers) laminated composite curved panel, D_{str1}, displaying $P_{VCT} = 29.6$ kN vs. the real buckling load $P_{exp.} = 29.5$ kN.

One should note the non-destructive nature of the predicted buckling using VCT ($P_{VCT} > P_{exp.}$ and in close proximity). Also, the stable

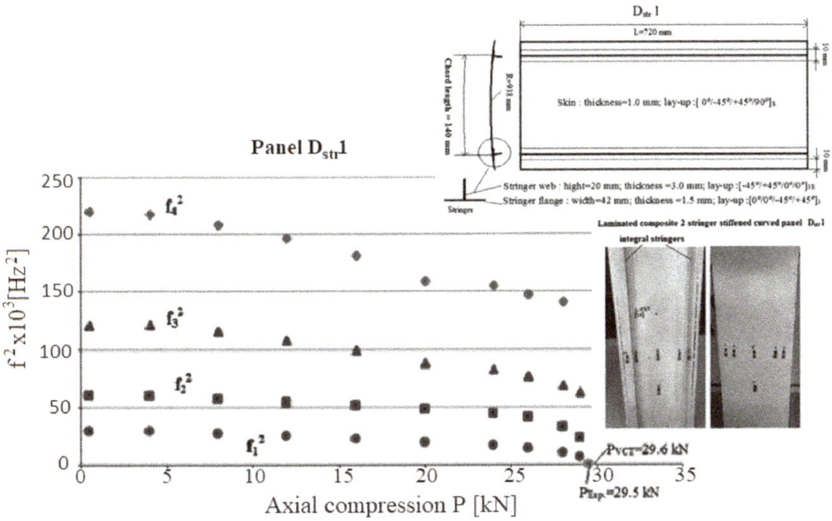

Fig. 5.15. Panel D_{str1} — Frequency squared vs. applied axial compression — experimental results to determine the first buckling load: $P_{VCT} = 29.6\,kN$ compared to $P_{exp.} = 29.5\,kN$ (adapted from [39]).

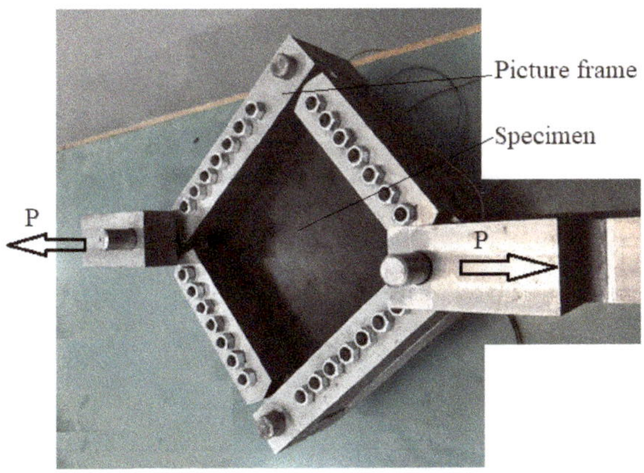

Fig. 5.16. Picture-frame apparatus to test square laminated composite plates under pure shear.

postbuckling behavior of a panel enables safe recording of natural frequencies at buckling without ruining the specimen.

To assess the applicability of the VCT to plates under pure shear, tests were initiated at the Krumbein Aerospace Structures Laboratory at Technion, I.I.T., Israel using a picture frame presented in Fig. 5.16. The frame has four hinges at it corners. Application of tensile load at two opposite corners (see Fig. 5.16) would induce compression in the other two corners, which will induce shear loads along the perimeter of the square plate. At a

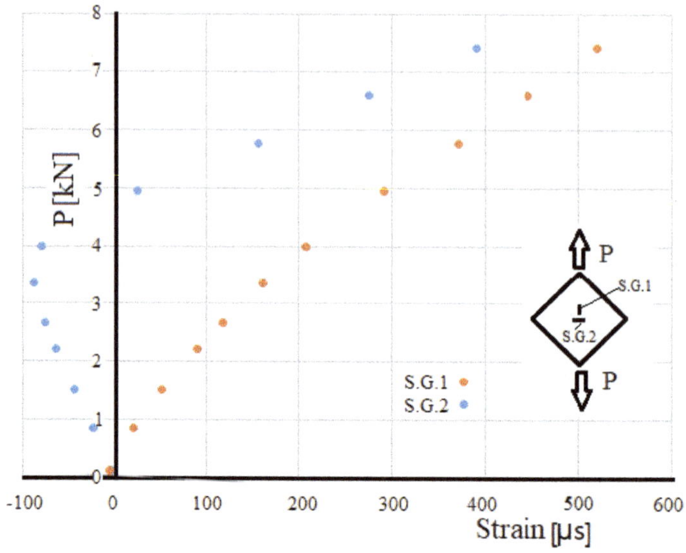

Fig. 5.17. Plate-1: Experimental compression and tensions strains vs. applied tensile loads.

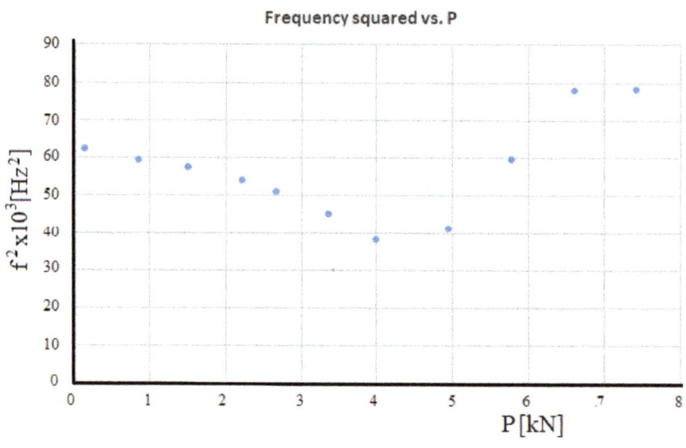

Fig. 5.18. Plate-1: Experimental first natural squared frequency vs. applied tensile loads.

certain load, the plate would buckle under shear. Using the modal hammer method, described above, natural frequencies were recorded as a function of applied load. Two strain gages were bonded to monitor the axial and lateral behavior of the plate (Fig. 5.17). The strain gages readings as a function of the applied load are presented in Fig. 5.17.

In parallel, the natural frequencies were also recorded. The tests results are presented in Fig. 5.18, where the first frequency squared is plotted vs.

the applied load. The frequency squared is monotonically reducing as the load is increased. Note that the frequency squared vs. the shear load shows a non-linear relations ship as reported in [35]. However at a certain load ($P = 4.6$ kN) the trend is changing direction, and the frequency squared start to climb up, while the load is still increasing. This tendency is also shown in Fig. 5.17 for strain gage 2. At $P = 4.6$ kN, the stain reading changes from negative (compressive strain) to positive (tensile strain) values. This value is considered the buckling load of the plate under shear, which can be calculated to be $V_{cr} = P/\sqrt{2} = 4.6/\sqrt{2} = 3.25\,\text{kN}$. It can be stated, that for a plate under shear loads, the VCT predicts very well the buckling load, as compared to experimental strain gages readings (see [38]).

Beside the above presented experimental results, the results presented in [40–43] display similar trend of reduction of the frequency squared with the increasing of the compression load. It is interesting to quote a recent experimental study presented by Shahgholian-Ghahfarokhi *et al.* [44] on composite sandwich plates with iso-grid cores. The application of the VCT for those specimens was done using a modified method (drawing a graph of $(1-p)^2$ vs. $(1-f^2)$), which will be described in the next shells chapter. The VCT results showed a very good agreement to actually performed tests, thus again showing the applicability and the high reliability of the method.

5.4. Cylindrical Shells

The last basic structure, a cylindrical shell (see Fig. 5.19) will be next discussed.

Buckling and natural frequencies of cylindrical shells had been widely treated in the literature (for example [45–68]). The specimens included stiffened and unstiffened, isotropic or laminated composite cylindrical shells under various boundary conditions, with and without initial geometric imperfections. However, a great discrepancy is found between experimental and numerical predictions for axially loaded cylindrical shells, with the experimental points being consistently lower than the predicted ones. This can be attributed to initial geometric imperfections, boundary conditions and load eccentricity found on the experimental specimens and not correctly taken into account when calculating the predicted buckling load. This leads to a well-known report, NASA SP 8007 [47] which introduced the concept of "knock-down factor", γ, defined as

$$\gamma \equiv P/P_{cr} = 1 - 0.902\left(1 - e^{-\frac{1}{16}\sqrt{\frac{R}{h}}}\right) \quad \text{for} \quad \frac{R}{h} < 1500, \tag{5.88}$$

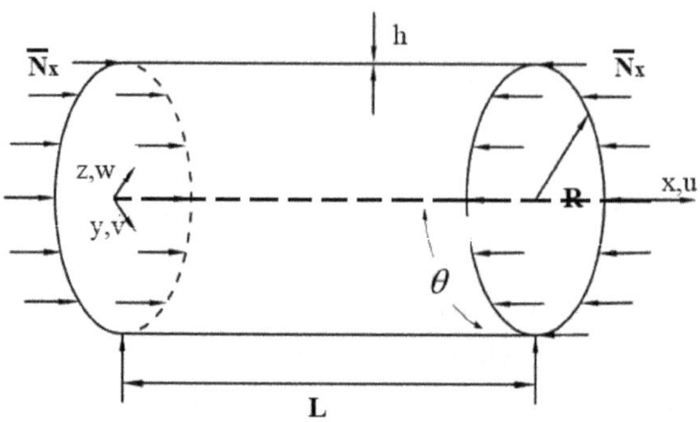

Fig. 5.19. A schematic drawing of a cylindrical shell loaded in compression.

where R and h are the radius and the thickness of the shell, respectively. Note that the value of the knock-down factor can reduce the axial carrying capacity of the shell by up to 70–80% (as a function of the ratio R/t). This is one of the drives behind the application of the VCT to shell-type structures.

The three governing equations for an isotropic cylindrical shell, with a radius R, thickness h, Young's modulus E and Poisson's ratio υ, while the three displacements are u, v and w (see Fig. 5.19), based on Donnell's equations (see [48]) can be written as, where $\bar{N}_x$ is considered positive in tension,

$$\frac{\partial^2 u}{\partial x^2}+\frac{(1-\upsilon)}{2R^2}\frac{\partial^2 u}{\partial \theta^2}+\frac{(1+\upsilon)}{2R}\frac{\partial^2 v}{\partial x\partial\theta}-\frac{\upsilon}{R}\frac{\partial w}{\partial x}=0,$$
$$\frac{(1+\upsilon)}{2R}\frac{\partial^2 u}{\partial x\partial\theta}+\frac{(1-\upsilon)}{2}\frac{\partial^2 v}{\partial x^2}+\frac{1}{R^2}\frac{\partial^2 v}{\partial\theta^2}-\frac{1}{R^2}\frac{\partial w}{\partial\theta}=0,$$
$$\upsilon R\frac{\partial u}{\partial x}+\frac{\partial v}{\partial\theta}-w-\left(\frac{h^2R^2}{12}\frac{\partial^4 w}{\partial x^4}+\frac{h^2}{6}\frac{\partial^4 w}{\partial x^2\partial\theta^2}+\frac{h^2}{12R^2}\frac{\partial^4 w}{\partial\theta^4}\right)$$
$$+\frac{(1-\upsilon^2)R^2\bar{N}_x}{Eh}\frac{\partial^2 w}{\partial x^2}=\frac{(1-\upsilon^2)R^2\rho}{E}\frac{\partial^2 w}{\partial t^2}. \quad (5.89)$$

Assuming a cylindrical shell with two simply supported SS3[c] type ends, namely

$$v(0)=v(L)=0,\quad w(0)=w(L)=0,$$
$$N_x(0)=N_x(L)=\frac{Eh}{(1-\upsilon^2)}\left[\frac{\partial u}{\partial x}+\upsilon\left(\frac{\partial v}{R\partial\theta}-\frac{w}{R}\right)\right]=0,$$
$$M_x(0)=M_x(L)=-D\left[\frac{\partial^2 w}{\partial x^2}+\upsilon\frac{\partial^2 w}{R^2\partial\theta^2}\right]=0, \quad (5.90)$$

[c]There are four types of simply supported boundary conditions: SS1: $w=M_x=N_x=N_{\grave{e}}=0$, SS2: $w=M_x=u=N_{\grave{e}}=0$, SS3: $w=M_x=N_x=v=0$, SS4: $w=M_x=u=v=0$ and four types of clamped boundary conditions: C1: $w=w_{,x}=N_x=N_{\grave{e}}=0$, C2: $w=w_{,x}=u=N_{\grave{e}}=0$, C3: $w=w_{,x}=N_x=v=0$, C4: $w=w_{,x}=u=v=0$.

and displacement components

$$
\begin{aligned}
u(x,\theta,t) &= A_{mn}\cos(mx)\cos(n\theta)e^{i\omega_{mn}t},\\
v(x,\theta,t) &= B_{mn}\sin(mx)\sin(n\theta)e^{i\omega_{mn}t},\\
w(x,\theta,t) &= C_{mn}\sin(mx)\cos(n\theta)e^{i\omega_{mn}t},
\end{aligned}
\tag{5.91}
$$

one can find the solution for the governing equations (5.90) by back substituting the expressions for the displacement in Eq. (5.90), yielding

$$
\begin{aligned}
&-A_{mn}\left(m^2+\frac{(1-\upsilon)}{2R^2}n^2\right)+B_{mn}\frac{(1+\upsilon)}{2R}mn-C_{mn}\frac{\upsilon}{R}m=0,\\
&A_{mn}\frac{(1+\upsilon)}{2R}mn-B_{mn}\left[\frac{(1-\upsilon)}{2}m^2+\frac{1}{R^2}n^2\right]+C_{mn}\frac{1}{R^2}n=0\\
&-A_{mn}\upsilon Rm+B_{mn}n\\
&-C_{mn}\left[1+\left(\frac{h^2R^2}{12}m^4+\frac{h^2}{6}m^2n^2+\frac{h^2}{12R^2}n^4\right)\right.\\
&\left.+\frac{\left(1-\upsilon^2\right)R^2\bar{N}_x}{Eh}m^2-\frac{\left(1-\upsilon^2\right)R^2\rho}{E}\omega^2\right]=0.
\end{aligned}
\tag{5.92}
$$

To obtain an unique solution, the determinant of the coefficients A_{mn}, B_{mn} and C_{mn} should be zero. This leads to the following relationship between the compression load $\bar{N}_x$ and the frequency squared ω^2 as a function of the longitudinal half waves m and the circumferential ones, n and all the other parameters of the shell:

$$
\begin{aligned}
&\frac{\left(1-\upsilon^2\right)R^2\rho}{E}\cdot\frac{(1-\upsilon)\left(m^2R^2+n^2\right)^2}{2}\omega^2\\
&-\frac{\left(1-\upsilon^2\right)m^2R^2}{Eh}\cdot\frac{(1-\upsilon)\left(m^2R^2+n^2\right)^2}{2}\bar{N}_x\\
&-\frac{(1-\upsilon)\left(m^2R^2+n^2\right)^2}{2}-\frac{(1-\upsilon)\left(m^2R^2+n^2\right)^4}{2}\cdot\frac{h^2}{12R^2}\\
&+\frac{(1-\upsilon)}{2}\left[\upsilon^2m^4R^4+2m^2R^2n^2+n^4\right]=0.
\end{aligned}
\tag{5.93}
$$

The buckling load, $\bar{N}_{x,cr}$ and the natural frequencies at zero axial load, ω_{mn}^2 can be obtained from Eq. (5.93), yielding

$$\omega_{mn}^2 = +\frac{E}{(1-\upsilon^2)R^2\rho} + \frac{E(m^2R^2+n^2)^2}{(1-\upsilon^2)R^2\rho}\cdot\frac{h^2}{12R^2} - \frac{E\cdot[\upsilon^2m^4R^4+2m^2R^2n^2+n^4]}{(m^2R^2+n^2)^2(1-\upsilon^2)R^2\rho}, \tag{5.94}$$

$$\bar{N}_{x,cr} = -\left\{\begin{array}{c}\dfrac{E\cdot h}{(1-\upsilon^2)\,m^2R^2} + \dfrac{E\cdot h\cdot(m^2R^2+n^2)^2}{(1-\upsilon^2)\,m^2R^2}\cdot\dfrac{h^2}{12R^2} \\ -\dfrac{E\cdot h\cdot[\upsilon^2m^4R^4+2m^2R^2n^2+n^4]}{(m^2R^2+n^2)^2(1-\upsilon^2)m^2R^2}\end{array}\right\}.$$

Taking into account the definition of the critical parameters in Eq. (5.93), Eq. (5.93) can be written in the following compact form

$$\frac{\omega^2}{\omega_{mn}^2} + \frac{\bar{N}_x}{\bar{N}_{x,cr}} = 1 \quad \text{or} \quad \frac{f^2}{f_{mn@\text{zeroload}}^2} + \frac{P}{P_{cr}} = 1$$

$$\Rightarrow f^2 + p = 1 \tag{5.95}$$

where

$$p = \frac{\bar{N}_x}{\bar{N}_{x,cr}} = \frac{P}{P_{cr}} \quad \text{and} \quad f^2 = \frac{\omega^2}{\omega_{mn}^2} = \frac{f^2}{f_{mn@\text{zeroload}}^2} \tag{5.96}$$

Therefore, it can be concluded that also for cylindrical shells, the natural frequency squared depends linearly with the applied axially compressed load, as depicted before for beams, columns and plates.

However, as it was demonstrated before for plates, test results do not show the relation presented in Eq. (5.95). A typical graph of the natural frequency squared vs. the applied axial compression (see Fig. 5.20) presents an experimental linear regression, however the test buckling load is much lower than the extrapolation to zero of the linear line. Trying to apply the VCT using a linear regression line would result in a large error for the prediction of the *in-situ* buckling load, with the P_{VCT} being much larger than the $P_{\text{exp.}}$.

Many efforts had been invested to try to apply the VCT in a reliable way. This can be found in the early studies of Souza *et al.* [54–56]. Using the experimental data of the Technion, I.I.T., in [50–51] he presented a new approach to a non-destructively prediction of the *in-situ* buckling load of aluminum stringer–stiffened cylindrical shells. Based on their assumption that a compressed cylindrical shell can be approximated by a strut and an arch interconnected at their centers, and loaded axially along the strut. The relation between the non-dimensional axial compression, p (see Eq. (5.96)), and

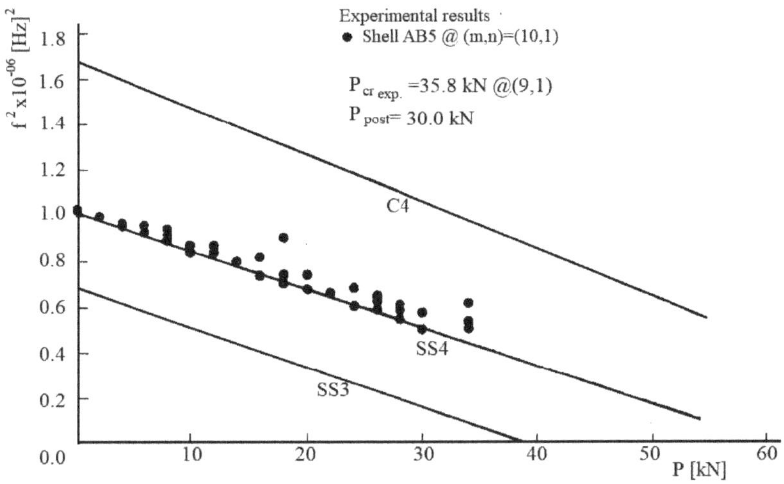

Fig. 5.20. Shell AB5: experimental results — natural frequency squared vs. axial compression for an aluminum stringer-stiffened cylindrical shell (adapted from [53]).

the non-dimensional natural frequency, f (see Eq. (5.96)), has the following form:

$$(1-p)^2 + (1-\xi^2)(1-f^4) = 1, \tag{5.97}$$

where ξ is the "experimental" knock-down factor based on the results of the test, at relatively low loads. The procedure starts with the acquisition of the natural frequencies at zero axial load. Then the load is increased and the non-dimensional frequency f is calculated for each load step, P which is normalized by the numerical buckling load P_{cr} to yield the variable p. At about 60% of the calculated buckling load the test is stopped, a straight line is drawn, starting at point $[(1-f^4)=0, (1-p)^2=1]$ till point $[\xi^2, (1-f^4)=1]$. The value of ξ^2 is determined as the cross point between the oblique line and the vertical line from $(1-f^4)=1$. The results for shell RO-34 are presented in Fig. 5.21, to yield $\xi^2 = 0.09$. Then the predicted buckling load will be calculated as

$$P_{\text{pred.}} = (1-\xi)P_{cr}. \tag{5.98}$$

For shell RO-34 the value of Eq. (5.98) is $P_{\text{pred.}} = (1-0.3)64.09 = 44.86\,\text{kN}$ in comparison with $P_{\text{exp.}} = 42.20\,\text{kN}$. The predicted buckling value is 6.3% over the experimental one, showing its nondestructive nature of the present method ($P_{\text{pred.}} > P_{\text{exp.}}$).

In the last decade, a high number of manuscripts, dealing with the application of VCT to laminated composite shells, under axial compression [57–70] had been published. While Jansen *et al.* [57] tried to develop a new

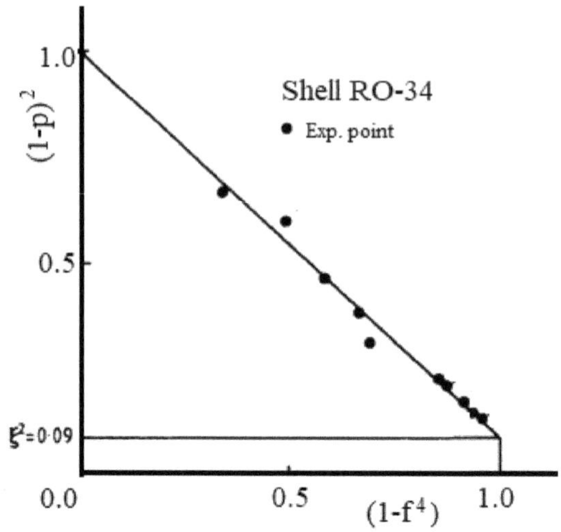

Fig. 5.21. Shell RO-34: Application of Souza's approach [54] — experimental results — natural frequency squared vs. axial compression for an aluminum stringer-stiffened cylindrical shell (adapted from [54]).

numerical approach to take into account the influence of the initial geometric imperfections on the natural frequencies of axially loaded cylindrical shells, all other publications [58–70] presented also experimental results. All of them are based on the approach proposed by Arbelo *et al.* [58–60]. The empirical approach is based on the modification of the method presented by Souza *et al.* [54]. The basic relation $1 - p = f^2$ is rewritten to yield the following relationship:

$$1 - p = f^2 = 1 - (1 - f^2) \Rightarrow (1 - p)^2 = [1 - (1 - f^2)]^2. \tag{5.99}$$

Then a graph of $(1 - p)^2$ vs. $(1 - f^2)$ is constructed as schematically depicted in Fig. 5.22. Following the definitions of p and f, presented in Eq. (5.96), a best fit second-order equation is approximated based on the experimental points. Based on Fig. 5.22, the fitted second-order polynomic curve of the measured experimental natural frequencies would have the following expression:

$$(1 - p)^2 = \alpha(1 - f^2)^2 + \beta(1 - f^2) + \chi \tag{5.100}$$

with the values of the constants A, B and C being determined by the best fit process. Finding the minimal point of the second order equation presented in Eq. (5.100) yields

$$(1 - f^2)_{\min} = -\frac{\beta}{2\alpha} \rightarrow (1 - p)^2_{\min} = \chi - \frac{\beta^2}{4\alpha} \equiv \xi^2 \tag{5.101}$$

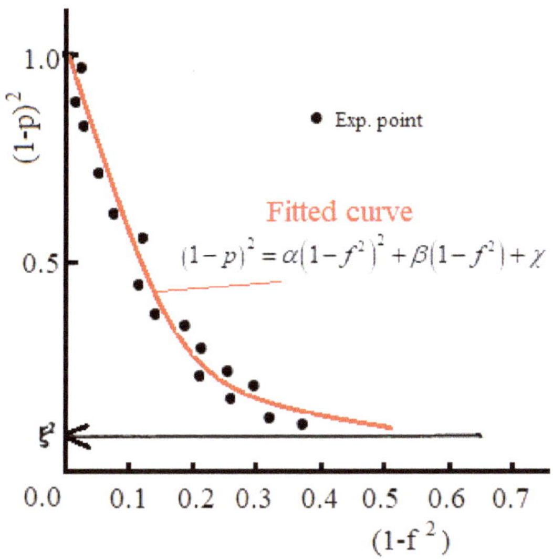

Fig. 5.22. Schematic determination of the predicted buckling load using VCT — according to Arbelo *et al.* [58–60] method.

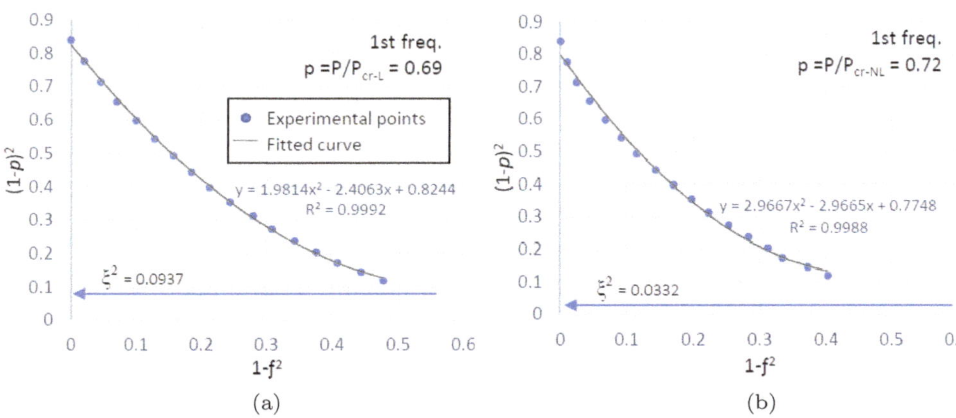

Fig. 5.23. Classical Laminated Shell — $(1-p)^2$ vs. $(1-f^2)$: (a) First Natural frequency for experimental results till $P/P_{cr-L} = 0.69$;, (b) First Natural Frequency for experimental results till $P/P_{cr-NL} = 0.72$ (adapted from [64]).

with ξ being the "knock-down factor" which represents the drop in the shell load carrying capacity, like Souza *et al.* [54] method, presented above, leading to the prediction of the *in-situ* tested specimen using the following form:

$$P_{\text{pred.}} = (1-\xi)P_{cr}. \tag{5.102}$$

Typical results for the application of Arbelo *et al.* [58–60] method is presented in Fig. 5.23, for a classical laminated shell described in [64].

More details about the excitation method can be found in Appendix D. The value of P_{cr-L} mentioned in the caption of Fig. 5.23 is 420 kN, while $P_{cr-NL} = 325$ kN. The predicted buckling load based on points up to $p = P/P_{cr-L} = 0.69$ is 291 kN while for $p = P/P_{cr-NL} = 0.72$, $P_{\text{pred.}} = 266$ kN, with the experimental value being 303 kN. Without doubt, using a linear code to predict the buckling load, P_{cr-L}, yields a precise buckling load for the *in-situ* laminated composite cylindrical shell, within 4% of the experimental one ($\rho = 291/303 = 0.96$). Note that the VCT predicted value of the buckling load is less, but within 4% of the experimental one, clearly representing its nondestructive nature of the method.

The same tendency of accurate, non-destructive buckling predictions are also reported by the other publications [58–63, 65–68], surely displaying the successful application of VCT to laminated composite cylindrical shells. It is important to note, that Franzoni *et al.* [65], succeeded to give an analytical meaning of the experimental method suggested by Arbelo *et al.* [58–60]. It is claimed in [65] that, due to initial geometric imperfections, the natural frequency of an isotropic cylindrical shell cannot reach zero at buckling. It is shown that the minimum value of the applied load squared, $(1-p)^2$ occurs at the point where a drop in the axially load-carrying capacity would appear due to the initial geometric imperfections, represented by $\xi^2 = (1 - P_{\text{imp}}/P_{cr})^2$. Therefore, minimizing the function presented by Eq. (5.100), namely $(1-p)^2 = \alpha(1-f^2)^2 + \beta(1-f^2) + \chi$, finding the ξ^2 and then calculating the predicted buckling value $P_{\text{pred.}} = (1-\xi)P_{cr}$ (see Eq. (5.102)), as suggested by the empirical method of Arbelo *et al.* [58–60] fits the theoretical model suggested by Franzoni *et al.* [65].

5.5. The Influence of Initial Geometric Imperfections

The influence of initial geometric imperfections on both the buckling load and natural frequencies for plates and cylindrical shells had been widely investigated and published in the literature. Typical references for isotropic and anisotropic plates can be found in [74–80] while for isotropic and anisotropic cylindrical shells in [81–90]. The researchers in [75, 76] report "a significant increase in the vibration frequencies may occur for imperfection amplitudes in the order of a fraction of the plate thickness, even in the absence of in-plane forces". In [79] it is stated that as a function of the applied compression, the frequencies of the laminated plates with various shapes of initial imperfections may be significantly higher than those for perfect plates, yielding typical schematic graphs as presented in Fig. 5.24.

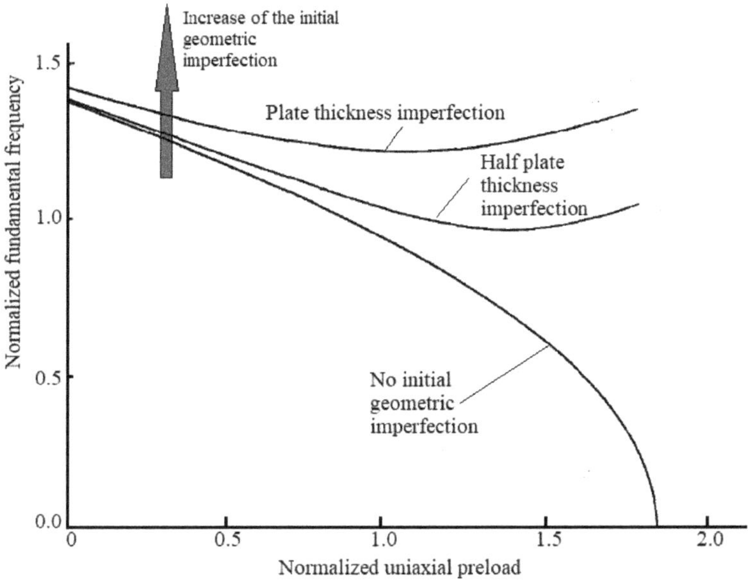

Fig. 5.24. Non-dimensional circular frequency squared vs. non-dimensional axial compression — a schematic graph based on [79].

Note that the graph in Fig. 5.24 is similar to Fig. 5.11, further strengthening the Massonnet [36] claim regarding the strong influence of the initial geometric imperfections on the natural frequencies of circular plates, while extending it to rectangular plates.

Note that Ilanko and Dickinson [77, 78] present in their manuscripts similar results as those displayed in Fig. 5.24, with experimental results fitting well to the numerical ones. The numerical results were obtained using the Rayleigh-Ritz method.

Kapania and Yang [80] present in their extensive review numerical procedures to study the influence of initial geometric imperfections on isotropic and laminated thin plates, quoting the studies of Ilanko and Dickinson [77, 78] and Hui and Leissa [75] and presenting similar results as being quantitatively displayed in Fig. 5.24.

As stated above, the influence of the initial geometric imperfections on the buckling and natural frequencies of circular cylindrical shells had been well investigated as one can see in the typical references [81–90] brought in the present review. In a recent chapter, (Chapter 10, from [31]), Jansen addressed this topic. He presented a general theory to investigate the effect of imperfections on stability and vibrations of shells, and in particular composite cylindrical shells. Using perturbation analysis, and applying various types of geometric imperfections he succeeded in predicting the sudden drop

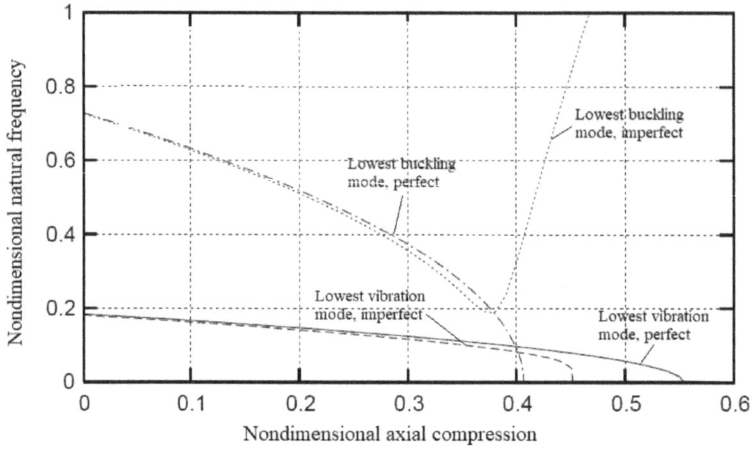

Fig. 5.25. The influence of axial load on linearized frequency of anisotropic shell (Booton's shell [81]: non-dimensional axisymmetric imperfection amplitude = $\xi_1 = -0.04$, nondimensional asymmetric imperfection amplitude = $\xi_2 = 0.05$ (adapted from [31]).

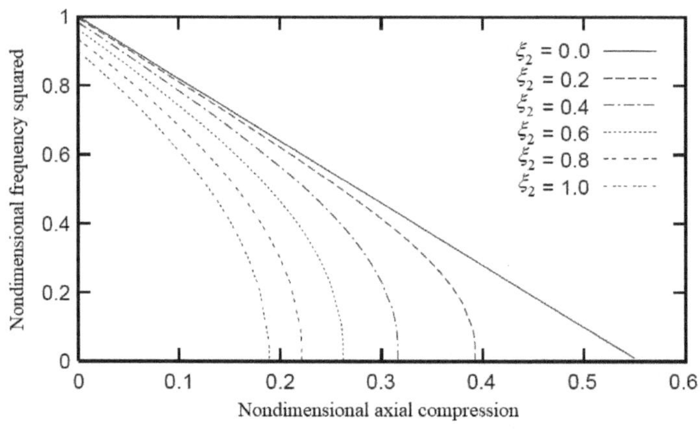

Fig. 5.26. The influence of axial load on linearized frequency of anisotropic shell with asymmetric imperfection (Booton's shell [81], lowest vibration mode: $\frac{W}{h} = \xi_2 \sin\frac{\pi x}{L}\cos\frac{6}{R}(y - 0.002x)$ (adapted from [31])).

of the frequency squared due to those imperfections using the shell data provided in [81] (see Table 5.2) as can be seen in Figs. 5.25–5.27 (the imperfection amplitudes were normalized by the shell thickness). The study presents results for various types of geometric imperfections with consistent behavior throughout the investigated parameters.

To sum up this section, it can be stated that the initial geometric imperfections are the main drive for the "peculiarities" encountered when testing plates or cylindrical shells for natural frequencies under loads.

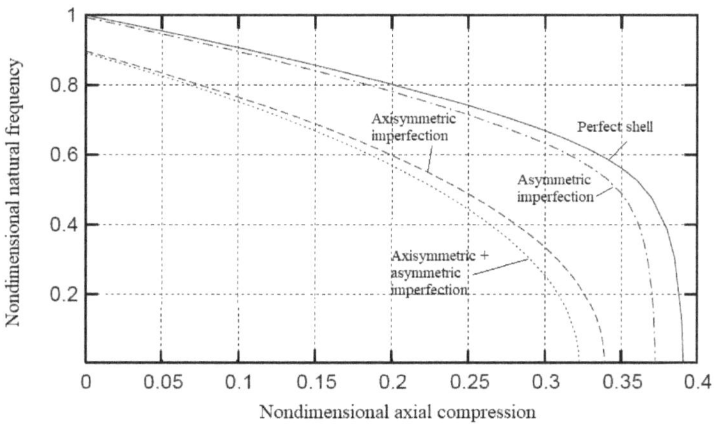

Fig. 5.27. The influence of imperfections and axial loading on the linearized frequency of anisotropic shell (Booton's shell [81]: imperfection mode: $\frac{W}{h} = -0.25\cos\frac{2m\pi x}{L} + 0.25[w_1(x)\cos n\theta + w_2(x)\cos n\theta]$ (adapted from [31])).

Table 5.2. Booton's shell data [81].

Shell geometry	R (radius) = 2.67 in = 0.06782 m
	L (length) = 3.776 in = 0.09591m
Laminate geometry	Three layers (the numbering starts at the outside of the shell)
	$h_1 = h_2 = h_3$ (layer thickness) = 0.0089 in = 0.000226 m
	$\theta_1 = 30°, \theta_1 = 0°, \theta_1 = -30°$ (layer orientation)
Layer properties	Glass epoxy (composite material)
	$E_{11} = 5.83 \cdot 10^6$psi = 40.196 GPa (major modulus of elasticity)
	$E_{22} = 2.42 \cdot 10^6$psi = 16.685 GPa (minor modulus of elasticity)
	$\upsilon_{12} = 0.363$ (major Poisson's ratio)
	$G_{12} = 6.68 \cdot 10^5$psi = 4.6056 GPa (shear modulus of elasticity)

5.6. Conclusions and Recommendations

The Vibration Correlation Technique (VCT) is today applied to all basic thin-walled structures, like columns (beams), plates, panels and shells yielding very good predictions for the experimental buckling load.

Recent publications provided new procedures to apply the VCT to cylindrical shells yielding an effective nondestructive method.

Changing the normalizing load to a nonlinear buckling load or reducing the number of experimental points would lead to worst predictions, and therefore it is recommended to choose the linear value as the reference.

Consistent measurement of the natural frequencies and their associated mode shapes using advanced equipment, like POLYTEC PSV 500 scanning laser vibrometer [63] will provide consistent and repetitive results. This conclusion and recommendation are based on the successful use of the scanning laser vibrometer by various researchers as reported in [58–68].

The lack of the capability to measure the mode shapes should not prevent the application of the described non-destructive method. The measured mode shapes serve only to monitor the correct path of the experimental natural frequencies as a function of the compression load.

Using a loudspeaker as a means to excite the tested shell proved to be a reliable source of excitation (although located outside the specimen) yielding consistent and repeatable results as described and successfully used in [64], although other means like modal hammer (successfully applied in [39, 44]) and are also viable.

It is recommended to have at least 15 experimental point to reach a reliable buckling prediction, using the VCT.

For cases where the lowest natural frequency cannot be detected, and the prediction is done using the second lowest frequency, the buckling prediction will be less accurate (90% of the experimental buckling load, compared to 96% when using the lowest natural frequency [64]), however the non-destructive nature of the VCT is preserved. One should note that according to data presented in [58–68], the lowest natural frequency will provide the best prediction using the VCT when applied to cylindrical shells. The same is true also for columns or beams under compressive loads, as the buckling mode is similar to the first natural frequency mode. In case of plates, the buckling and the natural frequency modes are dependent on the aspect ratio of the plate, and the VCT requirement of a similarity between the buckling and natural frequency mode might lead to a best buckling prediction when using higher modes (although this was still not reported in the literature). The results presented in [39, 44] to predict the buckling load of compressed plates using VCT, show that the use of the first natural frequency provided the best buckling prediction in comparison with the use of higher frequency modes (see details in [39]).

Although the method reached its relatively maturity, it is advised that more tests results should be provided mainly for plates under shear loading to construct a sound database for the application of VCT as a non-destructive method also for this type of loading. In addition, it is advised to test other shapes of shells under increasing loads and investigate the application of the VCT.

References

[1] Singer, J., Arbocz, J. and Weller, T. *Buckling experiments: Experimental Methods in Buckling of Thin-Walled Structures. Shells Built-Up Structures, Composites and Additional Topics*, Vol. 2, Chapter 15. John Wiley & Sons, Inc., New York, 2002, 1109 p.

[2] Lurie, H., Lateral vibrations as related to structural stability, Ph.D. thesis, California, Institute of Technology, Pasadena, CA, USA, 1950, 97 p.

[3] Lurie, H., Effective end restraint of columns by frequency measurements, *J. Aeronautical Sci.,* Vol. 18, 1951, pp. 556–557.

[4] Lurie, H., Lateral vibrations as related to structural stability, *J. Appl. Mech. ASME*, Vol. 19, 1952, pp. 195–204.

[5] Abramovich, H. and Kalnins, K., The vibration correlation technique revisited *Proc. 11th Int. Symp. Vib. Contin. Syst.* (Llanberis, Great Britain), 2017, pp. 1–3. http://www.isvcs.org/ISVCS11-Proceedings.pdf.

[6] Bokaian, A., Natural frequencies of beams under tensile axial loads, *J. Sound Vibration*, Vol. 142(3), 1990, pp. 481–498.

[7] Bokaian, A., Natural frequencies of beams under compressive axial loads, *J. Sound Vibration*, Vol. 126(1), 1988, pp. 49–65.

[8] Dhanaraj, R. and Palaninathan, Free vibration of initially stressed composite laminates, *J. Sound Vibration*, Vol. 142(3), 1990, pp. 365–378.

[9] Abramovich, H., *Advanced Aerospace Materials — Aluminum-Based and Composite Materials*, 2019, Chapter 7, Walter de Gruyter GmbH, Berlin/Boston, 308 p.

[10] Abramovich, H., Natural frequencies of Timoshenko beams under compressive axial loads, *J. Sound Vibration*, Vol. 157(1), 1992, pp. 183–189.

[11] Abramovich, H., Thermal buckling of cross-ply composite laminates using a first-order shear deformation theory, *Composite Struct.*, Vol. 28, 1994, pp. 201–213.

[12] Abramovich, H., A new insight on vibrations and buckling of a cantilevered beam under a constant piezoelectric actuation, *Composite Struct.,* Vol. 93, 2011, pp. 1054–1057.

[13] Abramovich, H., Axial stiffness variation of thin walled laminated composite beams using piezoelectric patches — A new experimental insight, *Int. J. Aeronautics and Aerospace Res.*, 3(2), 2016, pp. 97–105.

[14] Johnson, E.E. and Goldhammer, B.F., The determination of the critical load of a column or stiffened panel in compression by the vibration method, *Proc. Soc. Exp. Stress Analy.,* Vol. 11(1), 1953, pp. 221–232.

[15] Jacobson, M.J., and Wenner, M.L., Predicting buckling loads from vibration data, *Exp. Mech.,* Vol. 8(10), 1968, pp. 35N–38N.

[16] Balch, Jr., R.L., An experimental nondestructive method for determining the buckling load of a column with arbitrary end conditions, M.Sc. thesis, Oregon State College, Oregon, USA, June 1959, 45 p.

[17] Jubb, J.E.M., Phillips, I.G. and Becker, H., Interrelation of structural stability, stiffness, residual stress and natural frequency, *J. Sound Vibration*, Vol. 39(1), 1975, pp. 121–134.

[18] Belvin, W.K., Analytical and experimental vibration and buckling characteristics of a pretensioned stayed column, *J. Spacecraft*, Vol. 21(5), 1984, pp. 456–462, doi.org/10.2514/3.25680.

[19] Segal, A. and Baruch, M., A nondestructive dynamic method for the determination of the critical load of elastic columns, *Exp. Mech.*, Vol. 20(8), 1980, pp. 285–288.

[20] Coulter, B.A. and Miller, R.E., Vibration and buckling of beam-columns subjected to non-uniform axial loads, *Int. J. Numer. Methods Eng.*, Vol. 23, 1986, pp. 1739–1755.
[21] Lake, M.S. and Mikulas, Jr., M.M., Buckling and vibration analysis of a simply supported column with a piecewise constant cross, section, NASA TP 3090, March 1991, 18 p.
[22] Plaut, R.H., and Virgin, L.N. Use of frequency data to predict buckling. *J. Eng. Mech.*, 116(10), 1990, pp. 2330–2335.
[23] Virgin, L.N., Plaut, R.H. Effect of axial load on forced vibrations of beams, *J. Vibration*, Vol. 168(3), 1993, 395–405.
[24] Virgin, L.N., *Vibration of Axially Loaded Structures*, 2007, Cambridge University Press, 2007, 351 p.
[25] Chailleux, A., Hans, Y. and Verchery, G., Experimental study of the buckling of laminated composite columns and plates, *Int. J. Mech. Sci.*, Vol. 17, 1975, pp. 489–498.
[26] Tasopoulos, A.A., The relation between the buckling load factor and the fundamental frequency of a structure, *Structures Congress 2005, ASCE Conf. Proc.*, Vol. 171, New York, April 20–24, 2005, 17 p.
[27] Fässler, S.,Axial load determination using modal analysis, Rapport TVBK, Avdelningen för Konstructionsteknik, Lunds Tekniska Högskola, Lunds University, Sweden, 2014, 129 p.
[28] Carpinteri, A. and Paggi, M., A theoretical approach to the interaction between buckling and resonance instabilities, *J. Eng. Math.*, Vol. 78, 2013, pp. 19–35, doi:10.1007/s10665-011-9478-0.
[29] Pagani, A., Augello, R., and Carrera, E., Virtual vibration correlation technique (VCT) for nonlinear analysis of metallic and composite structures, *Proc. Int. Mechanical Engineering Congress & Exposition, IMECE* 2018, Pittsburg, USA, IMECE2018-86852, Nov. 09–15, 2018, 6 p, doi:10.1115/IMECE2018-86674.
[30] Żmuda-Trzebiatowski, Ł., Investigation of natural frequencies of axially loaded thin-walled columns, *MATEC Web of Conferences*, Vol. 219, 2018, Paper Id:02018, 8 p., doi:10.1051/matecconf/201821902018.
[31] Abramovich, H., *Stability and Vibrations of Thin Walled Composite Structures*, Woodhead Publishing Limited, 2017, 778 p.
[32] Kennedy, D., Williams, F.W. and Anderson M.S., Buckling and vibration analysis of laminated panels using VICONOPT, *J. Aerospace Eng.*, Vol. 7(3), 1994, pp. 245–262.
[33] Williams, F.W. and Anderson, M.S., Buckling and vibration analysis of shear-loaded prismatic plate assemblies with supporting structures, utilizing symmetric or repetitive cross-sections, in *Aspects of the Analysis of Plate Structures — A Volume in Honour of W.H. Wittrick*, D.J. Dawe, R.W. Horsington, A.G. Kamtekar and G.H.Little, (eds.,), Oxford University Press, Oxford, 1985, pp. 51–71.
[34] O'Leary, O.J., Williams, F.W. and Kennedy, D., Optimum stiffened panel design with fundamental frequency constraint, *Thin-Walled Struct.*, Vol. 39, 2001, pp. 555–569.
[35] Kennedy, D. and Lo, K.I., Critical buckling predictions for plates and stiffened panels from natural frequency measurements, *J. Phys. Conf. Ser.* Vol. 1106, 2018, Article Id 012018, 8 p.
[36] Massonnet, Ch., Le voilement des plaques planes sollicitees dans leur plan (Buckling of plates), *Final Report of the 3rd Congress of the International Association for Bridge and Struct., Eng.*, Liege, Belgium, September 1948, pp. 291–300.
[37] Abramovich, H., Gil, J., Grunwald, A. and Rosen, A., Vibrations and buckling of radially loaded circular plates, *TAE 332*, Department of Aeronautical Engineering, Technion, Israel Institute of Technology, Haifa, Israel, July 1978, 43 p.

[38] Abramovich, H., The Vibration Correlation Technique – A Reliable Nondestructive Method to Predict Buckling Loads of Thin Walled Structures, Thin Walled Structures, Vol. 159, Feb. 2021, Paper Id: 107308, 17 p.

[39] Abramovich, H., Govich, D., and Grunwald, A. Buckling prediction of panels using the vibration correlation technique, *Prog. Aerospace Sci.*, Vol. 78, 2015, pp. 62–73.

[40] Giaedi, A.A., A study on the effect of stress level on the vibration frequency of structures, Ph.D. thesis, Faculty of Engineering, University of Glasgow, Department of Aeronautics and Fluid-Mechanics, 1986, 206 p.

[41] Mandal, P., Prediction of buckling load from vibration measurements. In Drew H.R., and Pellegrino S. (eds) *New Approaches to Structural Mechanics, Shells and Biological Structures. Solid Mechanics and Its Applications*, Vol. 104. Springer, Dordrecht, 2002, doi:10.1007/978-94-015-9930-6_15.

[42] Singhatanadgid, P. and Sukajit, P., Determination of buckling load of rectangular plates using measured vibration data, *Proc. SPIE*, Vol. 7375, 2008, 737532, doi: 10.1117/12.839274.

[43] Singhatanadgid, P. And Sukajit, P., Experimental determination of the buckling load of rectangular plates using vibration correlation technique, *Struct. Eng. Mech.*, Vol. 7(3), 2011, pp. 31–349, doi:10.12989/sem.2011.37.3.331.

[44] Shahgholian-Ghahfarokhi, D., Aghaei-Ruzbahani, M. And Rahimi, G., Vibration correlation technique for the buckling load prediction of composite sandwich plates with iso-grid cores, *Thin-Walled Struct.*, Vol. 142, 2019, pp. 392–404.

[45] Mao, R. and Williams, F.W., Post-critical behavior of orthotropic circular cylindrical shells under time dependent axial compression, *J. Sound Vibration*, Vol. 201(3), 1998, pp. 307–327.

[46] Radhakrishnan, R., Prediction of buckling strengths of cylindrical shells from their natural frequencies, *Earthquake Eng. Struct. Dynam.*, Vol. 2, 1973, pp. 107–115.

[47] Weingarten, V.I. and Peterson, J.P., Buckling of thin-walled circular cylinders, NASA SP-8007 August 1968, Revised September 1986.

[48] Amabili, M., *Nonlinear Vibrations and Stability of Shells and Plates*, Cambridge University Press, 2008, 374 p.

[49] Rosen, A. and Singer J., Vibrations of axially loaded stiffened cylindrical shells, *J. Sound Vibration*, Vol. 34(3), 1974, pp. 357–378.

[50] Singer, J. and Rosen, A., Design criteria for buckling and vibration of imperfect stiffened shells, on I*CAS Proceedings*, Dexter, R.R. and Singer J. (eds.), Weizmann Science Press of Israel, Jerusalem, 1974, pp. 495–517.

[51] Rosen, A. and Singer J., Effect of axisymmetric imperfections on the vibrations of cylindrical shells under axial compression, *AIAA J.* Vol. 12(7), 1974, pp. 995–997, doi:10.2514/3.49393.

[52] Abramovich, H. and Singer, J., Correlation between vibrations and buckling of cylindrical shells under external pressure and combined loading, *Israel J. Techno.,* Vol. 16(1–2), 1978, pp. 34–44.

[53] Singer, J. and Abramovich, H., Vibration correlation techniques for definition of practical boundary conditions in stiffened shells, *AIAA J.*, Vol. 17(7), 1979, pp. 762–769.

[54] Souza, M.A., Fok, W.C. and Walker, A.C. Review of experimental techniques for thin-walled structures liable to buckling. Part I — Neutral and unstable buckling, *Exp. Tech.,* Vol. 7(9), 1983, pp. 21–25.

[55] Souza, M.A., Fok, W.C. and Walker, A.C. Review of experimental techniques for thin-walled structures liable to buckling — Part II: Stable buckling, *Exp. Tech.,* Vol. 7(10), 1983, pp. 36–39.

[56] Souza, M.A. and Assaid, L.M.B. A new technique for the prediction of buckling loads from nondestructive vibration tests. *Exp. Mech.,* 1991, Vol. 31(2), pp. 93–97.
[57] Jansen, E.L., Abramovich, H. and Rolfes, R., The direct prediction of buckling loads of shells under axial compression using VCT — A revisited and upgraded approach, *ICAS 2014, 29th Cong. Int. Council of the Aeronautical Sciences*, St. Petersburg, Russia, Sept. 7–12, 2014.
[58] Arbelo, M.A., de Almeida, S.F.M., Donadon, M.V., Rett, S.R., Degenhardt, R., Castro, S.G.P., Kanins, K. and Ozolins, O., Vibration correlation technique for the estimation of real boundary conditions and buckling load of unstiffened plates and cylindrical shells, *Thin-Walled Struct., 2014*, Vol. 79, 2014, pp. 119–128.
[59] Arbelo M A, Kalnins K, Ozolins O, Skukis, E., Castro, S.G.P. and Degenhardt, R., Experimental and numerical estimation of buckling load on unstiffened cylindrical shells using a vibration correlation technique, *Thin-Walled Struct.*, 2015, Vol. 94, pp. 273–279.
[60] Kalnins, K., Arbelo, M.A., Ozolins, O., Skukis, E., Castro, G.P. and Degenhardt, R., Experimental nondestructive test for estimation of buckling load on unstiffened cylindrical shells using vibration correlation technique, *Shock Vibration*, Vol. 2015, 2015, Article Id. 729684, 8 p.
[61] Skukis, E., Ozolins, O., Kalnins, K. and Arbelo, M.A., Experimental test for estimation of buckling load on unstiffened cylindrical shells by vibration correlation technique. *Procedia Eng.,* Vol. 172, 2017, pp. 1023–1030.
[62] Skukis, E., Ozolins, O., Andersons, J. *et al.* Applicability of the vibration correlation technique for estimation of the buckling load in axial compression of cylindrical isotropic shells with and without circular cutouts. *Shock Vibration*, Vol. 2017, Article ID 2983747, 2017, 15 p., https://doi.org/10.1155/2017/2983747.
[63] Shahgholian-Ghahfarokhi, D. and Rahimi, G. Buckling load prediction of grid-stiffened composite cylindrical shells using the vibration correlation technique, *Composite Sci. Techn.*, Vol. 167, 2018, pp. 470–481.
[64] Labans, E., Abramovich, H. and Bisagni, C. An experimental vibration-buckling investigation on classical and variable angle tow composite shells under axial compression. *J. Sound Vibration*, 449, (2019), pp. 315–329.
[65] Franzoni, F., Degenhardt, R., Albus, J. and Arbelo, M.A., Vibration correlation technique for predicting the buckling load of imperfection-sensitive isotropic cylindrical shells: An analytical and numerical verification. *Thin-Walled Struct.,* Vol. 140, 2019, pp. 236–247.
[66] Franzoni, F., Odermann, F., Labans, E., Bisagni, C., Arbelo, M.A. and Degenhardt, R., Experimental validation of the vibration correlation technique robustness to predict buckling of unstiffened composite cylindrical shells. *Composite Struct.,* Vol. 224, 2019, 111107.
[67] Franzoni, F., Odermann, F., Wilckens, D., Kalnins, K., Arbelo, M.A. and Degenhardt, R., Assessing the axial buckling load of a pressurized orthotropic cylindrical shell through vibration correlation technique. *Thin-Walled Struct.,* Vol. 137, 2019, pp. 353–366.
[68] Shahgholian-Ghahfarokhi, D. and Rahimi, G. H., Prediction of the critical buckling load of stiffened composite cylindrical shells with lozenge grid based on the nonlinear vibration analysis; *Modares Mech. Eng.,* Vol. 18(4), 2018, pp. 135-143.
[69] Ilanko, S., Monterrubio, L.E. and Mochida, Y., *The Rayleigh-Ritz Method for Structural Analysis*, John Wiley & Sons, 230 p, Chapter 2, 2014.
[70] Galef, A.E., Bending frequencies of compressed beams, *J. Acoust. Soc. Am.*, Vol. 44, 1968, 643 p.

[71] Amba-Rao, C.L., Effect of end conditions on the lateral frequencies of uniform straight columns, *Acoust. Soc. America*, Vol. 42, 1967, pp. 900–901.
[72] Valle, J., Fernández, D. and Madrenas, J., Closed-form equation for natural freqeuncies of beams under full range of axial loads modeled with a spring-mass system, *Int. J. Mech. Sci.,* Vols. 153–154, 2019, pp. 380–390.
[73] Guédé, Z. and Elishakoff, I., Apparently first closed-form solutions for inhomogeneous vibrating beams under axial loading, *Proc. Math. Phys. Eng. Sci. Royal Soc.*, Vol. 457, (2007), 2001, pp. 623–649.
[74] Silver, R.L.,The effect of initial curvature on the parametric vibration of rectangular plates subjected to an in-plane sinusoidal load, Ph.D. Dissertation, Syracuse University, Syracuse, NY, USA, 1972.
[75] Hui, D. and Leissa, A.W., Effects of geometric imperfections on vibrations of biaxially compressed rectangular flat plates, *J. Appl. Mech.,* Vol. 50(4a), 1983, pp. 751–756.
[76] Hui, D., Effects of geometric imperfections on large-amplitude vibrations of rectangular plates with hysteresis damping, *J. Appl. Mech.,* Vol. 51, 1984, pp. 217–220.
[77] Ilanko, S. And Dickinson, S.M., The vibration and post-buckling of geometrically imperfect, simply supported rectangular plates under uni-axial loading — Part I:Theoretical approach, *J. Sound Vibration*, Vol. 118(2), 1987, pp. 313–336.
[78] Ilanko, S. and Dickinson, S.M., The vibration and post-buckling of geometrically imperfect, simply supported rectangular plates under uni-axial loading — Part II :Experimental investigation, *J. Sound Vibration*, Vol. 118(2), 1987, pp. 337–351.
[79] Hui, D., Effects of geometric imperfections on frequency-load interaction of biaxially compressed antisymmetric angle ply rectangular plates, *J. Appl. Mech.,* Vol. 52, 1985, pp. 155–162.
[80] Kapania, R.K. and Yang, T.Y., Buckling, postbuckling and nonlinear vibrations of imperfect plates, *AIAA J.*, Vol. 25(10), 1987, pp. 1338–1346.
[81] Booton, M., Buckling of imperfect anisotropic cylinders under combined loading, UTIAS Report 203, 1976.
[82] Abramovich, H., Singer, J. and Yaffe, R., Initial imperfection measurements of stiffened shells and buckling predictions, *Israel J. Techno.,* Vol. 17, 1979, pp. 324–338.
[83] Singer, J. and Prucz, J., Influence of initial geometrical on vibrations of axially compressed stiffened cylindrical shells, *J. Sound Vibration*, Vol. 80(1), 1982, pp. 117–143.
[84] Abramovich, H., Yaffe, R. and Singer, J., Evaluation of stiffened shell characteristics from imperfection measurements, *J. Strain Anal.,* Vol. 22(1), 1987, pp. 17–23.
[85] Abramovich, H., Weller, T. and Yaffe, R., Application of a modified Donnell technique for the determination of critical loads of imperfect plates, *Computers and Struct.,* Vol. 37(4), 1990, pp. 463–469.
[86] Abramovich, H., Singer, J. and Weller, T., The influence of initial imperfections on the buckling of stiffened cylindrical shells under combined loading, in *Buckling of Shells Structures, On Land, in the Sea and in the Air*, J.F. Julien, (ed.,) 538 p. CRC Press, 9th Sept. 1991, *Proc. Int. Colloquium Buckling of Shell structures, on Land, in the Sea and in the Air*, Lyon, France, 17 Sept. 1991, 10 p.
[87] Singer, J. and Abramovich, H., The development of shell imperfection measurement techniques, *Thin-Walled Struct.,* 23, 1995, pp. 379–398.
[88] Abramovich, H., Singer, J. and Weller, T., The influence of repeated buckling on the geometrical imperfections of stiffened cylindrical shell under combined loading, *Int. J. Non-Linear Mech.,* Vol. 37, 2002, pp. 577–588, 2002.
[89] Jansen, E.L., The effect of geometric imperfections on the vibrations of anisotropic cylindrical shells, *Thin Walled Struct.,* Vol. 45, 2007, pp. 274–282.
[90] Arbocz, J. *et al.*, *Space Engineering- Buckling of Structures Handbook*, ECSS (European Cooperation for Space Standardization), ECSS-E-HB-32-24A, 24th March 2010, 463 p.

Appendix A

Table A.1. The data for Fig. 1.

$\frac{P}{P_{cr}}$	$(\omega/\omega_1)^2$	
	S–S	C–C
0.0	1.0	1.0
0.1	0.9	0.902735
0.2	0.8	0.804965
0.3	0.7	0.70667
0.4	0.6	0.607795
0.5	0.5	0.508312
0.6	0.4	0.40818
0.7	0.3	0.307342
0.8	0.2	0.205744
0.9	0.1	0.103297
1.0	0.0	0.0

Appendix B

Free harmonic vibration of an isotropic axially compressed column is governed by the following differential equation:

$$EIw_{,xxxx} + Pw_{,xx} + \rho Aw_{,tt} = 0, \tag{B.1}$$

where w is the lateral deflection, of the column, A is its cross-section and ρ the mass per unit volume. Division by EI yields:

$$w_{,xxxx} + k^2 w_{,xx} + c^2 w_{,tt} = 0, \tag{B.2}$$

where

$$k^2 = \frac{P}{EI}, \quad c^2 = \frac{\rho A}{EI}. \tag{B.3}$$

Instead of solving the differential equation presented by Eq. (B.2), as presented in Eqs. (5.5)-(5.9), we shall assume harmonic vibrations having the following expression:

$$w(x,t) = W(x) \sin \omega t. \tag{B.4}$$

Substituting Eq. (B.4) into Eq. (B.2) yields

$$W_{,xxxx} + k^2 W_{,xx} - c^2 \omega^2 W = 0. \tag{B.5}$$

For constant EI and simply supported column, the boundary conditions are

$$W(0) = W(0)_{,xx} = W(L) = W(L)_{,xx}. \tag{B.6}$$

The solution for Eq. (B.5) has the following form (which also satisfies the boundary conditions, Eq. (B.6)):

$$W = C \sin \frac{\pi x}{L}. \tag{B.7}$$

Substitution of Eq. (B.7) into Eq. (B.5) leads to the following form:

$$\omega^2 = \frac{\pi^2}{\rho A L^2}[P_E - P] \text{ where } P_E = \frac{\pi^2 EI}{L^2}. \tag{B.8}$$

Division of both sides of Eq. (B.8) by P_E yields the following expression:

$$\left(\frac{\omega}{\omega_1}\right)^2 = 1 - \frac{P}{P_E}, \tag{B.9}$$

where

$$\omega_1^2 = \frac{\pi^4 EI}{\rho A L^4}. \tag{B.10}$$

For the clamped–clamped case, the assumed buckling shape of the column would be

$$W = \widetilde{C}\left(1 - \cos \frac{2\pi x}{L}\right). \tag{B.11}$$

The expression in Eq. (B.11) satisfies the boundary conditions for a clamped column at both ends, namely

$$W(0) = W_{,x}(0) = W(L) = W_{,x}(L) = 0. \tag{B.12}$$

Using the Rayleigh's principle, one can obtain an upper limit to the fundamental frequency of the compressed column, by assuming the static mode given by Eq. (B.11) (see also [2]). Equating the total potential energy to the kinetic energy of the vibrating column (for simplicity, we assume an isotropic material) and taking the clamped-clamped case, yields the following expression:

$$\frac{1}{2}\int_0^L EI(\Phi_{,xx})^2 dx - \frac{P}{2}\int_0^L (\Phi_{,x})^2 dx = \frac{\omega^2}{2}\int_0^L \rho A \Phi^2 dx, \tag{B.13}$$

where Φ is the exact mode for a clamped–clamped loaded column having a length L. Note that the relationship presented in Eq. (B.13) is the exact one, only if the mode Φ is the correct mode of vibration in the presence of axial compression. In case we assume an approximated mode, as the one presented in Eq. (B.11), and demanding that the mode amplitude is normalized to unity, the equality changes into the following inequality:

$$\int_0^L (V_{,xx})^2 dx - \frac{P}{EI}\int_0^L (V_{,x})^2 dx \geq \frac{\omega^2 \rho A}{EI}$$

or

$$1 \geq \frac{\omega^2 c^2}{\int_0^L (V_{,xx})^2 dx} + \lambda^2 \frac{\int_0^L (V_{,x})^2 dx}{\int_0^L (V_{,xx})^2 dx}, \tag{B.14}$$

where

$$\lambda^2 = \frac{P}{EI}, \quad c^2 = \frac{\rho A}{EI}. \tag{B.15}$$

As already stated above, Eq. (B.14) gives an upper-level approximation for the relationship between the critical buckling load and the natural frequency of the compressed column. For the clamped–clamped column we get the following result.

The normalized amplitude $\widetilde{C}$ of the assumed mode W(x) has the following form:

$$\int_0^L \widetilde{C}^2 \left(1 - \cos\frac{2\pi x}{L}\right)^2 dx = 1 \Rightarrow \quad \widetilde{C} = \sqrt{\frac{2}{3L}}. \tag{B.16}$$

Performing the various mathematical operations in Eq. (B.14), we get the following inequality:

$$1 \geq \frac{\omega^2 c^2}{\frac{16\pi^4}{3L^4}} + \lambda^2 \frac{\frac{4\pi^2}{3L^2}}{\frac{16\pi^4}{3L^4}}$$

$$\Rightarrow 1 \geq \frac{3\rho A \omega^2 L^4}{16\pi^4 EI} + \frac{PL^2}{4\pi^2 EI}. \tag{B.17}$$

However, the buckling load and the lower natural frequency for a clamped–clamped column are given as

$$P_{cr} = \frac{4\pi^2 EI}{L^2}, \quad \omega_{n=1^2} = (22.4)^2 \frac{EI}{\rho A L^4} \tag{B.18}$$

Using Eq. (B.18), the inequality in Eq. (B.17) can be written as

$$1 \geq 0.96582687\frac{\omega^2}{\omega_{n=1}^2} + \frac{P}{P_{cr}} \quad \text{or} 1.03538225\left(1 - \frac{P}{P_{cr}}\right) \geq \frac{\omega^2}{\omega_{n=1}^2} \tag{B.19}$$

Figure B.1 presents the upper-level curve and the probable curve for a clamped–clamped (C–C) compressed column, and a column with simply supported boundaries at both ends (SS–SS). For the C–C column, we have an

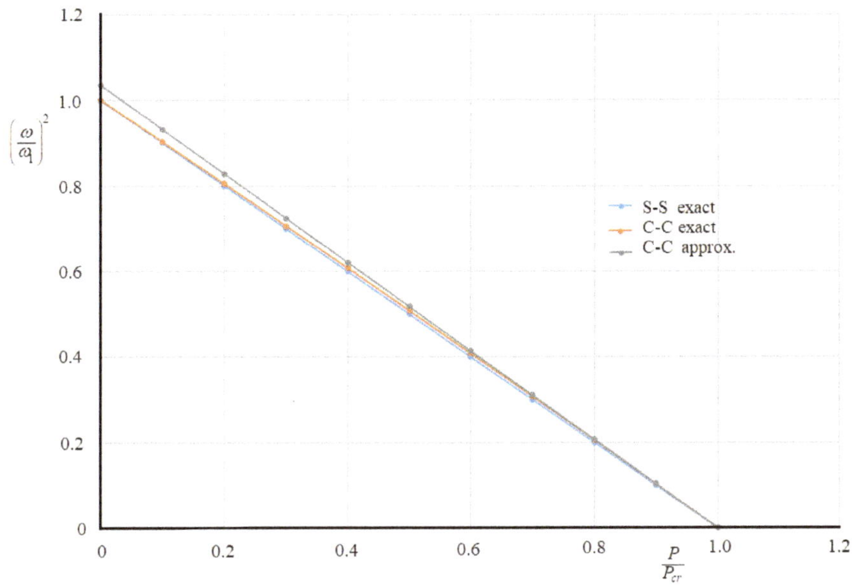

Fig. B.1. Non-dimensional circular frequency squared vs. non-dimensional axial compression for the assumed mode approach — a thin-walled column.

upper-level curve reaching above $P/P_{cr} = 1$, however, the probable curve should have $P/P_{cr} = 1$ at $\omega^2/\omega_{n=1}^2 = 0$ and $P/P_{cr} = 0$ at $\omega^2/\omega_{n=1}^2 = 1$. For the SS–SS column, as the buckling mode is exactly the mode of the compressed column with same boundary conditions, the exact linear relationship is shown in Fig. B.1.

It is interesting to note that in 1968, Galef [70] also suggested to use the Rayleigh principle to correlate the frequency squared to the applied compressive load, for clamped–free and clamped cases beam cases. His letter to the editor was submitted, after Amba-Rao [71] published his letter to editor in which he solved analytically the problem of first natural frequency vs. the applied compressive load for clamped–clamped, clamped–hinged (simply supported), hinged–hinged and clamped–free beam cases. The graph presented in [70] displays the nondimensional natural frequency as a function of the nondimensional axial compression (and not the frequency squared vs. axial compression) with frequency extrapolated to zero for nondimensional loads ranging from 0.9 to 1. Galef [70] then suggested that the graphs presented by Amba-Rao [71] follow the well-known relationship $\omega = \omega_0\sqrt{(1 - P/P_{cr})}$, where ω is the natural frequency in the presence of axial compression, ω_0 is the natural frequency of the uncompressed beam, P is the compressive load and P_{cr} is the Euler buckling load.

In a recent study, Valle *et al.* [72] provided solutions for various types of boundary conditions for beams, by modeling the cases as a spring-mass system. The results presented are for beams under axial tension, although the authors claim that their method is applicable also for compressive loads.

Guédé, and Elishakoff [73] addressed the topic of nonhomogeneous vibrating beams under axial compression and presented closed form solutions for the linear relationship between the square of the natural frequency and the axial load ratio, for various boundary conditions. The boundary conditions investigated were: clamped–free, simply supported at both ends, clamped–clamped and clamped–simply supported. The method used, included the assumption of various polynomial functions to describe the distribution of the mass and the rigidity EI along the beam and solution of the exact equation of motion. The authors included a comparison to Galerkin method results yielding a the same linear relationship between the frequency squared and the non-dimensional axial compression load. It is also claimed in [73] that the linear relationship is valid for all four types of boundary conditions for inhomogeneous beams in contrast to uniform beams, where the exact linear relationship is true only for columns with simply supported boundary conditions at both ends.

Appendix C: Southwell's Plot

Figure C.1 presents schematically the behavior of a column under axial compression. As can be seen, there is a great difference between an ideal column and a tested one. The question is what valid comparison can be made between theory and experiment.

Ayrton and Perry [1] were the first to introduce imperfection in the loading and in the form of the column, yielding the following relation [4]:

$$y_1 = \frac{C}{1 - \frac{P}{P_{\text{Euler}}^{\text{cantilever}}}} \quad \text{where} \quad P_{\text{Euler}}^{\text{cantilever}} = \frac{\pi^2 EI}{4L^2} \tag{C.1}$$

where

C is = initial geometric imperfection of a loaded cantilever strut;
y_1 is = the ordinate of the strut center line in the deformed configuration;
E is =Young's modulus;
I is = second moment of cross-sectional area;
L is = length of the strut; and
P is = axial compressive load.

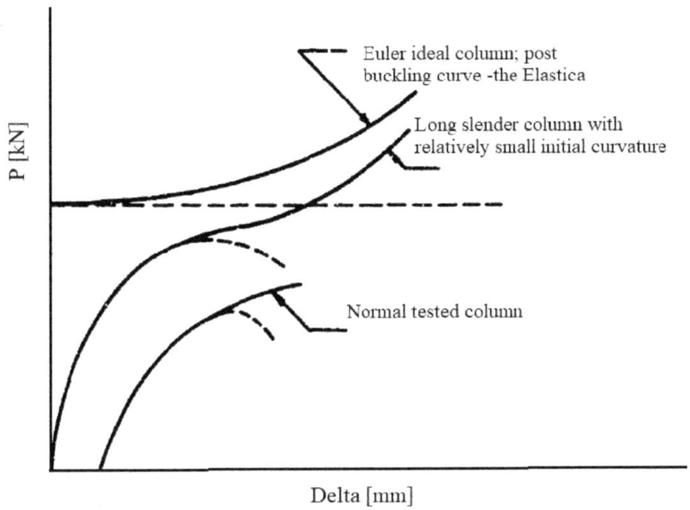

Fig. C.1. Schematic behavior of compressed columns.

Based on Eq. (C.1), the central deflection of any strut on any boundary conditions is, as follows:

$$\delta = \frac{C \cdot P}{P_{\text{Euler}} - P} \Rightarrow \frac{1}{\delta} = \frac{P_{\text{Euler}}}{C \cdot P} - \frac{1}{C},$$

$$\text{if } x = \frac{1}{\delta}, y = \frac{1}{P} \text{ then}$$

$$x = \frac{P_{\text{Euler}}}{C} y - \frac{1}{C} \text{ or } y = \frac{C}{P_{\text{Euler}}} x + \frac{1}{P_{\text{Euler}}}. \tag{C.2}$$

Then the inverse of the central deflection, $1/\delta$ is linearly connected to $1/P$ with a slope C/P_{Euler}. Another formulation stemming from Eq. (C.1) is

$$\delta = \frac{C \cdot P}{P_{\text{Euler}} - P} \Rightarrow \delta \frac{(P_{\text{Euler}} - P)}{P} = C,$$

$$\text{if } \bar{x} = \frac{\delta}{P}, \bar{y} = \delta \text{ then}$$

$$P_{\text{Euler}} \cdot \bar{x} - \bar{y} = C \text{ or } \bar{y} = P_{\text{Euler}} \cdot \bar{x} - C \tag{C.3}$$

It is clear that the slope of the linear equations is the Euler critical load, while the intersection with the $\bar{y}$ axis (namely δ axis) is the effective initial geometric imperfection.

Southwell [2] also treated the question posed at the beginning of the chapter. He assumed a simply supported compressed strut, as presented in Fig. C.2.

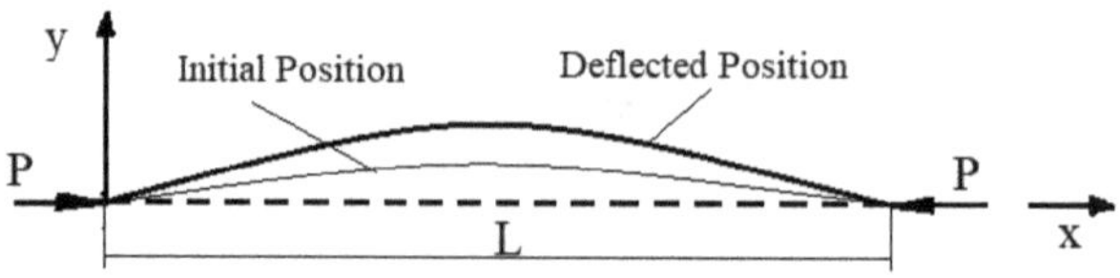

Fig. C.2. Schematic strut assumed by Southwell [2].

The governing equation, for the strut in the presence of initial geometric imperfection, y_0, is

$$EI\left(\frac{d^2y}{dx^2}-\frac{d^2y_0}{dx^2}\right)+Py=0\Rightarrow\frac{d^2y}{dx^2}+\lambda^2y=\frac{d^2y_0}{dx^2}$$

$$\text{where}\ \lambda^2=\frac{P}{EI}. \tag{C.4}$$

If we assume that, the initial imperfection and the lateral deflection have the following forms

$$y=\sum_{m=1}^{\infty}Y_m\sin\frac{m\pi x}{L}\quad\text{and}\quad y_0=\sum_{m=1}^{\infty}\bar{Y}_m\sin\frac{m\pi x}{L}, \tag{C.5}$$

substitution in Eq. (C.4) leads to

$$Y_m=\frac{\bar{Y}_m}{1-\frac{\mathrm{P}}{\mathrm{EI(m^2\pi^2/L^2)}}}\rightarrow\frac{Y_m}{\bar{Y}_m}=\frac{1}{1-\frac{\mathrm{P}}{m^2P_{\mathrm{Euler}}}}. \tag{C.6}$$

The deflection of the strut would be

$$y=\sum_{m=1}^{\infty}\bar{Y}_m\left(\frac{1}{1-\frac{\mathrm{P}}{m^2P_{\mathrm{Euler}}}}\right)\sin\frac{m\pi x}{L}. \tag{C.7}$$

The maximal deflection would appear at the middle of the strut, therefore its expression is given by substituting $x = L/2$ in Eq. (C.7), yielding

$$\delta\equiv Y_1-\bar{Y}_1-Y_3+\bar{Y}_3+\cdots. \tag{C.8}$$

As Southwell wrote in his manuscript [2] "$\cdots$ if P is a fairly considerable fraction of P_{Euler}", Eq. (C.6) can rewritten as

$$\delta=\frac{\bar{Y}_1}{1-\frac{P}{P_{\mathrm{Euler}}}}-\bar{Y}_1. \tag{C.9}$$

Equation (C.9) can be modified to the following expression:

$$\delta=\frac{\bar{Y}_1}{1-\frac{\mathrm{P}}{\mathrm{P_{Euler}}}}-\bar{Y}_1=\bar{Y}_1\left(\frac{1}{1-\frac{\mathrm{P}}{\mathrm{P_{Euler}}}}-1\right)=\bar{Y}_1\frac{\frac{\mathrm{P}}{\mathrm{P_{Euler}}}}{1-\frac{\mathrm{P}}{\mathrm{P_{Euler}}}}$$

$$\Rightarrow\delta\frac{\mathrm{P_{Euler}}}{\mathrm{P}}-\delta=\bar{Y}_1. \tag{C.10}$$

The basic generic equation: $\delta = \dfrac{C \cdot P}{P_{Euler} - P}$

where

δ = mid span elastic deflection; P = applied compressive load;

C = constant; $P_{Euler} = \dfrac{\pi^2 EI}{L^2}$ = Euler buckling load.

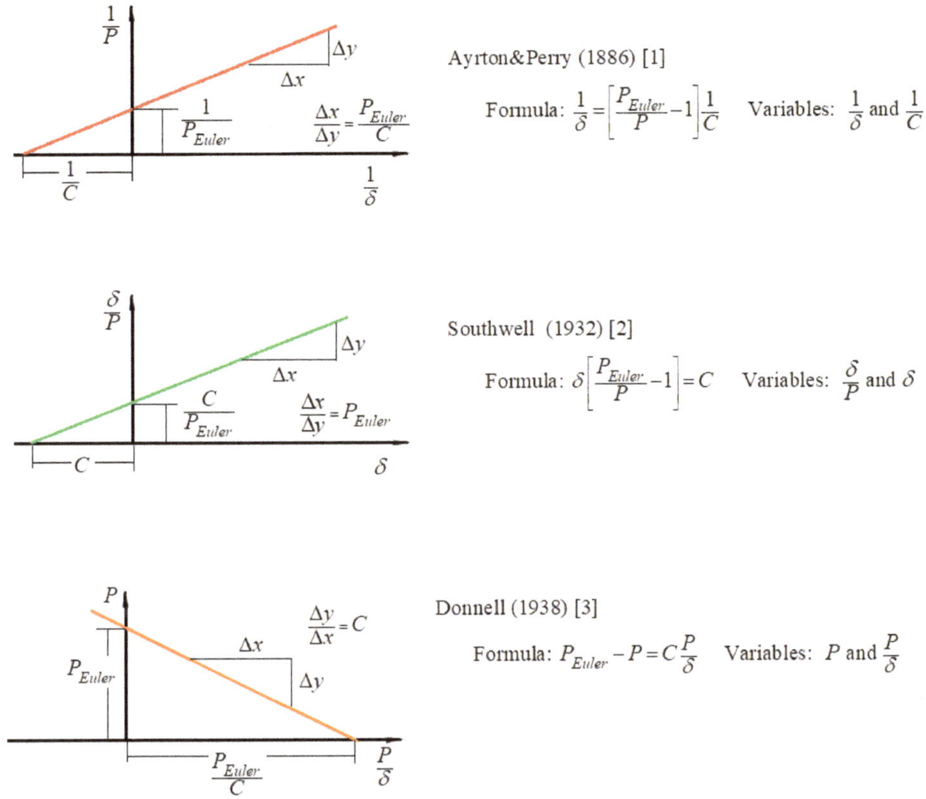

Fig. C.3. Schematic linear representations strut of Ayrton and Perry[1], Southwell [2] and Donnell [3] models.

Southwell used Eq. (C.10) to plot the ratio of the displacement to the applied load, δ/P vs. the displacement, δ. Donnell [3] used a third derivative of Eq. (C.9) and plotted the load vs. the ratio of the load to displacement. The differences among, the Southwell, the Donnell and the Ayrton and Perry representations [4, 5] are clearly presented in Fig. C.3.

Note that, although developed for a strut under axial compression, scientists are using the Southwell method also for other types of loading leading to buckling as described in [6].

References

[1] Ayrton, W. E. and Perry, J., On struts, *The Engineer*, 62, 1886, December 10th, pp. 464–465 and 1886 December 24th, pp. 513–515.
[2] Southwell, R. V., On the analysis of experimental observations in problems of elastic stability, *Proc. Royal Soc., A*, Vol. 135, 1932, pp. 601–616.
[3] Donnell, L.H., On the application of Southwell's method for the analysis of buckling tests, *Stephen Timoshenko 60th Anniversary Volume*, MacMillan Book Company, 1938, pp. 27–38.
[4] Horton, W.H., Cundari, F. L. and Johnson, R. W., Applicability of the Southwell plot to the interpretation of test data obtained from stability studies of elastic column and plate structures, USAAVLABS Technical Report 69-32, November 1971.
[5] Singhal, M.K., Studies on the elastic stability of bodies, Ph.D. thesis, presented to the Faculty of the Division of Graduate Studies and Research, School of Aerospace Engineering, Georgia Institute of Technology, August 1973, 159 p.
[6] Kalkan, İ., Application of Southwell method on the analysis of lateral torsional buckling tests on reinforced concrete beams, *Int. J. Eng. Res. Develop.*, Vol. 2(1), 2010, pp. 58–66.

Appendix D: Novel Excitation Methods

A good excitation method would enable to acquire the natural frequencies of a vibrating structure and lead to consistent results when applying VCT.

Normally, an electromagnetic shaker (Fig. D.1) would provide the necessary driving force to excite the specimen, as a function of frequency

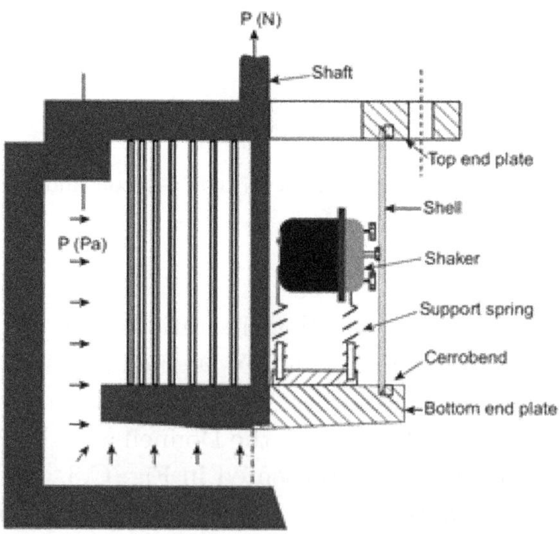

Fig. D.1. An electromagnetic shaker.

generator [1]. The response of the specimen can be recorded using a microphone and the resonance can be detected using Lissajous curves, with the excitation voltage and the specimen response voltage being supplied to the two axes of an oscilloscope, see Fig. D.2.

Exciting the specimen using a loudspeaker, would yield and ellipse, while using a shaker the Lissajous curve would be an eight (Fig. D.2(b)), due to the "push-pull" movement of the shaker which is twice the response frequency.

Another important topic for the successful application of VCT is the mode shapes of the tested specimen. When using a microphone, its voltage can be used to draw the mode shape of the vibrating specimen. Using a single accelerometer, to detect the response of the vibrating structure, will not enable the recording of the modes, only the vibrating frequency.

A novel excitation method is the use of modal hammer as described in [2]. The method uses a special dedicated hammer to slightly impact the specimen, and its response is recorded using an accelerometer bonded on the tested specimen and connected to NI DAQ system (see Fig. D.3). The excitation system is composed of an impact hammer (adapted from National Instruments—NI), a single accelerometer and a NI DAQ system with a Me Scope Modal Software, to detect the excited natural frequencies and their respective mode shapes of the tested panel. The output is in the form of Frequency Response Function (FRF) as shown in Fig. D.4. The determination of the natural frequencies is done by searching peaks on the Imaginary FRF chart, see Fig. D.4.

The most sophisticated way to excite a given structure is the use of a laser as described in [3].

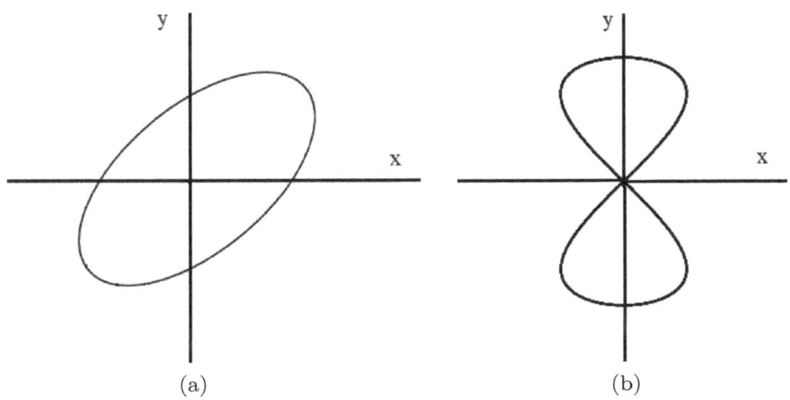

Fig. D.2. Lissajous curves: (a) an ellipse; (b) an eight curve.

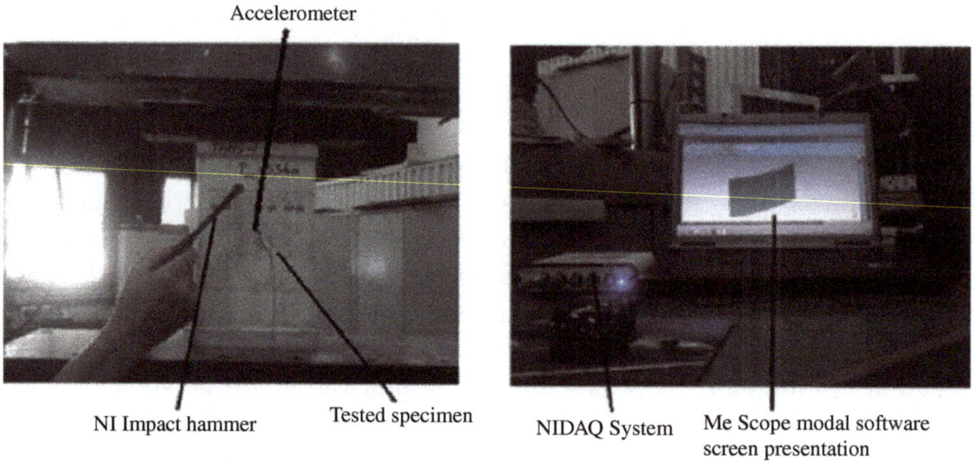

Fig. D.3. The modal hammer excitation test set-up (adapted from [2]).

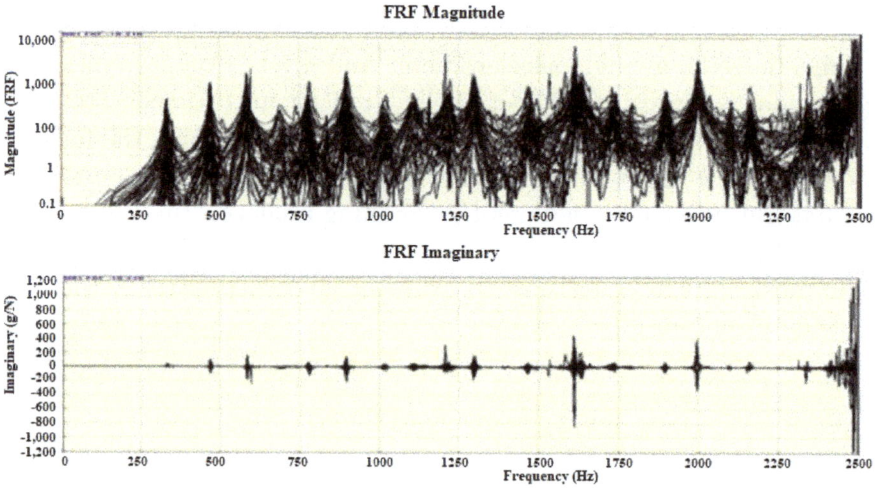

Fig. D.4. A typical output using NI modal hammer and its DAQ system (adapted from [2]).

The test series were performed on laminated composite shell at the laboratory of Aerospace Structures and Materials of the Delft University of Technology in The Netherlands. The test set-up assembly for static buckling test and VCT analysis is depicted in Figs. D.5(a) and D.5(b). It includes an MTS 3500 kN servo-hydraulic loading rig to provide the static compressive loads and the POLYTEC PSV 500 scanning laser vibrometer. The POLYTEC set-up can measure the natural frequencies and the vibration mode shapes

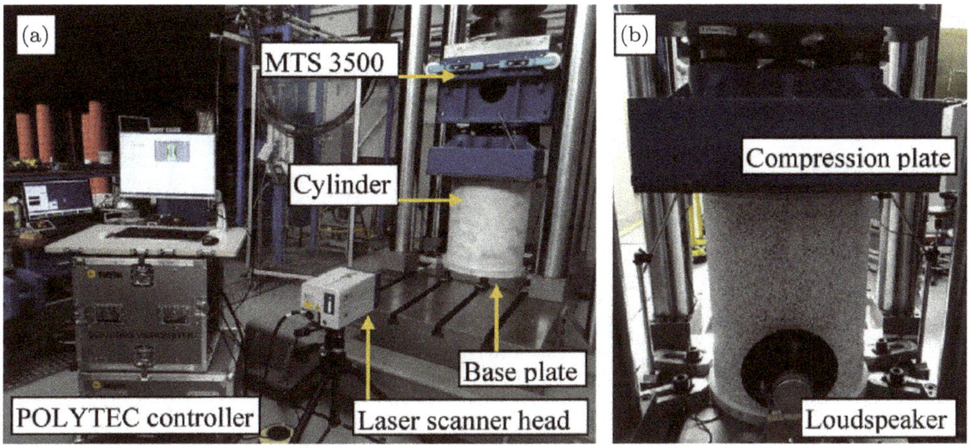

Fig. D.5. The test set-up: (a) frontal view; (b) rear view (adapted from [3]).

Experimental measured natural frequencies of the classical laminated shell.

Load [kN]	1st Frequency [Hz]	2nd Frequency [Hz]	3rd Frequency [Hz]	4th Frequency [Hz]
5	220.6	228.1	238.1	261.2
20	243.1	250.6	267.5	295.0
35	243.7	253.1	266.2	300.6
50	241.2	251.8	263.1	301.2
65	238.1	250.0	260.0	300.6
80	235.0	247.5	256.8	301.2
95	231.2	244.4	253.7	298.7
110	227.5	241.2	250.0	296.2
125	223.7	238.1	246.2	294.3
140	220.0	234.3	242.5	291.2
155	216.2	230.6	238.1	288.7
170	211.8	226.8	233.8	285.6
185	206.8	223.1	230.0	273.7
200	202.6	218.7	225.0	268.7
215	197.5	214.3	220.6	263.1
230	192.5	210.0	215.6	256.2
245	187.5	206.6	210.0	248.1
260	181.8	200.6	–	237.5
275	176.2	195.6	–	223.8

Fig. D.6. Typical experimental results for a classical laminated shell (adapted from [3]).

of the tested cylinders. It consists of a laser scanning head, data acquisition unit and a control unit, as shown in Fig. D.5. A loudspeaker located at the rear part of the tested specimen and controlled via an amplifier by the POLYTEC system within a predefined frequency spectrum excited the tested cylinders. The measurements were conducted using a 100–400 Hz frequency sweep, the range being fixed to include most of the low natural frequencies. The laser vibrometer is capable of covering approximately 160 degrees of the cylinder surface with high fidelity grid containing 450 points. Although to reduce the time of the scan, the scanned area was reduced to roughly 100 grid points, without reducing the measurements accuracy. After calibration of the scanning process, a typical measurement would include the following steps: (a) the axial compressive load level is applied, (b) the loudspeaker

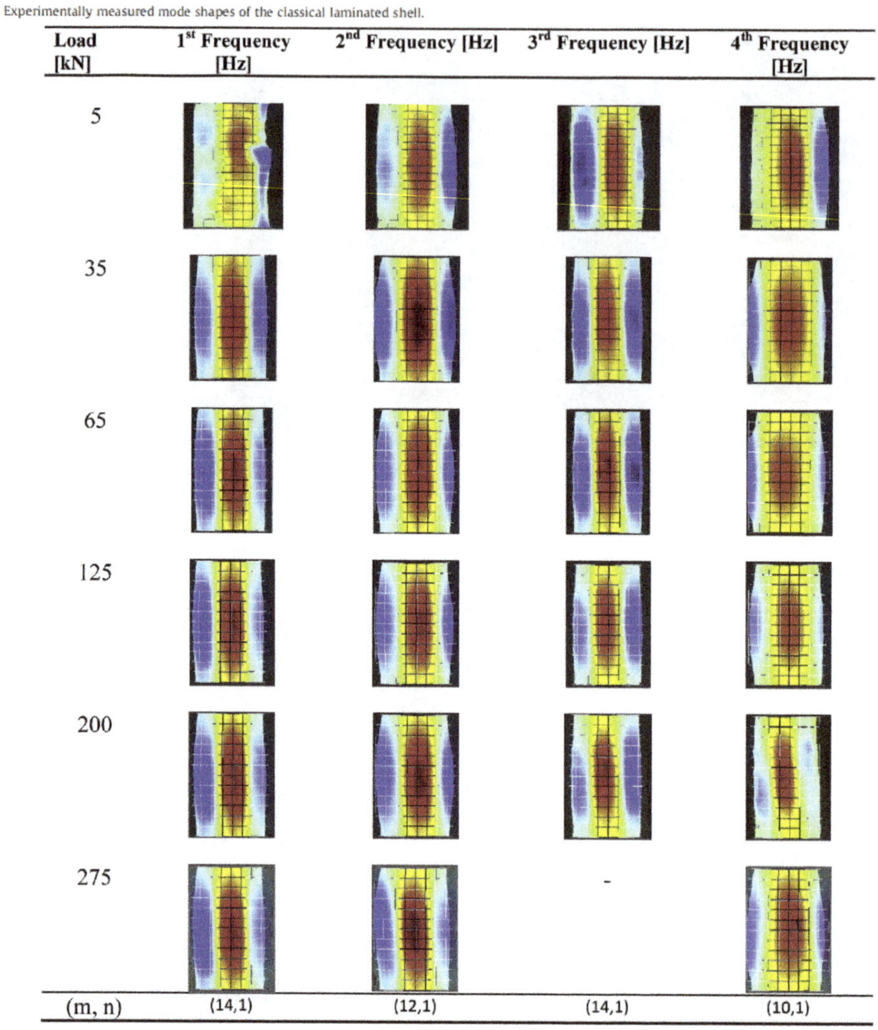

Fig. D.7. Typical experimental modal shapes for a classical laminated shell (adapted from [3]).

is exciting the cylinder on a predefined frequency sweep range, while the laser scanning head is measuring the cylinder response at each predefined grid point, scanning it repeatedly five times (c), a Fast Fourier's Transform (FFT) of the cylindrical shell response presenting the natural frequencies within the predefined frequency range.

Figures D.6 and D.7 present typical experimental results and the associated mode shapes, respectively.

References

[1] Abramovich, H., Singer, J. and Weller, T., Repeated buckling and its influence on the geometrical imperfections of stiffened cylindrical shells under combined loading, *Int. J. Non-Linear Mech.*, Vol. 37, 2002, pp. 577–588.
[2] Abramovich H, Govich D, Grunwald A. Buckling prediction of panels using the vibration correlation technique, *Prog. Aerospace Sci.*, Vol. 78, 2015, pp. 62–73.
[3] Labans E, Abramovich H, Bisagni C. An experimental vibration-buckling investigation on classical and variable angle tow composite shells under axial compression, *J. Sound and Vibration*, 2019, Vol. 449, pp. 315–329.

Appendix E: Summary of Axial Compression vs. Natural Frequencies Relationship for Thin-Walled Structures

Table E.1. Columns governing equations and load vs. natural frequency relationships.

Case	Governing equation	Load vs. natural frequency relationship
An isotropic column, axially compressed	$\mathrm{EI}\frac{\partial^4 \mathrm{w(x,t)}}{\partial \mathrm{x}^4} + \mathrm{P}\frac{\partial^2 w(x,t)}{\partial x^2} + \rho \mathrm{A}\frac{\partial^2 \mathrm{w(x,t)}}{\partial \mathrm{t}^2} = 0$	$\left(\frac{\omega}{\omega_n}\right)^2 + \frac{P}{P_{cr}} = 1$
An isotropic column, axially tensioned	$\mathrm{EI}\frac{\partial^4 w(x,t)}{\partial x^4} - P\frac{\partial^2 w(x,t)}{\partial x^2} + \rho A\frac{\partial^2 w(x,t)}{\partial t^2} = 0$	$1 + \frac{P}{P_{cr}} = \left(\frac{\omega}{\omega_\mathrm{n}}\right)^2$
An isotropic column axially compressed on a Winkler-type foundation	$\mathrm{EI}\frac{\partial^4 w(x,t)}{\partial x^4} + P\frac{\partial^2 w(x,t)}{\partial x^2} + \mathrm{k}_W \cdot w(x,t) + \rho A\frac{\partial^2 w(x,t)}{\partial t^2} = 0$	$\left(\frac{\omega}{\omega_\mathrm{n}}\right)^2 + \frac{P}{P_{cr}} = 1$
A symmetric laminated composite column under compression using CLT	$\mathrm{D}_{11}\frac{\partial^4 w(x,t)}{\partial x^4} + \bar{N}_{xx}\frac{\partial^2 w(x,t)}{\partial x^2} + I_2\frac{\partial^4 w(x,t)}{\partial t^2 \partial x^2} - \mathrm{I}_0\frac{\partial^2 w(x,t)}{\partial t^2} = 0$	$\left(\frac{\omega}{\omega_\mathrm{n}}\right)^2 + \frac{\bar{N}_{xx}}{\left(\bar{N}_{xx}\right)_{cr}} = 1$
A non-symmetric laminated composite column under compression using CLT	$\mathrm{D}_{11}\frac{\partial^4 w(x,t)}{\partial x^4} + \bar{N}_{xx}\frac{\partial^2 w(x,t)}{\partial x^2} + \mathrm{I}_2\frac{\partial^4 w(x,t)}{\partial t^2 \partial \mathrm{x}^2} - \mathrm{I}_0\frac{\partial^2 w(x,t)}{\partial t^2} = 0$	$\left(\frac{\omega}{\omega_\mathrm{n}}\right)^2 + \frac{\bar{N}_{xx}}{\left(\bar{N}_{xx}\right)_{cr}} = 1$
A symmetric laminated composite column under compression using FOSDT	$\mathrm{A}_{11}\left(\frac{\partial^2 u}{\partial x^2} + \frac{\partial^3 w}{\partial x^3}\right) = \mathrm{I}_0\frac{\partial^2 u}{\partial t^2}$ $\mathrm{KA}_{55}\left(\frac{\partial^2 w}{\partial x^2} + \frac{\partial \phi}{\partial x}\right) - \overline{N}_{xx}\frac{\partial^2 w}{\partial x^2} = \mathrm{I}_0\frac{\partial^2 w}{\partial t^2}$ $\mathrm{D}_{11}\frac{\partial^2 \phi}{\partial x^2} - \mathrm{KA}_{55}\left(\frac{\partial w}{\partial x} + \phi\right) = \mathrm{I}_2\frac{\partial^2 \phi}{\partial t^2}$	$\left(\frac{\omega}{\omega_1}\right)^2 + \frac{\bar{N}_{xx}}{\left(\bar{N}_{xx}\right)_{cr}} = 1$

(*Continued*)

Table E.1. (*Continued*)

Case	Governing equation	Load vs. natural frequency relationship
A non-symmetric laminated composite column under compression using FOSDT	$\mathrm{A}_{11}\left(\frac{\partial^2 u}{\partial x^2}+\frac{\partial^3 w}{\partial x^3}\right)+\mathrm{B}_{11}\frac{\partial^2 \phi}{\partial x^2}=\mathrm{I}_0\frac{\partial^2 u}{\partial t^2}+\mathrm{I}_1\frac{\partial^2 \phi}{\partial t^2}$ $\mathrm{KA}_{55}\left(\frac{\partial^2 w}{\partial x^2}+\frac{\partial \phi}{\partial x}\right)-\overline{N}_{xx}\frac{\partial^2 w}{\partial x^2}=\mathrm{I}_0\frac{\partial^2 w}{\partial t^2}$ $\mathrm{B}_{11}\left(\frac{\partial^2 u}{\partial x^2}+\frac{\partial^3 w}{\partial x^3}\right)+\mathrm{D}_{11}\frac{\partial^2 \phi}{\partial x^2}-\mathrm{KA}_{55}\left(\frac{\partial w}{\partial x}+\phi\right)=\mathrm{I}_2\frac{\partial^2 \phi}{\partial t^2}+\mathrm{I}_1\frac{\partial^2 u}{\partial t^2}$	$\left(\frac{\omega}{\omega_1}\right)^2+\frac{\bar{N}_{xx}}{\left(\bar{N}_{xx}\right)_{cr}}=1$
An isotropic column, axially compressed due to thermal loading	$\mathrm{EI}\frac{\partial^4 w(x,t)}{\partial x^4}+(\mathrm{A}\cdot\mathrm{E}\cdot\alpha\cdot\Delta\mathrm{T})\frac{\partial^2 w(x,t)}{\partial x^2}+\rho\mathrm{A}\frac{\partial^2 w(x,t)}{\partial t^2}=0$	$\left(\frac{\omega}{\omega_n}\right)^2+\frac{\Delta\mathrm{T}}{(\Delta\mathrm{T})_{cr}}=1$

Table E.2. Columns schematic drawings.

Case	Schematic drawing of the case
An isotropic column, axially compressed	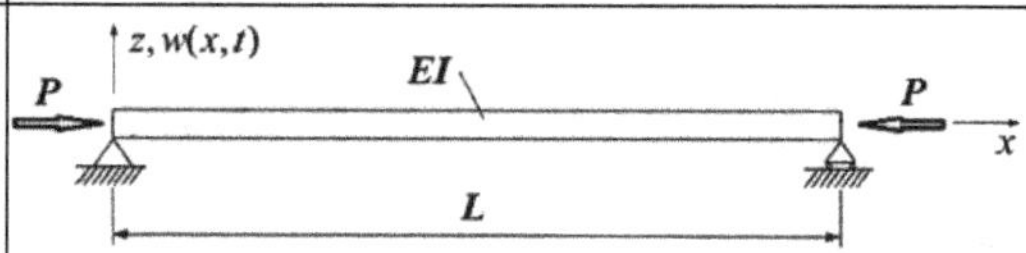
An isotropic column, axially tensioned	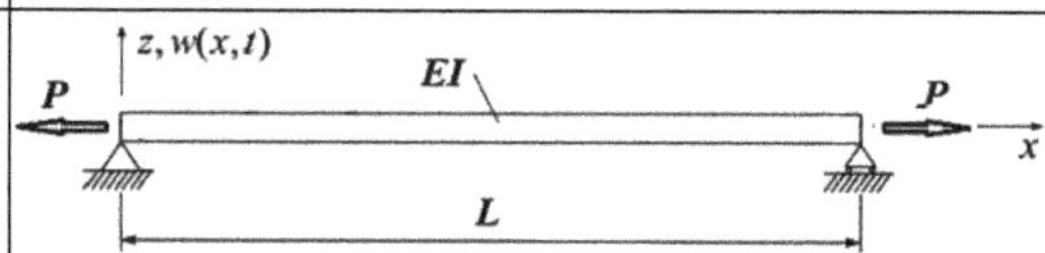
An isotropic column axially compressed on a Winkler-type foundation	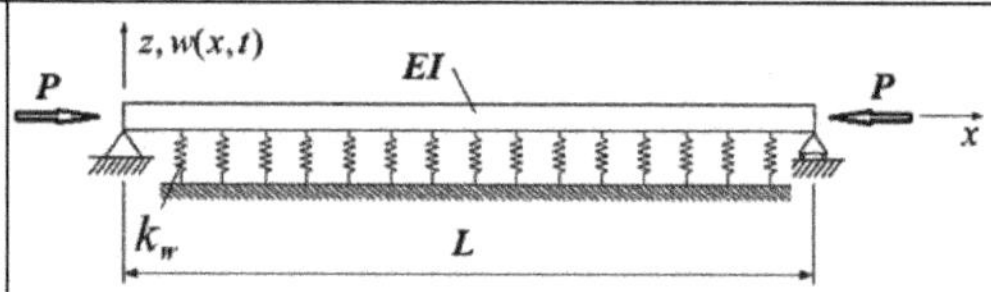
A symmetric laminated composite column under compression using CLT	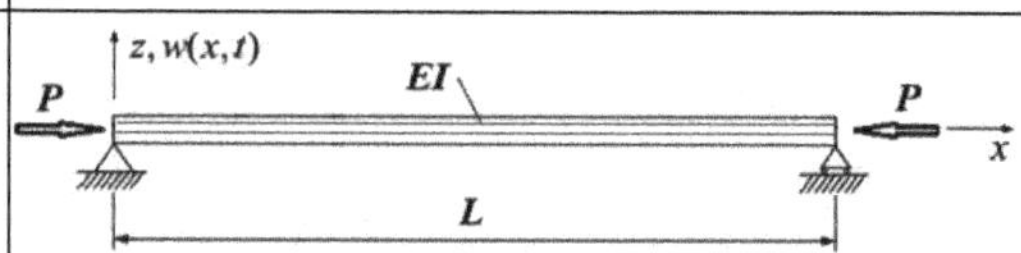
A non-symmetric laminated composite column under compression using CLT	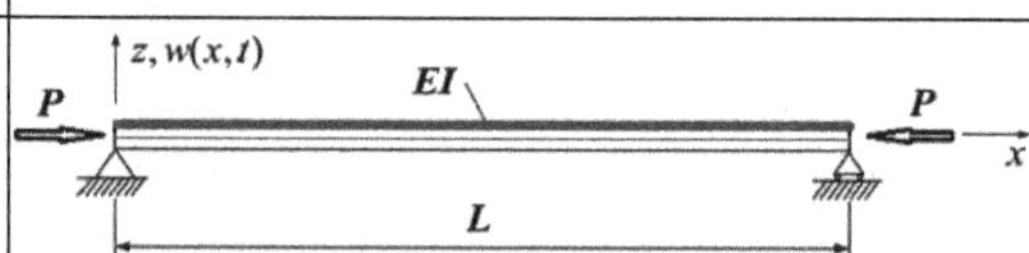

Table E.2. (*Continued*)

Case	Schematic drawing of the case
A symmetric laminated composite column under compression using FOSDT	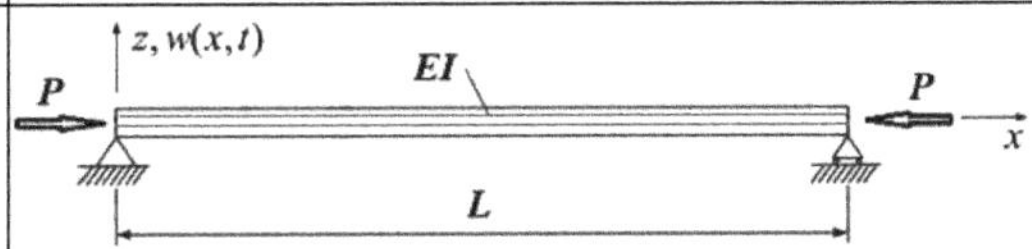
A non-symmetric laminated composite column under compression using FOSDT	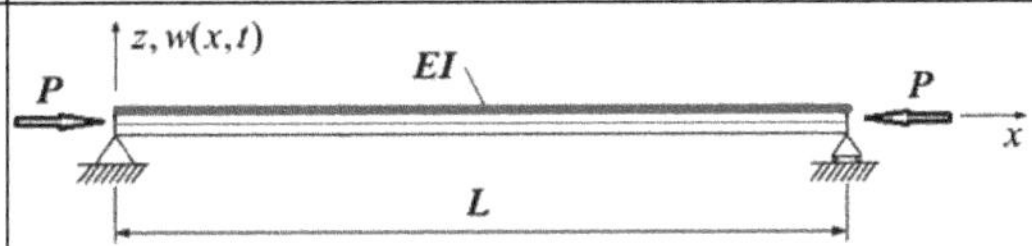
An isotropic column, axially compressed due to thermal loading	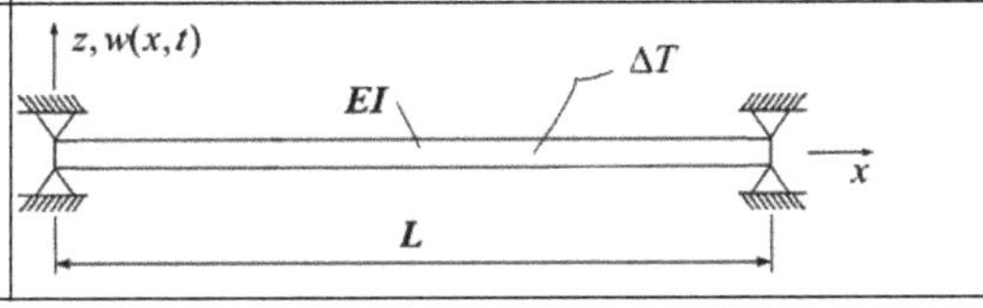

Table E.3. Table E3 Columns critical buckling loads and unloaded natural frequencies.

Case	P_{cr}, or $\left(\bar{N}_{xx}\right)_{\mathrm{cr}}$	ω_n^2
An isotropic column, axially compressed	$\frac{n^2\pi^2 EI}{L^2}$	$\frac{n^2\pi^4 EI}{\rho A L^4}$
An isotropic column, axially tensioned	$\frac{n^2\pi^2 EI}{L^2}$	$\frac{n^2\pi^4 EI}{\rho A L^4}$
An isotropic column axially compressed on a Winkler type foundation	$EI\frac{n^2\pi^2}{L^2}\left[1+\frac{k_W\cdot L^4}{n^4\pi^4 EI}\right]$	$\frac{n^4\pi^4 EI}{\rho A L^4}\left[1+\frac{k_W\cdot L^4}{n^4\pi^4 EI}\right]$
A symmetric laminated composite column under compression using CLT	$\left(\bar{N}_{xx}\right)_{cr} = D_{11}\frac{n^2\pi^2}{L^2}$	$\frac{n^4\pi^4 D_{11}}{L^4 I_0}\cdot\frac{1}{\left[1+\frac{I_2}{I_0}\left(\frac{n\pi}{L}\right)^2\right]}$
A non-symmetric laminated composite column under compression using CLT	$\left(D_{11}-\frac{B_{11}^2}{A_{11}}\right)\frac{n^2\pi^2}{L^2}$	$\frac{n^4\pi^4}{L^4 I_0}\cdot\frac{\left(D_{11}-B_{11}^2/A_{11}\right)}{\left[1+\frac{I_2}{I_0}\left(\frac{n\pi}{L}\right)^2\right]}$
A symmetric laminated composite column under compression using FOSDT*	$\frac{D_{11}n^2\pi^2}{L^2\left(1+\frac{n^2\pi^2 D_{11}}{L^2 KA_{55}}\right)}$	$\frac{n^4\pi^4 D_{11}}{L^4 I_0}\cdot\frac{1}{\left[1+\left(\frac{n\pi}{L}\right)^2\left(\frac{I_2}{I_0}+\frac{D_{11}}{KA_{55}}\right)\right]}$

(*Continued*)

Table E.3. (*Continued*)

A non-symmetric laminated composite column under compression using FOSDT	$\frac{(D_{11}-B_{11}^2/A_{11})n^2\pi^2}{L^2\left(1+\frac{n^2\pi^2(D_{11}-B_{11}^2/A_{11})}{L^2KA_{55}}\right)}$	$\frac{n^4\pi^4}{L^4I_0}\cdot\frac{(D_{11}-B_{11}^2/A_{11})}{\left[1+(\frac{n\pi}{L})^2\left(\frac{I_2}{I_0}+\frac{(D_{11}-B_{11}^2/A_{11})}{KA_{55}}\right)\right]}$
An isotropic column, axially compressed due to thermal loading	$(\Delta T)_{cr}=\frac{n^2\pi^2 I}{L^2\alpha A}$	$\frac{n^2\pi^4 EI}{\rho AL^4}$

Note: $K = 5/6$ for a rectangular cross-section. The accurate value of the shear coefficient is $\kappa = 10(1+\upsilon)/(12+11\upsilon)$ for a rectangular cross section and $\kappa = 6(1+\upsilon)/(7+6\upsilon)$ for a solid circular cross section, where υ is the Poisson's ratio.

Table E.4. Plates and shells governing equations and load vs. natural frequency relationships.

Case	Governing equation	Load vs. natural frequency relationship
A flat isotropic plate, compressed in the x and y directions*	$D\left[W_{,xxxx}(x,y,t) +2W_{,xxyy}(x,y,t)+W_{,yyyy}(x,y,t)\right] +\rho h W_{,tt}(x,y,t)=N_xW_{,xx}(x,y,t) +N_yW_{,yy}(x,y,t)$	$\left(\frac{\omega}{\hat{\omega}_{mn}}\right)^2+\frac{N}{N_{xcr}^{mn}}+\frac{N}{N_{ycr}^{mn}}=1$
A flat isotropic plate, under pure shear	$D\left[W_{,xxxx}(x,y,t)+2W_{,xxyy}(x,y,t) +W_{,yyyy}(x,y,t)\right] +\rho hW_{,tt}(x,y,t)=2N_{xy}W_{,xy}(x,y,t)$	No closed form solutions is available for this case. A highly non-linear relation between shear load and natural frequency squared is reported in [35].
A flat composite plate, compressed in the x and y directions using CLT**	$D_{11}W_{,xxxx}(x,y,t)+2(D_{12}+2D_{66})) W_{,xxyy}(x,y,t) +D_{22}W_{,yyyy}(x,y,t)+N_xW_{,xx}(x,y,t) +N_yW_{,\mathrm{yy}}(x,y,t) +\mathrm{I}_0W_{,tt}(x,y,t)-\mathrm{I}_2\left[W_{,tt_{xx}}(x,y,t) +W_{,tt_{yy}}(x,y,t)\right]=0$	$\left(\frac{\omega}{\hat{\omega}_{mn}}\right)^2+\frac{N}{N_{xcr}^{mn}}+\frac{N}{N_{ycr}^{mn}}=1$
An isotropic cylindrical shell axially compressed (Donnell's equations)	$\frac{\partial^2u}{\partial x^2}+\frac{(1-\upsilon)}{2\mathrm{R}^2}\frac{\partial^2u}{\partial\theta^2}+\frac{(1+\upsilon)}{2\mathrm{R}}\frac{\partial^2v}{\partial x\partial\theta}-\frac{\upsilon}{\mathrm{R}}\frac{\partial w}{\partial x}=0$ $\frac{(1+\upsilon)}{2\mathrm{R}}\frac{\partial^2u}{\partial x\partial\theta}+\frac{(1-\upsilon)}{2}\frac{\partial^2v}{\partial x^2}+\frac{1}{\mathrm{R}^2}\frac{\partial^2v}{\partial\theta^2} -\frac{1}{\mathrm{R}^2}\frac{\partial w}{\partial\theta}=0$ $\upsilon\mathrm{R}\frac{\partial u}{\partial x}+\frac{\partial v}{\partial\theta}-w -\left(\frac{h^2\ \mathrm{R}^2}{12}\frac{\partial^4w}{\partial x^4}+\frac{h^2}{6}\frac{\partial^4w}{\partial x^2\partial\theta^2}+\frac{h^2}{12R^2}\frac{\partial^4w}{\partial\theta^4}\right)$ $\frac{(1-\upsilon^2)\mathrm{R}^2\bar{N}_x}{Eh}\frac{\partial^2w}{\partial x^2}=\frac{(1-\upsilon^2)R^2\rho}{E}\frac{\partial^2w}{\partial t^2}$	$\left(\frac{\omega}{\hat{\omega}_{mn}}\right)^2+\frac{N}{N_{xcr}^{mn}}=1$

Note: $^*D=Eh^3/\left[12\left(1-\upsilon^2\right)\right]$; $^{**}D_{ij}=\frac{1}{3}\sum_{k=1}^{N}\overline{Q}_{ij}^{(k)}\left(z_{k+1}^3-z_k^3\right)$, $\overline{Q}_{ij}^{(k)}$ being the lamina stiffness after transformation $\begin{Bmatrix} I_0 \\ I_2 \end{Bmatrix}=\int_{-h/2}^{+h/2}\begin{Bmatrix} 1 \\ z^2 \end{Bmatrix}\cdot\rho\cdot dz$.

Table E.5. Plates and shell schematic drawings.

<table>
<tr><th>Case</th><th>Schematic drawing of the case</th></tr>
<tr><td>A flat isotropic plate, compressed in the x and y directions</td><td>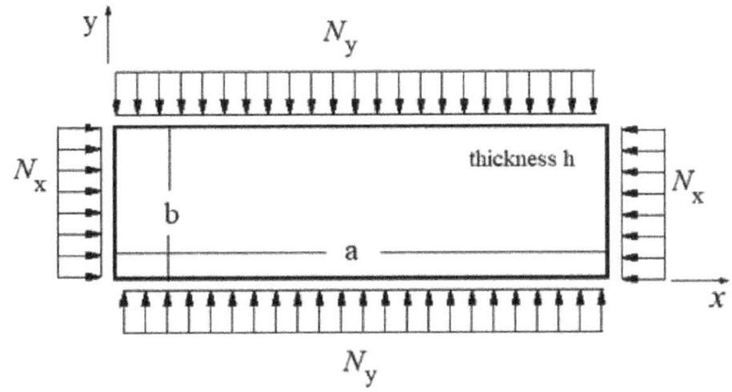
</td></tr>
<tr><td>A flat isotropic plate, under pure shear</td><td>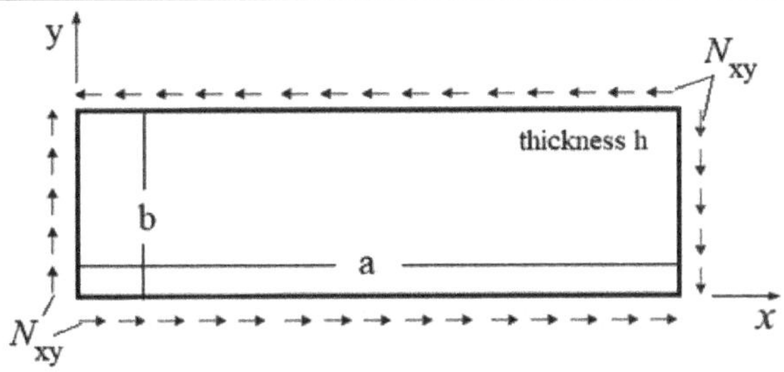
</td></tr>
<tr><td>A flat composite plate, compressed in the x and y directions using CLT</td><td>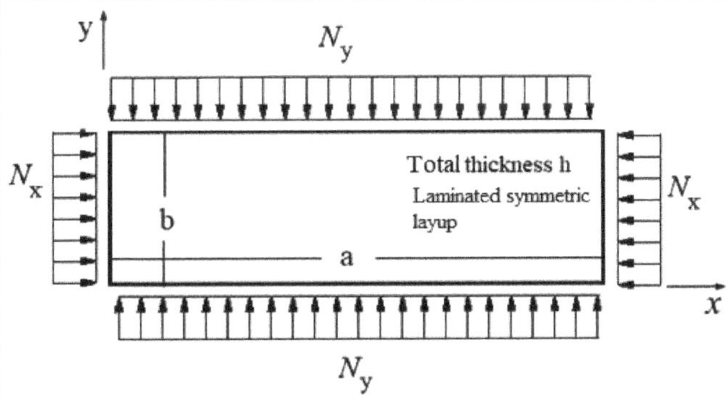
</td></tr>
<tr><td>An isotropic cylindrical shell axially compressed</td><td>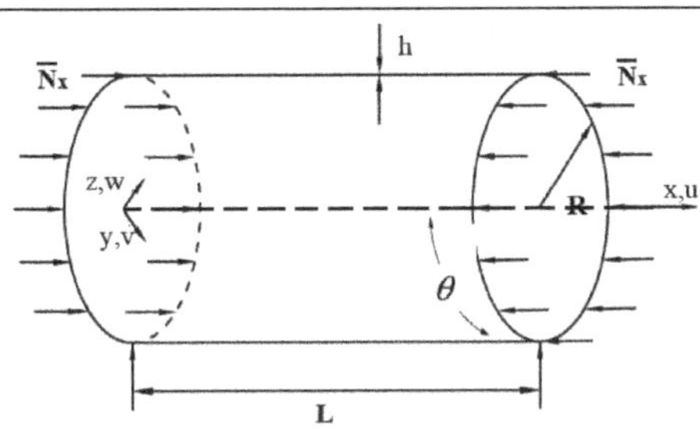
</td></tr>
</table>

Table E.6. Plates and shell critical buckling loads and unloaded natural frequencies.

Case	$(\bar{N}_x)_{cr}$	$(\bar{N}_y)_{cr}$	$\hat{\omega}^2_{mn}$
A flat isotropic plate, compressed in the x and y directions (simply supported all around)	$D\pi^2 \frac{a^2}{m^2}\left[\left(\frac{m}{a}\right)^2+\left(\frac{n}{b}\right)^2\right]^2$	$D\pi^2 \frac{b^2}{n^2}\left[\left(\frac{m}{a}\right)^2+\left(\frac{n}{b}\right)^2\right]^2$	$D\frac{\pi^4}{\rho h}\left[\left(\frac{m}{a}\right)^2+\left(\frac{n}{b}\right)^2\right]^2$
A flat isotropic plate, under pure shear* (simply supported all around)	$(\bar{N}_{xy})_{cr} = k\frac{\pi^2 D}{b^2}$ For a square plate $k = 9.34$**		$D\frac{\pi^4}{\rho h}\left[\left(\frac{m}{a}\right)^2+\left(\frac{n}{b}\right)^2\right]^2$
A flat isotropic plate, compressed in the x and y directions using CLT (simply-supported all around)	$\frac{D_{11}\left(\frac{m\pi}{a}\right)^4+D_{22}\left(\frac{n\pi}{b}\right)^4}{\left(\frac{m\pi}{a}\right)^2} + \frac{2(D_{12}+2D_{66})\left(\frac{m\pi}{a}\right)^2\left(\frac{n\pi}{b}\right)^2}{\left(\frac{m\pi}{a}\right)^2}$	$\frac{D_{11}\left(\frac{m\pi}{a}\right)^4+D_{22}\left(\frac{n\pi}{b}\right)^4}{\left(\frac{n\pi}{b}\right)^2} + \frac{2(D_{12}+2D_{66})\left(\frac{m\pi}{a}\right)^2\left(\frac{n\pi}{b}\right)^2}{\left(\frac{n\pi}{b}\right)^2}$	$\frac{D_{11}\left(\frac{m\pi}{a}\right)^4+D_{22}\left(\frac{n\pi}{b}\right)^4}{I_0+I_2\left[\left(\frac{m\pi}{a}\right)^2+\left(\frac{n\pi}{b}\right)^2\right]} + \frac{2(D_{12}+2D_{66})\left(\frac{m\pi}{a}\right)^2\left(\frac{n\pi}{b}\right)^2}{I_0+I_2\left[\left(\frac{m\pi}{a}\right)^2+\left(\frac{n\pi}{b}\right)^2\right]}$
	$(\bar{N}_x)_{cr}$		$\hat{\omega}^2_{mn}$
An isotropic cylindrical shell axially compressed (Donnell's Equations)	$\left\{\frac{E\cdot h}{(1-\upsilon^2)m^2R^2} + \frac{E\cdot h\cdot(m^2R^2+n^2)^2}{(1-\upsilon^2)m^2R^2}\cdot\frac{h^2}{12R^2} - \frac{E\cdot h\cdot[\upsilon^2m^4R^4+2m^2R^2n^2+n^4]}{(m^2R^2+n^2)^2(1-\upsilon^2)m^2R^2}\right\}$		$\frac{E}{(1-\upsilon^2)R^2\rho} + \frac{E(m^2R^2+n^2)^2}{(1-\upsilon^2)R^2\rho}\cdot\frac{h^2}{12R^2} - \frac{E\cdot[\upsilon^2m^4R^4+2m^2R^2n^2+n^4]}{(m^2R^2+n^2)^2(1-\upsilon^2)R^2\ rho}$

Note: *No closed form solutions available for the shear critical load $(\bar{N}_{xy})_{cr}$ (see [1] and Ch. 4 of [2])
**Timoshenko, S.P. and Gere, J.M., *Theory of elastic stability*, 2nd Edition, 1985, International Student Edition, McGraw-Hill Book Company, Inc. 541, p. page 382.

References

[1] Kennedy, D. and Lo, K.I., Critical buckling predictions for plates and stiffened panels from natural frequency measurements, *J. Phy. Conf. Ser.*, Vol. 1106, 2018, Article Id 012018, 8 p.

[2] Abramovich, H., *Stability and Vibrations of Thin Walled Composite Structures*, ©Woodhead Publishing Limited, 778 p.

Chapter 6

Behavior of Circular and Annular Plates

6.1. Introduction

The behavior of isotropic circular plates is analytically solved for buckling under compressive loads. Experimental validation is presented and the influence of the in-plane load on the natural frequencies are investigated in the framework of VCT.

Then the behavior of isotropic annular plates is presented leading to expressions for their in-plane buckling loads.

Orthotropic circular and annular plates under compressive loads are discussed, followed by vibrations of isotropic annular plates.

6.1.1. *Stress distribution in a radially compressed circular plate*

Figure 6.1 presents the geometry and the loading type of a flat circular plate with a radius a.

The stress function, Φ, for an axisymmetric case of a circular plate is obtained from the following equation (see [1]):

$$\left(\frac{d^2}{dr^2}+\frac{1}{r}\frac{d}{dr}\right)\left(\frac{d^2\Phi}{dr^2}+\frac{1}{r}\frac{d\Phi}{dr}\right)=\frac{d^4\Phi}{dr^4}+\frac{2}{r}\frac{d^3\Phi}{dr^3}-\frac{1}{r^2}\frac{d^2\Phi}{dr^2}+\frac{1}{r^3}\frac{d\Phi}{dr}=0. \tag{6.1}$$

The solution for Eq. (6.1) can be written as

$$\Phi = A_1\log r + A_2 r^2\log r + A_3 r^2 + A_4. \tag{6.2}$$

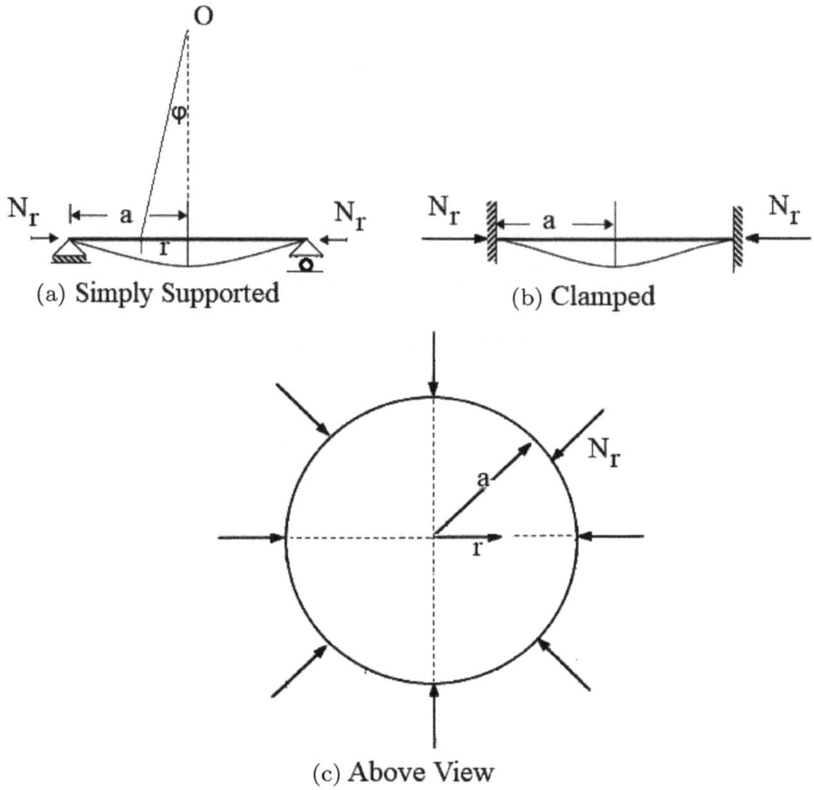

Fig. 6.1. The geometry of the circular plate and its various variables: (a) a simply supported case; (b) a clamped case; (c) above view of the plate.

Based on the stress function presented in Eq. (6.2) one can obtain the stresses, yielding

$$\sigma_r \equiv \frac{1}{r}\frac{\partial\Phi}{\partial r} + \frac{1}{r^2}\frac{\partial^2\Phi}{\partial\theta^2} = \frac{1}{r}\frac{d\Phi}{dr} = \frac{A_1}{r^2} + A_2(1 + 2\log r) + 2A_3,$$

$$\sigma_\theta \equiv \frac{\partial^2\Phi}{\partial r^2} = \frac{d^2\Phi}{dr^2} = -\frac{A_1}{r^2} + A_2(3 + 2\log r) + 2A_3,$$

$$\tau_{r\theta} \equiv -\frac{1}{r}\frac{\partial^2\Phi}{\partial r\partial\theta} + \frac{1}{r^2}\frac{\partial\Phi}{\partial\theta} = 0. \qquad (6.3)$$

To avoid a singularity at the origin of the plate (at its center), the constants A_1 and A_2 in Eq. (6.3) must vanish leading to the following results (for the case of no body forces)

$$\sigma_r = \sigma_\theta = \text{const.} \qquad (6.4)$$

6.2. Buckling of a Circular Plate

The equation leading to the axisymmetric buckling of a circular plate, as shown in Fig. 6.1 is presented in [1] as

$$r^2\frac{d^2\varphi}{dr^2} + r\frac{d\varphi}{dr} - \varphi = -\frac{Qr^2}{D}, \tag{6.5}$$

where φ is the angle between the axis of symmetry and a normal to the mid-plane of the deformed circular plate (see Fig. 6.1(a)), Q is the shear force per unit length and D, the flexural rigidity of the plate, is defined as

$$D = \frac{Eh^3}{12(1-\upsilon^2)}, \tag{6.6}$$

where E is the Young's modulus, h the thickness of the plate and υ is the Poisson's constant.

As no normal forces are acting on the circular plate, the following relationship exists:

$$Q = N_r \cdot \varphi. \tag{6.7}$$

Defining the following variables:

$$\frac{N_r}{D} \equiv \rho^2, \quad b \equiv \rho r, \tag{6.8}$$

where N_r is the compressive force per unit length (see Fig. 6.1(c)), transforms Eq. (6.5) into the following form:

$$b^2\frac{d^2\varphi}{db^2} + b\frac{d\varphi}{db} + (b^2-1)\varphi = 0. \tag{6.9}$$

The general solution of Eq. (6.9) (see [1]) is

$$\varphi = A_1 \cdot J_1(b) + A_2 \cdot Y_1(b), \tag{6.10}$$

where A_1 and A_2 are constants to be found using the boundary conditions of the problem and $J_1(b)$ and $Y_1(b)$ are Bessel functions of first and second kinds, respectively.

To satisfy the symmetry condition at the center of the plate ($r = 0 \rightarrow b = 0$) the angle φ must be zero. As the function $Y_1(b)$ goes to infinity as b tends to zero, the constant A_2 must vanish ($A_2 = 0$).

For the clamped case, ones obtains

$$\varphi_{r=a} = 0 \Rightarrow J_1(b) = J_1(\rho a) = 0 \tag{6.11}$$

leading to the smallest root of Eq. (6.11) to be $\rho a = 3.832$. Substituting the value of the smallest root in the first term of Eq. (6.8) yields

$$\frac{N_r}{D} \equiv \rho^2 \Rightarrow (N_r)_{cr} = \rho^2 \cdot D = \frac{(3.832)^2 D}{a^2} \approx \frac{14.684D}{a^2}. \tag{6.12}$$

In the case of a compressed circular plate resting on a simply supported perimeter edge, ones needs to enforce the zero bending moment, namely

$$\left(\frac{d\varphi}{dr} + \upsilon\frac{\varphi}{r}\right)_{r=a} = 0 \Rightarrow \rho a J_0(\rho a) - (1-\upsilon)J_1(\rho a) = 0,$$

while

$$\frac{dJ_1(b)}{db} = J_0(b) - \frac{J_1(b)}{b}, \tag{6.13}$$

where J_0 is the zero order Bessel function. Assuming $\upsilon = 0.3$, the smallest root of the transcendental Eq. (6.13) would be $\rho a = 2.05$. Substituting the value of the smallest root in the first term of Eq. (6.8) yields

$$\frac{N_r}{D} \equiv \rho^2 \Rightarrow (N_r)_{cr} = \rho^2 \cdot D = \frac{(2.05)^2 D}{a^2} \approx \frac{4.20D}{a^2}. \tag{6.14}$$

6.3. The Initial Geometric Influence

Actual plates are never perfect due to the presence of initial geometric imperfection, leading to the appearance of additional terms on the right-hand side of Eq. (6.5). In this case, the problem ceases to be an eigenvalue problem.

The initial geometric imperfections lead to an increasing transverse deflection due to the increase of the radial load. A coupling between the equations of equilibrium in the lateral direction and in the in-plane direction is found. The von-Karman equations for Cartesian system of coordinates can be written as (see [1]) as

$$\frac{\partial^4\Phi}{\partial x^4} + 2\frac{\partial^4\Phi}{\partial x^2\partial y^2} + \frac{\partial^4\Phi}{\partial y^4} = E\left[\left(\frac{\partial^2 w}{\partial x\partial y}\right)^2 - \frac{\partial^2 w}{\partial x^2}\frac{\partial^2 w}{\partial y^2}\right],$$

$$\frac{\partial^4 w}{\partial x^4} + 2\frac{\partial^4 w}{\partial x^2\partial y^2} + \frac{\partial^4 w}{\partial y^4} = \frac{h}{D}\left(p + \frac{\partial^2\Phi}{\partial y^2}\frac{\partial^2 w}{\partial x^2} + \frac{\partial^2\Phi}{\partial x^2}\frac{\partial^2 w}{\partial y^2} - 2\frac{\partial^2\Phi}{\partial x\partial y}\frac{\partial^2 w}{\partial x\partial y}\right), \tag{6.15}$$

where Φ is the Airy function, E is the Young's modulus, h is the plate's thickness, p is the applied pressure and D is the flexural rigidity of the plate (see Eq. (6.6)). Note that the equations presented in (6.15) assume a perfect plate without initial geometric imperfections and they are coupled, nonlinear, partial fourth-order differential equations.

For the case of a rectangular plate with initial geometric imperfection, denoted w_o, Eqs. (6.15) will have the following form:

$$\frac{\partial^4\Phi}{\partial x^4}+2\frac{\partial^4\Phi}{\partial x^2\partial y^2}+\frac{\partial^4\Phi}{\partial y^4}=E\left\{\left(\frac{\partial^2(w+w_o)}{\partial x\partial y}\right)^2-\frac{\partial^2(w+w_o)}{\partial x^2}\frac{\partial^2(w+w_o)}{\partial y^2}\right.$$
$$\left.-\left[\left(\frac{\partial^2 w_o}{\partial x\partial y}\right)^2-\frac{\partial^2 w_o}{\partial x^2}\frac{\partial^2 w_o}{\partial y^2}\right]\right\}$$
$$\frac{\partial^4 w}{\partial x^4}+2\frac{\partial^4 w}{\partial x^2\partial y^2}+\frac{\partial^4 w}{\partial y^4}=\frac{h}{D}\left(p+\frac{\partial^2\Phi}{\partial y^2}\frac{\partial^2(w+w_o)}{\partial x^2}+\frac{\partial^2\Phi}{\partial x^2}\frac{\partial^2(w+w_o)}{\partial y^2}\right.$$
$$\left.-2\frac{\partial^2\Phi}{\partial x\partial y}\frac{\partial^2(w+w_o)}{\partial x\partial y}\right). \tag{6.16}$$

Dealing with a circular plate with initial geometric imperfections, with no lateral pressure, $p=0$, Eqs. (6.16) will be written as (using polar coordinates, r and θ):

$$\left(\frac{d^2}{dr^2}+\frac{1}{r}\frac{d}{dr}\right)\left(\frac{d^2\Phi}{dr^2}+\frac{1}{r}\frac{d\Phi}{dr}\right)$$
$$=E\left\{-\frac{d^2(w+w_o)}{dr^2}\frac{1}{r}\frac{d(w+w_o)}{dr}+\frac{d^2w_o}{dr^2}\frac{1}{r}\frac{dw_o}{dr}\right\},$$
$$\left(\frac{d^2}{dr^2}+\frac{1}{r}\frac{d}{dr}\right)\left(\frac{d^2w}{dr^2}+\frac{1}{r}\frac{dw}{dr}\right)$$
$$=\frac{h}{D}\left\{\frac{1}{r}\frac{d\Phi}{dr}\frac{d^2(w+w_o)}{dr^2}+\frac{d^2\Phi}{dr^2}\frac{1}{r}\frac{d(w+w_o)}{dr}\right\}. \tag{6.17}$$

The next step is to integrate both equations (Eqs. (6.17)) with zero constants due to the integration. This leads to

$$\frac{1}{E}\int r\left(\frac{d^2}{dr^2}+\frac{1}{r}\frac{d}{dr}\right)\left(\frac{d^2\Phi}{dr^2}+\frac{1}{r}\frac{d\Phi}{dr}\right)dr$$
$$=\frac{1}{E}\int r\left\{\frac{d^2}{dr^2}\left(\frac{d^2\Phi}{dr^2}+\frac{1}{r}\frac{d\Phi}{dr}\right)+\frac{1}{r}\frac{d}{dr}\left(\frac{d^2\Phi}{dr^2}+\frac{1}{r}\frac{d\Phi}{dr}\right)\right\}dr$$
$$=\frac{1}{E}\int\left\{r\frac{d}{dr}\left[\frac{d}{dr}\left(\frac{d^2\Phi}{dr^2}+\frac{1}{r}\frac{d\Phi}{dr}\right)\right]+\frac{d}{dr}\left(\frac{d^2\Phi}{dr^2}+\frac{1}{r}\frac{d\Phi}{dr}\right)\right\}dr$$
$$=\frac{1}{E}\int\left\{\frac{d}{dr}\left[r\cdot\frac{d}{dr}\left(\frac{d^2\Phi}{dr^2}+\frac{1}{r}\frac{d\Phi}{dr}\right)\right]\right\}dr$$
$$=\frac{1}{E}\cdot r\cdot\frac{d}{dr}\left(\frac{d^2\Phi}{dr^2}+\frac{1}{r}\frac{d\Phi}{dr}\right),$$

$$\int\left\{-\frac{d^2(w+w_o)}{dr^2}\cdot\frac{d(w+w_o)}{dr}+\frac{d^2w_o}{dr^2}\cdot\frac{dw_o}{dr}\right\}dr$$
$$=\int\frac{d}{dr}\cdot\frac{1}{2}\left\{-\left[\frac{d(w+w_o)}{dr}\right]^2+\left[\frac{dw_o}{dr}\right]^2\right\}dr$$
$$=\frac{1}{2}\left\{-\left[\frac{d(w+w_o)}{dr}\right]^2+\left[\frac{dw_o}{dr}\right]^2\right\}=-\frac{1}{2}\left(\frac{dw}{dr}\right)^2-\frac{dw}{dr}\cdot\frac{dw_o}{dr},$$
$$\frac{D}{h}\int r\left(\frac{d^2}{dr^2}+\frac{1}{r}\frac{d}{dr}\right)\left(\frac{d^2w}{dr^2}+\frac{1}{r}\frac{dw}{dr}\right)dr$$
$$=\frac{D}{h}\int r\left\{\frac{d^2}{dr^2}\left(\frac{d^2w}{dr^2}+\frac{1}{r}\frac{dw}{dr}\right)+\frac{1}{r}\frac{d}{dr}\left(\frac{d^2w}{dr^2}+\frac{1}{r}\frac{dw}{dr}\right)\right\}dr$$
$$=\frac{D}{h}\int\left\{r\frac{d}{dr}\left[\frac{d}{dr}\left(\frac{d^2w}{dr^2}+\frac{1}{r}\frac{dw}{dr}\right)\right]+\frac{d}{dr}\left(\frac{d^2w}{dr^2}+\frac{1}{r}\frac{dw}{dr}\right)\right\}dr$$
$$=\frac{D}{h}\int\left\{\frac{d}{dr}\left[r\cdot\frac{d}{dr}\left(\frac{d^2w}{dr^2}+\frac{1}{r}\frac{dw}{dr}\right)\right]\right\}dr$$
$$=\frac{D}{h}\cdot r\cdot\frac{d}{dr}\left(\frac{d^2w}{dr^2}+\frac{1}{r}\frac{dw}{dr}\right),$$
$$\int\left\{\frac{d\Phi}{dr}\frac{d^2(w+w_o)}{dr^2}+\frac{d^2\Phi}{dr^2}\frac{d(w+w_o)}{dr}\right\}dr=\int\frac{d}{dr}\cdot\left\{\frac{d\Phi}{dr}\frac{d(w+w_o)}{dr}\right\}dr$$
$$=\frac{d\Phi}{dr}\frac{d(w+w_o)}{dr}=\frac{d\Phi}{dr}\left(\frac{dw}{dr}+\frac{dw_0}{dr}\right) \tag{6.18}$$

Rearranging Eqs. (6.18) leads to

$$\frac{1}{E}\cdot\frac{1}{r}\cdot\frac{d}{dr}\left(\frac{d^2\Phi}{dr^2}+\frac{1}{r}\frac{d\Phi}{dr}\right)=-\frac{1}{r^2}\frac{1}{2}\left(\frac{dw}{dr}\right)^2-\frac{1}{r^2}\frac{dw}{dr}\cdot\frac{dw_o}{dr},$$
$$\frac{D}{h}\cdot\frac{1}{r}\cdot\frac{d}{dr}\left(\frac{d^2w}{dr^2}+\frac{1}{r}\frac{dw}{dr}\right)=\frac{1}{r}\frac{d\Phi}{dr}\left(\frac{1}{r}\frac{dw}{dr}+\frac{1}{r}\frac{dw_o}{dr}\right). \tag{6.19}$$

Introducing the following new non-dimensionalized functions, instead of the unknown functions, in Eqs. (6.19), Φ, w and w_o, namely

$$\Pi=-\frac{1}{A^2\eta^2E}\cdot\left(\frac{1}{r}\cdot\frac{d\Phi}{dr}\right),$$
$$\mathrm{K}=-\frac{a}{A^2\eta}\cdot\left(\frac{1}{r}\cdot\frac{dw}{dr}\right),$$
$$\mathrm{K}_o=-\frac{a}{A^2\eta}\cdot\left(\frac{1}{r}\cdot\frac{dw_o}{dr}\right), \tag{6.20}$$

where the various constants appearing in Eqs. (6.20) are defined as follows:

$$\eta = \frac{1}{\sqrt{12(1-\upsilon^2)}} \cdot \frac{h}{a}, \qquad \alpha = \frac{Ar}{a} \tag{6.21}$$

with A being an arbitrary constant, a the radius of the circular plate and υ is the Poisson's ratio of the material. Having defined all the new variables, Eqs. (6.19) will have the following new form:

$$\begin{aligned} \frac{1}{\alpha^3} \cdot \frac{d}{d}\left(\alpha^3 \cdot \frac{d\Pi}{d\alpha}\right) &= \frac{1}{2} \cdot \mathrm{K}^2 + \mathrm{K} \cdot \mathrm{K}_0, \\ \frac{1}{\alpha^3} \cdot \frac{d}{d}\left(\alpha^3 \cdot \frac{d\mathrm{K}}{d\alpha}\right) &= -\Pi \cdot (\mathrm{K} + \mathrm{K}_0). \end{aligned} \tag{6.22}$$

Assuming a parabolic initial geometric imperfection, having the form (like in [2])

$$w_o = B \cdot \eta \cdot a \cdot \left(1 - \frac{r^2}{a^2}\right), \tag{6.23}$$

we get

$$\mathrm{K}_0 = \frac{2\mathrm{B}}{A^2}. \tag{6.24}$$

The expression for the constant B can be shown to be

$$B = \frac{\varphi_0}{\gamma h} \quad \text{where } \varphi_0 = B \cdot \eta \cdot a = B \cdot \gamma \cdot h, \tag{6.25}$$

and φ_0 is the initial deflection at the center of the circular plate.

The two main unknowns, Π and K presented in Eqs. (6.22), are described as series

$$\begin{aligned} \Pi &= \sum_{i=0}^{\infty} \pi_i \cdot \alpha^{2i}, \\ \mathrm{K} &= \sum_{i=0}^{\infty} k_i \cdot \alpha^{2i}. \end{aligned} \tag{6.26}$$

Substitution of Eqs. (6.26) in Eqs. (6.22) and equating coefficients of the same will yield to the following recursive equations:

$$\begin{aligned} 2 \cdot i \cdot (2 \cdot i + 2) \cdot \pi_i &= \frac{1}{2} \sum_{m+n=i-1}^{\infty} k_m \cdot k_n + \frac{2B}{A^2} k_{i-1}, \\ 2 \cdot i \cdot (2 \cdot i + 2) \cdot k_i &= -\frac{1}{2} \sum_{m+n=i-1}^{\infty} \pi_m \cdot k_n - \frac{2B}{A^2} \pi_{i-1}. \end{aligned} \tag{6.27}$$

For the case of clamped boundary conditions on the perimeter of the circular plate, the following expressions can be written

$$w(r=a)=0, \quad \frac{dw(r=a)}{dr}=0,$$
$$\sigma_r=\frac{1}{r}\cdot\frac{d\Phi}{dr}=\frac{N_r}{h}=p_r. \tag{6.28}$$

For a non-dimensionalized case, Eqs. (6.28) will have the following form:

$$\mathrm{K}(\alpha=A)=0, \quad \Pi(\alpha=A)=\Pi_e, \text{ where}$$
$$\Pi_e=\frac{p_r}{A^2\eta^2E}. \tag{6.29}$$

The determination of the constant A, for the case of clamped circular case is done using the first condition in Eqs. (6.29) yielding

$$\mathrm{K}=0 \quad \text{for } \alpha=A \text{ or}$$
$$0=k_0+k_1\cdot A^2+k_2\cdot A^4+k_3\cdot A^6+k_4\cdot A^8+k_5\cdot A^{10}+\cdots. \tag{6.30}$$

A will be the smallest root of the transcendental equation presented above. Once the unknown A was determined, the value of the function Π at $\alpha=A$ can be calculated and the compression applied on the perimeter of the circular plate can be written to be

$$p_r=A^2\cdot\eta^2\cdot E\cdot\Pi_{\alpha=A}. \tag{6.31}$$

The critical compression leading to the buckling of a clamped circular plate will then be written as

$$(p_r)_{cr}=14.68\cdot\eta^2\cdot E. \tag{6.32}$$

The non-dimensional loading will then be written as

$$\frac{p_r}{(p_r)_{cr}}=\frac{N_r}{(N_r)_{cr}}=\frac{A^2\cdot\Pi_{\alpha=A}}{14.68} \tag{6.33}$$

The displacement at the center of the circular plate can be calculated to yield the following expression:

$$\varphi=\eta a\cdot\int_0^A \mathrm{K}\cdot\alpha\cdot d\alpha=\eta a\cdot\left(k_0\cdot\frac{A^2}{2}+k_1\cdot\frac{A^4}{4}+k_2\cdot\frac{A^6}{6}+k_3\cdot\frac{A^8}{8}+\cdots\right)$$
$$\text{or}$$
$$\frac{\varphi}{\eta a}=\frac{\varphi}{\gamma h}=k_0\cdot\frac{A^2}{2}+k_1\cdot\frac{A^4}{4}+k_2\cdot\frac{A^6}{6}+k_3\cdot\frac{A^8}{8}+\cdots. \tag{6.34}$$

Note that the number of terms in Eq. (6.34) depends on the non-dimensional compression ratio. For $p_r/(p_r)_{cr}$, around 1, one should use terms up to the power of 10, while for $p_r/(p_r)_{cr} = 2$, the power should increase to 20.

The boundary conditions for a circular plate simply supported at $r = a$ will be

$$w(r = a) = 0 \quad \frac{d^2w(r = a)}{dr^2} + \frac{\upsilon}{r}\frac{dw(r = a)}{dr} = 0,$$
$$\sigma_r = \frac{1}{r} \cdot \frac{d\Phi}{dr} = \frac{N_r}{h} \equiv p_r, \tag{6.35}$$

where the second boundary condition, the moment, will be written as

$$\sum_{i=0}^{\infty} (2 \cdot i + 1 + \upsilon) \cdot \mathrm{K}_i \cdot A^{2 \cdot i} = 0. \tag{6.36}$$

Using the definition for Π_e from Eq. (6.29), we can arrive at a similar expression, like the one presented in Eq. (6.33), but for simply supported boundary conditions

$$\frac{p_r}{(p_r)_{cr}} = \frac{N_r}{(N_r)_{cr}} = \frac{A^2 \cdot \Pi_{\alpha=A}}{4.20}. \tag{6.37}$$

The deflection at the center of the circular plate and its non-dimensional form are given by Eq. (6.34).

6.4. Results

Some typical results, based on the equations presented above, will be next presented, with more results to be found in [4].

A computer code was written to find the relation between central plate deflection and the applied radial load. The code is based on the recursive relations presented in Eqs. (6.27), with the series being calculated to their 20th term, while the variable A is an arbitrary constant (see [4]). The convergence of the series is determined by using Eqs. (6.26) and (6.36). The iterations start with the assumption of the variables π_0 and k_0 (namely the first terms for each series). Then the value of Π_e for a given value of the $p_r/(p_r)_{cr}$ ratio is calculated using Eq. (6.29). Then the code would update the value of variables π_0 and k_0 until a prescribed convergence is achieved. Once the last variable of the series k_i is obtained, we use Eq. (6.34) to calculate the non-dimensional deflection $\varphi/\gamma h$. This is done for each assumed ratio of the non-dimensional pressure $p_r/(p_r)_{cr}$. Typical results can be seen in Figs. 6.2 and 6.3.

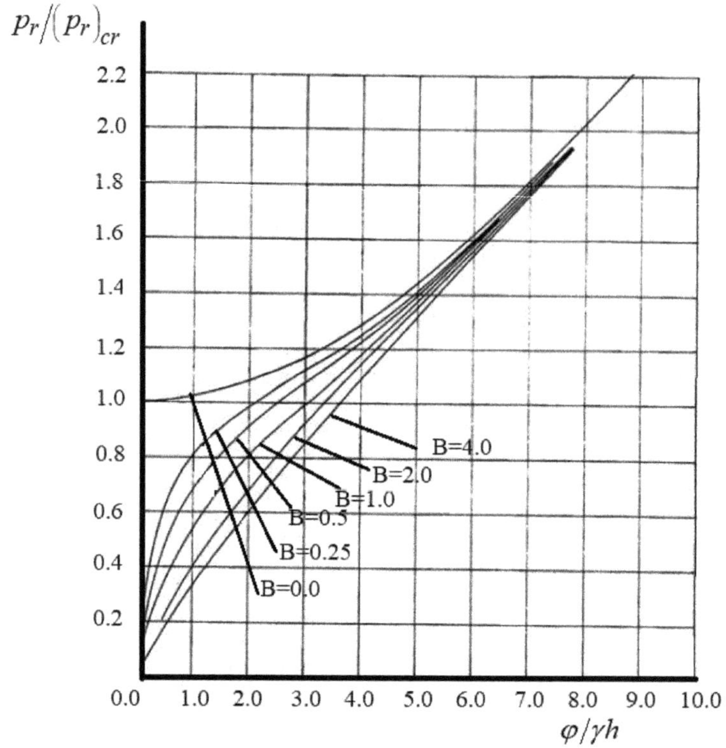

Fig. 6.2. The non-dimensional pressure as a function of the non-dimensional deflection for a clamped circular plate (adapted from [2]).

6.5. The Experimental Set-Up

Six circular plates under uniform compressive pressure on their circumference were tested. Out-of-plane deflections and natural frequencies were measured and the data was stored using a PC.

The plates were made from Aluminum Alloy AL-7075-T6, having the following properties: Young's modulus, $E = 75\,\text{GPa}$, Poisson's ratio, $\nu = 0.3$. The diameter of the plate was taken as $2 * a = 200\,\text{mm}$ and the thickness, $h = 1.06\,\text{mm}$ (see [4]). The loading system and the boundary conditions are presented in Figs. 6.4–6.6. Thirty segments enabled the application of the pressure on the circumference of the specimens (more details in [4]).

A hand pump enabled the supply of the pressure, which was measured by a pressure transducer.

The displacement were measured by a LVDT (Linear Variable Differential Transformer) mounted on a special bridge (see #5 in Fig. 6.4), thus enabling the recording of both radial and circumferential out-of-plane deflections.

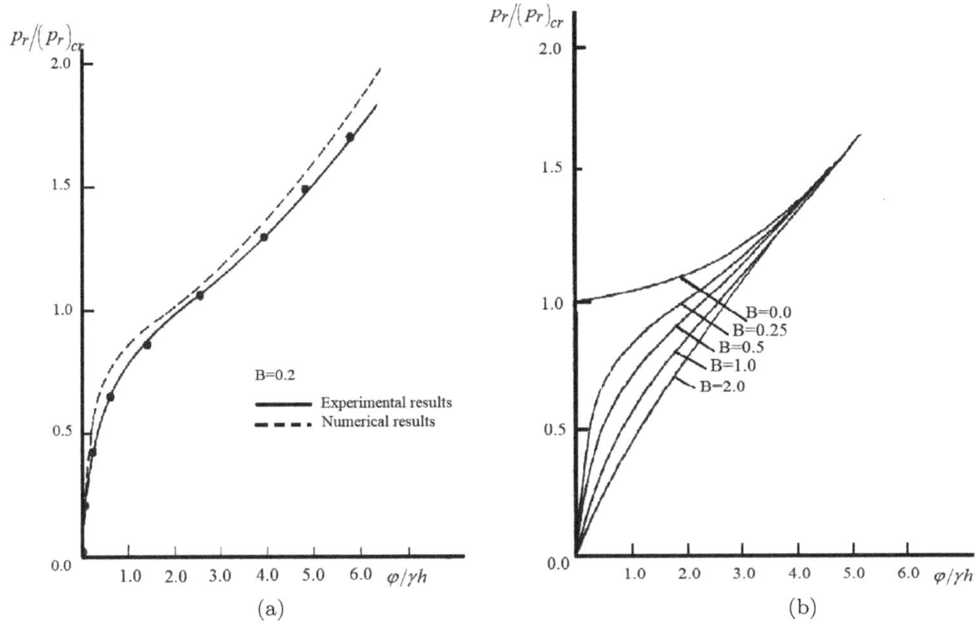

Fig. 6.3. The non-dimensional pressure as a function of the non-dimensional deflection for a simply supported circular plate: (a) experimental vs. numerical results for initial imperfection $B = 0.2$ and (b) the influence of the initial imperfection B on the post-buckling plate behavior (see [4]).

Fig. 6.4. The test set-up.

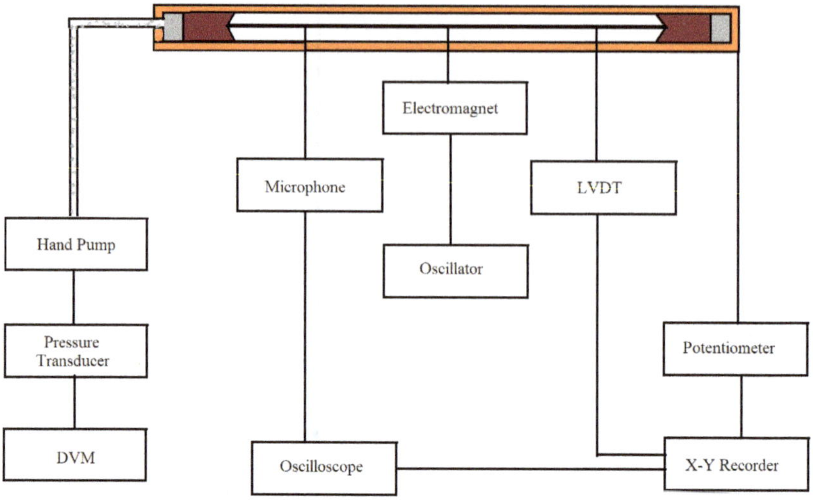

Fig. 6.5. Block diagram of the experimental set-up.

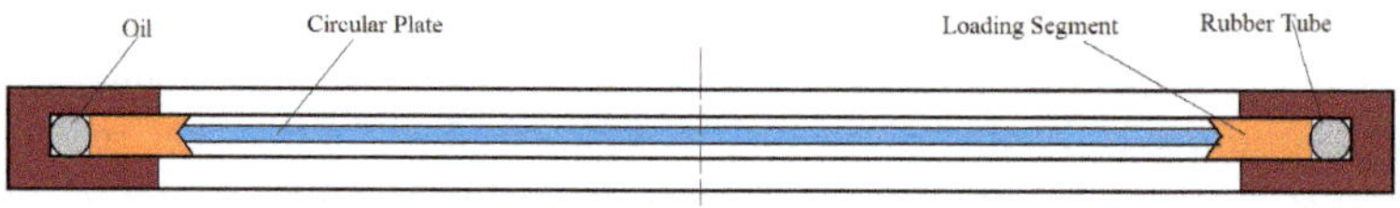

Fig. 6.6. A schematic drawing of the hydraulic loading system.

The excitation of the plate was done using and an electromagnet (see Fig. 6.4) and the driving frequency was supplied by an oscillator and read on a digital counter. Note that the core of the electromagnet was bonded to the specimen causing it to vibrate. The resonance was detected using Lissajous shapes on an oscilloscope, and the plate response was measure with a LVDT for frequencies up to 100 Hz.

Figure 6.7 presents the experimental results for the dynamic part of the tests. It presents the reduction of the non-dimensional squared natural frequency as a function of the non-dimensional circumferential pressure. Massonnet [3, 4] suggested in his manuscript that a non-destructive method can be used to predict the buckling a circular plate, by following the reduction of the lowest natural frequency with increasing compressive pressure. However, he noted that the natural frequency of the plate would not be reduced to zero at its buckling load. Instead, due to initial geometric imperfections, at a certain compressive load, the natural frequencies will increase again, as displayed in Fig. 6.7. This point can be used to predict

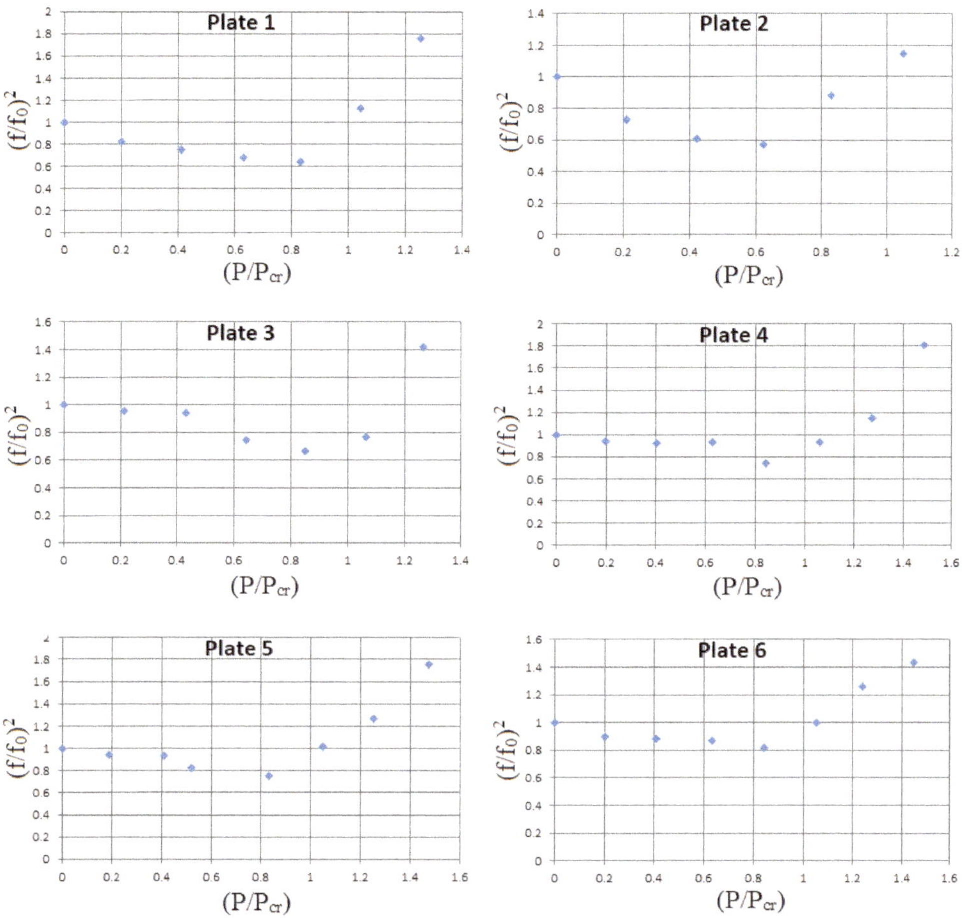

Fig. 6.7. The experimental squared of the non-dimensional frequency vs. the non-dimensional lateral pressure — Plates 1–6.

the actual buckling load of the plate. This method to predict buckling loads of a structure is also known by the acronym VCT (Vibration Correlation Technique) [5].

The summary of the experimental results are presented in Table 6.1. Note that the experimental buckling pressure was visually determined.

Comparing the predicted to the experimental buckling pressures, as displayed in Table 6.1, yields excellent results, with the VCT-based buckling pressure being equal or slightly below than the actual buckling pressure, exhibiting its non-destructive nature. Note that due to the stable post-buckling behavior of a plate, one can reach until the buckling load and beyond, without damaging the specimen.

Table 6.1. Theoretical, experimental and non-destructive predicted (VCT) buckling pressures.

Plate #	Theoretical buckling pressure (kPa)	Experimental buckling pressure (kPa)	Predicted buckling pressure (kPa)
1	324.0	275.8	275.76
2	324.0	204.1	201.33
3	324.0	275.8	275.78
4	324.0	275.8	272.34
5	324.0	275.8	269.48
6	324.0	275.8	272.89

6.6. Radially Compressed Annular Plates

Figure 6.8 presents a general annular plate with an inner radius of b and outer radius of a being compressed by an inner compression p_i and outer compression of p_o.

As presented in [6], the static von Karman equations for a general thin elastic isotropic plate in polar coordinates are

$$D\left[\frac{\partial^4 w}{\partial r^4}+\frac{2}{r}\frac{\partial^3 w}{\partial r^3}-\frac{1}{r^2}\frac{\partial^2 w}{\partial r^2}+\frac{2}{r^2}\frac{\partial^4 w}{\partial r^2\partial\theta^2}\right]$$

$$+D\left[\frac{1}{r^3}\frac{\partial w}{\partial r}-\frac{2}{r^3}\frac{\partial^3 w}{\partial r\partial\theta^2}+\frac{1}{r^4}\frac{\partial^4 w}{\partial\theta^4}+\frac{4}{r^4}\frac{\partial^2 w}{\partial\theta^2}\right]=\frac{\partial^2 w}{\partial r^2}\left(\frac{1}{r}\frac{\partial\Phi}{\partial r}+\frac{1}{r^2}\frac{\partial^2\Phi}{\partial\theta^2}\right)$$

$$+\frac{\partial^2\Phi}{\partial r^2}\left(\frac{1}{r}\frac{\partial w}{\partial r}+\frac{1}{r^2}\frac{\partial^2 w}{\partial\theta^2}\right)-2\left(\frac{1}{r}\frac{\partial^2 w}{\partial r\partial\theta}-\frac{1}{r^2}\frac{\partial w}{\partial\theta}\right)\left(\frac{1}{r}\frac{\partial^2\Phi}{\partial r\partial\theta}-\frac{1}{r^2}\frac{\partial\Phi}{\partial\theta}\right),$$

$$\frac{1}{Eh}\left[\frac{\partial^4\Phi}{\partial r^4}+\frac{2}{r}\frac{\partial^3\Phi}{\partial r^3}-\frac{1}{r^2}\frac{\partial^2\Phi}{\partial r^2}+\frac{2}{r^2}\frac{\partial^4\Phi}{\partial r^2\partial\theta^2}\right]$$

$$+\frac{1}{Eh}\left[\frac{1}{r^3}\frac{\partial\Phi}{\partial r}-\frac{2}{r^3}\frac{\partial^3\Phi}{\partial r\partial\theta^2}+\frac{1}{r^4}\frac{\partial^4\Phi}{\partial\theta^4}+\frac{4}{r^4}\frac{\partial^2\Phi}{\partial\theta^2}\right]$$

$$=-\frac{\partial^2 w}{\partial r^2}\left(\frac{1}{r}\frac{\partial w}{\partial r}+\frac{1}{r^2}\frac{\partial^2 w}{\partial\theta^2}\right)+\left(\frac{1}{r}\frac{\partial^2 w}{\partial r\partial\theta}-\frac{1}{r^2}\frac{\partial w}{\partial\theta}\right)^2. \tag{6.38}$$

For an axisymmetric case Eq. (6.38) is simplified to

$$D\left[\frac{d^4 w}{dr^4}+\frac{2}{r}\frac{d^3 w}{dr^3}-\frac{1}{r^2}\frac{d^2 w}{dr^2}+\frac{1}{r^3}\frac{dw}{dr}\right]=\frac{1}{r}\frac{d^2 w}{dr^2}\frac{d\Phi}{dr}+\frac{1}{r}\frac{dw}{dr}\frac{d^2\Phi}{dr^2},$$

$$\frac{1}{Eh}\left[\frac{d^4\Phi}{dr^4}+\frac{2}{r}\frac{d^3\Phi}{dr^3}-\frac{1}{r^2}\frac{d^2\Phi}{dr^2}+\frac{1}{r^3}\frac{d\Phi}{dr}\right]=-\frac{1}{r}\frac{\partial^3 w}{\partial r^3}. \tag{6.39}$$

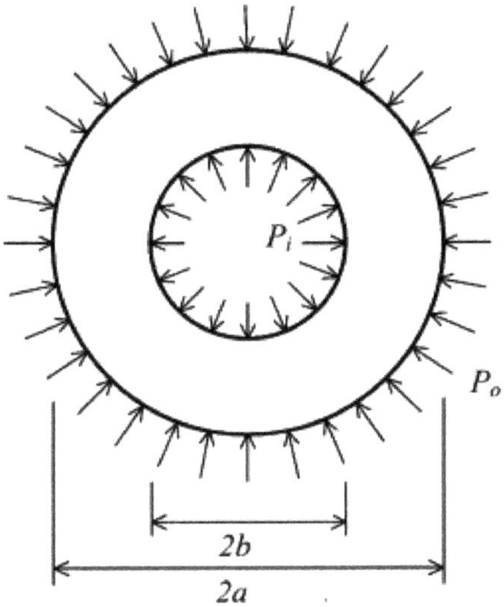

Fig. 6.8. Schematic drawing of a loaded annular plate.

For an annular plate (see Fig. 6.8) having an inner and outer radii b, a, respectively, the governing equation can be written for a general in-plane loadings, without normal loads (see also [6])

$$D\nabla^4 w = N_{rr}\frac{\partial^2 w}{\partial r^2} + N_{\theta\theta}\left(\frac{1}{r}\frac{\partial w}{\partial r} + \frac{1}{r^2}\frac{\partial^2 w}{\partial \theta^2}\right) - 2N_{r\theta}\left(\frac{1}{r^2}\frac{\partial w}{\partial \theta} - \frac{1}{r}\frac{\partial^2 w}{\partial r \partial \theta}\right),$$

where

$$\nabla^4 w = \left(\frac{\partial^2}{\partial r^2} + \frac{1}{r}\frac{\partial}{\partial r} + \frac{1}{r^2}\frac{\partial^2}{\partial \theta^2}\right)\left(\frac{\partial^2}{\partial r^2} + \frac{1}{r}\frac{\partial}{\partial r} + \frac{1}{r^2}\frac{\partial^2}{\partial \theta^2}\right) w,$$

$$N_{rr} \equiv h\sigma_{rr}, N_{\theta} \equiv h\sigma_{\theta\theta}, N_{r\theta} \equiv h\sigma_{r\theta}. \tag{6.40}$$

For the plate depicted in Fig. 6.8, and denoting the applied loads at the inner and outer boundaries as $\sigma_{rr,@r=a} = p_0 = -\lambda$, $\sigma_{rr,@r=b} = p_i = -\lambda, N_{r\theta} = 0$, $\lambda > 0$, Eq. (6.40) can be re-written as

$$D\nabla^4 w = \frac{\partial^2 w}{\partial r^2} - \lambda h\left(\frac{1}{r}\frac{\partial w}{\partial r} + \frac{1}{r^2}\frac{\partial^2 w}{\partial \theta^2}\right)$$

$$= -\lambda h\left(\frac{\partial^2 w}{\partial r^2} + \frac{1}{r}\frac{\partial w}{\partial r} + \frac{1}{r^2}\frac{\partial^2 w}{\partial \theta^2}\right) = -\lambda h \nabla^2 w \quad \text{therefore}$$

$$\nabla^4 w + \frac{\lambda h}{D}\nabla^2 w = 0. \tag{6.41}$$

Defining the following non-dimensional terms:

$$\rho = \frac{r}{a}, \quad \xi = \frac{b}{a}, \quad \xi < \rho < 1. \tag{6.42}$$

One can suggest a solution for Eq. (6.41) in the following form

$$w(\rho, \xi) = \Theta_n(\rho) \cos n\theta, \quad 0 \leq \theta \leq 2\pi. \tag{6.43}$$

Substituting Eq. (6.42) into Eq. (6.41) yields

$$\nabla_n^4 \Theta_n(\rho) + \frac{\lambda h a^2}{D} \nabla_n^2 \Theta_n(\rho) = 0, \quad \text{where}$$

$$\nabla_n^2 = \frac{d^2}{d\rho^2} + \frac{1}{\rho}\frac{d}{d\rho} - \frac{n^2}{\rho^2}. \tag{6.44}$$

The solution for Eq. (6.44) can be written in terms of Bessel functions,[a] depending on the plate boundary conditions. It can be presented as

$$\Theta_0(\rho) = A_0 J_0\left[\sqrt{\frac{\lambda h r}{D}}\right] + B_0 Y_0\left[\sqrt{\frac{\lambda h r}{D}}\right] + C_0$$

$$+ D_0 \ln \rho \quad \text{for } n = 0, \ \rho = \frac{r}{a} \quad \text{and}$$

$$\Theta_n(\rho) = A_n J_n\left[\sqrt{\frac{\lambda h r}{D}}\right] + B_n Y_n\left[\sqrt{\frac{\lambda h r}{D}}\right]$$

$$+ C_0 \rho^n + D_0 \rho^{-n} \quad \text{for } n = 1, 2, \ldots, \ \rho = \frac{r}{a}, \tag{6.45}$$

where J_n, Y_n are the Bessel functions of the first and second kind, respectively, while A_0, B_0, C_0, D_0 and A_n, B_n, C_n, D_n are arbitrary constants of integration to be determined from the imposed boundary condition.

Next we shall present some of Yamaki's [7] results, who is considered to present the most general solutions.

(a) The circumferences at $r = a$ and $r = b$ are clamped (more details on the boundary conditions can be found in Appendix A):
For this case, the boundary conditions are $w = dw/d\rho = 0$. Imposing the boundary conditions on the general solution (Eq. (6.45)) yields for a fixed n four homogenous linear equations in A_n, B_n, C_n, D_n. Vanishing of the determinant of the coefficients matrix leads to the critical buckling stress, λ_{cr}.
(b) The circumference at $r = a$ is clamped, while at $r = b$ is simply supported:

[a]Bessel functions, originally discovered by Daniel Bernoulli, were later generalized by Friedrich Bessel (22 July 1784–17 March 1846, a German astronomer, mathematician, physicist and geodesist) are canonical solutions $y(x)$ of the Bessel differential equation $x^2 \frac{d^2y}{dx^2} + x\frac{dy}{dx} + (x^2 - \alpha^2)y = 0$, where α is an arbitrary complex number.

For this case, the boundary conditions are as follows:

$$w(r=a)=0, \quad \frac{dw(r=a)}{\partial \rho}=0,$$

while

$$w(r=b)=0, \quad \frac{\partial^2 w(r=b)}{\partial \rho^2}+\frac{\upsilon}{\rho}\left[\frac{\partial w(r=b)}{\partial \rho}+\frac{1}{\rho}\frac{\partial^2 w(r=b)}{\partial \theta^2}\right]=0.$$

As in (a), demanding the vanishing of the determinant of the coefficients matrix leads to the critical buckling stress, λ_{cr}.

(c) The circumference at $r=a$ is clamped, while at $r=b$ is free:
For this case, the boundary conditions are as follows:

$$w(r=a)=0, \quad \frac{dw(r=a)}{\partial \rho}=0,$$

while

$$\frac{\partial^2 w(r=b)}{\partial \rho^2}+\frac{\upsilon}{\rho}\left[\frac{\partial w(r=b)}{\partial \rho}+\frac{1}{\rho}\frac{\partial^2 w(r=b)}{\partial \theta^2}\right]=0,$$

$$\frac{\partial}{\partial \rho}\left[\frac{\partial^2 w(r=b)}{\partial \rho^2}+\frac{1}{\rho}\frac{\partial w(r=b)}{\partial \rho}+\frac{1}{\rho^2}\frac{\partial^2 w(r=b)}{\partial \theta^2}\right] + \frac{(1-\upsilon)}{\rho}\frac{\partial}{\partial \theta}\left[\frac{1}{\rho}\left(\frac{\partial w\,(r=b)}{\partial \rho}\right)\right]+D\frac{\partial w(r=b)}{\partial \rho}=0$$

As in (a) and (b), demanding the vanishing of the determinant of the coefficients matrix leads to the critical buckling stress, λ_{cr}. See Appendix B for buckling coefficients for a radial loaded annular plate with inner free boundary conditions and clamped boundary conditions for the external circumference.

(d) The circumference at $r=a$ is simply supported, while at $r=b$ is free:
For this case, the boundary conditions are as follows:

$$w(r=b)=0, \quad \frac{\partial^2 w(r=b)}{\partial \rho^2}+\frac{\upsilon}{\rho}\left[\frac{\partial w(r=b)}{\partial \rho}+\frac{1}{\rho}\frac{\partial^2 w(r=b)}{\partial \theta^2}\right]=0,$$

while

$$\frac{\partial^2 w(r=b)}{\partial \rho^2}+\frac{\upsilon}{\rho}\left[\frac{\partial w(r=b)}{\partial \rho}+\frac{1}{\rho}\frac{\partial^2 w(r=b)}{\partial \theta^2}\right]=0,$$

$$\frac{\partial}{\partial \rho}\left[\frac{\partial^2 w(r=b)}{\partial \rho^2}+\frac{1}{\rho}\frac{\partial w(r=b)}{\partial \rho}+\frac{1}{\rho^2}\frac{\partial^2 w(r=b)}{\partial \theta^2}\right] + \frac{(1-\upsilon)}{\rho}\frac{\partial}{\partial \theta}\left[\frac{1}{\rho}\left(\frac{\partial w(r=b)}{\partial \rho}\right)\right]+D\frac{\partial w(r=b)}{\partial \rho}=0.$$

As in (a) and (b), demanding the vanishing of the determinant of the coefficients matrix leads to the critical buckling stress, λ_{cr}. See Appendix B for buckling coefficients for a radial loaded annular plate with inner free boundary conditions and simply supported boundary conditions for the external circumference.

A more detailed example of how to calculate the buckling load of an annular plate is presented in Appendix C.

6.7. Buckling of Orthotropic Circular Plates

The orthotropic circular plate, having a radius a and a thickness h has the following stress-strain relations:

$$\begin{aligned}
&\sigma_{rr} = E_{rr}\varepsilon_{rr} + \upsilon_{r\theta}E_{\theta\theta}\varepsilon_{\theta\theta} \equiv Q_{11}\varepsilon_{rr} + Q_{12}\varepsilon_{\theta\theta},\\
&\sigma_{\theta\theta} = \upsilon_{r\theta}E_{rr}\varepsilon_{rr} + E_{\theta\theta}\varepsilon_{\theta\theta} \equiv Q_{12}\varepsilon_{rr} + Q_{22}\varepsilon_{\theta\theta},\\
&\tau_{r\theta} = G_{r\theta}\gamma_{r\theta} \equiv Q_{66}\gamma_{r\theta}, \quad \text{where}\\
&\upsilon_{r\theta}E_{\theta\theta} = \upsilon_{r\theta}E_{rr}\varepsilon_{rr} \Rightarrow Q_{21} = Q_{12}.
\end{aligned} \tag{6.46}$$

The general buckling equation for an orthotropic circular plate in plane loaded can be presented within the classical lamination theory as (see also [8])

$$\begin{aligned}
&\frac{h^3}{12}\left[Q_{11}\frac{\partial^4 w}{\partial r^4} + 2\frac{(Q_{12}+2Q_{66})}{r^2}\frac{\partial^4 w}{\partial r^2\partial\theta^2} + \frac{Q_{22}}{r^4}\frac{\partial^4 w}{\partial\theta^4} + 2\frac{Q_{11}}{r^3}\frac{\partial^3 w}{\partial r^3}\right]\\
&\quad + \frac{h^3}{12}\left[-2\frac{(Q_{12}+2Q_{66})}{r^3}\frac{\partial^3 w}{\partial r\partial\theta^2} - \frac{Q_{22}}{r^2}\frac{\partial^2 w}{\partial r^2}\right.\\
&\quad \left. + 2\frac{(Q_{12}+2Q_{66}+Q_{22})}{r^4}\frac{\partial^2 w}{\partial\theta^2} + \frac{Q_{22}}{r^3}\frac{\partial w}{\partial r}\right]\\
&= -N_r\frac{\partial^2 w}{\partial r^2} - N_\theta\left(\frac{1}{r}\frac{\partial w}{\partial r} + \frac{1}{r^2}\frac{\partial^2 w}{\partial\theta^2}\right) - 2N_{r\theta}\left(\frac{1}{r^2}\frac{\partial w}{\partial\theta} - \frac{1}{r}\frac{\partial^2 w}{\partial r\partial\theta}\right).
\end{aligned} \tag{6.47}$$

For the axisymmetric radial case, the pre-buckling loads will be [8]

$$N_r = h\sigma_0\left(\frac{r}{a}\right)^{k-1}, \quad N_r = kh\sigma_0\left(\frac{r}{a}\right)^{k-1}, \quad N_{r\theta} = 0. \tag{6.48}$$

where k is an orthotropic constant defined below.

According to [8], Eq. (6.47) is re-casted in a nondimensional form $\rho = r/a$ yielding

$$\begin{aligned}
&\frac{\partial^4 \widehat{w}}{\partial \rho^4} + 2\frac{(\alpha_1 + 2\alpha_3)}{r^2}\frac{\partial^4 \widehat{w}}{\partial \rho^2 \partial \theta^2} + \frac{k^2}{\rho^4}\frac{\partial^4 \widehat{w}}{\partial \theta^4} + \frac{2}{\rho}\frac{\partial^3 \widehat{w}}{\partial \rho^3} - 2\frac{(\alpha_1 + 2\alpha_3)}{\rho^3}\frac{\partial^3 \widehat{w}}{\partial \rho \partial \theta^2}\\
&\quad - \frac{k^2}{\rho^2}\frac{\partial^2 \widehat{w}}{\partial \rho^2} + 2\frac{(\alpha_1 + k^2 + 2\alpha_3)}{\rho^4}\frac{\partial^2 \widehat{w}}{\partial \theta^2} + \frac{k^2}{\rho^3}\frac{\partial \widehat{w}}{\partial \rho}\\
&\quad + \frac{12a^2}{Q_{11}h^3}\left[N_\rho \frac{\partial^2 \widehat{w}}{\partial \rho^2} + N_\theta\left(\frac{1}{\rho}\frac{\partial \widehat{w}}{\partial \rho} + \frac{1}{\rho^2}\frac{\partial^2 \widehat{w}}{\partial \theta^2}\right)\right.\\
&\quad \left. + 2N_{r\theta}\left(\frac{1}{\rho^2}\frac{\partial \widehat{w}}{\partial \theta} - \frac{1}{\rho}\frac{\partial^2 \widehat{w}}{\partial \rho \partial \theta}\right)\right] = 0 \quad \text{where}\\
&\alpha_1 = \frac{Q_{12}}{Q_{11}}, \quad \alpha_3 = \frac{Q_{66}}{Q_{11}}, \quad k^2 = \frac{Q_{22}}{Q_{11}}.
\end{aligned} \tag{6.49}$$

The non-dimensional deflection ($\widehat{w} = w/a$) can be assumed to have the following form [8]:

$$\widehat{w} = W(\rho)\cos m\theta. \tag{6.50}$$

Then, substituting the proposed deflection function in Eq. (6.49) leads to an ordinary differential equations having variable coefficients:

$$\begin{aligned}
&\frac{d^4 W}{d\rho^4} + \frac{2}{\rho}\frac{d^3 W}{d\rho^3} - \frac{\lambda^2}{\rho^2}\frac{\partial^2 W}{\partial \rho^2} + \frac{\lambda^2}{\rho^3}\frac{\partial W}{\partial \rho} + \frac{\beta^2}{\rho^4}W\\
&\quad + \gamma^{k+1}\left[\rho^{k-1}\frac{\partial^2 W}{\partial \rho^2} + k\rho^{k-2}\frac{\partial W}{\partial \rho} - km^2\rho^{k-3}W\right] = 0, \quad \text{where}\\
&\lambda^2 = 2m^2(\alpha_1 + 2\alpha_3) + k^2,\\
&\beta^2 = m^4 k^2 - 2m^2(\alpha_1 + k^2 + 2\alpha_3),\\
&\gamma^{k+1} = h\sigma_0 \frac{12a^2}{Q_{11}h^3}.
\end{aligned} \tag{6.51}$$

For the symmetric buckling case, $m = 0$, we get from Eq. (6.51)

$$\begin{aligned}
&\frac{d^4 W}{d\rho^4} + \frac{2}{\rho}\frac{d^3 W}{d\rho^3} - \frac{k^2}{\rho^2}\frac{\partial^2 W}{\partial \rho^2} + \frac{k^2}{\rho^3}\frac{\partial W}{\partial \rho} + \frac{\beta^2}{\rho^4}W\\
&\quad + \gamma^{k+1}\left[\rho^{k-1}\frac{\partial^2 W}{\partial \rho^2} + k\rho^{k-2}\frac{\partial W}{\partial \rho}\right] = \left(\frac{d}{d\rho} + \frac{1}{\rho}\right)\\
&\quad \times \left(\frac{d^3 W}{d\rho^3} + \frac{1}{\rho}\frac{d^{32} W}{d\rho^2} - \frac{k^2}{\rho^2}\frac{\partial W}{\partial \rho} + \gamma^{k+1}\rho^{k-1}\frac{\partial W}{\partial \rho}\right) = 0.
\end{aligned} \tag{6.52}$$

The solution for Eq. (6.52) has the following form [8–10]:

$$W(\rho) = C_1 \int_0^{\rho} S_{\delta,\varphi}(\eta\rho^{\kappa})d\rho + C_2 \sum_{n=0}^{\infty} J_{(\varphi+2n+1)}(\eta\rho^{\kappa})$$

$$+ C_3 \sum_{n=0}^{\infty} J_{(-\varphi+2n+1)}(\eta\rho^{\kappa}) + C_4, \quad \text{where}$$

$$\delta = \frac{1-k}{1+k}, \quad \varphi = \frac{2k}{1+k}, \quad \eta = \frac{2\gamma^{\kappa}}{K+1}, \quad \kappa = \frac{k+1}{2},$$

$$S_{\mu_1,\mu_2}(z) = \text{Lommel's function [10].} \tag{6.53}$$

Note that for the isotropic case ($k = 1$) Eq. (6.53) reduces to

$$W(\rho) = C_1 \log(\rho) + C_2 J_0(\gamma\rho) + C_3 Y_0(\gamma\rho) + C_4, \tag{6.54}$$

which is well known in the literature.

A quick view on the solution presented by Eq. (6.53), reveals that the constants C_1 and C_2 should vanish to ensure a finite deflection at the center of the circular plate. Then, Eq. (6.53) will obtain the following expression:

$$\widehat{w}(\rho,\theta) = \left[C_2 \sum_{n=0}^{\infty} J_{(\varphi+2n+1)}(\eta\rho^{\kappa}) + C_4 \right] \cos m\theta. \tag{6.55}$$

Imposing the boundary conditions on Eq. (6.55) would lead to the eigenvalue of the problem.

For the clamped case, these boundary conditions would be displacement and slope at the external circumference ($r = a$) equal to zero leading to

$$J_{\varphi}(\eta) = 0, \tag{6.56}$$

while for the simply supported case (displacement and the radial moment vanish at the external edge) the characteristic buckling equation will be

$$(k + \alpha_1)J_{\varphi}(\eta) - \kappa\varphi J_{\varphi+1}(\eta) = 0. \tag{6.57}$$

One should note that substituting in Eqs. (6.56) and (6.57), $k = 1$ and $\alpha_1 = v$ will provide the solution for the isotropic circular plates.

Figures 6.9 and 6.10 present the buckling coefficient for the symmetric case for a clamped and simply supported circular plate.

The asymmetric case of orthotropic circular plates is presented in detail in [8], including adequate equations.

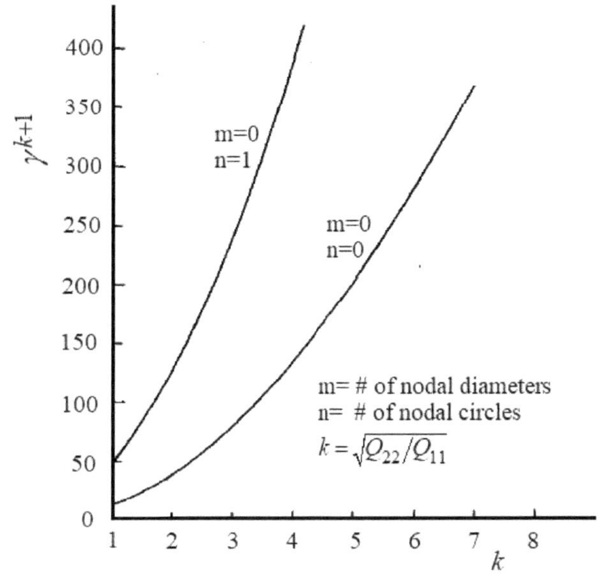

Fig. 6.9. Clamped circular orthotropic plate — buckling coefficient for symmetric buckling case.

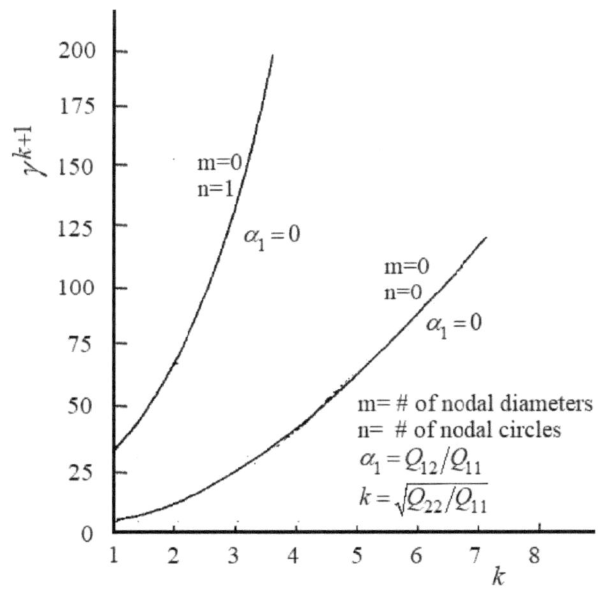

Fig. 6.10. Simply supported circular orthotropic plate — buckling coefficient for symmetric buckling case.

6.8. Buckling of Orthotropic Annular Plates

Figure 6.11 presents a thin polar orthotropic annular plate, under uniform inplane loadings, p_{in} along the inner circumference at radius b, and p_{out} along the outer edge at radius a. The uniform thickness of the plate is h. As presented in [11], the prebuckling membrane stresses can be written as

$$\sigma_r = \alpha r^{k-1} + \beta r^{-k-1},$$

$$\sigma_\theta = \alpha k r^{k-1} + \beta k r^{-k-1},$$

$$\sigma_{r\theta} = 0, \quad \text{where}$$

$$k = \sqrt{\frac{E_\theta}{E_r}} = \sqrt{\frac{D_\theta}{D_r}}, D_r = \frac{E_r h^3}{12\,(1 - \upsilon_r \upsilon_\theta)}, D_\theta = \frac{E_\theta h^3}{12(1 - \upsilon_r \upsilon_\theta)},$$

$$\alpha = -\frac{p_{\text{out}} a^{k+1} - p_{in} b^{k+1}}{a^{2k} - b^{2k}}, \beta = \frac{(p_{\text{out}} b^{k-1} - p_{in} a^{k-1})(ab)^{k+1}}{a^{2k} - b^{2k}}, \quad (6.58)$$

where E_r and E_θ are the Young's moduli in the radial and tangential directions, respectively.

The buckling shape of the annular plate will experience m circumferential waves, leading to the following expression for the plate deflection

$$W(r, \theta) = W_m(r) \cos m\theta. \quad (6.59)$$

The solution is obtained by applying the Rayleigh–Ritz method. The strain energy V and the work T being done by the loading at the boundaries can be

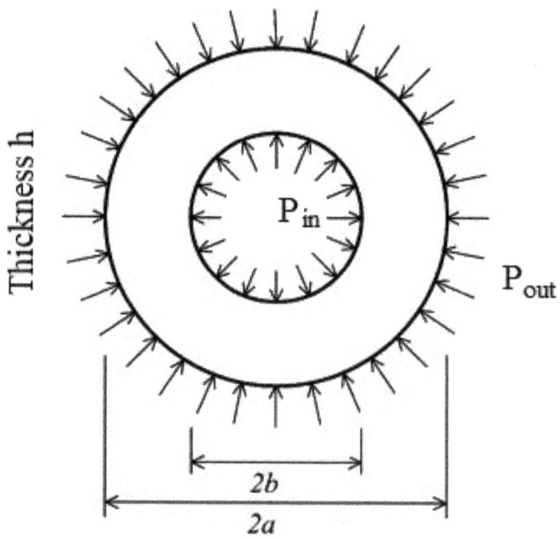

Fig. 6.11. Orthotropic annular plate under compressive loadings.

written as

$$V = \frac{\pi}{2}(1+\varepsilon_{0m})\int_b^a \left[D_r \left(\frac{d^2W_m}{dr^2}\right)^2 + 2\frac{\widetilde{D}}{r}\frac{d^2W_m}{dr^2}\left(\frac{dW_m}{dr} - \frac{m^2}{r}W_m\right)\right] rdr$$

$$+\frac{\pi}{2}(1+\varepsilon_{0m})\int_b^a \left[\frac{D_\theta}{r^2}\left(\frac{dW_m}{dr} - \frac{m^2}{r}W_m\right)^2\right.$$

$$\left. + 4D_{r\theta}\frac{m^2}{r^2}\left(\frac{dW_m}{dr} - \frac{W_m}{r}\right)^2\right] rdr,$$

$$T = \frac{\pi}{2}(1+\varepsilon_{0m})\,h\left\{\int_b^a \left[\sigma_r\left(\frac{dW_m}{dr}\right)^2 + \sigma_\theta\frac{m^2}{r^2}\left(\frac{m}{r}W_m\right)^2\right] rdr\right\},$$

where

$$\widetilde{D} = \upsilon_\theta D_r = \upsilon_r D_\theta, \quad D_{r\theta} = \frac{Gh^3}{12}, \quad G = \text{shear modulus}$$

$$\varepsilon_{00} = 1 \quad \text{and} \quad \varepsilon_{0m} = 0 \quad \text{for } m \neq 0. \tag{6.60}$$

The application of the Rayleigh–Ritz method requires to express the radial mode shape to be written as

$$W_m(r) = \sum_{i=1}^{\infty} \phi_i \Delta_i(r), \tag{6.61}$$

where $\Delta_i(r)$ are admissible functions satisfying the problem's boundary conditions and $\phi_i(i = 1, 2, 3\ldots)$ are constants to be determined from the stationary condition of the total potential $\delta(V + T) = 0$.

6.9. Vibrations of Isotropic Annular Plates

Vogel and Skinner [12] have dealt the topic of natural vibrations of isotropic annular flat plates in detail, presenting frequency determinant for various boundary conditions.

The governing equation for the free vibration of an annular plate, having a thickness of $2h$, an external and inner radii a and b, respectively, in term of the displacement w of a point along the neutral plane of the plate is written as [12]

$$D\left[\frac{\partial^2}{\partial r^2} + \frac{1}{r}\frac{\partial}{\partial r} + \frac{1}{r^2}\frac{\partial^2}{\partial \theta^2}\right]^2 w + \rho\frac{\partial^2 w}{\partial t^2} = 0, \tag{6.62}$$

where ρ is the mass density of the material and $D = Eh^2/[3(1-\upsilon^2)]$.

The assumed deflection form has the following form:

$$w(r,\theta) = W(r)\sin(n\theta - \theta)e^{i\omega t}. \tag{6.63}$$

Substitution of Eq. (6.63) into Eq. (6.62) leads to

$$D\left[\frac{d^2}{dr^2} + \frac{1}{r}\frac{d}{dr} + \frac{n^2}{r^2}\right]^2 W - \lambda^4 W = 0, \quad \text{where}$$

$$\lambda^4 = 3(1-\upsilon^2)\frac{\rho\omega^2}{Eh^2} \tag{6.64}$$

The general solution for Eq. (6.64) is given by [12] as

$$W(r) = A_1 J_n(\lambda r) + A_2 Y_n(\lambda r) + A_3 I_n(\lambda r) + A_4 K_n(\lambda r), \tag{6.65}$$

where J_n and Y_n are Bessel functions and I_n and K_n are modified Bessel functions.

Imposing the boundary conditions of a given annular plate will lead to the determination of the natural frequencies, by demanding the vanishing of the coefficients (A_1, A_2, A_3 and A_4) determinant.

Another approach is to solve the problem by applying the Rayleigh–Ritz method, as is presented in [13]. A one-term Rayleigh–Ritz deflection mode was assumed in [13]

$$w(r,\theta,t) = A_1(r-a)\cos(n\theta)e^{i\omega t}, \tag{6.66}$$

where A_1 is an arbitrary constant, θ is a circumferential angle and ω is the angular frequency. For an axisymmetric loading, the kinetic energy T and the static potential V can be written as

$$T = \frac{\rho h}{2}\int_b^a\int_0^{2\pi}\left[\frac{\partial w(r,\theta,t)}{\partial t}\right]^2 r\,dr\,d\theta = \frac{\rho h\omega^2}{2}\int_b^a\int_0^{2\pi}[w(r,\theta)]^2 r\,d\theta\,dr,$$

$$V = \frac{1}{2}\int_b^a\int_0^{2\pi}\left[\sigma_r h\left(\frac{\partial w}{\partial r}\right)^2 + \frac{\sigma_\theta h}{r^2}\left(\frac{\partial w}{\partial \theta}\right)^2\right.$$

$$\left. + D\left(\frac{\partial^2 w}{\partial r^2} + \frac{1}{r}\frac{\partial w}{\partial r} + \frac{1}{r^2}\frac{\partial^2 w}{\partial \theta^2}\right)^2\right] r\,d\theta\,dr$$

$$+\frac{D}{2}\int_b^a\int_0^{2\pi}\left[-2(1-\upsilon)\frac{\partial^2 w}{\partial r^2}\left(\frac{1}{r}\frac{\partial w}{\partial r} + \frac{1}{r^2}\frac{\partial^2 w}{\partial \theta^2}\right)\right.$$

$$\left. + 2(1-\upsilon)\frac{\partial^2 w}{\partial r^2}\left(\frac{1}{r}\frac{\partial^2 w}{\partial r\partial\theta} - \frac{1}{r^2}\frac{\partial w}{\partial \theta}\right)^2\right] r\,d\theta\,dr, \tag{6.67}$$

where

$$\sigma_r = \frac{pa^2}{h(a^2-b^2)} + \frac{1}{r^2}\frac{pa^2b^2}{h(a^2-b^2)}, \quad \sigma_\theta = \frac{pa^2}{h(a^2-b^2)} - \frac{1}{r^2}\frac{pa^2b^2}{h(a^2-b^2)},$$

and a, b are the outer and inner radii, respectively and p is an uniform compression per unit length applied on the outer circumference. Substitution of the expression for w (Eq. (6.66)) and equating $T = V$ yields

$$\begin{aligned}
&\beta^2 3\left(1-\upsilon^2\right)\left(\frac{1}{12} - \frac{\Delta^4}{4} + \frac{2\Delta^3}{3} - \frac{\Delta^2}{2}\right) \\
&\quad = \lambda\left[\frac{1}{2} - n^2 - \frac{n^2\left(1+\Delta^2\right)-\Delta^2}{1-\Delta^2}\ln\Delta\right] \\
&\qquad - \left[\left(1+n^2\right)^2\ln\Delta + 2\frac{\Delta-1}{\Delta}\left(1-n^2\right)n^2 + \left(1-\upsilon+\frac{n^2}{2}\right)n^2\frac{\Delta^2-1}{\Delta^2}\right],
\end{aligned}$$

where

$$\beta \equiv 4\omega\sqrt{\frac{\rho a^4}{Eh^2}}, \quad \lambda \equiv \frac{pa^2}{D}, \quad \Delta \equiv \frac{b}{a}. \tag{6.68}$$

Substituting $\lambda \equiv \frac{pa^2}{D} = 0$ yields the characteristic equation for the plate natural frequencies, namely

$$\begin{aligned}
&\beta^2 3(1-\upsilon^2)\left(\frac{1}{12} - \frac{\Delta^4}{4} + \frac{2\Delta^3}{3} - \frac{\Delta^2}{2}\right) \\
&\quad = -\left[(1+n^2)^2\ln\Delta + 2\frac{\Delta-1}{\Delta}\left(1-n^2\right)n^2\right. \\
&\qquad \left. + \left(1-\upsilon+\frac{n^2}{2}\right)n^2\frac{\Delta^2-1}{\Delta^2}\right].
\end{aligned} \tag{6.69}$$

Substituting $\beta \equiv 4\omega\sqrt{\frac{\rho a^4}{Eh^2}} = 0$ yields the characteristic equation for the plate buckling,

$$\begin{aligned}
&\lambda\left[\frac{1}{2} - n^2 - \frac{n^2(1+\Delta^2)-\Delta^2}{1-\Delta^2}\ln\Delta\right] \\
&\quad = \left[(1+n^2)^2\ln\Delta + 2\frac{\Delta-1}{\Delta}(1-n^2)n^2\right. \\
&\qquad \left. + \left(1-\upsilon+\frac{n^2}{2}\right)n^2\frac{\Delta^2-1}{\Delta^2}\right].
\end{aligned} \tag{6.70}$$

The results presented in [13] proved to be very accurate for $n = 0, 1$, while for $n = 3, 4$ the accuracy is only about 10% (compared to 1.5% before).

References

[1] Timoshenko, S.P. and Gere, J.M., *Theory of Elastic Stability*, 2nd Edition, McGraw-Hill International Book Company, Singapore, 1963, 539 p.

[2] Massonnet, Ch., Voilment les plaques solicitée dans leur plan, *Final Report of the 3rd Congress of the International Association for Bridge and Structural Engineering*, Liège, Belgium, September 1948, pp. 291–300.

[3] Massonnet, Ch., Les relations entre les modes normaux de vibration et la stabilité des systems elastiques, *Bull. Cours Lab., d'Essais Contruction Genie Civil et d'Hydraulique Fluviale*, Vol. 1(1 and 2), 1940.

[4] Abramovich, H.,Gil, J., Grunwald, A. and Rosen, A., Vibrations and buckling of radially loaded circular plates, *TAE* 332, July 1978, Department of Aeronautical Engineering, Technion, Israel Institute of Technology, Haifa, Israel, 43 p.

[5] H. Abramovich, *Stability and Vibrations of Thin Walled Composite Structures*, Elsevier Ltd, 2017, United Kingdom, 778 p.

[6] Bloom, F. and Coffin, D., *Handbook of Thin Plate Buckling and Postbuckling*, Chapman & Hall/CRC, 2001, 786p.

[7] Yamaki, S., Buckling of a thin annular plate under uniform compression, *Trans. ASME J. Appl. Mech.* Vol. 25(3), 1958, pp. 267–273.

[8] Swamidas, A.S.J. and Kunukkasseril, V.X., Buckling of orthotropic circular plates, *AIAA J.*, Vol. 11(12), 1973, pp. 1633–1636.

[9] Kamke, E., *Differential Gleichungen Losungmethoden und Losungen*, 3rd Edition Chelsea, New York, 1959, pp. 437–442.

[10] Erdelyi, A., Magnus, W., Oberhettinger, F. and Tricomi, F.G., *Higher Transcendental Functions*, Vol. 2, McGraw-Hill, New York, 1953, 40p.

[11] Ramaiah, G.K., Buckling of polar orthotropic annular plates under uniform inplane compressive forces, *Trans. ASME J. Appl. Mech.*, Vol. 48, 1981, pp. 643–653.

[12] Vogel, S.M. and Skinner, D.W., Natural frequencies of transversely vibrating uniform annular plates, *Trans. ASME J. Appl. Mech.* Vol. 32(4), 1965, pp. 926–931.

[13] Rosen, A. and Libai, A., Transverse vibrations of compressed annular plates, *J. Sound Vibration* Vol. 40(1), 1975, pp. 149–153.

Appendix A: Plate Boundary Conditions in Polar Coordinates-General Expressions

For a given w displacement the linear strains in a polar coordinate system are defined as

$$\varepsilon_{rr} = -z\frac{\partial^2 w}{\partial r^2},$$

$$\varepsilon_{\theta\theta} = -z\left(\frac{1}{r}\frac{\partial w}{\partial r} + \frac{1}{r^2}\frac{\partial^2 w}{\partial \theta^2}\right),$$

$$\varepsilon_{r\theta} = -z\left(-\frac{1}{r^2}\frac{\partial w}{\partial \theta} + \frac{1}{r}\frac{\partial^2 w}{\partial r \partial \theta}\right), \tag{A.1}$$

where ε_{rr} and $\varepsilon_{\theta\theta}$ are the strains in the r, θ directions, respectively, while $\varepsilon_{r\theta}$ is the shear strain.

The moments and the shear resultants are defined as

$$M_{rr} = D\left[\frac{\partial^2 w}{\partial r^2} + \upsilon\left(\frac{1}{r}\frac{\partial w}{\partial r} + \frac{1}{r^2}\frac{\partial^2 w}{\partial \theta^2}\right)\right],$$

$$M_{\theta\theta} = D\left[\left(\frac{1}{r}\frac{\partial w}{\partial r} + \frac{1}{r^2}\frac{\partial^2 w}{\partial \theta^2}\right) + \upsilon\frac{\partial^2 w}{\partial r^2}\right],$$

$$M_{r\theta} = -D(1-\upsilon)\left[-\frac{1}{r^2}\frac{\partial w}{\partial \theta} + \frac{1}{r}\frac{\partial^2 w}{\partial r \partial \theta}\right],$$

$$V_r = D\frac{\partial}{\partial r}\left[\frac{\partial^2 w}{\partial r^2} + \left(\frac{1}{r}\frac{\partial w}{\partial r} + \frac{1}{r^2}\frac{\partial^2 w}{\partial \theta^2}\right)\right],$$

$$V_\theta = D\frac{1}{r}\frac{\partial}{\partial \theta}\left[\left(\frac{1}{r}\frac{\partial w}{\partial r} + \frac{1}{r^2}\frac{\partial^2 w}{\partial \theta^2}\right) + \frac{\partial^2 w}{\partial r^2}\right]. \tag{A.2}$$

The plate governing equation will be written as

$$\left[\frac{\partial^2}{\partial r^2} + \frac{1}{r}\frac{\partial}{\partial r} + \frac{1}{r^2}\frac{\partial^2}{\partial \theta^2}\right]^2 w = -\frac{q(r,\theta)}{D}, \quad \text{where}$$

$$D = \frac{Eh^3}{12(1-\upsilon^2)}, \tag{A.3}$$

and q is the vertical loading per unit area.

The geometric boundary conditions would be the vertical displacement, w and the slope, dw/dr, while the natural boundary conditions would be the moment M and the shear force V (the relevant expressions are given in Eq. (A.2)).

In an axisymmetric case (no dependence on θ) the expressions, written above, are simplified to yield

$$M_{rr} = D\left[\frac{\partial^2 w}{\partial r^2} + \upsilon\frac{1}{r}\frac{\partial w}{\partial r}\right], M_{\theta\theta} = D\left[\frac{1}{r}\frac{\partial w}{\partial r} + \upsilon\frac{\partial^2 w}{\partial r^2}\right],$$

$$M_{r\theta} = 0, \quad V_r = D\frac{\partial}{\partial r}\left[\frac{\partial^2 w}{\partial r^2} + \frac{1}{r}\frac{\partial w}{\partial r}\right], \quad V_\theta = 0,$$

$$\left[\frac{d^2}{dr^2} + \frac{1}{r}\frac{d}{dr}\right]^2 w = \frac{1}{r}\frac{d}{dr}\left\{r\frac{d}{dr}\left[r\frac{d}{dr}\left(r\frac{dw}{dr}\right)\right]\right\} = -\frac{q(r)}{D} \tag{A.4}$$

For the simple case of uniform load q, four times integration with respect to r, applied on the last expression in Eq. (A.4), leads to

$$w(r) = -\frac{qr^4}{64D} + \alpha\frac{r^2}{4}(\ln r - 1) + \beta\frac{r^2}{4} + \gamma, \ln r + \delta, \tag{A.5}$$

where α, β, γ and δ are constants to be defined by the relevant boundary conditions of the plate.

Table B.1. κ vs. the ratio b/a (adapted from [1, 2]).

	Simply supported on the outer circumference, free at the inner circumference									
b/a	0	0.1	0.2	0.3	0.4	0.5	0.6	0.7	0.8	0.9
κ	0.35	0. 33	0.30	0.27	0.23	0.21	0.19	0.18	0.17	0.16
	Clamped on the outer circumference, free at the inner circumference									
b/a	0	0.1	0.2	0.3	0.4	0.5				
κ	1.22	1.17	1.11	1.21	1.48	2.07				

Appendix B: Buckling of an Annular Plate Under Uniform External Radial Compression

The annular plate has an inner radius b, and an outer radius a, with h being its thickness. The inner circumference is free, while the outer one is either simply supported or clamped. The expression for the critical buckling stress is given as (see also [1, 2])

$$\sigma_{r,cr} = \kappa \frac{E}{1 - v^2} \left(\frac{h}{a} \right)^2 \tag{B.1}$$

E is the Young's modulus, while v is the Poisson's ratio. The value of the coefficient κ depends on the ratio b/a as displayed in the Table B.1.

References

[1] Timoshenko, S.P. and Gere, J.M., *Theory of Elastic Stability*, 2nd Edition, International Student Edition, McGraw-Hill Book Co., 1963, Singapore, 541 p.

[2] Roark, R.J. and Young, W.C., *Formulas for stress and strain*, 5th Edition, International Student Edition, McGraw-Hill Book Co., Singapore, 1976, 624 p.

Appendix C: Buckling of an Annular Plate — A Detailed Example

The case consists of a thin circular annular plate, with a clamped outer circumference $r = a$, while the inner circumference $r = b$ is free. The uniform loading is applied on the outer circumference. This example was dealt in [1–3].

As presented in Section 6.6, the governing equation of the plate has the following form [1]:

$$\begin{aligned} D&\left[\frac{\partial^4 w}{\partial r^4}+\frac{2}{r}\frac{\partial^3 w}{\partial r^3}-\frac{1}{r^2}\frac{\partial^2 w}{\partial r^2}+\frac{2}{r^2}\frac{\partial^4 w}{\partial r^2\partial\theta^2}+\frac{1}{r^3}\frac{\partial w}{\partial r}\right.\\ &\left.-\frac{2}{r^3}\frac{\partial^3 w}{\partial r\partial\theta^2}+\frac{1}{r^4}\frac{\partial^4 w}{\partial\theta^4}+\frac{4}{r^4}\frac{\partial^2 w}{\partial\theta^2}\right]\\ &=D\nabla^4 w=N_r\frac{\partial^2 w}{\partial r^2}+N_\theta\left[\frac{1}{r}\frac{\partial w}{\partial r}+\frac{1}{r^2}\frac{\partial^2 w}{\partial r^2}\right], \end{aligned} \tag{C.1}$$

where

$$N_r=-N_0\frac{a^2}{a^2-b^2}\left(1-\frac{b^2}{r^2}\right),\quad N_\theta=-N_0\frac{a^2}{a^2-b^2}\left(1+\frac{b^2}{r^2}\right).$$

Note that the shear load is zero ($N_{r\theta}=0$) and N_0 is the uniform compression at the boundary.

Equation (C.1) can be re-casted in a more compact form to yield

$$\begin{aligned} &D\nabla^4 w+\lambda\left[\frac{1}{b^2}+\frac{1}{r^2}\right]\nabla^2 w=\frac{2\lambda}{r^2}\frac{\partial^2 w}{\partial r^2},\ \text{where}\\ &\nabla^4 w=\nabla^2\nabla^2 w,\quad \nabla^2 w=\left(\frac{\partial^2}{\partial r^2}+\frac{1}{r}\frac{\partial}{\partial r}+\frac{1}{r^2}\frac{\partial^2}{\partial\theta^2}\right)w,\\ &\lambda=N_0\frac{a^2b^2}{D(a^2-b^2)}. \end{aligned} \tag{C.2}$$

The boundary conditions for the present problem read

$$\begin{aligned} &w=0\quad\text{and}\quad\frac{\partial w}{\partial r}=0\quad\text{at } r=a,\\ &\frac{\partial^2 w}{\partial r^2}+\upsilon\left(\frac{1}{r}\frac{\partial w}{\partial r}+\frac{1}{r^2}\frac{\partial^2 w}{\partial\theta^2}\right)=0,\quad\text{and}\\ &\frac{\partial}{\partial r}\left(\frac{\partial^2 w}{\partial r^2}+\frac{1}{r}\frac{\partial w}{\partial r}+\frac{1}{r^2}\frac{\partial^2 w}{\partial\theta^2}\right)\\ &+\frac{1-\upsilon}{r}\frac{\partial}{\partial\theta}\left(\frac{1}{r}\frac{\partial^2 w}{\partial r\partial\theta}-\frac{1}{r^2}\frac{\partial w}{\partial\theta}\right)=0\quad\text{at } r=b. \end{aligned} \tag{C.3}$$

As suggested in [1, 2], the solution for Eq. (C.2) has the following form:

$$w(r,\theta)=\Gamma_n(r)\cos(n\theta),\quad n=0,1,2,3,\ldots \tag{C.4}$$

The suggested solution is substituted in Eq. (C.2), using the following variable transformation $H = \sqrt{\lambda}(r/b)$ leading to

$$\frac{d^4\Gamma_n}{dH^4} + \frac{2}{H}\frac{d^3\Gamma_n}{dH^3} - \left(\frac{1+2n^2+\lambda}{H^2} - 1\right)\frac{d^2\Gamma_n}{dH^2} + \left(\frac{1+2n^2+\lambda}{H^3} + \frac{1}{H}\right)\frac{d\Gamma_n}{dH}$$
$$+ \left[\frac{n^2\left(n^2-\lambda-4\right)}{H^4} - \frac{n^2}{H^2}\right]\Gamma_n = 0, \quad n = 0, 1, 2, 3, \ldots \tag{C.5}$$

Note that for $n = 0$, Eq. (C.5) simplifies to

$$\frac{d^3\Psi}{dH^3} + \frac{2}{H}\frac{d^2\Psi}{dH^2} - \left(\frac{1+\lambda}{H^2} - 1\right)\frac{d\Psi}{dH} + \left(\frac{1+\lambda}{H^3} + \frac{1}{H}\right)\Psi = 0,$$
$$\text{where } \Psi \equiv \frac{d\Gamma_0}{dH}. \tag{C.6}$$

Equation (C.6) has to fulfill the last three boundary conditions presented in Eq. (C.3) (the first is inherently fulfilled), namely

$$\Psi = 0 \quad \text{at } H = \sqrt{\lambda}\left(\frac{a}{b}\right)$$
$$\frac{d\Psi}{dH} + \frac{\upsilon}{H}\Psi = 0, \quad \text{and}$$
$$\frac{d}{dH}\left[\frac{d\Psi}{dH} + \frac{1}{H}\Psi\right] = 0 \quad \text{at } H = \sqrt{\lambda}\left(\frac{a}{b}\right). \tag{C.7}$$

The form of a possible solution satisfying the last boundary condition given in Eq. (C.7) is given by

$$\Psi = \alpha J_i(H) + \beta J_{-i}(H), \tag{C.8}$$

where $i = \sqrt{\lambda + 1}$, $J_i(H)$ and $J_{-i}(H)$ are Bessel functions of the first kind and α and β are arbitrary constants to be determined from the first two boundary conditions given in Eq. (C.7), leading to a determinant for the non-dimensional buckling load,

$$\begin{vmatrix} J_i(H) & J_{-i}(H) \\ J_{i-1}(H) + \dfrac{\upsilon - i}{\sqrt{\lambda}} J_i(H) & J_{-i-1}(H) + \dfrac{\upsilon + i}{\sqrt{\lambda}} J_{-i}(H) \end{vmatrix}$$
$$= \begin{vmatrix} J_i\left[\sqrt{\lambda}\left(\dfrac{a}{b}\right)\right] & J_{-i}\left[\sqrt{\lambda}\left(\dfrac{a}{b}\right)\right] \\ J_{i-1}(\sqrt{\lambda}) + \dfrac{\upsilon - i}{\sqrt{\lambda}} J_i(\sqrt{\lambda}) & J_{-i-1}(\sqrt{\lambda}) + \dfrac{\upsilon + i}{\sqrt{\lambda}} J_{-i}(\sqrt{\lambda}) \end{vmatrix} = 0. \tag{C.9}$$

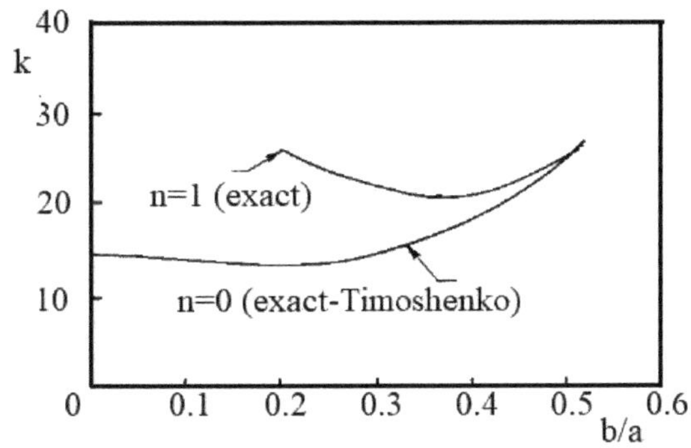

Fig. C.1. Buckling coefficients for exact solutions obtained in [1] (adapted from [1, 2]).

The expression for the critical buckling load is presented in a generic form as

$$N_{0,cr} = k\frac{D}{a^2}, \quad \text{where } k = \lambda\left[\left(\frac{a}{b}\right)^2 - 1\right]. \tag{C.10}$$

Figure C.1 presents the buckling coefficient, k, as a function of the aspect ratio b/a of the plate.

To obtain a solution for $n = 1$ (having one wave around the plate circumference), one has to substitute $n = 1$ in Eq. (C.5) leading to

$$\frac{d^4\Gamma_1}{dH^4} + \frac{2}{H}\frac{d^3\Gamma_1}{dH^3} - \left(\frac{3+\lambda}{H^2} - 1\right)\frac{d^2\Gamma_1}{dH^2} + \left(\frac{3+\lambda}{H^3} + \frac{1}{H}\right)\frac{d\Gamma_1}{dH}$$
$$-\left[\frac{3+\lambda}{H^4} + \frac{1}{H^2}\right]\Gamma_1 = 0. \tag{C.11}$$

The way to solve Eq. (C.11) is difficult and cumbersome, however we shall present the way as is proceeded in [1]. First a variable transformation is used: $\Gamma_1 \equiv H\gamma$ *and* $\Psi = d\gamma/dH$, leading to a compact form of Eq. (C.11)

$$\frac{d^2\Psi}{dH^2} + \frac{3}{H}\frac{d\Psi}{dH} + \left[1 + \frac{1-\varepsilon^2}{H^2}\right]\Psi - \frac{\mu}{H^3} = 0,$$

where

$$\varepsilon^2 = \lambda + 4 \quad \text{and} \quad \mu = \text{arbitrary constant}. \tag{C.12}$$

The general solution for Eq. (C.12) has the following form (see [1]):

$$\Psi = \frac{1}{H}[\alpha J_i(H) + \beta J_{-i}(H) + \mu R_{-1,i}(H)],$$

where

$$R_{-1,i}(H) = \frac{\pi}{2\sin(i\pi)}\left[J_i(H)\int\frac{J_{-i}(H)}{H}dH - J_{-i}(H)\int\frac{J_i(H)}{H}dH\right] \tag{C.13}$$

where α and β are arbitrary constants to be determined from the imposed boundary conditions. Writing $\Psi(H)$ in terms of the variable $\Gamma_1(H)$ and substituting in Eq. (C.13) yields

$$H\frac{d\Gamma_1}{dH} - \Gamma_1 = H[\alpha J_i(H) + \beta J_{-i}(H) + \mu R_{-1,i}(H)]. \tag{C.14}$$

The boundary conditions presented in Eq. (C.3) transform into

$$\begin{aligned}
&\Gamma_1 = 0 \quad \text{and} \quad \frac{d\Gamma_1}{dH} = 0 \quad \text{at } H = \sqrt{\lambda}\frac{a}{b} \\
&\frac{d^2\Gamma_1}{dH^2} + \frac{\upsilon}{H^2}\left(H\frac{d\Gamma_1}{dH} - \Gamma_1\right) = 0, \quad \text{and} \\
&\frac{d^3\Gamma_1}{dH^3} + \frac{1}{H^2}\frac{d^2\Gamma_1}{dH^2} - \frac{3-\upsilon}{H^2}\frac{d\Gamma_1}{dH} + \frac{3-\upsilon}{H^3}\Gamma_1 = 0 \quad \text{at } H = \sqrt{\lambda}.
\end{aligned} \tag{C.15}$$

Imposing the boundary conditions from Eq. (C.15) into the general solution, Eq. (C.14) leads to the following equations, which are casted in a matrix form:

$$\begin{bmatrix} J_i\left(\sqrt{\lambda}\frac{a}{b}\right) & J_{-i}\left(\sqrt{\lambda}\frac{a}{b}\right) \\ (1+\upsilon-i)J_i(\sqrt{\lambda}) + \sqrt{\lambda}J_{i-1}(\sqrt{\lambda}) & (1+\upsilon+i)J_{-i}(\sqrt{\lambda}) + \sqrt{\lambda}J_{-i-1}(\sqrt{\lambda}) \end{bmatrix}$$
$$\times \begin{Bmatrix} \alpha \\ \beta \end{Bmatrix} = \begin{Bmatrix} 0 \\ 0 \end{Bmatrix}$$

$$\text{and } \mu = 0. \tag{C.16}$$

Demanding that the coefficients determinant in Eq. (C.16) would vanish, yields the results presented in Fig. C.1.

References

[1] Majumdar, S. Buckling of a thin annular plate under uniform compression, *AIAA J.*, Vol. 9(9), 1971, pp. 1701–1707.
[2] Timoshenko, S.P. and Gere, J.M., *Theory of Elastic Stability*, 2nd Edition, International Student Edition, McGraw-Hill Book Co., 1963, Singapore, 541 p.
[3] Roark R.J. and Young, W.C., *Formulas for Stress and Strain*, 5th Edition, International Student Edition, McGraw-Hill Book Co., 1976, Singapore, 624 p.

Chapter 7

Dynamic Buckling of Thin-Walled Structures

7.1. Introduction

This chapter will deal with what is called in the literature "dynamic" buckling of columns, plates and shells (metal and composite materials). First, the term "dynamic" buckling will be presented and defined followed by examples from the literature. Then the equations of motions for columns and plates will be presented and numerical and numerical and experimental results of tests performed on columns, plates and shells will be highlighted.

Many references can found in the literature regarding dynamic buckling of thin-walled structures (see [1–50]. The topic of applying an axially time dependent load onto a column, a basic thin-walled structure, thus inducing lateral vibrations and eventually causing the buckling of the column was studied for many years. This behavior is also called *vibration buckling*, as proposed by Lindberg [11]. As it is described in his fundamental report [11], the axial oscillating load might lead to unacceptable large vibrations amplitudes at a critical combination of the frequency and amplitude of the axial load and the inherent damping of the column. This behavior is presented in Fig. 7.1(a) (adapted from [11]), where an oscillating axial load induces bending moments that cause lateral vibrations of the column. As described in [11] the column will laterally vibrate at large amplitudes when the loading frequency will be twice the natural lateral bending frequency of the column. The term used by Lindberg, *vibration buckling* presents some kind of similarity to vibration resonance. However, in the case of vibration resonance the applied load is in the same direction as the motion, namely in our case, lateral to the column, and the resonance will occur when the loading frequency equals the natural frequency of the column. This type of vibration

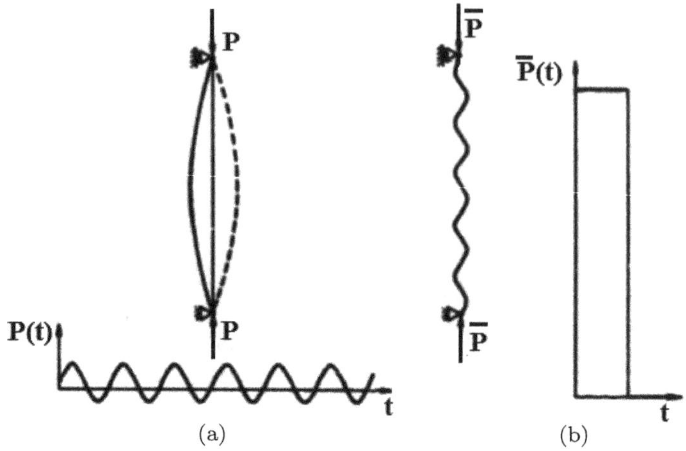

Fig. 7.1. (a) Buckling under parametric resonance; (b) pulse-type buckling.

buckling was called by Lindberg as: *dynamic stability of vibrations induced by oscillating parametric loading.* This type of resonance is also called in the literature as *parametric resonance* (see an application of this type of dynamic stability in Simitses book [17] and in [39]).

Another type of vibration type is often called *pulse buckling*, where the structure will be deformed to unacceptably large amplitudes as a result of a transient response of the structure to the dynamic axially applied load [11]. One should note that the sudden applied load might cause a permanent deformation due to plastic response of the column, a snap to a larger post-buckling deformation or simply a return to its undeform state. This is pictured in Fig. 7.1(b) (adapted from [11]) where the response of the column to a sudden short-time axial load is shown.

One should note that buckling will occur when an unacceptably large deformation or stress is encountered by the column. The column can withstand a large axial load before reaching the buckling condition provided the load duration is short enough. Under an intense, short duration axial load, the column would buckle into a very high-order mode as shown in Fig. 7.1(b). Lindberg [11] claims that pulse buckling falls under the following mathematical definition: *dynamic response of structural systems induced by time-varying parametric loading.* Throughout this chapter, the *pulse buckling* will be equivalent to dynamic buckling.

As stated above, the dynamic buckling of thin-walled structures has been widely addressed in the literature. It started with the famous paper by Budiansky and Roth [1], thru Hegglin's report on dynamic buckling of columns [2] and continued with Budiansky and Hutchinson [3] and

Hutchinson and Budiansky [4] in the mid-1960. Then more structures have been addressed as presented in typical references [5–38] and [40–43].

One of the most intriguing and challenging thing is to define a criterion to clearly define the critical load causing the structure to buckle under the subjected pulse loading. As presented by Kubiak [31] and also by Ari Gur [10, 18, 21] and others in [16, 19, 23–25] a new term is introduced called *DLF* (*Dynamic Load Factor*) to enable the use of the dynamic buckling criteria. It is defined as

$$\text{DLF} \equiv \frac{\text{Pulse Buckling Amplitude}}{\text{Static Buckling Amplitude}} \equiv \frac{(P_{cr})_{\text{dyn}}}{(P_{cr})_{\text{static}}}. \tag{7.1}$$

According to Kubiak [31], the most popular criterion had been proposed by Volmir [7] for plates subjected to in-plane pulse loading. As quoted in [31], Volmir proposed the following criterion: "*Dynamic critical load corresponds to the amplitude of pulse load (of constant duration) at which the maximum plate deflection is equal to some constant value k (k-half or one plate thickness).*" Another, very widely used criterion has been formulated and proposed by Budiansky and Hutchinson [3], based on an earlier work [1] and latter extended [4]. Originally, the criterion was formulated for shell-type structures but was used also for columns and plates. The criterion claims that: "*Dynamic stability loss occurs when the maximum deflection grows rapidly with the small variation of the load amplitude*". This criterion is schematically presented in Fig. 7.2(b), where $R(\lambda, t)$ is the response of the simply nonlinear model assumed in [3] and presented also in Fig. 7.2(a),

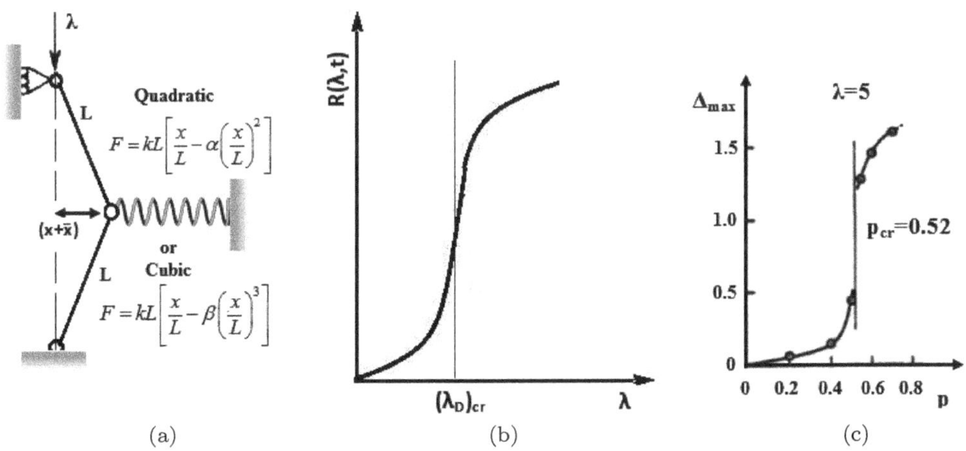

Fig. 7.2. (a) The nonlinear model (adapted from [3]); (b) the Budiansky and Hutchinson (B&H) schematic criterion (from [3]); (c) the application of the B&H criterion to axisymmetric dynamic buckling of clamped shallow spherical shells (adapted from [1]).

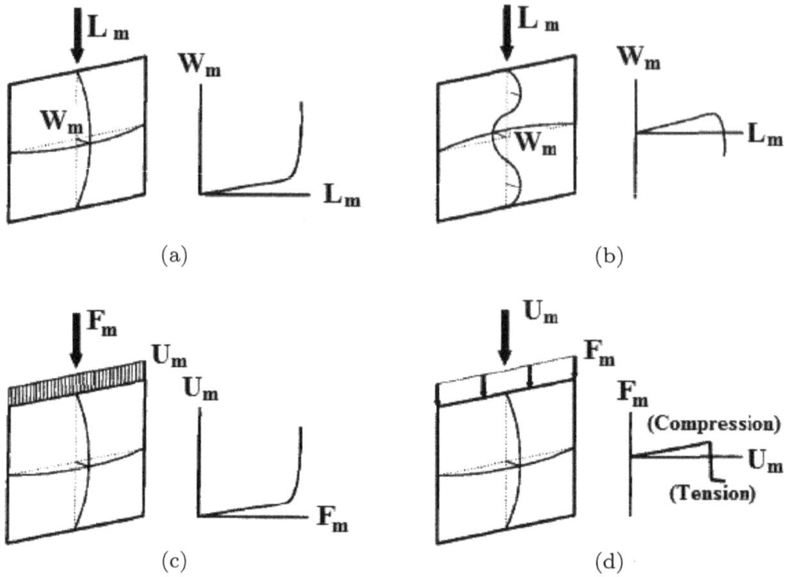

Fig. 7.3. Buckling criteria (adapted from [3]): (a) the first criterion; (b) the second criterion; (c) the third criterion; and (d) the fourth criterion.

and λ is the non-dimensional applied dynamic pulse-type compressive load. Figure 7.2(c), presents the application of the criterion for the axisymmetric dynamic buckling of clamped shallow spherical shells as presented in [1].

Other dynamic buckling criteria were suggested and applied by Ari-Gur and Simonetta in their manuscript [21]. They formulated four criteria presented schematically in Figs. 7.3(a)–7.3(d). The first criterion (see Fig. 7.3(a)) correlates the maximum lateral deflection, W_m to the pulse intensity defined as L_m. The first criterion is stated as: "*Buckling occurs when, for a given pulse shape and duration, a small increase in the pulse intensity causes a sharp increase in the rate of growth of the peak lateral deflection*" [21]. The authors claim that this criterion can be used to both displacement and force loading types for a wide range of pulse frequencies, however for very short pulse durations the results might be misleading. The reason for this is connected to the characteristics of the out-of-plane deflections which, for short pulse duration in the vicinity of buckling loads, turn out to have short wavelength patterns which are associated with smaller peak deflections. Therefore, the authors present the second criterion (Fig. 7.3(b)), which comes to answer the deficiency of the first criterion, and is suitable to patterns of short wavelength deflection shapes. It claims that "*dynamic buckling occurs when a small increase in the pulse intensity causes a decrease in the peak lateral deflection*" and is relevant to only impulsive loads and may be used in complimentary to the first criterion. The last two criteria presented in [21] are connected the intensity of the applied load vs. the maximum response of the loaded edge, say at $x = 0$. According to the third buckling criterion (Fig. 7.3(c)), "*buckling occurs when a small increase in the force intensity* F_m *causes a sharp increase in the peak longitudinal displacement* U_m *at* $x = 0$". The buckling phenomenon under pulse loading is due to the diminishing of the structural resistance to an in-plane compressive load when the dynamic lateral deflections grow rapidly. The fourth buckling criterion, associated to a displacement pulse (Fig. 7.3(d)), states that "*buckling occurs when a small increase in the pulse displacement intensity* U_m *causes a transition of the peak reaction force* F_m *at* $x = 0$ *from compression to tension*". This transition would come true, when the tensile force needed to keep the deforming structure at the prescribed U_m would be greater than the maximum compression at its loaded edge.

Another interesting criterion was suggested by Petry and Fahlbush [23], claiming that the Budiansky–Hutchinson criterion is very conservative for structures with stable post-buckling equilibrium path because it does not take into account load carrying-capacity of the structure. The criterion proposed is based on stress analysis and claims that: "a stress failure occurs if the effective stress σ_E exceeds the limit stress of the material, σ_L; a dynamic response caused by an impact is defined to be dynamically stable if $\sigma_E \leq \sigma_L$ is fulfilled at every time everywhere in the structure". Using this criterion, which is claimed to be practical also for ductile materials (by using the yield stress σ_Y instead of σ_L), the DLF (see Eq. (7.1)) is modified to

$$\mathrm{DLF} \equiv \frac{(N_F)_{\mathrm{dyn.}}}{(N_F)_{\mathrm{static}}} \tag{7.2}$$

where $(N_F)_{\mathrm{dyn.}}$ and $(N_F)_{\mathrm{static}}$ are defined as the dynamic failure and static failure loads, respectively.

7.2. Dynamic Buckling of Columns

7.2.1. *Dynamic buckling of columns using CLT*

The column differential equations of motion under time-dependent axial compression, as presented in Fig. 7.4, are given by

$$N_{x,x} = I_0\ddot{u} - I_1\ddot{w}_{,x}, \tag{7.3}$$

$$[M_{x,x} + N_x \cdot w_{,x} + I_2 \cdot \ddot{w}_{,x}]_{,x} = I_0\ddot{w} + I_1\ddot{u}_{,x}, \tag{7.4}$$

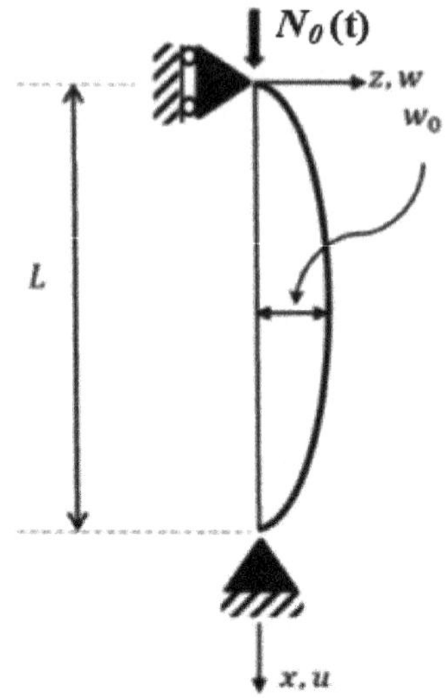

Fig. 7.4. A column with geometric initial imperfection, w_0, subjected to time-dependent axial compression ($N_0(t)$ is the axial compression at $x = 0$).

where

$$(I_0; I_1; I_2) = \rho \int_{-h/2}^{h/2} (1; z; z^2) dz \tag{7.5}$$

and $()_{,x}$ is the first partial differentiation with respect to x, $\ddot{u}, \ddot{w}$ are the second partial differentiation of the axial and lateral displacements with respect to time, t, h is the total thickness of the column and ρ is the relevant densities for each layer in the laminate.

Assuming that the force and moment resultants can be written as a function of the strains and curvatures, according to the CLT approach we obtain

$$\begin{Bmatrix} N_x \\ M_x \end{Bmatrix} = \begin{bmatrix} A_{11} & B_{11} \\ B_{11} & D_{11} \end{bmatrix} \begin{Bmatrix} \varepsilon_x \\ \kappa_x \end{Bmatrix}, \tag{7.6}$$

where

$$(A_{11}; B_{11}; D_{11}) = \int_{-h/2}^{h/2} \overline{Q}_{11} (1; z; z^2) dz \tag{7.7}$$

and $\overline{Q}_{11}$ is the plane stress reduced stiffness for each layer after transformation (for an exact expression, see Chapter 2). The strain-displacement (u longitudinal displacement, w lateral displacement) relationships are as follows:

$$\begin{Bmatrix} \varepsilon_x \\ \kappa_x \end{Bmatrix} = \begin{Bmatrix} u_{,x} + \frac{1}{2}(w_{,x}^2 - w_{0,x}^2) \\ -(w - w_0)_{,xx} \end{Bmatrix}. \tag{7.8}$$

Substitution of Eqs. (7.6)–(7.8) into Eqs. (7.3) and (7.4), while neglecting rotary inertia, $I_2 = 0$, yields

$$A_{11}(u_{,xx} + w_{,x} \cdot w_{,xx} - w_{0,x} \cdot w_{0,xx}) \\ -B_{11}(w - w_0)_{,xxx} = I_0\ddot{u} - I_1\ddot{w}_{,x}, \tag{7.9}$$

$$A_{11}\left[(u_{,x} \cdot w_{,x})_{,x} + \frac{3}{2}w_{,x}^2 \cdot w_{,xx} - w_{0,x}\left(w_{0,xx} \cdot w_{,x} - \frac{1}{2}w_{0,x} \cdot w_{,xx}\right)\right] \\ + B_{11}[u_{,xx} + w_{0,xx}(w - w_o)_{,x}]_{,x} - D_{11}[w - w_o]_{,xxxx} \\ = I_0\ddot{w} + I_1\ddot{u}_{,x}. \tag{7.10}$$

Assuming a symmetric laminate, $B_{11} = I_1 = 0$, then Eqs. (7.9) and (7.10) are simplified to yield

$$A_{11}(u_{,xx} + w_{,x} \cdot w_{,xx} - w_{0,x} \cdot w_{0,xx}) = I_0\ddot{u}, \tag{7.11}$$

$$A_{11}\left[(u_{,x} \cdot w_{,x})_{,x} + \frac{3}{2}w_{,x}^2 \cdot w_{,xx} - w_{0,x}\left(w_{0,xx} \cdot w_{,x} - \frac{1}{2}w_{0,x} \cdot w_{,xx}\right)\right] \\ -D_{11}[w - w_o]_{,xxxx} = I_0\ddot{w}. \tag{7.12}$$

One should note that all the terms in Eqs. (7.11) and (7.12) are a function of both x and t.

The associated boundary conditions (pinned–pinned or simply supported) for the symmetric laminate are

$$\text{Lateral deflection: } w(0,t) = w(L,t) = 0, \tag{7.13}$$

$$\text{Bending moment: } D_{11} \cdot w_{,xx}(0,t) = D_{11} \cdot w_{,xx}(L,t) = 0, \tag{7.14}$$

$$\text{Axial force: } A_{11} \cdot u_{,x}(0,t) = N_x(0,t) = -N_0(t) = -N_0 \sin\frac{\pi t}{T}, \tag{7.15}$$

$$\text{Axial displacement: } u_{,x}(L,t) = 0, \tag{7.16}$$

while the initial conditions assume that $w(x,0) = w_0$, $w_{,x}(x,0) = u(0,0) = \dot{u}(0,0) = 0$.

Note that Eqs. (7.11) and (7.12) can be used for an isotropic column, by letting $A_{11} = EA$, the axial stiffness (E = Young's modulus and A = cross-section of the column) and the bending stiffness, $D_{11} = Eh^3/[12(1 - \upsilon^2)]$.

The solution for the equations of motion, (7.11)–(7.12), can be obtained using energy methods, like the Galerkin method, using the boundary and initial conditions presented in Eqs. (7.13)–(7.16). To apply the Galerkin approach, one should assume displacements which satisfy the boundary conditions of the problem, namely for the present case, the following functions will be assumed:

$$w(x,t) = A(t)\sin\frac{\pi x}{L}, \quad w_0(x) = A_0 \sin\frac{\pi x}{L},$$
$$u(x,0) = B(t)\cos\frac{\pi x}{L}, \quad N_0(t) = N_0 \sin\frac{\pi t}{T}. \tag{7.17}$$

Multiplying Eq. (7.11) by the expression for $u(x,\ 0)$ while Eq. (7.12) is multiplied by $w(x,t)$ and integrating both from $x = 0$ to $x = L$ will exclude the x dependence, yielding two nonlinear time-dependent equations which have to be solved numerically.

For each value of N_0, the response of the beam w and u can be calculated for an assumed initial geometric imperfection, $w_0(x,t) = A_0 \sin\frac{\pi x}{L}$, where A_0 is a known amplitude (usually percentage of the column's thickness), for a constant time period T.

A typical drawing is presented in Fig. 7.5 (adapted from [16]). Calculating the dynamic buckling for each value of the $2T/T_b$ and dividing it by the static buckling load, will yield the DLF for the tested or calculated case (see Eq. (7.1)).

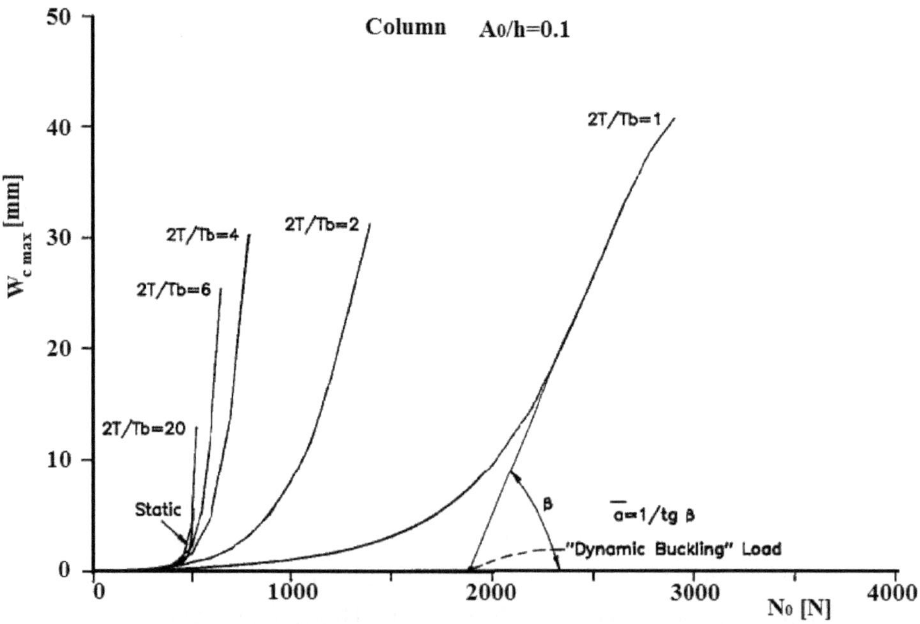

Fig. 7.5. The influence of the loading duration (T) on the column's response (adapted from [16]), T_b first natural period of the column, A_0 is the amplitude of the initial geometric imperfection.

7.2.2. *Dynamic buckling of columns using FSDT*

To solve the dynamic buckling of columns (see Fig. 7.4) using the FSDT we shall assume the following displacement field:

$$\begin{aligned} u_x(x,z,t) &= u(x,t) + z\phi_x(x,t), \\ u_z(x,z,t) &= w(x,t) - w_0(x), \end{aligned} \tag{7.18}$$

where the assumed variables u and w are displacements in the x and z directions of a point on the mid-plane of the plate (namely at $z = 0$), respectively, and ϕ_x is a rotation about the x-axis, while w_0 is the initial geometric imperfection of the column. The equations of motion can be written as (see Chapters 2 and 4):

$$\begin{aligned} &N_{x,x} = I_0\ddot{u} + I_1\ddot{\phi}_x, \\ &Q_{x,x} + [N_x \cdot (w_{,x} - w_{o,x})]_{,x} = I_0, \ddot{w}, \\ &M_{x,x} - Q_x = I_2\ddot{\phi}_x + I_1\ddot{u}. \end{aligned} \tag{7.19}$$

In Eq. (7.19), Q_x is the shear force or the transverse force resultant defined as

$$Q_x = K\int_{-h/2}^{+h/2} \tau_{xy} \cdot dz, \tag{7.20}$$

where K is the shear correction coefficient computed by equating the strain energy due to transverse shear stresses to the strain energy due to true transverse shear as calculated by a 3D elasticity theory while τ_{xy} is the transverse shear stress. For a homogenous beam having a rectangular cross-section, $K = 5/6$ (see Chapter 3). The force and moment resultants as a function of the displacements are given by

$$\begin{Bmatrix} N_x \\ M_x \\ Q_x \end{Bmatrix} = \begin{bmatrix} A_{11} & B_{11} & 0 \\ B_{11} & D_{11} & 0 \\ 0 & 0 & KA_{55} \end{bmatrix} \begin{Bmatrix} u_{,x} + \frac{1}{2}(w_{,x}^2 - w_{0,x}^2) \\ \phi_{x,x} \\ \phi_x + w_{,x} - w_{0,x} \end{Bmatrix}. \tag{7.21}$$

As was assumed in Section 7.2.1, the strains have the following form:

$$\begin{Bmatrix} \varepsilon_x \\ \gamma_{xz} \end{Bmatrix} = \begin{Bmatrix} u_{x,x} + \frac{1}{2}(w_{,x}^2 - w_{0,x}^2) \\ u_{x,z} + u_{z,x} \end{Bmatrix} = \begin{Bmatrix} u_{,x} + \frac{1}{2}(w_{,x}^2 - w_{0,x}^2) + z\phi_{x,x} \\ \phi_x + w_{,x} - w_{0,x} \end{Bmatrix} \tag{7.22}$$

Substituting Eq. (7.21) into Eq. (7.19) yields the following equations of motion expressed by the three assumed displacements, $u(x,t)$, $w(x,t)$,

$\phi_x(x,t)$ and the known initial geometric imperfection, $w_0(x)$ (see also a similar derivation in [24]):

$$\begin{aligned}
&A_{11}(u_{,xx} + w_{,x} \cdot w_{,xx} - w_{0,x} \cdot w_{0,xx}) + B_{11}\phi_{x,xx} = I_0\ddot{u} + I_1\ddot{\phi}_x,\\
&[A_{11}(u_{,xx} + w_{,x} \cdot w_{,xx} - w_{0,x} \cdot w_{0,xx}) + B_{11}\phi_{x,xx}](w_{,x} - w_{0,x})\\
&\quad + \left[A_{11}\left(u_{,x} + \frac{1}{2}w_{,x}^2 - \frac{1}{2}w_{0,x}^2\right) + B_{11}\phi_{x,x}\right](w_{,xx} - w_{0,xx})\\
&\quad + KA_{55}(\phi_{x,x} + w_{,xx} - w_{0,xx}) = I_0\ddot{w},\\
&\quad + B_{11}(u_{,xx} + w_{,x} \cdot w_{,xx} - w_{0,x} \cdot w_{0,xx}) + D_{11}\phi_{x,xx}\\
&\quad - KA_{55}(\phi_x + w_{,x} - w_{0,x}) = I_2\ddot{\phi}_x + I_1\ddot{u}.
\end{aligned} \tag{7.23}$$

For a symmetric layup, $B_{11} = I_1 = 0$, then Eqs. (7.23) are simplified to yield

$$\begin{aligned}
&A_{11}(u_{,xx} + w_{,x} \cdot w_{,xx} - w_{0,x} \cdot w_{0,xx}) = I_0\ddot{u},\\
&[A_{11}(u_{,xx} + w_{,x} \cdot w_{,xx} - w_{0,x} \cdot w_{0,xx})](w_{,x} - w_{0,x})\\
&\quad + \left[A_{11}\left(u_{,x} + \frac{1}{2}w_{,x}^2 - \frac{1}{2}w_{0,x}^2\right)\right](w_{,xx} - w_{0,xx})\\
&\quad + KA_{55}(\phi_{x,x} + w_{,xx} - w_{0,xx}) = I_0\ddot{w},\\
&D_{11}\phi_{x,xx} - KA_{55}(\phi_x + w_{,x} - w_{0,x}) = I_2\ddot{\phi}_x.
\end{aligned} \tag{7.24}$$

As in the previous section, one should note that all the terms in Eqs. (7.24) are a function of both x and t.

The associated boundary conditions (pinned–pinned or simply supported) for the symmetric laminate are

$$\text{Lateral deflection: } w(0,t) = w(L,t) = 0, \tag{7.25}$$

$$\text{Bending moment: } D_{11} \cdot \phi_{x,x}(0,t) = D_{11} \cdot \phi_{x,x}(L,t) = 0, \tag{7.26}$$

$$\text{Axial force: } A_{11} \cdot u_{,x}(0,t) = N_x(0,t) = -N_0(t) = -N_0 \sin\frac{\pi t}{T}, \tag{7.27}$$

$$\text{Axial displacement: } u_{,x}(L,t) = 0, \tag{7.28}$$

while the initial conditions assume that $w(x,0) = w_0$, $w_{,x}(x,0) = u(0,0) = \dot{u}(0,0) = 0$.

As before, the analytic solution of the equations of motion cannot be obtained, therefore energy methods, like Galerkin method, are suggested. Suitable assumed solutions for the case of simply supported boundary

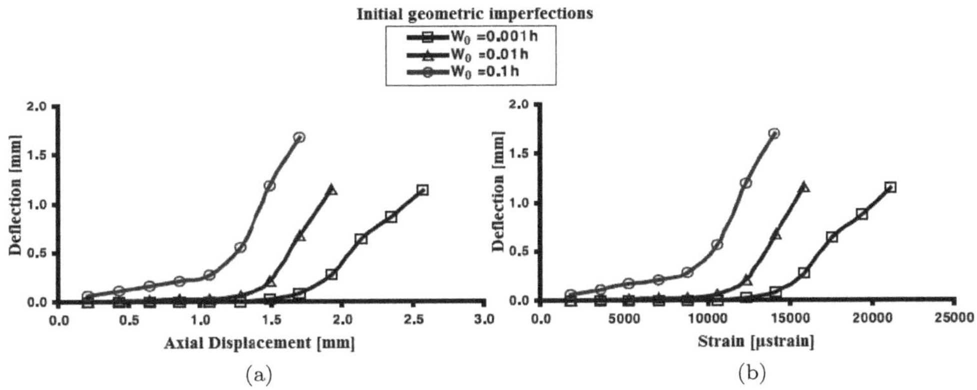

Fig. 7.6. The dynamic pulse-buckling response of a column ($L = 150$[mm], Width $=$ 20[mm]) for various initial geometric imperfections: (a) mid-span deflection vs. maximal axial displacement at the impacted end; (b) mid-span deflection vs. compressive strain at the neutral axis of the column mid-span (adapted from [24]).

conditions are as follows:

$$u(x,0) = A(t)\cos\frac{\pi x}{L}, \quad w(x,t) = B(t)\sin\frac{\pi x}{L}, \quad \phi_x(x,t) = C(t)\cos\frac{\pi x}{L},$$
$$w_0(x) = A_0\sin\frac{\pi x}{L}, \quad N_0(t) = N_0\sin\frac{\pi t}{T}. \tag{7.29}$$

The procedure described before in Section 7.2 is again employed to obtain the response of the column to a pulse-type loading. As can be seen in Fig. 7.6, based on the results presented in [24], the response of the column has a similar behavior to what had been presented in Fig. 7.5. Note that in Fig. 7.6(b), the compressive strain is a linear function of the applied axial compression load, therefore it can be compared with what it is presented in Fig. 7.5.

7.3. Dynamic Buckling of Plates

The present derivation for the dynamic buckling of orthotropic rectangular plates under uniaxial loading is based on Ekstrom's fundamental study [8] presented already in 1973 using the models developed by Lekhnitskii for anisotropic plates [44]. Let us assume a simply supported rectangular orthotropic plate uniaxially loaded as presented in Fig. 7.7. We choose the natural axes of the material to coincide with the coordinate axes leading to

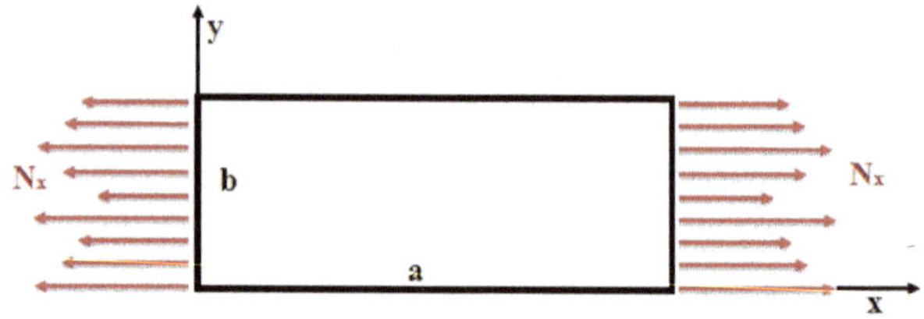

Fig. 7.7. A thin orthotropic plate uniaxial loaded.

the following in-plane stress–strain relations

$$\begin{Bmatrix} \sigma_x \\ \sigma_y \\ \tau_{xy} \end{Bmatrix} = \frac{1}{1-\upsilon_{xy}\cdot\upsilon_{yx}} \begin{bmatrix} E_x & \upsilon_{xy}E_y & 0 \\ \upsilon_{yx}E_x & E_y & 0 \\ 0 & 0 & (1-\upsilon_{xy}\cdot\upsilon_{yx})G_{xy} \end{bmatrix} \begin{Bmatrix} \varepsilon_x \\ \varepsilon_y \\ \gamma_{xy} \end{Bmatrix}$$

$$\equiv \begin{bmatrix} Q_{11} & Q_{12} & 0 \\ Q_{21} & Q_{22} & 0 \\ 0 & 0 & Q_{66} \end{bmatrix} \begin{Bmatrix} \varepsilon_x \\ \varepsilon_y \\ \gamma_{xy} \end{Bmatrix}, \tag{7.30}$$

$$\begin{Bmatrix} \varepsilon_x \\ \varepsilon_y \\ \gamma_{xy} \end{Bmatrix} = \begin{bmatrix} \frac{1}{E_x} & -\frac{\upsilon_{xy}}{E_x} & 0 \\ -\frac{\upsilon_{yx}}{E_y} & \frac{1}{E_y} & 0 \\ 0 & 0 & \frac{1}{G_{xy}} \end{bmatrix} \begin{Bmatrix} \sigma_x \\ \sigma_y \\ \tau_{xy} \end{Bmatrix}, \tag{7.31}$$

and

$$Q_{12} = Q_{21} \Rightarrow \upsilon_{yx} = \frac{E_x}{E_y}\upsilon_{xy} \tag{7.32}$$

The dynamic buckling problem can be solved assuming out-of-plane initial geometric imperfections (w_0), and therefore the strains will be assumed to have the following expressions:

$$\begin{Bmatrix} \varepsilon_x \\ \varepsilon_y \\ \gamma_{xy} \end{Bmatrix} = \begin{Bmatrix} \frac{\partial u}{\partial x} + \frac{1}{2}\left[\left(\frac{\partial w}{\partial x}\right)^2 - \left(\frac{\partial w_0}{\partial x}\right)^2\right] \\ \frac{\partial v}{\partial y} + \frac{1}{2}\left[\left(\frac{\partial w}{\partial y}\right)^2 - \left(\frac{\partial w_0}{\partial y}\right)^2\right] \\ \frac{\partial u}{\partial x} + \frac{\partial v}{\partial y} + \frac{\partial^2 w}{\partial x \partial y} - \frac{\partial^2 w_0}{\partial x \partial y} \end{Bmatrix}, \tag{7.33}$$

where $w(x, y, t)$ is the total out-of-plane, $w_0(x, y)$ is the initial out-of-plane geometric displacement and $u(x, y, t)$ and $v(x, y, t)$ are the in-plane displacements in the x and y directions, respectively.

The compatibility equation for the stress function $F(x, y, t)$ and the plate equations of motion (see [8, 44]) can be written as

$$\frac{1}{E_y} \cdot \frac{\partial^4 F}{\partial x^4} + \frac{1}{E_x} \cdot \frac{\partial^4 F}{\partial y^4} + 2\left(\frac{1}{2G_{xy}} - \frac{\upsilon_{xy}}{E_x}\right)$$
$$= \left(\frac{\partial^2 w}{\partial x \partial y}\right)^2 - \left(\frac{\partial^2 w_0}{\partial x \partial y}\right)^2 + \frac{\partial^4 w_0}{\partial x^2 \partial y^2} - \frac{\partial^4 w}{\partial x^2 \partial y^2}, \tag{7.34}$$

$$D_x \frac{\partial^4 (w - w_0)}{\partial x^4} + 2(D_1 + 2D_{xy}) \frac{\partial^4 (w - w_0)}{\partial x^2 \partial y^2} + D_y \frac{\partial^4 (w - w_0)}{\partial y^4}$$
$$= h\left[\frac{\partial^2 w}{\partial x^2} \cdot \frac{\partial^2 F}{\partial y^2} - 2\frac{\partial^2 w}{\partial x \partial y} \cdot \frac{\partial^2 F}{\partial x \partial y} + \frac{\partial^2 w}{\partial y^2} \cdot \frac{\partial^2 F}{\partial x^2} - \rho \frac{\partial^2 w}{\partial t^2}\right], \tag{7.35}$$

where, h is the plate thickness and ρ is the plate density. The flexural stiffnesses of the plate are defined as follows:

$$D_x \equiv \frac{E_x h^3}{12(1 - \upsilon_{xy}\upsilon_{yx})}, \quad D_y \equiv \frac{E_y h^3}{12(1 - \upsilon_{xy}\upsilon_{yx})}, \quad D_{xy} = \frac{G_{xy} h^3}{12},$$
$$D_1 \equiv \frac{E_x \upsilon_{yx} h^3}{12(1 - \upsilon_{xy}\upsilon_{yx})} = \frac{E_y \upsilon_{xy} h^3}{12(1 - \upsilon_{xy}\upsilon_{yx})}. \tag{7.36}$$

The out-of-plane boundary conditions of the present problem, assuming that the initially straight edges of the plate remain straight after buckling can be written as

$$@x = 0, a: \ w = w_0 = 0 \quad \text{and} \quad \frac{\partial^2 w}{\partial x^2} = \frac{\partial^2 w_0}{\partial x^2} = 0,$$
$$@y = 0, b: \ w = w_0 = 0 \quad \text{and} \quad \frac{\partial^2 w}{\partial y^2} = \frac{\partial^2 w_0}{\partial y^2} = 0. \tag{7.37}$$

The in-plane boundary conditions have no restrain at $y = 0, b$ and in the x direction

$$@x = 0, a: \quad \frac{1}{b}\int_0^b \sigma_x dy = \frac{1}{b}\int_0^b \frac{\partial^2 F}{\partial y^2} dy = -P,$$
$$@y = 0, b: \quad \frac{1}{a}\int_0^a \sigma_y dx = \frac{1}{a}\int_0^a \frac{\partial^2 F}{\partial x^2} dx = 0, \tag{7.38}$$

where P is the compression load acting on the plate.

The assumed solution for the w, and the initial geometric imperfection, w_0, have the following forms:

$$w(x, y, t) = A(t) \sin\frac{m\pi x}{a} \sin\frac{n\pi y}{b}, \tag{7.39}$$

$$w_o(x, y, t) = A_o \sin\frac{m\pi x}{a} \sin\frac{n\pi y}{b}. \tag{7.40}$$

One should note that the expressions in Eqs. (7.39) and (7.40) satisfy the boundary conditions assumed above (Eq. (7.37)). Substituting those equations in the compatibility equation leads to the following relationship between the stress function F and the displacements

$$\frac{1}{E_y} \cdot \frac{\partial^4 F}{\partial x^4} + \frac{1}{E_x} \cdot \frac{\partial^4 F}{\partial y^4} + 2\left(\frac{1}{2G_{xy}} - \frac{\upsilon_{xy}}{E_x}\right)$$
$$= \frac{m^2 n^2 \pi^4}{2a^2 b^2}\left[\cos\frac{2m\pi x}{a} + \cos\frac{2n\pi y}{b}\right](A^2 - A_0^2). \tag{7.41}$$

The solution for Eq. (7.41), which also satisfies the in-plane boundary conditions, Eq. (7.38), can be written as

$$F(x, y, t) = (A^2 - A_0^2)\left[\frac{a^2 n^2 E_y}{32 b^2 m^2}\cos\frac{2m\pi x}{a} + \frac{b^2 m^2 E_x}{32 a^2 n^2}\cos\frac{2n\pi y}{b}\right] - \frac{P}{2}y^2. \tag{7.42}$$

Equation (7.42) is then inserted in Eq. (7.32) yielding

$$\frac{\pi^4}{h}\left[D_x\frac{m^4}{a^4} + 2(D_1 + 2D_{xy})\frac{m^2 n^2}{a^2 b^2} + D_y\frac{n^4}{b^4}\right][A(t) - A_0] = \frac{m^2\pi^2}{a^2}P(t)\cdot A(t)$$
$$-\rho\frac{d^2 A(t)}{dt^2} + \frac{\pi^4}{8}\left[\frac{m^4}{a^4}E_x\cos\frac{2n\pi y}{b} + \frac{n^4}{b^4}E_y\cos\frac{2m\pi x}{a}\right][A^2(t) - A_0^2]A(t). \tag{7.43}$$

Note that Eq. (7.43) is a nonlinear equation containing functions of the variables x, y and t. To eliminate the x, y dependency, the Galerkin method is applied, which demands the multiplication of both sides of the equation by $\sin\frac{m\pi x}{a}\sin\frac{n\pi y}{b}dxdy$ and integrating over the middle plane of the plate yielding the following nonlinear time-dependent expression

$$\frac{d^2 A(t)}{dt^2} + \frac{\pi^4}{\rho h}\left[D_x\frac{m^4}{a^4} + 2(D_1 + 2D_{xy})\frac{m^2 n^2}{a^2 b^2} + D_y\frac{n^4}{b^4}\right][A(t) - A_0]$$
$$-\frac{m^2\pi^2}{a^2\rho}P(t)\cdot A(t) + \frac{\pi^4}{24\rho}\left[\frac{m^4}{a^4}E_x + \frac{n^4}{b^4}E_y\right]\left[A^2(t) - A_0^2\right]A(t) = 0, \tag{7.44}$$

where m and n are both odd integers. The time dependence of the term $P(t)$ is chosen according to its time dependence, namely for a step function $P(t) = P_0$ (constant), for an impulse $P(t) = P_0\delta(t)$ (where (t) is the Kronecker delta.[a]) or any other time functions. Equation (7.44) can be non-dimensionalized (assuming $n = 1$) to yield the following expression (see discussion in [8]):

$$\frac{d^2\xi}{d\tau^2} + S\left[\frac{m^4 + 2R_{12}m^2\beta^2 + R_{22}\beta^4}{4}(\xi - \xi_0) - m^2\tau\xi \right.$$
$$\left. + \frac{(1 - \upsilon_{xy} \cdot \upsilon_{yx})}{8}(m^4 + R_{22}\beta^4)(\xi^2 - \xi_0^2)\xi\right] = 0. \quad (7.45)$$

where

$$P_1 \equiv \frac{4D_x\pi^2}{a^2h}, \quad \xi \equiv \frac{A}{h}, \quad \xi_0 \equiv \frac{A_0}{h}, \quad \beta \equiv \frac{a}{b}, \quad \tau \equiv \frac{P}{P_1} = \frac{c \cdot t}{P_1},$$
$$R_{12} \equiv \frac{D_1 + 2D_{xy}}{D_x}, \quad R_{22} \equiv \frac{D_y}{D_x} = \frac{E_y}{E_x}, \quad S \equiv \frac{P_1^3\pi^2}{c^2a^2\rho}. \quad (7.46)$$

Equation (7.45) is solved using numerical methods for integration of non-linear equations, like the famous Runge-Kutta methods.[b] Solutions for the response of the plate for various S values are presented in [8].

It is interesting to note that Eq. (7.44) can be used to obtain the static plate buckling load and its natural frequencies, as well as the expression for large plate deflections.

Assuming the static case for a perfect plate,

$$\frac{d^2A(t)}{dt^2} = 0 \quad \text{and} \quad A_0 = 0 \quad (7.47)$$

and neglecting high-order terms like A^3, Eq. (7.44) yields

$$\frac{\pi^2}{h}\left[D_x\frac{m^4}{a^4} + 2(D_1 + 2D_{xy})\frac{m^2n^2}{a^2b^2} + D_y\frac{n^4}{b^4}\right] - \frac{m^2}{a^2}P = 0. \quad (7.48)$$

The critical buckling load of the plate is obtained from Eq. (7.48) by assuming $n = 1$, namely

$$P_{cr} = \frac{\pi^2 D_x}{b^2h}\left[\left(\frac{mb}{a}\right)^2 + \frac{2(D_1 + 2D_{xy})}{D_x} + \frac{D_y}{D_x}\left(\frac{a}{mb}\right)^2\right]$$
$$\Rightarrow (P_{cr})_{\min.} \text{ occurs when } \frac{a}{b}\sqrt{\frac{D_y}{D_x}} = \text{integer}. \quad (7.49)$$

[a]The function is 1 if the variables are equal, and 0 otherwise, namely:

$$\delta_{ij} = \begin{cases} 0 & \text{if } i \neq j, \\ 1 & \text{if } i = j. \end{cases}$$

[b]Runge–Kutta methods are a family of implicit and explicit iterative methods (including the Euler Methods routine), used in temporal discretization for the approximate solutions of ordinary differential equations developed approximately. in 1900 by C. Runge and M. W. Kutta.

Neglecting high-order terms like A^3, and assuming a perfect plate ($A_0 = 0$) Eq. (7.44) is written as

$$\frac{d^2A(t)}{dt^2} + \frac{\pi^4}{\rho h}\left[D_x\frac{m^4}{a^4} + 2(D_1 + 2D_{xy})\frac{m^2n^2}{a^2b^2} + D_y\frac{n^4}{b^4}\right]A(t)$$
$$-\frac{m^2\pi^2}{a^2\rho}P(t)\cdot A(t) = 0. \tag{7.50}$$

The second term in Eq. (7.50), is the square of the natural frequency of the perfect plate, namely

$$\omega_{mn}^2 = \frac{\pi^4}{\rho h}\left[D_x\frac{m^4}{a^4} + 2(D_1 + 2D_{xy})\frac{m^2n^2}{a^2b^2} + D_y\frac{n^4}{b^4}\right]. \tag{7.51}$$

Finally, assuming the static case, the expression for the postbuckling behavior for a perfect plate with large deflections is obtained from Eq. (7.44), namely

$$\left\{\frac{\pi^4}{h}\left[D_x\frac{m^4}{a^4} + 2\left(D_1 + 2D_{xy}\right)\frac{m^2n^2}{a^2b^2} + D_y\frac{n^4}{b^4}\right] - \frac{m^2\pi^2}{a^2}PA\left(t\right)\right\}$$
$$+\frac{\pi^4}{24}\left[\frac{m^4}{a^4}E_x + \frac{n^4}{b^4}E_y\right]A^3 = 0. \tag{7.52}$$

Equation (7.50) shows that the lateral deflection would be zero until the critical buckling load given by Eq. (7.49) is reached. After that point, non-zero deflections are possible while the load-deflection relation is cubic.

For further references dealing with the behavior of plates under dynamic buckling, the reader is referred to [14–16, 18, 19, 21, 23, 31, 35].

The definition of the buckling loads for plates, both for a static and dynamic case is one of the often issues to be solved consistently. To assist the definition of the buckling load for a plate, a method is described in Appendix A, which can be easily applied for experimental and numerical data consisting of deflections vs. applied load.

7.4. Dynamic Buckling of Thin-Walled Structures-Numerical and Experimental Results

The dynamic buckling of thin-walled structures has been dealt in depth also from the experimental point of view, beside various calculations using different approaches. Figure 7.8(a) presents a shallow clamped spherical cap loaded by a sudden rectangular pressure q applied at time $t = 0$ and held constant for a given period of time and then suddenly removed (as described

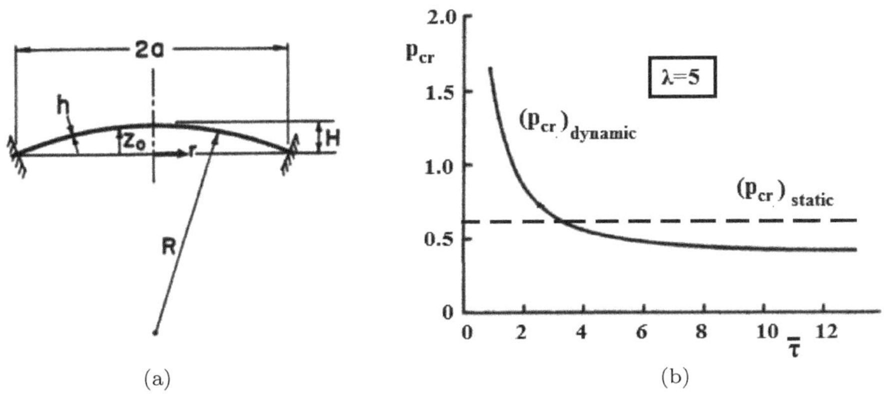

Fig. 7.8. (a) A clamped shallow spherical cap and (b) variation of the non-dimensional critical pressure with non-dimensional load duration (adapted from [1]).

in [1]). Typical numerical results are then shown in Fig. 7.8(b), showing the reduction of the critical non-dimensional pressure with the increasing the duration of the applied pressure (see other results in [1]). It is interesting to note that the static critical load, that from a certain non-dimensional time duration (in this case from approximately $\overline{\tau} = 3$) the static critical pressure is above the dynamic critical pressure, whereas for short-time duration the dynamic critical load is higher than the static one, as is normally known for short duration loads. The various non-dimensional terms used in Fig. 7.8 were defined in [1] as

$$\overline{\tau} = \frac{ct}{R}, \quad c = \sqrt{\frac{E}{\rho}}, \quad p = \frac{q}{q_0},$$

$$\lambda = 2[3(1-\upsilon^2)]^{1/4}\left(\frac{H}{h}\right)^{1/2}, \quad q_0 = \frac{2E}{[3(1-\upsilon^2)]^{1/2}}\left(\frac{h}{R}\right)^{1/2}, \quad (7.53)$$

where E is the Young's modulus, υ is the Poisson's ratio and ρ is the density.

Similar results were reported in [23], where an isotropic plate (Fig. 7.9(a)) is sudden loaded by a rectangular impulse and the response of the plate is presented using the term DLF as being defined in Eq. (7.1) as a function of the non-dimensional time duration, where T_s = the impact period and T_p = the natural period of the plate. As shown by other researchers, in the vicinity of $T_s/T_p = 1$ the term DLF is less that unity, implying that the static load is higher than the dynamic one. Shifting this period ratio towards 0.5 shows a rapid increase in the value of DLF, reaching a maximal value of 3.6 in the vicinity of $T_s/T_p = 0.1$ (approx.), namely as expected the plate can withstand high dynamic compressive stresses (3.6 the static buckling

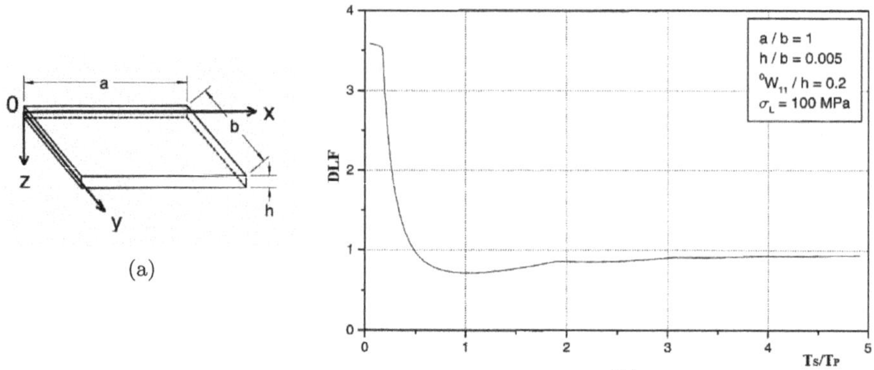

Fig. 7.9. (a) An isotropic flat plate and (b) variation of DLF with the non-dimensional load duration (adapted from [23]).

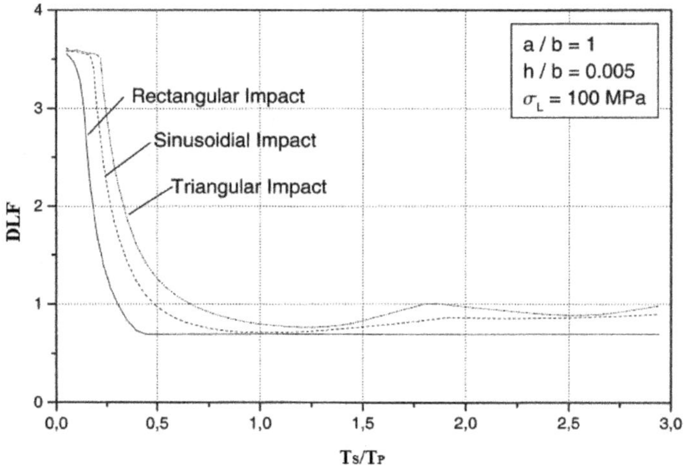

Fig. 7.10. Variation of DLF with the non-dimensional load duration for three types of impulse (adapted from [23]).

load of the plate) provided the time duration is very short. Impacting the plate with sinusoidal or triangular impulses displays the same behavior as the rectangular one (see Fig. 7.10 taken from [23]).

Experimental results for laminated composite plates under impulse type uniaxial loading obtained by dropping masses and measuring the response using a back-to-back strain gages bonded in the center of the plate is described in [19]. Figure 7.11 presents the experimental variation of the DLF term defined as

$$DLF \equiv \frac{(\varepsilon_{cr})_{\text{dyn.}}}{(\varepsilon_{cr})_{\text{st.}}} \tag{7.54}$$

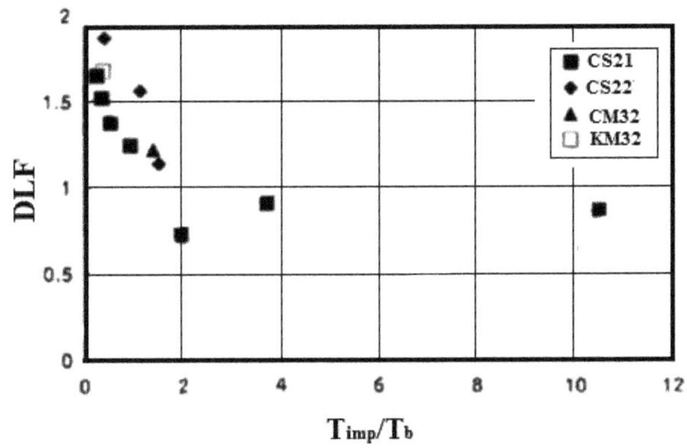

Fig. 7.11. Variation of DLF with the non-dimensional load duration — experimental results (adapted from [19]).

Table 7.1. The properties of the tested plates (from [19]).

Specimen	Material	Lay-up	Total number of layers	Total thickness (measured) [mm]
CS21	Graphite-Epoxy HT-T300 (Torey)	$(\pm 45°, \pm 45°, \pm 45°)_{sym}$	12	1.63
CS22	Graphite-Epoxy HT-T300 (Torey)	$(\pm 45°, \pm 45°, \pm 45°)_{sym}$	12	1.63
CM32	Graphite-Epoxy HT-T300 (Torey)	$(0°, \pm 45°, 90°, \pm 45°, 0°)$	9	1.125
KM32	Kevlar (Dupont)	$(0^0, \pm 45°, 90°, \pm 45°, 0°)$	9	1.125

as a function of the non-dimensional period ratio T_{imp}/T_b, where T_{imp} is the period of the impact as measured by the bonded strain gages and T_b is the lowest natural frequency of the plate.

As shown before, DLF less than one was measured for some values of non-dimensional period ratio, while for short periods, the DLF is higher than unity, reaching experimental values of up to 2. The properties of the specimens presented in Fig. 7.11 are summarized in Table 7.1, while the material properties are presented in Table 7.2 (using the data published in [19]). Table 7.3 presents additional data in the form of natural frequencies

Table 7.2. The material properties of the tested plates (from [19]).

Material	E_{11} [MPa]	E_{22} [MPa]	G_{12} [MPa]	ν_{12}	ν_{21}
Graphite-Epoxy HT-T300 (Torey)	122	8.55	3.88	0.32	0.022
Kevlar (Dupont)	70.8	5.5	2.05	0.34	0.026

Table 7.3. The experimental results- the tested plates (from [19]).

Specimen	AR $[a \times b]$ (mm^2)	f_{exp} [Hz]	$(\varepsilon_{cr})_{\text{st.}}$ [μs]
CS21	2 [150 × 300]	350	2600
CS22	2 [150 × 300]	250	2413
CM32*	1 [225 × 225]	393	90
KM32	1 [225 × 225]	186	400

*Nominal clamped boundary conditions.

and static buckling strains as obtained during the tests described in [19]. One should note that except plate CM32, which was on clamped boundary conditions all around the perimeter of the plate, all the other plates had simply supported boundary conditions along the four edges of the plate.

As the measured response of the plates was in the form of strains, measured by two back-to-back bonded strain gages, the compression strain ε_c and the bending strain ε_b were calculated[c] and the axial compression load and lateral out-of-plane deflection were replaced by the compression and bending strains, respectively.

The definition of the static load was performed using the modified Donell's approach, found to be the most appropriate for plates with initial geometric imperfections, w_0, (see a discussion in [45]), having the following expression:

$$\frac{\varepsilon_b}{\varepsilon_c} = \frac{\varepsilon_b + w_0}{(\varepsilon_{cr})_{st.} + a(\varepsilon_b^2 + 3\varepsilon_b w_0 + 2w_0^2)}. \tag{7.55}$$

Using experimental data in the form of ε_b vs. ε_c one can curve fit it using Eq. (7.53) to yield the static buckling strain $(\varepsilon_{cr})_{\text{st.}}$ together with the initial geometric imperfection w_0 and the constant a. Based on the work performed in [16], the dynamic buckling strain $(\varepsilon_{cr})_{\text{dyn.}}$ is determined by curve fitting the following equation to the experimental data (ε_b vs. ε_c)

$$\frac{\varepsilon_b}{\varepsilon_c} = \frac{\varepsilon_b + w_0}{(\varepsilon_{cr})_{\text{dyn.}} + a\varepsilon_b} \tag{7.56}$$

[c] $\varepsilon_c \equiv \frac{\varepsilon_1+\varepsilon_2}{2}, \varepsilon_b \equiv \frac{\varepsilon_1-\varepsilon_2}{2}$, where $\varepsilon_1, \varepsilon_2$ are the measured strains.

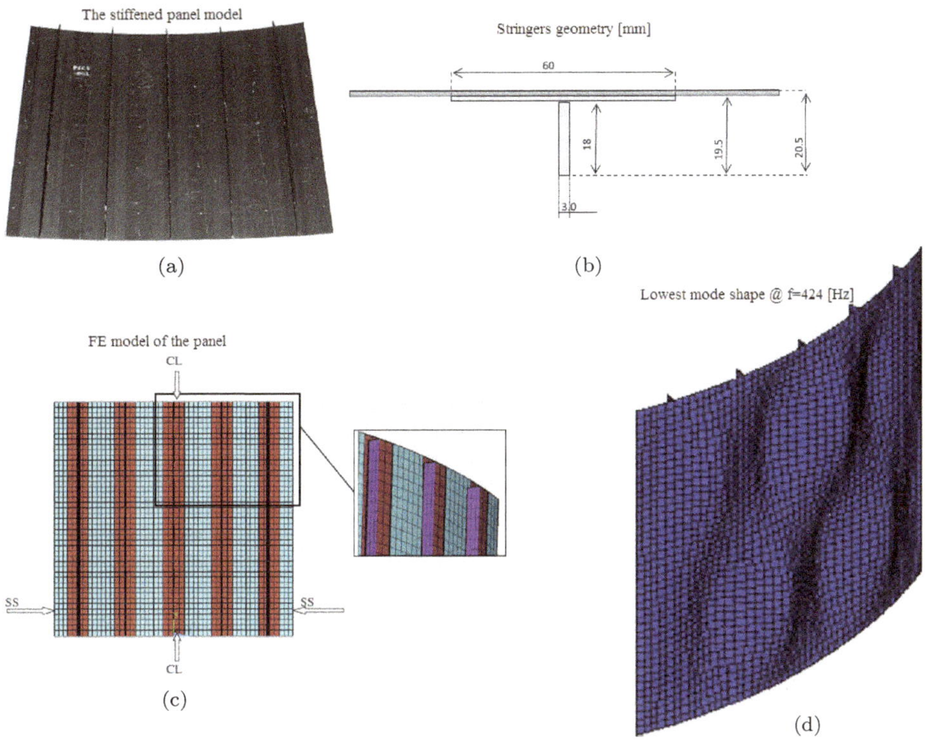

Fig. 7.12. Dynamic buckling investigation of a curved laminated composite stringer stiffened panel: (a) the stringered panel model; (b) geometric dimensions of the stringer; (c) the FE model; (d) the lowest mode shape of the stringered panel @ $f = 424$ [Hz] (adapted from [38]).

As for the static case, the two terms, the constant a and the initial geometric imperfection w_0, are also determined from the curve fitting process, yielding consistent results as presented in [19].

A dynamic buckling investigation was numerically performed in [38] on a curved laminated composite stringer stiffened panel. Details of the actual model, the dimensions of the stringer, the FE model and the mode shape at $f = 424$ [Hz] (the lowest frequency of the panel) are presented in Fig. 7.12.

The variation of the DLF (defined by Eq. (7.1)) vs. the non-dimensional load duration is shown in Fig. 7.13, where T is the period of the applied load and $T_b = 1/f_b = 2358.5$ [μsec], is the lowest natural period of the stringered panel ($f_b = 424$ [Hz]). As shown before for other structural cases, also for this stringered panel, the DLF is lower than unity in the vicinity of the lowest natural frequency of the specimen, returning to values above unity for very short periods, while for long periods of time, the DLF tends to unity.

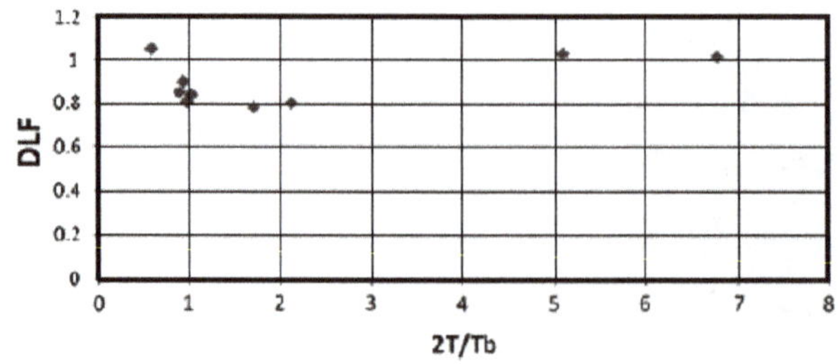

Fig. 7.13. Variation of DLF with the non-dimensional load duration — experimental results (adapted from [38]).

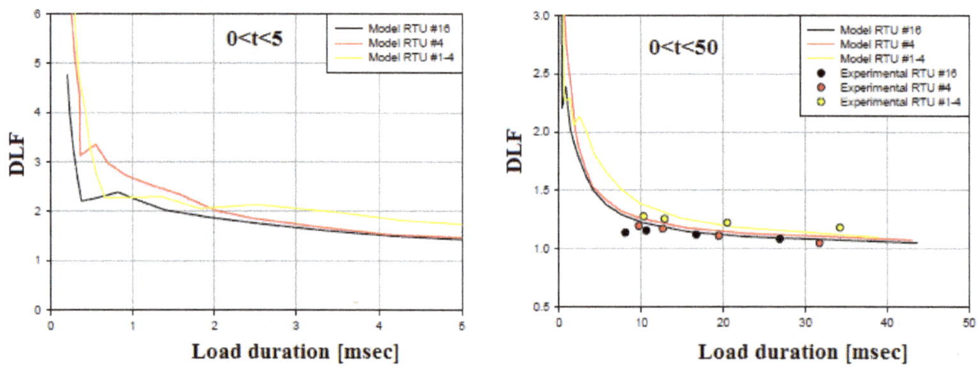

Fig. 7.14. Typical results from [37]: Variation of DLF with load duration — experimental and numerical results of specimens Type 1 (adapted from [37]).

Investigations of shells under dynamic type loading have been performed by many investigators, like in [20, 22, 25, 26, 29, 30, 37, 46, 47]. A comprehensive experimental and numerical investigation was performed in the Ph.D. thesis written by Eglis [37], in which composite cylinders were subjected to gradually and suddenly applied axial compression loads, including half-sine shaped pulse loads leading to the eventually dynamic buckling of the tested specimens. Figure 7.14 shows typical numerical and experimental results for the variation of the DLF with the loading period. The general trend presented in Fig. 7.14 is that the DLF is above unity for the tests performed on Type 1 shells [37]. The mode shapes for loading periods less, equal and above the lowest natural frequency (the largest period) are presented in Fig. 7.15. At $T < \tau/2$, the shell buckling modes are similar to the static buckling mode, but with a larger number of longitudinal waves, at load duration $T = \tau/2$ the buckling mode starts to transform to an axisymmetric mode while for $T > \tau/2$ leads to axisymmetric buckling mode together with a slight drop of DLF at a load duration of $T = 2a/c$ (a — shell length,

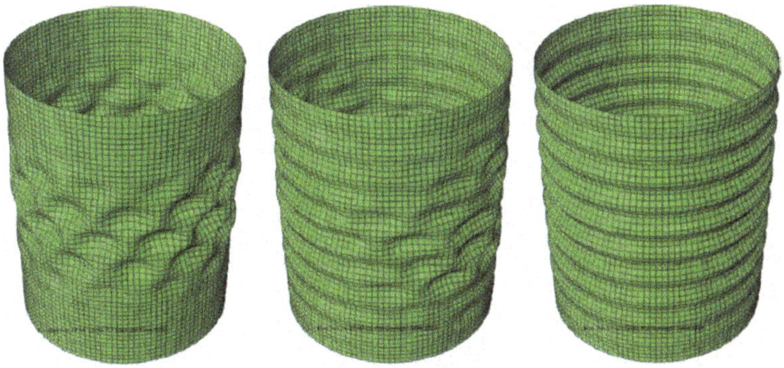

Fig. 7.15. Typical results from [37]: Variation of dynamic buckling mode shapes (not to scale) with load duration (τ represents the longest natural bending period of the shell) — numerical results of specimens Type 1 (adapted from [37]).

c — speed of sound in the shell). The natural periods for RTU #16, RTU #4 and RTU #1–4 tested shells (see Fig. 8.14) were τ = 2.92 [msec], 4.04 [msec] and 6.06 [msec], respectively (see [37]).

Similar results are presented in [46], where numerical and experimental investigations of a composite laminated cylindrical shell under impulsive loading are presented. The tested shell (see Fig. 7.16(a)) had a laminate of $[0°, -45°, +45°]$, was nominally clamped at its edges, with a length of 230 [mm] (between the clamped edges), a diameter of 250 [mm] and manufactured from graphite-epoxy with the following properties: E_{11} = 137.0 [GPa], E_{22} = 9.81 [GPa], G_{12} = 5.886 [GPa], and ν_{12} = 0.34, and 0.125 [mm] layer thickness. The dynamic loading was calculating using a dropping mass on the end plate with a mass of 32 [kg]. The calculated shell static mode shape is presented in Fig. 7.16(b), while the dynamic response for 746 [μsec] and 2170 [μsec] impact duration are shown in Figs. 7.16(c) and 7.16(d), respectively. As expected, the static and the dynamic modes have different patterns. The calculated DLF for numerical and experimental results is presented in Fig. 7.17 for various initial geometric imperfections. Although the experimental results do not go below DLF = 1, the numerical results show a DLF lower than unity in the vicinity of $T/T_b = 1$, for zero initial geometric imperfections, while for higher values of geometric imperfections, the DLF curve is higher the unity.

A recent research [47] presents similar results to those described in [46], for the behavior of composite cylindrical shells having laminates of $[0°/90°/0°/90°/90°/0°/90°/0°]$ and $[0°/0°/60°/-60°/-60°/60°/0°/0°]$ under various durations of load impulse. A step pulse was used for the compressive loading,

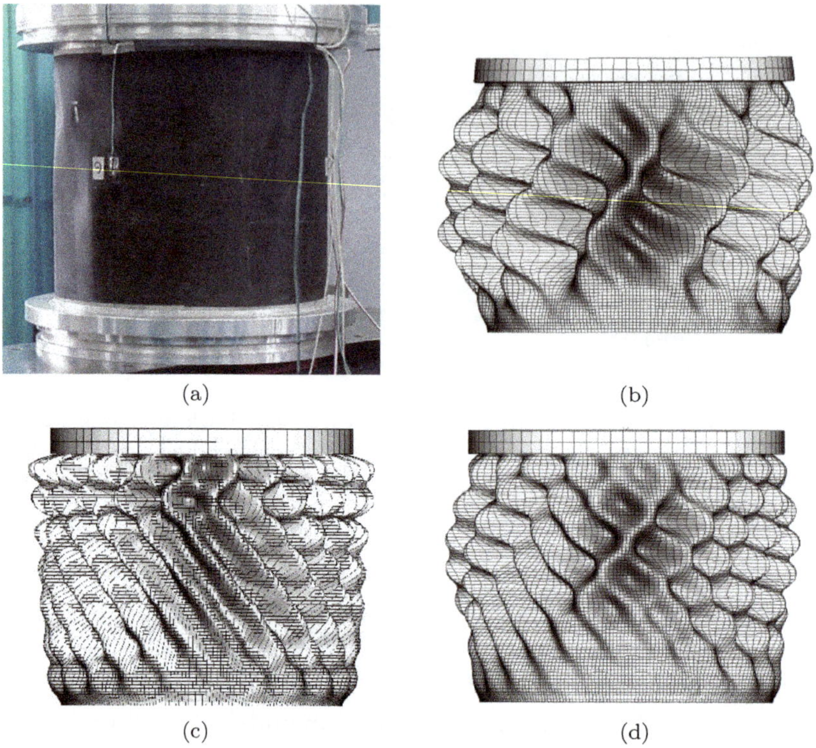

Fig. 7.16. Typical results from [46]: (a) the tested cylindrical shell; (b) the static mode shape; (c) dynamic mode shape @ 746 [μsec]; and (d) dynamic mode shape @ 2170 [μsec] (adapted from [46]).

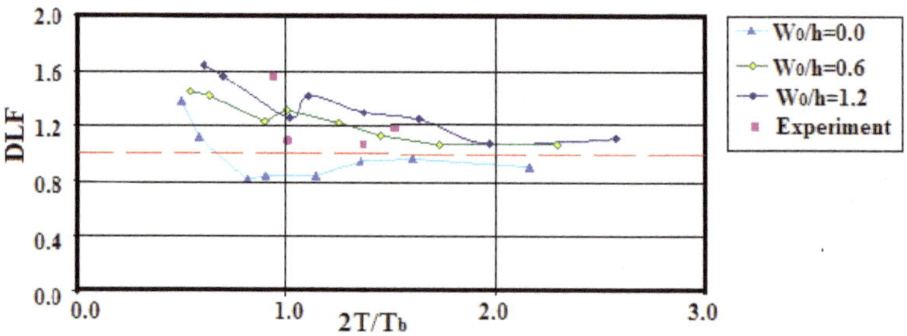

Fig. 7.17. DLF vs. non-dimensional impulse period ($T_b = 1/f_b$, where f_b is the lowest natural frequency of the shell) (adapted from [46]).

while calculations using the ABAQUS/Standard code yielded a lowest natural frequency of 427 [Hz] and a static linear buckling load of 97.19 [kN] for the [0°/90°/0°/90° /90°/0°/90°/0] laminate and 120.39 [kN] for the [0°/0° /60°/−60°/−60°/60°/0°/0°] laminate. ABAQUS/Explicit [48] was used for the dynamic buckling analysis of the shells under the impulsive loading.

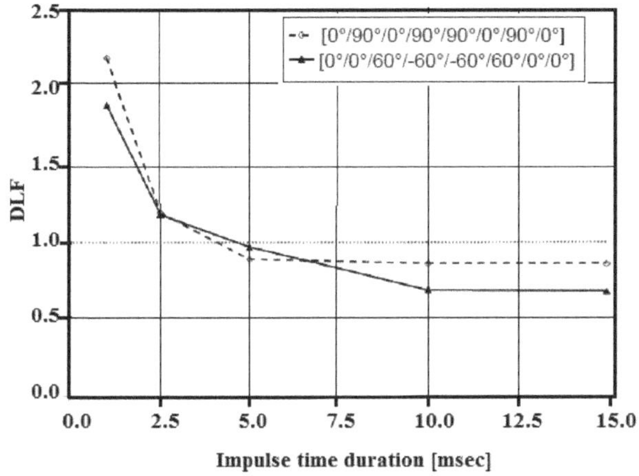

Fig. 7.18. DLF vs. non-dimensional impulse period ($T_b = 1/f_b$, where f_b is the lowest natural frequency of the shell) (adapted from [47]).

Figure 7.18 presents the DLF, calculated using Eq. (7.1), as a function of the period of the applied step impulse loading. For short periods, the DLF is higher than unity, while from $T = 5$ [msec] it drops below unity for both laminates. As the largest natural period can be calculated to 2.342 [msec], it again shows that in the vicinity of the lowest natural frequency the DLF might fall below unity also for laminated composite cylindrical shells.

A recent publication [49] investigates the DLF for various types of impulse loads (Fig. 7.19) applied axially to laminated composite cylindrical shells. The numerical study is using ABAQUS/Explicit [48] FE code for the dynamic calculations. The dynamic buckling of the shell was consistently determined using the following method: A relatively long period of $t = 60\text{T}$, where $T = 3.4\,\text{msec}$ was investigated to observe the shell response to the pulse loading, as the shell might vibrate and not buckle. The shell vertical displacement u_z was monitored. If the edge of the shell returns to its unloaded position ($u_z = 0$ at $t = 60\,\text{T}$) or oscillates around this position ($u_{z\text{-}ave} = 0$, where the term $u_{z\text{-}ave}$ is the average displacement of the vertical displacement at $t = 60\,\text{T}$) then the shell is stable. When the edge of the shell does not return to its unloaded initial position, $u_z \neq 0$ or it oscillates around a point which is not the initial position, $u_{z\text{-}ave} \neq 0$, then the shell is considered as buckled. The highest experienced load which leads to this response will be the dynamic buckling load of the shell. This criterion is displayed in Fig. 7.20. Two consecutive axial compression were applied, $P_{01} = 340\,\text{kN}$ and $P_{02} = 360\,\text{kN}$ for a trapezoidal time shape (Fig. 7.19). The response of the axial end shortening was drawn for $0 < t < 60\,\text{T}$, where

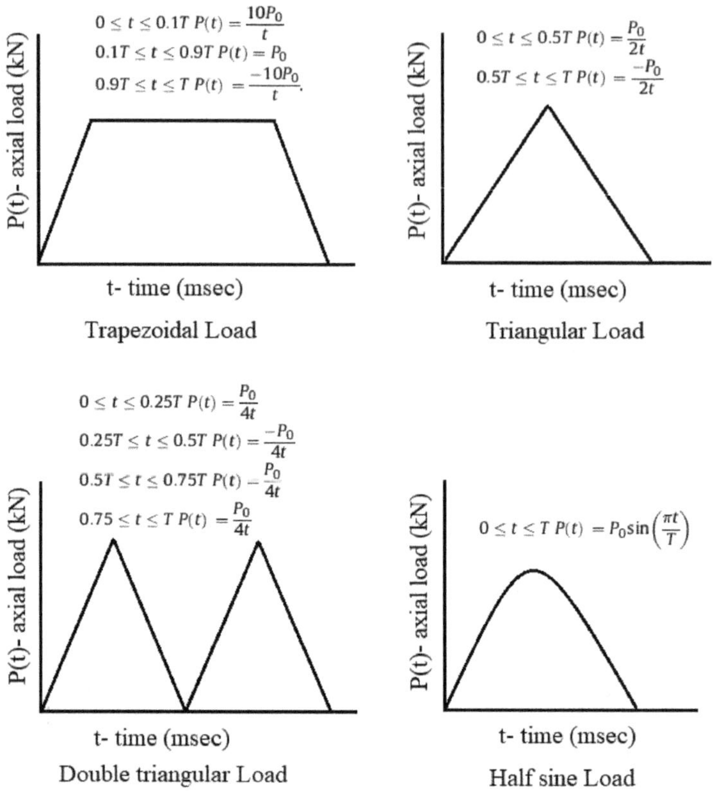

Fig. 7.19. Shapes of the impulse loads at their associated equations investigated in [48].

$T = T_b = 3.4038$ msec (T_b the lowest natural period of the shell). While for the first load ($P_{01} = 340$ kN) the shell response shows oscillations around the zero axial end shortening, for the second load ($P_{01} = 360$ kN) the average displacement is negative (Fig. 7.20(c), $u_{z\text{-}ave} = -0.07$ mm), meaning that the shell buckled dynamically. The influence of the initial geometric imperfection (taken as the static buckling shape of the shell multiplied by a factor less then unity) was investigated for the four time shapes and the dynamic buckling load was determined according to the above described criterion.

Typical results are presented in Figs. 7.21, 7.22 and Tables 7.4 and 7.5.

Note that increasing the initial geometric imperfections lowers the increases the value of DLF, as depicted in Table 7.4. From Table 7.5, one can see that only for a trapezoidal impulse shape, the DLF < 1.

To sum up this chapter, one draws the following conclusions:

(1) The dynamic buckling load was defined yielding a DLF index.
(2) A DLF < 1 was experienced for columns and plates in the vicinity of their first natural frequency.

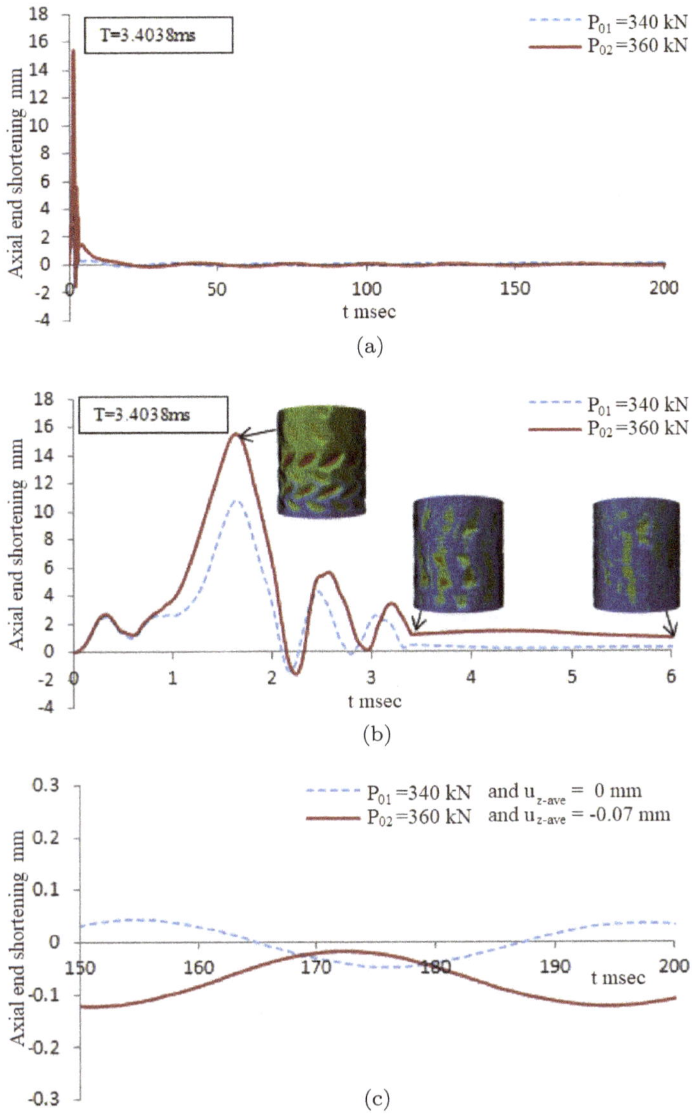

Fig. 7.20. Axial displacement for $T/T_b = 1$ ($T_b = 3.4038\,\text{msec}$): (a) shell response for $0 < t < 60\,\text{T}$; (b) shell response for $0 < t < 2\text{T}$; and (c) shell response for $50\,\text{T} < t < 60\,\text{T}$.

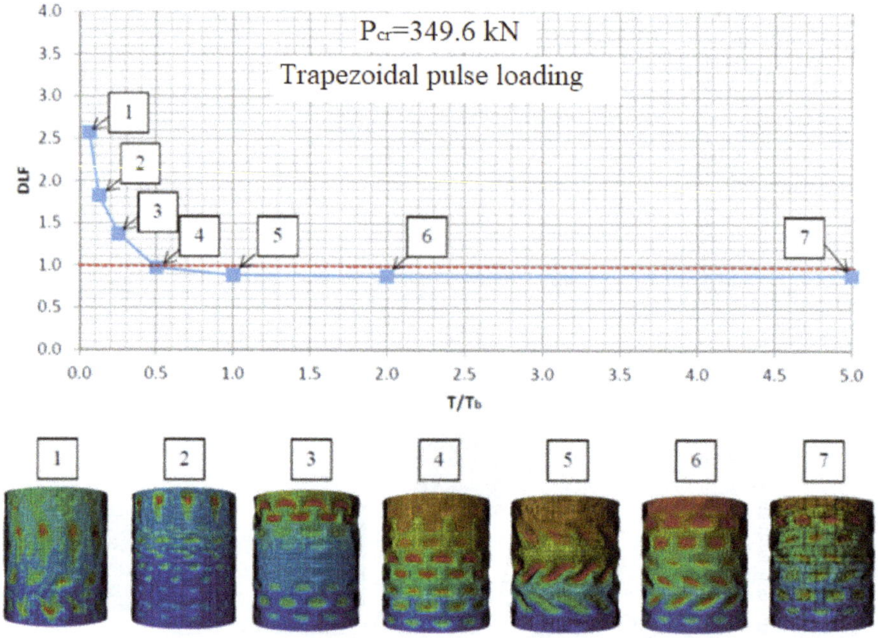

Fig. 7.21. DLF (Dynamic Load Factor) vs. load duration with an initial geometric imperfection $w_0 = 0.1t$.

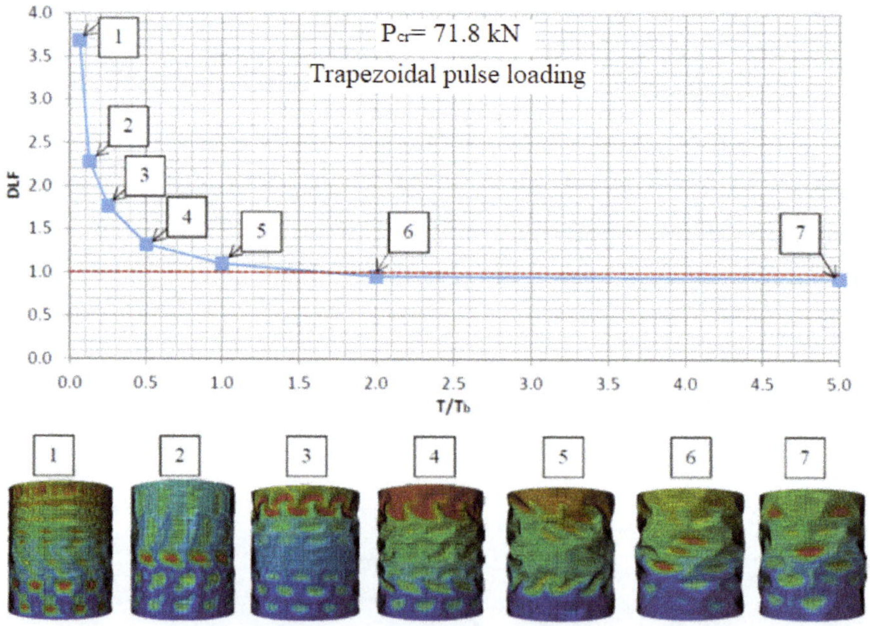

Fig. 7.22. DLF (Dynamic Load Factor) vs. load duration with an initial geometric imperfection $w_0 = 0.3t$.

Table 7.4. Dynamic buckling loads and their respective DLF for various load durations (trapezoidal time pulse) and initial geometric imperfections, w_0.

Initial Geometric Imperfection w_0	T/T_b	0.0625	0.125	0.25	0.5	1.0	2.0	5.0
$0.1t$[a]	DLF[b]	2.57	1.83	1.37	0.97	0.89	0.87	0.89
	Dynamic buckling load, kN	900	640	480	340	310	305	310
$0.2t$	DLF[c]	3.34	2.00	1.67	1.13	1.13	0.95	0.93
	Dynamic buckling load, kN	1000	600	500	340	340	285	280
$0.3t$	DLF[d]	3.68	2.28	1.77	1.33	1.11	0.96	0.94
	Dynamic buckling load, kN	1000	620	480	360	300	260	255

Note: [a]t = shell thickness = 1.448 mm; [b]static buckling load = 349.6 kN; [c]Static buckling load = 300.0 kN; [d]Static buckling load = 71.9 kN.

Table 7.5. Dynamic buckling loads and their respective DLF for various types of impulse shape at an initial geometric imperfections, $w_0 = 2t$ (t = shell thickness).

Type of impulse shape	T/T_b	0.0625	0.125	0.25	0.5	1.0	2.0	5.0
Trapezoidal	DLF[a]	3.34	2.00	1.67	1.13	1.13	0.95	0.93
	Dynamic buckling load, kN	1000	600	500	340	340	285	280
Triangular	DLF[a]	13.30	9.00	5.70	3.40	2.50	1.90	1.30
	Dynamic buckling load, kN	4000	2700	1700	1020	740	560	000
Double triangular	DLF[a]	13.70	9.30	5.70	4.00	2.50	1.70	1.30
	Dynamic buckling load, kN	4100	2800	1700	1200	760	520	400
Half sine	DLF[a]	10.00	6.00	4.30	2.80	2.10	1.50	1.20
	Dynamic buckling load, kN	3000	1800	1300	840	620	460	370

Note: [a]Static buckling load = 300.0 kN, Double triangular.

(3) A DLF < 1 was also found for cylindrical shells for a trapezoidal impulse loading.
(4) In view of those results, one should be aware that in some circumstances the dynamic buckling load might be less than the statical one, leading to the conclusion that not always a design for a thin-walled structure capable of withstanding static loads would cover also a dynamic loading.

References

[1] Budiansky, B. and Roth, R.S., Axisymmetric dynamic buckling of clamped shallow spherical shells, Collected Papers on Instability of Shell Structures, NASA TN-D-1510, 1962, pp. 597–606.
[2] Hegglin, B., Dynamic buckling of columns, SUDAER No. 129, June 1962, Department of Aeronautics & Astronautics, Stanford University, Stanford, CA, 55 p.
[3] Budiansky, B. and Hutchinson, J.W., Dynamic buckling of imperfection sensitive structures", *Proc. 11th Int. Congr. Applied Mechanics*, pp. 636–651, Berlin, 1964. H. Götler, ed., Springer-Verlag, 1966.
[4] Hutchinson, W.J. and Budiansky, B., Dynamic buckling estimates, *AIAA J.*, Vol. 4(3), 1966, pp. 525–530.
[5] Lock, M.H., A study of buckling and snapping under dynamic load, Air Force Report No. SAMSO-TR-68-100, Aerospace Report No. TR-0158(3240-30)-3, December 1967, 55 p.
[6] Burt, J.A., Dynamic buckling of shallow spherical caps subjected to a nearly axisymmetric step pressure load, Master Thesis, Naval Postgraduate School, Monterey, CA, September 1971, 75 p.
[7] Volmir, S.A., *Nonlinear Dynamics of Plates and Shells*, Science, Moscow, USSR, 1972.
[8] Ekstrom, R.E., Dynamic buckling of a rectangular orthotropic plate, *AIAA J.*, Vol. 11(12), 1973, pp. 1655–1659.
[9] Lee, L.H.N., Dynamic buckling of an inelastic column, *Int. J. Solids Struct.*, Vol. 17(3), 1981, pp. 271–279.
[10] Ari-Gur, J., Weller, T. and Singer J., Experimental and theoretical studies of columns under axial impact, *Int. J. Solids Struct.*, Vol. 18(7), 1982, pp, 619–641.
[11] Lindberg, H.E., Dynamic pulse buckling-theory and experiment, SRI International, DNA 6503H, Handbook, Menlo Park, CA, 1983.
[12] Lee, L.H.N. and Ettestad, K.L., Dynamic buckling of an ice strip by axial impact, *Int. J. Impact Eng.*, Vol. 1(4), 1983, pp. 343–356.
[13] Gary, G., Dynamic buckling of an elastoplastic column, *Int. J. Impact Eng.*, Vol. 1(4), 1983, pp. 357–375.
[14] Dannawi, M. and Adly. M., Constitutive equation, quasi-static and dynamic buckling of 2024-t3 plates — experimental result and analytical modelling, *J. Phys. Colloques*, Vol. 49(C3), 1988, pp. C3-575–C3-588.
[15] Birman, V., Problems of dynamic buckling of antisymmetric rectangular laminates, *Composite Struct.*, Vol. 12(1), 1989, pp. 1–15.
[16] Weller, T., Abramovich, H. and Yaffe, R., Dynamic buckling of beams and plates subjected to axial impact, *Computers Struct.*, Vol. 32(3–4), 1989, pp. 835–851.
[17] Simitses, G.J., *Dynamic Stability of Suddenly Loaded Structures*, Springer, New York, 290 p.

[18] Ari-Gur, J. and Hunt, D.H., Effects of anisotropy on the pulse response of composite panels, *Composite Eng. (now Composites Part B)*, Vol. 1(5), 1991, pp. 309–317.
[19] Abramovich, H. and Grunwald, A., Stability of axially impacted composite plates, *Composite Struct.*, Vol. 32(1–4), 1995, pp. 151–158.
[20] Schokker, A., Sridharan, S. and Kasagi, A., Dynamic buckling of composite shells, *Computers Struct.*, Vol. 59(1), 1996, pp. 43–53.
[21] Ari-Gur, J. and Simonetta, R., Dynamic pulse buckling of rectangular composite plates, *Composites Part B*, Vol. 28, 1997, pp. 301–308.
[22] Eslami, M.R., Shariyat, M. and Shakeri, M., Layerwise theory for dynamic buckling and postbuckling of laminated composite cylindrical shells, *AIAA J.*, Vol. 36(10), 1998, pp. 1874–1882.
[23] Petry, D. and Fahlbusch, G., Dynamic buckling of thin isotropic plates subjected to in-plane impact, *Thin-Walled Struct.*, Vol. 38(3), 2000, pp. 267–283.
[24] Zheng, Z. and Farid, T., Numerical studies on dynamic pulse buckling composite laminated beams subjected to an axial impact pulse, *Composite Struct.*, Vol. 56(3), 2002, pp. 269–277.
[25] Yaffe, R. and Abramovich, H., Dynamic buckling of cylindrical stringer stiffened shells, *Computers Struct.*, Vol. 81(9–11), 2003, pp. 1031–1039.
[26] Chamis, C.C. and Abumeri, G.H., Probabilistic dynamic buckling of smart composite shells, NASA/TM-2003-212710, 2003.
[27] Zhang, Z., Investigation on dynamic pulse buckling and damage behavior of composite laminated beams subject to axial pulse, Ph.D. thesis, Faculty of Engineering, Civil Engineering, Dalhousie University, Halifax, Nova Scotia, Canada, 2003, 228 p.
[28] Lindberg, H.E., *Little Book of Dynamic Buckling*, LCE Science/Software, September 2003.
[29] Bisagni, C., Dynamic buckling tests of cylindrical shells in composite materials, *24th Int. Congr. Aeronautical Sciences* (ICAS2004), 29 August–3 September, 2004, Yokohama, Japan.
[30] Bisagni, C., Dynamic buckling of fiber composite shells under impulsive axial compression, *Thin-Walled Struct.*, 43(3), 2005, pp. 499–514.
[31] Kubiak, T., Dynamic buckling of thin-walled composite plates with varying widthwise material properties, *Int. J. Solids Struct.*, Vol. 42, 2005, pp. 5555–5567.
[32] Ji, W. and Waas, A.M., Dynamic bifurcation buckling of an impacted column, *International J. Eng. Sci.*, Vol. 46(10), 2008, pp. 958–967.
[33] Kubiak, T., Dynamic buckling estimation for beam-columns with open cross-sections, Paper No. 13, *Proc. Twelfth Int. Conf. Civil, Structural and Environmental Engineering Computing*, B.H.V. Topping, L.F. Costa Neves and R.C. Barros, (Editors), 2009, Civil-Comp Press, Stirlingshire, Scotland.
[34] Jabareen, M. and Sheinman, I., Dynamic buckling of a beam on a nonlinear elastic foundation under step loading, *J. Mech. Mater. Struct.*, Vol. 4(7–8), 2009, pp. 1365–1373.
[35] Michalska, K.K., About some important parameters in dynamic buckling analysis of plated structures subjected to pulse loading, *Mech. Mech. Eng.*, Vol. 14(2), 2010, pp. 269–279.
[36] Mania, R.J., Membrane-flexural coupling effect in dynamic buckling of laminated columns, *Mech. Mech. Eng.*, Vol. 14(1), 2010, pp. 137–150.
[37] Eglitis, E., Dynamic buckling of composite shells, Ph.D. thesis, Riga Technical University, Faculty of Civil Engineering, Institute of Materials and Structures, 2011, Riga, Latvia, 172 p.
[38] Abramovich, H. and Less, H., Dynamic buckling of a laminated composite stringer-stiffened cylindrical panel", *Composite Part B*, Vol. 43(5), 2012, pp. 2348–2358.

[39] Chung, M., Lee, H.J., Kang, Y.C., Lim, W.-B., Kim, J. H., Cho, J.Y., Byun, W. Kim, S.J. and Park, S.-H., Experimental study on dynamic buckling phenomena for supercavitating underwater vehicle, *Int. J. Naval Arch. Ocean Eng.*, Vol. 4, 2012, pp. 183–198, http://dx.doi.org/10.2478/IJNAOE-2013-0089.
[40] Landa, A., The buckling resistance of structures subjected to impulsive type actions, Master Thesis, Norwegian University of Science and Technology, Department of Marine Technology NTNU Trondheim, Norway, February 2014, 104 p.
[41] Straume, J.G., Dynamic buckling of marine structures, Master Thesis, Norwegian University of Science and Technology, Department of Marine Technology NTNU Trondheim, Norway, June 2014, 148 p.
[42] Kuzkin, V.A., Structural model for the dynamic buckling of a column under constant rate compression, preprint, 2015, arXiv:1506.00427 [physics.class-ph].
[43] Mouhata, O. and Abdellatif, K., Dynamic buckling of stiffened panels, *Procedia Eng.*, Vol. 125, 2015, pp. 1001–1007.
[44] Lekhnitskii, S.G., *Anisotropic Plates*, 2nd Edition, Gordon and Breach, New York, 1968; translation from Russian.
[45] Abramovich, H., Weller, T. and Yaffe R., Application of a modified Donnell technique for the determination of critical loads of imperfect plates, *Comput. Struct.*, Vol. 37, 1990, pp. 463–469.
[46] Abramovich, H., Weller, T. and Pevzner, P., Dynamic buckling behavior of thin-walled composite circular cylindrical shells under axial impulsive loading, *Proc. AIAC-12, Twelfth Australian Int. Aerospace Congr.*, Melbourne, 19–22 March 2007
[47] Citra, V. and Priyadarsini, R.S., Dynamic buckling of composite cylindrical shells subjected to axial impulse, *Int. J. Scientific Eng. Res.*, Vol. 4(5), 2013, pp. 162–165.
[48] *ABAQUS/Explicit User's Manual*, Version 2017, ABAQUS Inc., USA, 2017.
[49] Zaczynska, M., Abramovich, H. and Bisagni, C., Parametric studies on the dynamic buckling phenomenon of a composite cylindrical shell under impulsive axial compression, *J. Sound Vibration*, Vol. 482, 2020, Paper Id: 115462, 17 p.

Appendix A: Calculation of the Critical Buckling Load of a Uniaxial Loaded Plate from Test Results

One of the issues facing a researcher is to correctly define the buckling load of a plate for the static and dynamic cases, based on experimental points obtained either by tests or numerically. The following method was already published and applied in [38] and is based on the work performed by Brown.[d] It was shown that for practical structural geometries and loading conditions the following equation holds:

$$\delta_i^2 - \delta_0^2 = \alpha^2 h^2 \theta = \alpha^2 h^2 \left[\frac{P_i}{P_{cr}} - 1 + \frac{\delta_0}{\delta_i} \right], \tag{A.1}$$

[d] Brown, V.L., Linearized least-squared technique for evaluating plate-buckling loads, *J. Eng. Mech.*, Vol. 116(5), 1990, pp. 1050–1057.

where δ_i is the lateral plate deflection due to corresponding in-plane loading P_i; δ_0 is initial the lateral plate deflection; h is the thickness of the plate and α is a constant accounting for the load configuration and boundary conditions. After some mathematical manipulations, Eq. (A.1) can be rewritten as

$$P_i\delta_i = A_1 + A_2\delta_i + A_3\delta_i^3, \tag{A.2}$$

where

$$A_1 = -P_{cr}\delta_0, \quad A_2 = P_{cr}\left[1 - \frac{\delta_0^2}{\alpha^2 h^2}\right], \quad A_3 = \frac{P_{cr}}{\alpha^2 h^2}. \tag{A.3}$$

Equation (A.3) is presented in a suitable form for the application of the Least Square Method to fit a curve to a series of given data points, P_i, δ_i. To do so, let us define the sum of the squares of the residuals for m data points as

$$\text{SUM} = \sum_{i=1}^{m} \left(P_i\delta_i - A_1 - A_2\delta_i - A_3\delta_i^3\right)^2. \tag{A.4}$$

Taking partial derivatives of Eq. (A.4) with respect to the coefficients A_1, A_2 and A_3 (see Eq. (A.3)) and equating them to zero (thus minimizing the error), yields the following system of equations written in a matrix form:

$$\begin{bmatrix} m & \sum_{i=1}^{m}\delta_i & \sum_{i=1}^{m}\delta_i^3 \\ \sum_{i=1}^{m}\delta_i & \sum_{i=1}^{m}\delta_i^2 & \sum_{i=1}^{m}\delta_i^4 \\ \sum_{i=1}^{m}\delta_i^3 & \sum_{i=1}^{m}\delta_i^4 & \sum_{i=1}^{m}\delta_i^6 \end{bmatrix} \begin{Bmatrix} A_1 \\ A_2 \\ A_3 \end{Bmatrix} = \begin{Bmatrix} \sum_{i=1}^{m} P_i\delta_i \\ \sum_{i=1}^{m} P_i\delta_i^2 \\ \sum_{i=1}^{m} P_i\delta_i^4 \end{Bmatrix}. \tag{A.5}$$

The three linear equations, presented in a matrix form in Eq. (A.5), are then solved for the three unknowns A_1, A_2 and A_3. Substituting them into Eq. (A.3) and the results in Eq. (A.2) leads the following cubic equation for the single unknown P_{cr}:

$$P_{cr}^3 - A_2 P_{cr}^2 - A_1^2 A_3 = 0. \tag{A.6}$$

Solving Eq. (A.6) will yield the buckling load of the plate based on the displacement–force curve, generated either by experiments or by numerical calculations.

Chapter 8

Introduction to Piezoelectricity

8.1. Introduction

Without doubt, it can be stated that the current research field carrying the name *Piezoelectricity*, was initiated and started by the French Curie brothers, Jacques Curie (1856–1941) and Pierre Curie (1859–1906) with their seminal studies [1, 2]. They discovered peculiar characteristics for some crystalline minerals like quartz, tourmaline, topaz, cane sugar and Rochelle salt. They found that the application of tensile and compression stresses would generate voltages of opposite polarity and proportional to the applied stresses. This phenomenon is called the *direct piezoelectric effect*, which is schematically presented in Figs. 8.1(a) and 8.1(b). The source of the word piezoelectricity comes from Greek and means electricity resulting from pressure (*piesi* means pressure in Greek). In the year following the discovery of the direct effect, 1881, another famous French-Luxembourgish physicist and inventor Gabriel Lippman (1845–1921) [3] predicted the existence of the *piezoelectric converse effect*, which is based on fundamental thermodynamic principles. By the end of 1881, the Curie brothers experimentally confirmed the existence of the piezoelectric converse effect. They found that using one of the voltage-generating crystals, and exposing it to an applied electric field, it would lengthened or shortened according to the polarity of the field, and this elongation/shortening is in proportion to electrical field strength, as being schematically shown in Figs. 8.2(a) and 8.2(b).

As stated above, piezoelectric materials exhibit two types of piezoelectric effects: the direct and the converse ones. The direct piezoelectric effect defines the piezoelectric material's capacity to transform the mechanical energy in the form of applied stress (or strain) into electrical energy while the converse effect describes its capacity to transform the applied electrical voltage into

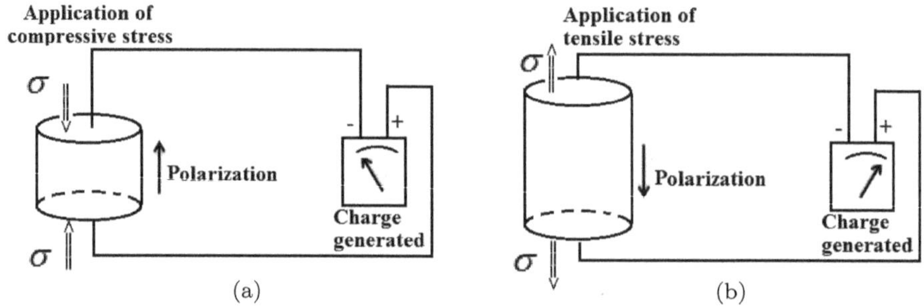

Fig. 8.1. The schematic direct piezoelectric effect: (a) at applied compressive stress and (b) at applied tensile stress.

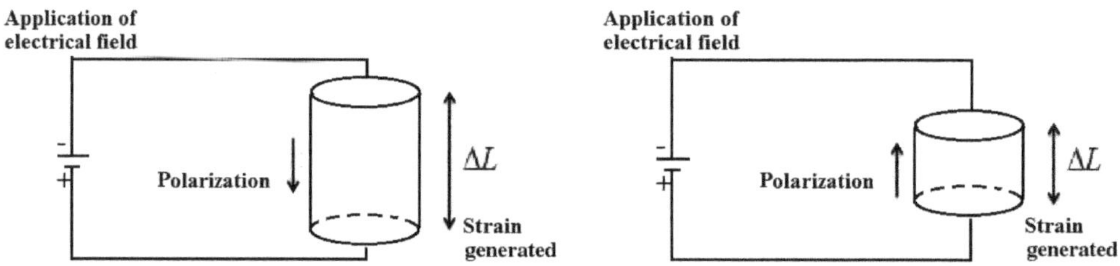

Fig. 8.2. The schematic converse piezoelectric effect: (a) at applied electrical field in the polarization direction; (b) at applied electrical field opposite the polarization direction.

mechanical energy in the form of strain. Therefore, the direct piezoelectric effect is responsible for the material's ability to function as a sensor and the converse piezoelectric effect is accountable for its ability to function as an actuator. Based on these properties, the piezoelectric material can be employed both as a sensor (using the direct piezoelectric effect) and as an actuator (using the converse piezoelectric effect). The coupled electromechanical behavior of piezoelectric materials can be derived and presented by two constitutive equations (see [4–6, 64, 66]). The direct piezoelectric effect has the following form:

$$D_i = d_{im}T_m + \widetilde{\varepsilon}_{ij}^{T}E_j, \tag{8.1}$$

while the converse piezoelectric effect has the following equation:

$$\varepsilon_k = S_{km}^{E}T_m + d_{jk}E_j, \tag{8.2}$$

where D_i represents the dielectric displacement in N*m/V or in C/m^2, ε_k is the strain vector, E_j is the applied electrical field vector in V/m, and T_m is the stress vector in N/m^2. d_{im} and d_{jk} are the piezoelectric coefficients in m/V or in C/N, $\tilde{\varepsilon}_{ij}^T$ is the dielectric permittivity at constant stress in N/V^2 or in F/m, and S_{km}^E is the elastic compliance matrix at constant applied electrical field in m^2/N. The compact form of Eqs. (8.1) and (8.2) can be expanded to obtain the following expressions

$$\begin{bmatrix} D_1 \\ D_2 \\ D_3 \end{bmatrix} = \begin{bmatrix} d_{11} & d_{12} & d_{13} & d_{14} & d_{15} & d_{16} \\ d_{21} & d_{22} & d_{23} & d_{24} & d_{25} & d_{26} \\ d_{31} & d_{32} & d_{33} & d_{34} & d_{35} & d_{36} \end{bmatrix} \begin{bmatrix} T_1 \\ T_2 \\ T_3 \\ T_4 \\ T_5 \\ T_6 \end{bmatrix} + \begin{bmatrix} \tilde{\varepsilon}_{11}^T & \tilde{\varepsilon}_{12}^T & \tilde{\varepsilon}_{13}^T \\ \tilde{\varepsilon}_{21}^T & \tilde{\varepsilon}_{22}^T & \tilde{\varepsilon}_{23}^T \\ \tilde{\varepsilon}_{31}^T & \tilde{\varepsilon}_{32}^T & \tilde{\varepsilon}_{33}^T \end{bmatrix} \begin{bmatrix} E_1 \\ E_2 \\ E_3 \end{bmatrix}, \tag{8.3}$$

$$\begin{bmatrix} \varepsilon_1 \\ \varepsilon_2 \\ \varepsilon_3 \\ \varepsilon_4 \\ \varepsilon_5 \\ \varepsilon_6 \end{bmatrix} = \begin{bmatrix} S_{11}^E & S_{12}^E & S_{13}^E & S_{14}^E & S_{15}^E & S_{16}^E \\ S_{21}^E & S_{22}^E & S_{23}^E & S_{24}^E & S_{25}^E & S_{26}^E \\ S_{31}^E & S_{32}^E & S_{33}^E & S_{34}^E & S_{35}^E & S_{36}^E \\ S_{41}^E & S_{42}^E & S_{43}^E & S_{44}^E & S_{45}^E & S_{46}^E \\ S_{51}^E & S_{52}^E & S_{53}^E & S_{54}^E & S_{55}^E & S_{56}^E \\ S_{61}^E & S_{62}^E & S_{63}^E & S_{64}^E & S_{65}^E & S_{66}^E \end{bmatrix} \begin{bmatrix} T_1 \\ T_2 \\ T_3 \\ T_4 \\ T_5 \\ T_6 \end{bmatrix} + \begin{bmatrix} d_{11} & d_{21} & d_{31} \\ d_{12} & d_{22} & d_{32} \\ d_{13} & d_{23} & d_{33} \\ d_{14} & d_{24} & d_{34} \\ d_{15} & d_{25} & d_{35} \\ d_{16} & d_{26} & d_{36} \end{bmatrix} \begin{bmatrix} E_1 \\ E_2 \\ E_3 \end{bmatrix}, \tag{8.4}$$

where the stress and the strain vectors are defined as (using the Voight notation)

$$T_1 = \sigma_{11}, \quad T_2 = \sigma_{22}, \quad T_3 = \sigma_{33},$$
$$T_4 = \tau_{23} = \tau_{32}, \quad T_5 = \tau_{31} = \tau_{13}, \quad T_6 = \tau_{12} = \tau_{21}, \tag{8.5}$$
$$\varepsilon_1 = \varepsilon_{11}, \quad \varepsilon_2 = \varepsilon_{22}, \quad \varepsilon_3 = \varepsilon_{33},$$
$$\varepsilon_4 = \varepsilon_{23} = \varepsilon_{32}, \quad \varepsilon_5 = \varepsilon_{31} = \varepsilon_{13}, \quad \varepsilon_6 = \varepsilon_{12} = \varepsilon_{21}. \tag{8.6}$$

Although the matrices in Eqs. (8.3) and (8.4) are fully populated for the general case, in most common piezoelectric materials like PZT (Lead Zirconate Titanate — Pb $[Zr_xTi_{1-x}]O_3$ $(0 \leq x \leq 1)$), due to the assumption that its piezo-ceramic material is isotropic in a plane perpendicular to the poling direction, several terms in the various matrices, presented above would become equal to zero or equal to each other. Therefore, for the PZT Eqs. (8.3) and (8.4) would be written as

$$\begin{bmatrix} D_1 \\ D_2 \\ D_3 \end{bmatrix} = \begin{bmatrix} 0 & 0 & 0 & 0 & d_{15} & 0 \\ 0 & 0 & 0 & d_{24} & 0 & 0 \\ d_{31} & d_{31} & d_{33} & 0 & 0 & 0 \end{bmatrix} \begin{bmatrix} T_1 \\ T_2 \\ T_3 \\ T_4 \\ T_5 \\ T_6 \end{bmatrix} + \begin{bmatrix} \tilde{\varepsilon}^T_{11} & 0 & 0 \\ 0 & \tilde{\varepsilon}^T_{22} & 0 \\ 0 & 0 & \tilde{\varepsilon}^T_{33} \end{bmatrix} \begin{bmatrix} E_1 \\ E_2 \\ E_3 \end{bmatrix}, \tag{8.7}$$

$$\begin{bmatrix} \varepsilon_1 \\ \varepsilon_2 \\ \varepsilon_3 \\ \varepsilon_4 \\ \varepsilon_5 \\ \varepsilon_6 \end{bmatrix} = [S^E_{km}] \begin{bmatrix} T_1 \\ T_2 \\ T_3 \\ T_4 \\ T_5 \\ T_6 \end{bmatrix} + \begin{bmatrix} 0 & 0 & d_{31} \\ 0 & 0 & d_{31} \\ 0 & d_{23} & d_{33} \\ 0 & d_{24} & 0 \\ d_{15} & 0 & 0 \\ 0 & 0 & 0 \end{bmatrix} \begin{bmatrix} E_1 \\ E_2 \\ E_3 \end{bmatrix}, \tag{8.8}$$

where the compliance matrix at constant applied electrical field, S^E_{km}, is given by

$$[S^E_{km}] = \begin{bmatrix} \frac{1}{E_{11}} & -\frac{\upsilon_{12}}{E_{11}} & -\frac{\upsilon_{13}}{E_{11}} & 0 & 0 & 0 \\ -\frac{\upsilon_{12}}{E_{11}} & \frac{1}{E_{11}} & -\frac{\upsilon_{13}}{E_{11}} & 0 & 0 & 0 \\ -\frac{\upsilon_{13}}{E_{11}} & -\frac{\upsilon_{13}}{E_{11}} & \frac{1}{E_{33}} & 0 & 0 & 0 \\ 0 & 0 & 0 & \frac{2(1+\upsilon_{13})}{E_{33}} & 0 & 0 \\ 0 & 0 & 0 & 0 & \frac{2(1+\upsilon_{13})}{E_{33}} & 0 \\ 0 & 0 & 0 & 0 & 0 & \frac{2(1+\upsilon_{12})}{E_{11}} \end{bmatrix}, \tag{8.9}$$

and the stiffness matrix at constant applied electrical field, C^E_{km} is then defined as

$$[C^E_{km}] = \begin{bmatrix} \frac{1-\upsilon_{13}^2}{E_{11}E_{33}\Theta} & \frac{\upsilon_{12}+\upsilon_{13}^2}{E_{11}E_{33}\Theta} & \frac{\upsilon_{13}+\upsilon_{12}\upsilon_{13}}{E_{11}E_{33}\Theta} & 0 & 0 & 0 \\ \frac{\upsilon_{12}+\upsilon_{13}^2}{E_{11}E_{33}\Theta} & \frac{1-\upsilon_{13}^2}{E_{11}E_{33}\Theta} & \frac{\upsilon_{13}+\upsilon_{12}\upsilon_{13}}{E_{11}E_{33}\Theta} & 0 & 0 & 0 \\ \frac{\upsilon_{13}+\upsilon_{12}\upsilon_{13}}{E_{11}E_{33}\Theta} & \frac{\upsilon_{13}+\upsilon_{12}\upsilon_{13}}{E_{11}E_{33}\Theta} & \frac{1-\upsilon_{13}^2}{E_{11}^2\Theta} & 0 & 0 & 0 \\ 0 & 0 & 0 & \frac{E_{33}}{2(1+\upsilon_{13})} & 0 & 0 \\ 0 & 0 & 0 & 0 & \frac{E_{33}}{2(1+\upsilon_{13})} & 0 \\ 0 & 0 & 0 & 0 & 0 & \frac{E_{11}}{2(1+\upsilon_{12})} \end{bmatrix}, \tag{8.10}$$

Table 8.1. Typical properties of piezoelectric materials.

Property	$BaTiO_3$	PZT-4	PZT-5A	PZT-5H	PZT-8	PVDF	PMN-6%PT	PMN-33%PT
d_{31} (10^{-12}C/N)	−78	−123	−171	−275	−97	−23	−1400	−520
d_{33} (10^{-12}C/N)	149	289	374	593	225	33	2400	2200
d_{15} (10^{-12}C/N)		496	584	741	330			
k_{33}	0.48	0.70	0.71	0.75	0.64	0.15	0.90	0.93
Curie temp. (°C)	115	328	365	193	300	100	145	100

where

$$\Theta \equiv \frac{(1+v_{12})(1-v_{12}-2v_{13}^2)}{E_{11}^2 E_{33}}, \tag{8.11}$$

and v_{12}, v_{12} are the Poisson's ratios and E_{11}, E_{33} are the principal Young's moduli in the directions 1(planar direction) and 3 (thickness direction).

To conclude this introductory part, the properties of common piezoelectric materials are summed up in Table 8.1 (see also [66, 83]).

The constant k_{33} appearing in Table 8.1 is called the piezoelectric coupling constant, relating the electrical field and the mechanical strain, both in the thickness direction, 3. The Curie temperature is the temperature at which the piezoelectric material "looses" its polarity, and PMN-6%PT and PMN-33%PT are synthetic single crystals.

It is instructive and interesting to show the complexity of the piezoelectric phenomenon, by shortly describing the strain versus the electrical field in typical piezoelectric materials, as depicted in Fig. 8.3. The behavior is for a ceramic perovskite material, PZT, which is also a ferroelectric material. The strains in both longitudinal and transverse modes exhibit hysteresis effects. The strain increases with the applied electric field initially linearly and later gets saturated. The strain is of the order of about 10^{-3} for an applied electrical field in the range of 1–1.5 kV for PZT, which is the best-known piezoelectric ceramic, used widely for transducer applications. For other piezoelectric materials, it is much less. When the applied electrical field is gradually reduced, the strain does not follow the same curve. The change in strain lags behind the change in the applied electrical field. When the electrical field is zero, a remnant strain is seen in the material. If the electrical field is increased in the reverse direction, the strain becomes zero at a particular negative electrical field. If the electrical field is now increased

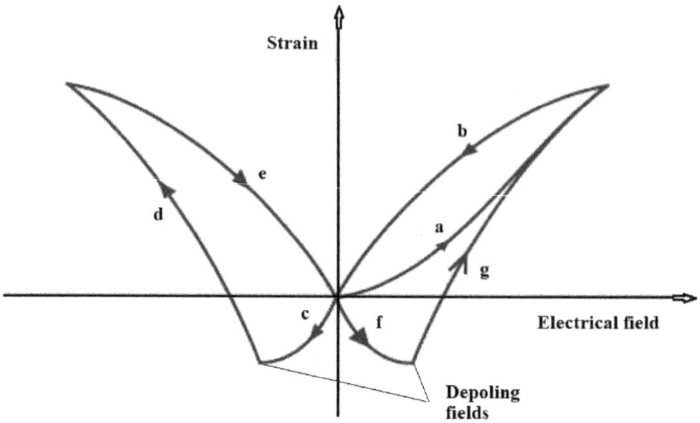

Fig. 8.3. Strain vs. applied electrical field for a typical piezoelectric material.

in the same direction, the strain increases and attains saturation. Thus, the strain versus the applied electrical field is a symmetric curve as shown in Fig. 8.3. A closed loop is formed when the electrical field in the reverse direction is again brought to zero and increased in the positive direction as shown in Fig. 8.3, while starting the curve at point *a* and following the paths *b*–*g*. Note also the two depoling fields presented on the hysteretic curve.

8.2. Piezoelectricity Books

Since its discovery by the Curie brothers in 1880, the knowledge on piezoelectricity had tremendously increased and a large number of books had been published [4–67]. These books contain various levels of knowledge regarding piezoelectric materials, their applications and how to formulate appropriate equations to investigate the behavior of smart structures. The books authored by Cady [4], Tiersten [5] and Yaffe *et al.* [6] are considered the rigorous mathematical references for the behavior of piezoelectric materials and their application. Additional books containing various levels of mathematical treatment for the piezoelectric effects are the books written by van Randeraat [7], Ikeda [8], Rosen *et al.* [9] and more recently, Katzir [10].

Other books would treat certain aspect of the topic, like ferroelectric transducers and sensors [11], piezoelectric ultrasonic transducers [12] and resonators built from piezoelectric material and their uses [13]. Other authors, like Tzou and Anderson [14], address the piezoelectricity under the title of *Intelligent Structural Systems*. Ultrasonic motors based on piezoelectric material are solely discussed in [15], while Uchino, one of the eminent scholars in the field of piezoelectricity, also addressed the piezoelectric actuators issue, enlarging it to ultrasonic motors [16].

Basic structural elements like plates and shells, made from piezoelectric material have also being investigated and published in a series of various books (see [17–19]).

A more applied book, discussing various issues of sensing by piezoelectric transducers, including acoustic emission and the associated electronic equipment, was authored by Gautschi [20].

Some of the manufacturers of piezoelectric materials, like APC, summarized their experience in the form of a book [21], enabling prospective customers to have basic knowledge about piezoelectricity.

Yang [22–26], another well-known scientist in the area of piezoelectricity, authors a series of books on this topic: he started with the introduction and basic equations for the piezoelectric effects [22], continued with the mechanical aspects of the topic [23] and analysis of piezoelectric devices [24] and concluding with special topics in piezoelectricity [25] and investigation of the dynamic behavior in the form of anti-plane motions of piezoceramics and acoustic wave devices [26].

Following the trend of various researchers to summarize their experience and knowledge in a particular field in the form of a book, Moheimani and Fleming [27] wrote a book addressing the issue of piezoelectric transducers to be used to control vibration and increase damping of structures. Similar books had been published, covering advanced dielectric piezoelectric materials outlining their properties, manufacturing process and their use as transducers [28] while Arnau [29] highlighted the application of piezoelectric transducers. Safari and Akdoğan [30] display in their book the use of piezoelectric and acoustic-type materials to generate transducers to lower/reduce noise.

Heywang *et al.* [31] claimed already in 2008 that the piezoelectric topic is already scientific mature to review its past and predict its future technology. Brockman [32] addresses the issue of adaptive fiber composites having piezoelectric fibers and presents the dynamic behavior of rotating devices made of adaptive fiber composites.

In 2010, Uchino [33] summarizes the topic of advanced piezoelectric materials, presenting the up-to-date equations of motion beside various technologies to fabricate offers. Without doubt, this book is one of the best in its scientific area and offers the reader a broad look on the topic. Another short book [34], also published in 2010, summarizes the processing and characterization of piezoelectric ceramic materials, ending with various applications for these materials. The same materials are the main issue in an additional book [35], this time addressing their structure, typical achieved properties and their uses.

More focused books, like the book authored by Jalili [36], in which the use of piezoelectric transducers to control vibrations of structures is presented for macro- micro- and nanoscale systems and Defaÿ [37], addressing the issue of thin films made from ferroelectric and piezoelectric materials to be used in microsystems, are also available on the market.

Erturk and Inman [38] were the first to summarize one of the important applications for piezoelectric materials — the harvesting of environmental parasitic mechanical energy and transforming it into useful electrical energy for various devices. Ilyas [39] recently treated the same issue in a short concise book.

Although it was apparent that the piezoelectric topic was well treated in those numerous books quoted above, starting 2012, another "wave" of new books covering various pin-pointed aspects of piezoelectric materials, appeared on the market. Those aspects included biomedical applications of piezoelectric-type nanomaterials [40], another book by Yang [41], this time covering the vibration of piezoelectric plates, mechanics of piezoelectricity [42], the use of piezoelectric-type materials and various devices in the medical sector [43] and dynamic fracture of piezoelectric made structures using the boundary integral equation method [44]. This list continues with a book authored by Levinson [45], dealing with piezoelectric accelerometers and their associated integral electronics, a more electronic treatment of piezoelectric transduces appearing in [46] and again a book dealing with piezoelectric materials and their applications [47].

In 2017, Giurgitiu [48] published a well-documented book on the use of piezoelectric wafer active sensors to health monitor various structures. This book is one of the best ones on this topic and is used by various scientists in their relevant research. The same topic of using piezoelectric materials in SHM (Structural Health Monitoring), energy harvesting and biomechanics is the subject of another recent book [49].

Apart of the above-described books, dealing with all the aspects of piezoelectricity, there is another series of books, under the general title of *Intelligent Structures* or *Smart Materials and Structures* (see for example [50–68]) in which the topic of piezoelectricity is one chapter in the book. It starts with the book authored by Chong *et al.* [50], the very basic book of Gandhi and Thompson [51] and the book by Guran and Inman [52], in which the intelligent structures are addressed together with wave motion and nonlinear mechanics.

Other books in these series, display various application for smart materials. The application of structures made of smart materials and their eventual application in mechanical and civil engineering sectors is dealt in [53], while [54] addresses the topic of smart structure (including piezoelectric materials) developing analytical tools for the design of those advanced structures. Gabert and Tzou [55] edited a book based on an IUTAM Symposium containing lectures on various smart structures, held in 2001. Suleman [56] summarized the course on the applications and related technologies of smart structures given, at CISM Udine in 2001, in a dedicated book.

Other aspects of smart/intelligent structures were also found their way in various books, like the dynamic analysis of smart structures [57, 62], and the use of advanced innovative structures built from smart materials for seismic applications [58].

Janocha [59] edited an interesting book on the application of smart structures in various sectors, casting the German name of *Adaptronics* for those structures involving both adaptive end electronic systems.

Carrera *et al.* [60] present classical and advanced theories to solve various problems of plates and shells built also from smart materials, while the behavior, mathematical representations and various applications are dealt in a number of books [61, 63–66]. Note that some of these books can serve also as text book for various undergraduate and graduate students [61, 63, 64, 66].

Finally, recent books, edited by Lopes *et al.* [67] and Araujo and Soares [68], present dynamic aspects of smart structures and highlight papers from *ECCOMAS* thematic conference on smart structures and materials held in 2017, respectively.

8.3. Harvesting

According to the above books references, quoted above, the piezoelectricity topic was well investigated and documented. However, the number of manuscripts dealing with this issue is only increasing in the recent years (starting 2014 till today).

One of the main issues addressed by researchers is energy harvesting using piezoelectric materials (see [69–113] as typical references). Figure 8.4 presents schematically the way the mechanical energy is harvested and transformed into electrical engineering, using the direct piezoelectric effect.

The literature presents two main modes of harvesting: the 31 mode generator (mainly harvesting from vibrations) and the 33 mode generator (direct pressure on the piezoelectric patch) as shown in the Fig. 8.5 (adapted from [66]).

Fig. 8.4. Energy harvesting system — A schematic view.

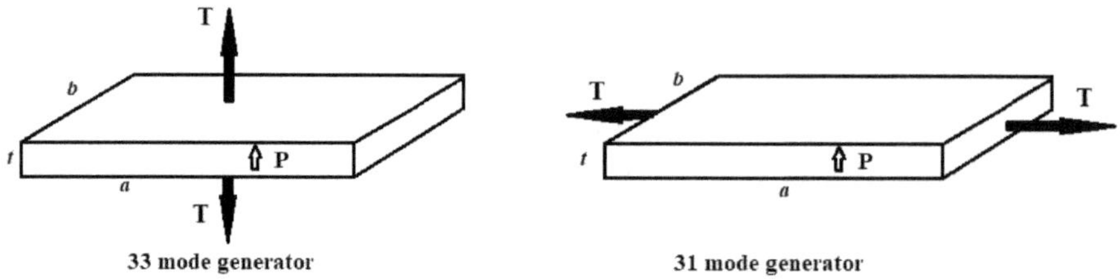

Fig. 8.5. Two modes of electrical energy generation — A schematic view.

In Fig. 8.5, the letter P stands for the polarity, its arrow showing its direction through the piezoelectric patch, T represents the mechanical applied stress, and the piezoelectric patch has a shape of a plate having $a * b * t$ (length $\times$ width $\times$ thickness). Assuming that the piezoelectric patch acts far from its natural frequency, it can be assumed to work as a capacitor (see for example [66, 71–73, 84]) and therefore the relationship among the mechanical, electrical and the geometric dimensions can be written for the 31 mode generator as

$$\begin{aligned} Q_{at\,V=0} &= d_{31} \cdot T_1 \cdot (a \cdot b), \\ V_{at\,Q=0} &= g_{31} \cdot T_1 \cdot t, \end{aligned} \tag{8.12}$$

while for the 33 mode generator the equations are similar

$$\begin{aligned} Q_{at\,V=0} &= d_{33} \cdot T_3 \cdot (a \cdot b), \\ V_{at\,Q=0} &= g_{33} \cdot T_3 \cdot t, \end{aligned} \tag{8.13}$$

where Q and V are the induced charge and voltage on the piezoelectric electrodes, respectively, T_1 and T_3 are the applied normal mechanical stresses in the directions 1 and 3, respectively (see also Eq. (8.5)) and d_{31} (or d_{33}) and g_{31} (or g_{33}) are the piezoelectric field constant and the piezoelectric voltage constant, respectively. To obtain the expression of the electrical energy, the charge is multiplied by the voltage and divided by two, yielding for the 31 mode generation the following equation:

$$U = \frac{Q \cdot V}{2} = \frac{d_{31} \cdot g_{31} \cdot T_1^2 \cdot (a \cdot b \cdot t)}{2} = \frac{d_{31} \cdot g_{31} \cdot T_1^2 \cdot Vol.}{2}, \tag{8.14}$$

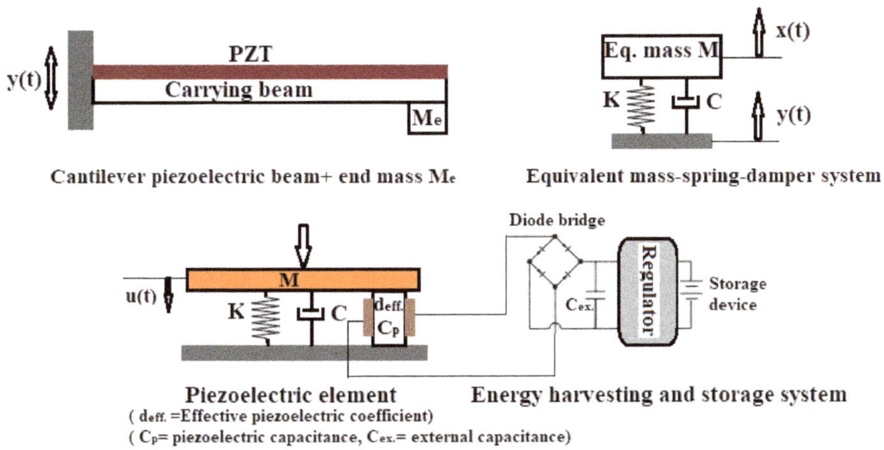

Fig. 8.6. A cantilever piezoelectric beam + end mass, its equivalent mass, spring, damper system and the harvester + energy storage view — A schematic view (adapted from [66]).

while for the 33 mode generation the expressions is

$$U = \frac{Q \cdot V}{2} = \frac{d_{33} \cdot g_{33} \cdot T_3^2 \cdot (a \cdot b \cdot t)}{2} = \frac{d_{33} \cdot g_{33} \cdot T_3^2 \cdot Vol.}{2}, \tag{8.15}$$

where *Vol.* is the volume of the piezoelectric plate.

It is clear from Eqs. (8.14) and (8.15) that to maximize the energy one would use a piezoelectric material having a maximal value for the expression $d_{33} \cdot g_{33} = \frac{d_{33}^2}{\varepsilon_{33}^E}$ (or $d_{31} \cdot g_{31} = \frac{d_{31}^2}{\varepsilon_{33}^E}$).

To perform the harvesting of vibrational energy, a device called in the literature *a harvester*, which normally consists of a cantilever-based beam equipped with piezoelectric patches or layers bonded either on one side of the carrying beam, or on both sides. To reduce the natural frequency of the piezo-beam, to the ambient frequency, a mass is added to the free end of the cantilever, called *the end mass*. The base of the cantilever is attached to the vibrating object from which the mechanical energy would then be scavenged into electrical energy using a suitable electrical circuit. The need of a carrying beam is either due to the brittleness of the piezoelectric material (has the characteristics of a ceramic material) in the case of using PZT or due to the flexibility when using PVDF materials. Figure 8.6 presents a typical cantilever piezoelectric cantilever beam, its equivalent one-degree of freedom system consisting of a mass, spring, and a damper, and the way the piezoelectric device is connected electrically, by using a diode bridge connected to the regulator and the storage system, which can be a battery, or a capacitor and even a super-capacitor.

More detailed derivations can be found in the book of Erturk and Inman [38], the book authored by Abramovich [66] and in [70–72, 76, 78].

Based on the derivation presented in detail in [78], the equation of motion for a cantilevered piezoelectric beam (having only one layer — see Fig. 8.6) using the Bernoulli–Euler theory (shear deformations are ignored) can be written as

$$D\frac{d^4W(x)}{dx^4} - m\omega^2 W(x) = 0 \tag{8.16}$$

where the external excitation $y(t) = Ae^{i\omega t}$ (see Fig. 8.6), where A is the excitation amplitude and ω is the excitation frequency, D is the bending stiffness of the cantilevered beam (including the piezoelectric layer), m is the total mass of the carrying beam and the piezoelectric layer and $W(x)$ is the lateral displacement of the beam. The solution of the Eq. (8.16) has the following general form:

$$W(x) = A_1\cosh(\beta x) + A_2\sinh(\beta x) + A_3\cos(\beta x) + A_4\sin(\beta x),$$

where

$$\beta^4 = \left(\frac{m\omega^2}{D}\right), \tag{8.17}$$

where A_1 — A_4 are constants to be imposed by the relevant boundary conditions of the problem, and x is the longitudinal coordinate. These boundary conditions can be written as

$$\text{at } x = 0: \; W(0) = A, \tag{8.18a}$$

$$\text{at } x = 0: \; \tfrac{dW(0)}{dx} = 0, \tag{8.18b}$$

$$\text{at } x = L: \; -D \cdot \tfrac{d^2W(L)}{dx^2}, + \left[E_{\text{piezo}} \cdot V \cdot \tfrac{(H+h)}{2} \cdot b\right] = 0, \tag{8.18c}$$

$$\text{at } x = L: \; D \cdot \tfrac{d^3W(L)}{dx^3} = -\omega^2 \cdot M_e \cdot W(L), \tag{8.18d}$$

where H and h are the thicknesses of the carrying beam and the piezoelectric layer, respectively and b is the width of the beam.

An additional equation connects the voltage generated V due to the piezoelectric layer to the other variables of this problem, yielding

$$V = \Lambda \cdot \beta \cdot [-A_1\sinh(\beta L) - A_2\cosh(\beta L) + A_3\sin(\beta L) - A_4\cos(\beta L)],$$

$$\text{where } \Lambda = \frac{E_{\text{piezo}} \cdot d_{31} \cdot (H+h)}{2 \cdot \left(\frac{1}{z_L} + \frac{1}{z_{\text{piezo}}}\right)} \tag{8.18e}$$

with z_L and Z_{piezo} being the external impedance (see Fig. 8.6) and the piezoelectric impedance, respectively, and E_{piezo} being the Young's modulus of the piezoelectric layer. Application of the boundary conditions (Eqs. 8.18(a)–8.18(e)) will yield the four constants A_1–A_4 and the generated voltage V. Note that Eqs. (8.18a)–(8.18d) can provide the natural frequencies of the beam with the piezoelectric layer, assuming $V = 0$.

With the general method of harvesting the vibration energy of various bodies using a piezoelectric harvester, being explained above, we can now review the various new manuscripts being published in the literature on this topic.

Caliò *et al.* [69] present a well-detailed review on piezoelectric energy harvesting solutions, updated for the year 2014. It includes emphasis on the material type, its operating modes (d_{31}, d_{33} and even d_{15}, the shear mode) and device configurations, from resonant to non-resonant devices and also to rotational solutions. The reviewed literature is compared based on power density and bandwidth. The review contains also various circuit solutions used for the conversion process. They state the d_{31} operating mode is the most used, while the d_{33} mode did not show the expected performances. Very low power densities, in the range of mW/cm^3, are reported for the references included in the review.

Uchino recently published three manuscripts [70–72], in which the piezoelectric harvesting is dealt in depth. While in [70] the topic is only reviewed and in [71] the merits of piezoelectric perovskites are described in detail, in [72] a detailed review is presented. It includes three major phases associated with piezoelectric energy harvesting: (1) mechanical–mechanical energy transfer, including mechanical stability of the piezoelectric transducer under large stresses, and mechanical impedance matching, (2) mechanical–electrical energy transduction, relating to the electro mechanical coupling factor in the composite transducer structure, and (3) electrical–electrical energy transfer, including electrical impedance matching, such as a direct current (DC/DC) conversion to transfer the energy to a rechargeable battery. This extended review includes a historical background of the piezoelectric energy harvesting, followed by pointing out several misconceptions by current researchers. The main part deals with step-by-step detailed energy flow analysis energy harvesting systems with PZT-based devices to provide comprehensive strategies on how to improve the efficiency of the harvesting system.

Wei and Jing [73] recently (217) presented a comprehensive review on energy harvesting, including the modelling and the realization of this topic using piezoelectric, electromagnetic and electrostatic approaches. The survey includes also energy harvesting using nonlinear characteristics to increase their output and broadened their bandwidth. The paper reviews three main and important vibration-to-electricity conversion mechanisms, namely piezoelectric, electromagnetic and electrostatic. For the piezoelectric conversion systems, the authors the equations of motion having the following form:

$$m \cdot \frac{d^2 z(t)}{dt^2} + d_m \cdot \frac{dz(t)}{dt} + k \cdot z(t) - \theta \cdot V(t) = 0 \tag{8.19}$$

$$\theta \cdot \frac{dz(t)}{dt} + C_p \cdot \frac{V(t)}{dt} + \frac{V(t)}{R_l} = 0 \tag{8.20}$$

leading to the following equation estimating the harvested power, from the vibrating harvester

$$P_{\text{out}} = \frac{m \cdot \varsigma_m \cdot \alpha \cdot \Omega^2 \cdot \omega^4}{\omega_n\{[1 - (1 + 2\varsigma_m \cdot \alpha) \cdot \Omega^2]^2 + [(2\varsigma_m + (1 + \kappa^2)\alpha)\Omega - \alpha \cdot \Omega^3]^2\}}, \tag{8.21}$$

where $z(t)$ is the coordinate of the vibrating mass, $V(t)$ is the voltage generated, m is the vibrating mass, ς_m is the non-dimensional damping factor, $\alpha = \omega_n \cdot C_p \cdot R_{\text{load}}$, ω_n is the natural frequency of the vibrating system (mass, m, + damper, d_m, + spring, k), C_p stands for the capacitance of the vibrating piezoelectric body, R_{load} represents the external electrical load, ω is the excitation frequency, $\Omega = \omega/\omega_n$and $\kappa^2 = \theta^2/(k \cdot C_p)$, the non-dimensional electromechanical coupling coefficient, where $\theta = -\varepsilon_{33} \cdot A_p/t_p$ (A_p and t_p being the area of the piezoelectric electrode and the thickness of the piezoelectric layer, respectively). As one of important factors to estimate the power output performance, the energy conversion efficiency for the three conversion mechanisms is also presented. Based on the various references reviewed in the survey, it turns up that the maximal measured power is 650 (μW) at a frequency of 1700 Hz with much less power at lower natural frequencies.

Another recent (2017) interesting review is the work of Ahmed *et al.* [74], in which the authors discuss the various approaches in vibration-based energy scavenging where piezoelectric materials are employed as the energy conversion medium. They classified the harvesting process into two major categories, according to the excitation type: an intermittent source defined as mechanical or the ambient vibration sources that do not have a constant or continuous amplitude of excitation and a continuous source. The review contains many examples of various harvesters quoting power generation mainly in the micro-watt region.

Tran *et al.* [75] had chosen to review harvesters working nonlinearly. The authors claim that using stochastic loading, or internal resonances, or being multi-degrees-of-freedom, or mechanical stoppers or parametric excitations, will lead to nonlinear behavior and thus enhance the energy harvesting. These techniques are capable of significantly enlarging the frequency bandwidth and, in some cases, increasing the amplitude response. The enhancement in performance, might bring to the development of advanced devices capable of harvesting energy more efficiently from ambient vibrations, although the authors do not quote the powered harvested using those nonlinear harvesters.

Priya *et al.* [76] address the topic of piezoelectric harvesting for micro-electro-mechanical-systems (MEMS). To achieve the target of battery-less wireless sensors, the authors claim that a piezo-MEMS having a volume of $42.5\,\text{mm}^3$ should be able to generate about $100\,\mu\text{W}$ continuous power from low-frequency ambient vibrations (less than $100\,\text{Hz}$). This extended review covers all the basics of harvesting using piezoelectric materials in the microscale dimensions. It discusses issues like power density and bandwidth of reported structures vibrating at low-frequency input. The paper includes description of the recent advancements in piezoelectric materials and resonator structures. Deposition of a crystalline overlayer on a crystalline substrate and grain texturing of piezoelectric materials as means to achieve a much higher energy conversion efficiency is also discussed. Lead-free piezoelectric thin films for embedded medical systems and their incorporation in MEMS processes are also reviewed. The authors claim in their survey that nonlinear resonating beams for wide bandwidth resonance would enable wide bandwidth and low-frequency operation of energy harvesters. Another important aspect dealt by the review is the management of the power circuit. To maximize the net energy harvested, the authors claim that it is essential for the power management circuit of a small-scale energy harvester to dissipate minimal power, and thus it requires special circuit design techniques and a simple maximum power point tracking scheme.

Avvari *et al.* [77] address in their review an aspect which is not always investigated in the literature. They discuss the long-term mechanical fatigue of a harvester, having the shape of a cantilever, by examining the strains at its root. The article presents an experimental investigation on the fatigue behavior of a cantilever piezoelectric energy harvesting beam at different base excitation levels. The experimental study is enhanced with an analytical formulation of the problem, to examine the strain levels and with finite element analysis formulation to model the piezoelectric energy harvesting beam with a macro-fiber composite piezoelectric transducer. A two-dimensional model was developed based on the three-dimensional model to investigate the crack propagation in the piezoelectric material bonded on the energy harvesting beam. The monitoring of the electro-mechanical progression of damage in the experimental specimens was performed using the impedance technique. The research provides an insight into the behavior of the piezoelectric energy

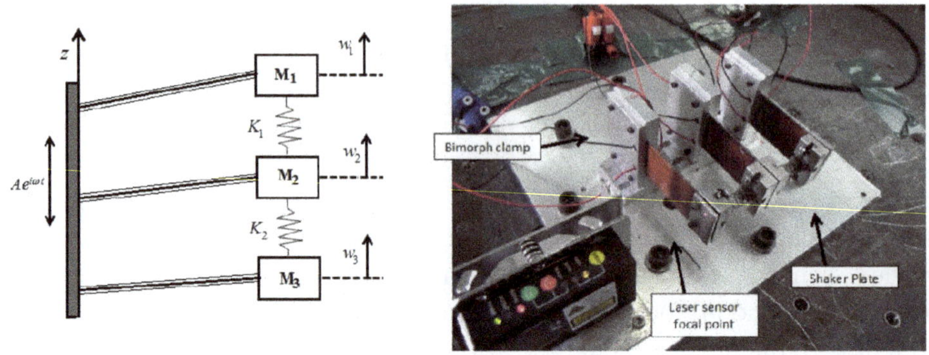

(a) (b)

Fig. 8.7. The three bimorphs system: (a) the schematic model, and (b) the experimental model (from [78]).

harvesting beam undergoing fatigue loading due to a uniform sinusoidal base excitation by analyzing the output voltage, resonant frequency, tip displacement, tip velocity, and impedance variations.

One of the inherent problems of single cantilever type harvester, namely its narrow bandwidth is treated in a very recent paper by Abramovich and Har-nes [78] by suggesting the use of three cantilevers with tip end masses, interconnected by springs at their free end (see Fig. 8.7), yielded an enhanced bandwidth for the system. An analytical model was developed (see Eqs. (8.16a)–(8.18e)) and its predictions were compared with experimental results and results published in the literature, yielding a very good correlation. A parameter investigation was then performed to evaluate the role of the problem parameters on the power output. Experimental results displayed the figure of 20 mW for the harvested power at the system first natural frequency and 5 mW at the second one.

Another interesting review authored by Todaro *et al.* [79] deal with piezoelectric MEMS vibrational harvesters reviewing its merits and trying to predict its future. This review presents the state of research on energy harvesters based on piezoelectric thin films focusing on devices at MEMS scale and exploiting MEMS technologies. It provides an overview on piezoelectric thin film materials exploited for energy harvesting including ferroelectric piezoceramics, textured polycrystalline non-ferroelectric piezoelectric materials and piezopolymers. Among ferroelectric ceramics, recent advances on thin films with giant piezoelectric coefficients and high performance lead free piezoelectric have been highlighted. According to the present status of research on energy harvesters, the performances of the reported devices are still lower (maximal power density being $86\,\mu W/mm^3$) for practical applications. It is pointed out that the main challenges are: power density and the bandwidth. To try and solve these issue, different solutions in terms of device architecture and alternative approaches have been proposed by the authors. Their conclusion is that a proper choice of the materials, new device architectures and strategies exploited for bandwidth and power density improvements, will be the future research on those devices.

Fang and Wang [80] present another aspect of using piezoelectric-based vibration-driven system this time not for harvesting, but for actuating small-size crawling-type locomotion robots with a known major drawback of limited stroke output that could severely constraint the system's locomotion performance. To advance the state of the art in this area, the authors propose two novel designs of piezoelectric vibration-driven locomotion systems. The first utilizes the resonant amplification concept, and the second explores the design of a bistable device. It is worth to note that these two designs can also be applied for harvesting energy from ambient vibrations.

Li [81] discuss in his short and concise review harvesting acoustic energies besides ambient vibrational energy using piezoelectric material. Acoustic energy harvesters generally are composed of two components: an acoustic resonator and a piezoelectric oscillator. The amplified acoustic wave inside the acoustic resonator drives the oscillation of piezoelectric element and generates electricity. This Helmholtz resonator was shown to produce $\sim 0.1\,\text{nW}$, however various improvements presented in the review quotes a higher power, $42\,\mu\text{W}$.

Narita and Fox [82] recently (2018) present an overview on energy harvesting, materials and systems. Three main categories are discussed: piezoelectric ceramics/polymers, magnetostrictive alloys, and magnetoelectric (ME) multiferroic composites. State-of-the-art harvesting materials and structures are presented with a focus on characterization, fabrication, modeling and simulation, and durability and reliability. Some perspectives and challenges for the future development of energy harvesting materials are also highlighted. They report the value of $5800\,\mu\text{W}$ (for a volume of 1,471 mm^3) for the power being generated by a PZT-5H knee joint. According to their review, this is the highest power generated by a piezoelectric device. It is worth mentioning also the enhanced 2018 review of Yang *et al.* [83], whose content was already used in the previous parts of the present survey.

Gafforelli *et al.* [84] present a theoretical study of a nonlinear piezoelectric harvester based on a simple model in which the piezoelectric coupling is treated as a linear dashpot added to a nonlinear Duffing oscillator, yielding a power generation being bounded by the mechanical damping for both linear and nonlinear harvesters, with the latter being spread out over a wider

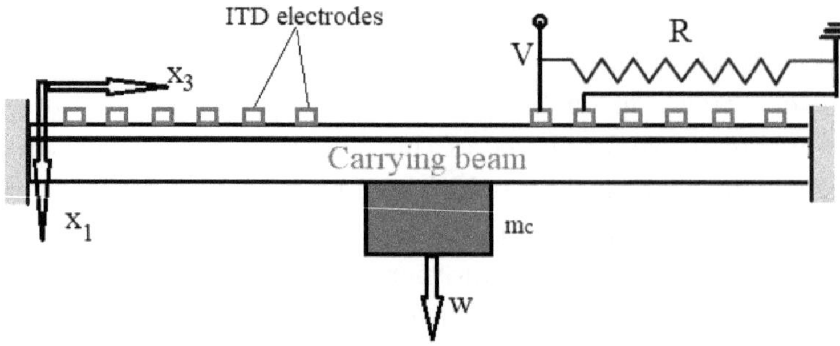

Fig. 8.8. Doubly clamped bridge-shaped piezolaminated beam with tip mass (adapted from [84]).

bandwidth. A more accurate model is then studied, using a bridge-shaped nonlinear harvester working in d_{33} mode as presented in Fig. 8.8. Note that the direct influence of an external load resistor is included to consider a more accurate nonlinear coupling effect instead of the linear electrical damping employed in the first model. The second model is suitable for predicting the response of a generic piezoelectric nonlinear harvester with arbitrary excitation frequency and external load.

Baur *et al.* [85] tackled the issue of using an organic flexible polymer composite (PVDF+ diamine-functionalized C60 fullerenes into P(VDF-TrFE) yielding a new composite that exceeds the piezoelectric properties of pure P(VDF-TrFE) by up to 29% at room temperature and 79% at elevated temperature of 140°C. The authors hope that the new developed composites may find application in areas that require flexible and sensitive piezoelectric materials that can withstand elevated temperatures. Another source of energy, in the form of vibrations induced due to aerodynamic phenomena like vortex-induced vibrations, flutter, buffeting, and galloping, and their harvesting using piezoelectric harvesters is discussed by Abdelkefi [86]. From an energy harvesting point of view, the harvester is placed in a flow field and excited to undergo large limit-cycle (LCO) oscillations (see a schematic view in Fig. 8.9(a)) or vortex-induced oscillations on a circular cylinder (see Fig. 8.9(b)) that can be converted to electrical energy using piezoelectric and/or electromagnetic transducers. The review presents a long list of studies which were dedicated to harvesting of vibrational energy due to aerodynamic processes, quoting a value of $P = 50\,\text{mW}$ for the maximal-induced electrical power from a galloping phenomenon.

Another review containing the use of piezoelectric polymers as functional materials for electromechanical transducers, smart materials and structures is brought up by Ramadan *et al.* [87]. It presents an overview of piezoelectric

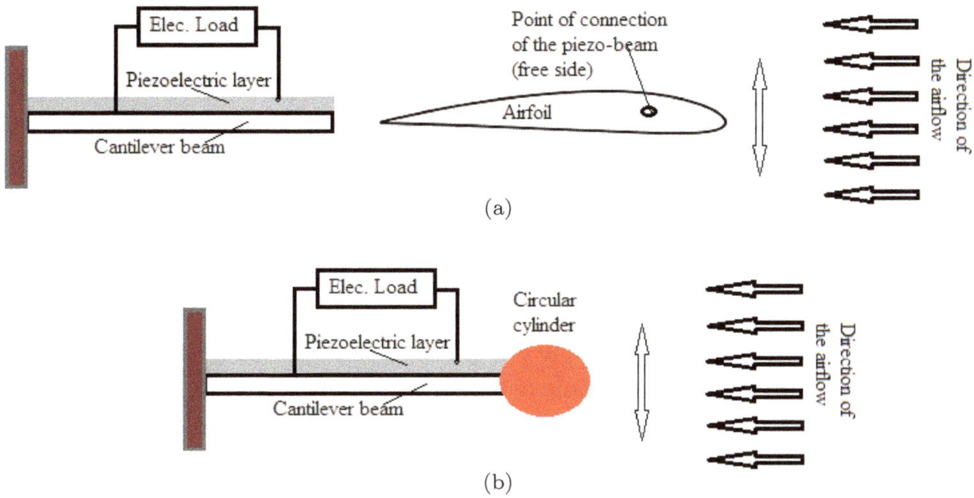

Fig. 8.9. A schematic drawing of harvesting using aerodynamic forces: (a) an airfoil undergoing LCO and (b) a circular cylinder under induced vortex oscillations (adapted from [86]).

polymers based on their operating principle like bulk piezoelectric polymers, piezocomposites and voided charged polymers. The state-of-the-art for year 2014, for those type of piezo-polymers is presented focusing on their fabrication techniques and material properties. A comparison between the different piezoelectric polymers and common inorganic piezoelectric materials (PZT, ZnO, AlN and PMN-PT) is also outlined in terms of piezoelectric properties. The use of these piezo-polymers in various electromechanical devices like tactile sensors, energy harvesters, acoustic transducers and inertial sensors is discussed. Wang *et al.* [88] is presenting the opposite aspect, namely the use of all-inorganic flexible piezoelectric energy harvester using a 2D mica. To demonstrate their capability, an all-inorganic, large scale, flexible piezoelectric $Pb(Zr_{0.52}Ti_{0.48})O_3$ energy harvester was fabricated yielding an outstanding performance (i.e., open-circuit voltage of 120 V, short-circuit current density of $150\,\mu A/cm^2$ and power density of 42.7 mW/cm^3).

Kumar and Chhabra [89] and Tian *et al.* [90] address in their reviews the topic of MEMS scale harvesters. While the authors in [89] tried to advocate the energy harvesters campaign to create completely self-powered devices and micro electro mechanical systems, the review in [90] outlines the knowledge to fabricate a piezoelectric energy harvester (PEH), the research progress made discussing piezoelectric materials, piezoelectric modes (d_{31}, d_{33} and d_{15}) of operation and energy harvester structures The authors present ways of optimizing the PEH's structure and matching it to the external excitation frequency, by adjusting the resonant frequency, frequency up-converting and broadening the presents bandwidth.

An interesting manuscript by Cha *et al.* [91] presents a study on energy harvesting using a piezoelectric biomimetic fish tail, modeled as a geometrically tapered beam with heterogeneous physical properties, undergoing large amplitude vibration in a viscous fluid. Fluid-structure interactions are described through a hydrodynamic function, which accounts for added mass and nonlinear hydrodynamic damping. The tail undulations caused the two piezoelectric composites (MFC-M4007P1-PES) having an active surface area of $40\times7\,\text{mm}^2$ to produce electrical energy, eventually to provide the power for a wireless communication module attached to the tail. As can be expected for such a piezoelectric device, having a limited volume of piezoelectric material, the reported power is only $0.18\,\mu$W.

The topic of flexible stretchable piezoelectric nanogenerators is another new topic dealing with harvesting. A few manuscripts [92–97] address this issue. While Lee *et al.* [92] highlight the research progress of transparent and flexible ZnO nanorods/nanowires, two-dimensional ZnO nanosheets, stretchable micro-patterned P(VDF-TrFE) polymer, $ZnSnO_3$ nanocubes-based piezoelectric NGs along with graphene and hydrophobic sponge structure-based triboelectric NGs, and their potential applications in powering portable electronics, Hu and Wang [93] claim that by introducing a new sandwich structure design, high performance piezoelectric nanogenerators can be achieved through a very simple fabrication process with good mechanical stability by utilizing ZnO nanowires. By further optimizing the nanomaterials' properties and device structure, the piezoelectric nanogenerators (at open circuit) voltage can be elevated to over 37 V, however this is expected to be much lower for a closed circuit. The use of a nanogenerator as a sustainable power source for self-powered system is then discussed and their use as active sensors for deformation detection, vortex capture and ambient wind-velocity detection is highlighted. In [94], Briscoe and Dunn review nanostructured piezoelectric energy harvesters, based on ZnO nanorods claiming that the nanostructures are crystallographically aligned and non-ferroelectric, and therefore do not require poling. However, recently other well-known materials have also been investigated for nanostructured energy harvesters including lead zirconate titanate (PZT) and barium titanate, with the potential for higher power outputs due to their higher piezoelectric coefficients. Dagdeviren *et al.* [95] join the other researches and present another extended review on flexible and stretchable nano devices by highlighting the essential mechanical to electrical conversion processes in those devices and systems, along with key considerations in their designs by presenting quantitative, experimentally validated mechanics models leading to guidelines in the selection of optimized configurations and materials choices, focusing on thin geometries, neutral mechanical plane construction and controlled buckling, while using options such as organic polymers, inorganic nanomaterials and various types of composites. Their review includes also representative applications in biomedicine, starting from devices for mechanical energy harvesting from natural motions of internal organs to sensors and actuators for the skin. Bera [96] also surveys the status of nanogenerators claiming that PZT nanowires that can be used for wearable and self-powered devices. A built nanogenerator could generate an output voltage of 6 V and produce a current of 45 leading to a power of 270 nW. This nanogenerator was built cost effectively and was demonstrated to light a commercial LCD and power a ZnO nanowire UV sensor for the quantitative detection of UV light. Another nanogenerator made of barium titanate produced an output current density of $0.19\,\mu\text{A/cm}^2$ and a power density of $\sim 7\,\text{mW/cm}^3$ showing a promising future for those devices. Jin *et al.* [97] discuss their research during which they had developed a polarization-free high-crystallization β-PVDF ($h\beta$ PVDF) based piezoelectric nanogenerator to be used as an acceleration sensor having a high sensitivity ($2.405\,\text{nA/(m/s}^2)$) and an excellent stability (97% remaining after 10000 cycles, presenting a promising future for the device.

Another aspect of the energy harvesting issue is presented by Rastegar [98], by addressing piezoelectric-based energy harvesting devices and their charge collection and storage electronics for use in very harsh environment of gun-fired munitions. The idea behind those harvesters is to capture energy from firing acceleration and vibratory motions during the flight and convert into electrical energy using piezoelectric-type harvesters. The review presents the application of the developed piezoelectric-based energy harvesting devices with event detection circuitry to the development of self-powered initiators or switching devices with full no-fire safety circuitry for protection against accidental drops, transportation vibration, and other similar low amplitude accelerations and/or high amplitude but short duration acceleration events is presented. The designs allow the use of a very small piezoelectric elements, thereby making such devices to be highly miniaturized. These devices can be readily hardened to withstand very high g firing setback accelerations in excess of 100,000 g and the harsh firing environments.

As described above, the harvesting of parasitic mechanical energy and its transformation into usable electrical energy using piezoelectric material

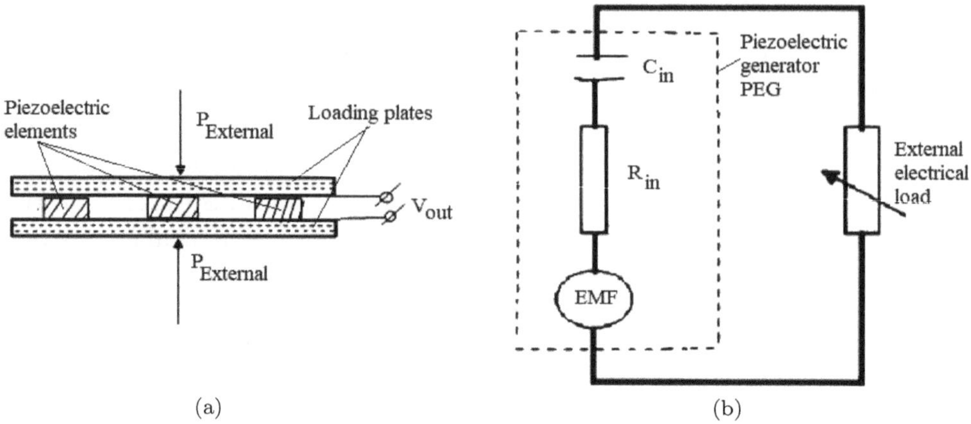

Fig. 8.10. The piezoelectric generator-PEG: (a) a schematic drawing of the mechanical layout and (b) the equivalent electrical circuit and its associated external electrical load (adapted from [99]).

operating in the d_{33} mode, led to a series of manuscripts [99–107] describing piezoelectric actuators embedded on driving roads, and scavenging the energy from passing motor traffic. The idea is to build a piezoelectric generator which will sense the passing vehicle, deform due to its weight and thus produce electrical energy due to the direct piezoelectric effect. The two early papers by Abramovich *et al.* [99, 100] report laboratory experiments on those generators (see Fig. 8.10) and discuss the high-power multi-element piezoelectric generators made from PZT ceramic under long-term cyclic external mechanical loading and showing that a safe-side stress level of 30 MPa can be used with a small reduction in the generated power as can be seen in Fig. 8.11. To compensate for the relatively low electric power, due to relatively reduced mechanical stresses applied on the PZT disks, one can increase the volume of the material used by placing layers of piezoelectric material one on top of the other, each subjected to the same mechanical stress. This will yield the required electric power from a safe given mechanical stress without reduction in its output. The follow-up manuscripts (see [101–107]) mainly review the state-of-the-art for piezoelectric road generators, discussing the prospects of realizing it, while presenting new results on this topic. Kim *et al.* [101] review what had been done on piezoelectric road generators and intend to perform tests on Georgia, USA highways, Duarte and Ferreira [102] also outline the development of energy harvesting technologies for road pavements, trying to divide them into various classes, while discussing and presenting a technical analysis and comparison for those technologies, using the results achieved with available prototypes. Kour and Charif [103] aims to

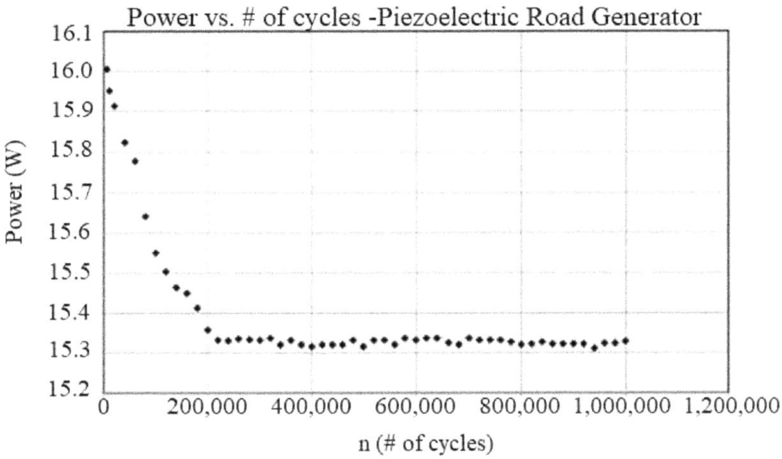

Fig. 8.11. The power output of a piezoelectric road generator loaded at 30 MPa at a frequency of 5 Hz (adapted from [99]).

assess the functionality of piezoelectricity in roads to use the energy due the moving vehicles. The energy is converted into electrical energy using piezoelectric technology to replace fossil fuel in streetlight applications, advocating this new technology claiming that piezoelectric road is a new energy evolution to provide a sustainable solution in terms of environment, economy, and social needs.

Papagiannakis *et al.* [104] present the development of several piezoelectric prototypes capable of harvesting energy from the action of traffic on highways. The amount of energy available for harvesting is explored through finite element simulations of the strain energy in pavements under moving tire loads. The prototypes developed involve various configurations of cylindrical and prismatic piezoelectric elements. Exploratory analysis shows that it is desirable to alternate the polarity of stacked piezoelectric elements and connect them in parallel to avoid generating unmanageably high voltages. The prototypes are tested in uniaxial compression under sinusoidal loading at a frequency of 10 Hz. Curves are fitted to the electrical power vs. stress laboratory data. For a single pass of a 44.48-kN truck tire load, the electrical power generated is estimated to be between 1.0 and 1.8 W, far from the experimental results presented in [99–100]. Yang *et al.* [105], in a recent manuscript (2017), present the results for a road generator using a stacked mode. The capability of their model to withstand repeated loading up to 150 kN, yielded only 100,000 cycling for the piezoelectric transducer, without reduction of the output power. The authors quote an open circuit voltage of 280 V, without mentioning the achieved power. It is worth mentioning that the amount of piezoelectric material used for the piezoelectric road generator was small, expecting a small output power. Another new paper (2018) on piezoelectric road generators is the one presented by Qabur and Alshammari [106] in which they review again the state-of-the-art for harvesting energy from roadways, claiming that PZT can be considered as the most efficient material due to its unique features according to many references and real or practical applications, that various factors such as geometry, thickness and structure affect the output of the piezoelectric process and that energy harvesters have the ability to be enhanced or improved by minimizing the mechanical and electrical losses of the harvesting stages in order to achieve the highest efficiency possible. They sum up by saying that the implementation of piezoelectric materials on roadways needs further research and a more comprehensive analysis of different data that are obtained from real life or practical applications. Finally, Walubita *et al.* [107] review in their article the state of the art in road energy harvesting technology, with a focus on piezoelectric systems, including an analysis of the impact of the technology from social and environmental standpoints. Overall, their literature findings indicate that the expansion of the roadway energy harvesting technology to a large practical scale is feasible, but such an undertaking should be wisely weighed from broader perspectives. Ultimately, the article provides a positive outlook of the potential contributions of road energy harvesting technologies to the ongoing energy and environmental challenges of human society. The article highlights an experimental study carried by the Texas A & M Transportation Institute (TTI) in which a prototype of a piezoelectric module, named Highway Sensing and Energy Conversion (HiSEC) was experimented in a laboratory at TTI. The HiSEC module is made up of three pairs of piezo disks stacked and connected with two diodes. The montage is thereafter enclosed in a metal case covered on the top with an impact cap. The maximal power obtained by this module was approximately 13.5 mW at a mechanical stress of 0.827 MPa (120 psi) and a frequency of 1 Hz. Note the small amount of the piezoelectric material, the low mechanical stress and the low frequency employed in their experiment, leading to low performances.

8.4. Other Aspects of Piezoelectricity

Other aspects of the piezoelectric materials are presented in the following references. Gripp and Rade [108] review the use of shunted piezoelectric transducers as effective vibration and noise tools. Although the topic was treated in the literature, the present review succeeds in presenting new ideas, like various types of shunt circuits, including the traditional passive resonant circuits, multimode resonant circuits, adaptive tuning circuits, switching circuits and negative capacitance. Yang *et al.* [109] developed an innovative piezoelectric brace, which provides both passive energy-dissipating and energy-harvesting capabilities — a passive suppression of structural vibrations and conversion of vibration energy into reusable electricity. The numerical results presented in the manuscript indicate that the piezoelectric braces passively dissipate energy through inclined oval-shaped hysteretic loops. The harvested energy is up to approximately 40% of the input energy.

Another research direction is presented by Zheng *et al.* [110] and by Quan *et al.* [111]. Both manuscripts deal with improving the properties of the piezoelectric materials, by excluding the lead from the synthesized piezoelectric material. While in [110] a survey of the unique advances in lead-free perovskites piezoelectric bulk materials, along with the descriptions of phase boundaries, domain configurations and piezoelectric effects is presented, the team in [111] presents another lead free piezoelectric ceramics, the $Bi_{0.5}(Na, K)_{0.5}TiO_3$ carrying the name BNKT, and having good piezoelectric, ferroelectric, and dielectric properties compared to PZT. The review is dedicated and focused on the role of dopants and secondary phase to piezoelectric properties of BNKT. It is claimed that the dopants are expected to change the crystal structure and create the vacancies which will cause the enhancement of the piezoelectric properties.

The use of piezoelectric materials in the biomedical sector is another topic dealt by various researchers and published in the literature (see [112–117]). They cover various aspects like: nano- piezoelectric devices — transducers for energy harvesting from natural motions of internal organs, sensors, and actuators for medical applications [112], piezoelectric based sensors [113], immune-sensors and DNA sensors [114], piezoelectric tools for bone surgery [115], the application of the piezoelectric technology as a surgical tool for precise removal of soft tissue and bone [116] and the application of piezoelectric materials to enhance tissue regeneration by acting as sensitive mechanic-electrical transduction systems leading to bone repair [117, 118].

The topic of piezoelectric motors was recently (2017) reviewed by Behera and Nemade in [119]. The authors survey ultrasonic motors and surface acoustic wave motors using resonant vibrations of piezoelectric substrates. The traditional ultrasonic motors using standing and traveling waves along with surface acoustic wave-driven motors are highlighted.

Vinogradov *et al.* [120] present the R&D of a peristaltic multiphase piezoelectric micro-pump, highlighting one of the current used of piezoelectric materials.

Jung *et al.* [121] deal with the use of micro-machined ultrasonic transducers (MUTs) as an alternative to conventional piezocomposite ultrasonic transducers, primarily due to the advantages that microelectromechanical systems provide. The manuscript various aspects of the piezoelectric MUTs including multi-electrode devices, partially clamped boundary conditions, and 3D (curved and domed types) MUTs, where the latter can provide an electromechanical coupling coefficient of up to 45%. The authors claim that arrays of piezoelectric MUTs can be implemented for various applications including intravascular ultrasound, fingerprint sensors, rangefinders in air, and wireless power supply systems, yielding an ideal solution for applications such as mobile biometric security (fingerprint sensors) and rangefinders due to their superior power efficiency and compact size.

Khan *et al.* [122] address the topic of piezoelectric thin films, highlighting their advantages such as a wide dynamic range and high sensitivity for high energy harvesters and large displacement and low power consumption for actuators. The review covers PZT based thin films, lead-free piezoelectric thin films, piezo-polymer films and cellulose-based electroactive paper (EAP).

Liu and Cohen [123] treat a small niche on the piezoelectric materials, by addressing piezoelectric material having negative longitudinal piezoelectric coefficients, thus contracting in the direction of an applied electric field. Using first-principles calculations, the authors demonstrated that several hexagonal ABC ferroelectrics possess significant negative longitudinal piezoelectric effects due to their strong ionic bonds associated with small effective charges and rigid potential energy surfaces. Moreover, ferroelectrics with negative longitudinal piezoelectric coefficients show anomalous pressure-enhanced ferroelectricity. The results presented might be used as design principles to aid the search for new piezoelectric materials for novel electromechanical device applications.

Calahorra and Kar-Narayan [124] address another piezoelectric niche by researching piezoelectricity arising from the non-centrosymmetric ZB structure in III–V semiconductors and exhibiting shear piezoelectricity (relative to the basic cubic cell vectors), with an identical value for all three orientations. The review addresses the challenges related to the study of piezoelectricity in III–V nanowires and the opportunities that lie therein in terms of device applications. Cui *et al.* [125] review in a recent article (from 2018) the new trend of two-dimensional layered materials with a non-centrosymmetric structure which exhibit great potential for nano-scale electromechanical systems and electronic devices. The review presents the preparation of these functional 2D layered materials, including exfoliation methods and vapor phase deposition growth, followed by a general introduction to various piezo/ferro-electric characterization methods. Typical 2D piezoelectric and ferroelectric materials and their electronic properties, together with their potential applications are highlighted. Garcia *et al.* [126] present another 2D material in the form of Iβ-nano-cellulose as a potentially new orthotropic 2D piezoelectric crystal. The authors claim that the order of magnitude estimated for the 2D Iβ-nano-cellulose piezoelectric response, $\sim$pm/V, ranks the material at the level of currently used piezoelectric energy generators and new artificial 2D designs.

Tuukkanen and Rajala [127] compiled a review on printable piezoelectric sensors. Various sensors, made of PVDF (polyvinylidenefluoride), cellulose nanofibril (CNF) and cellulose nanocrystal, are surveyed. Those materials are processed in a solution in combination with printed electrodes to obtain full printability of the sensors. A commercial PVDF film and an in-house fabricated CNF and CNC film are used as sensor materials. Conducting polymer, graphene and carbon nanotube (CNT) based inks are used as solution-processable electrode materials in the sensors, whereas conventional metallic electrodes are used as reference electrode material. It is claimed that the sensor sensitivity varies from 2 to 42 pC/N in the transverse direction, depending on set of the sensor and electrode materials used.

Marzeika *et al.* [128] address the issue of Langevin-type piezoelectric transducers, normally operating in the longitudinal mode by employing d_{33} vibrations of piezoceramic elements. Those transducers are widely used in various applications like high-accuracy positioning, ultrasonic welding, bonding, drilling, and others. Unlike those existing transducers, the team of researchers developed a novel design of a piezoelectric actuator based on two bending-type Langevin transducers. Two half-wavelength transducers are connected by a special aluminum plate. Piezoceramic rings with two-directional polarizations are used for bending oscillation excitation using an input voltage with a phase shifted by $\pi/2$ on the two different transducers. The forward and backward elliptical motion of the contact tip is controlled by changing phases of the electrical signals applied on the transducers. It was shown that a bidirectional motion of the slider can be achieved by counterchanging phases of the excitation signal on the different transducers. The developed experimental prototype actuator achieved a maximum output force equal to 8.8 N.

Finally, a recent review by Presas *et al.* [129] outlines the use of piezoelectric sensors in submerged systems. The review present newly developed transducers based on PZT type patches used as sensors/actuators for applications like modal analysis, active sound control (AVC), sound cancellation, energy harvesting and atomic force microscopes in physiologic mediums, all being submerged.

References

[1] Curie, J. and Curie, P., Développement, par pression, de l'électricité polaire dans les cristaux hémièdres faces 'inclinées, *C. R. Acad. Sci. Gén.*, Vol. 91, 1880, pp. 294–295.
[2] Curie, J. and Curie, P., Sur l'électricité polaire dans les cristaux hémièdres 'faces inclinées, *C. R. Acad. Sci. Gén.*, Vol. 91, 1880, pp. 383–386.
[3] Lippman, G., Principe de la conservation de l'électricité, ou second principe de la théorie des phénomènes électriques, *J. Phys. Theor. Appl.*, Vol. 10(1), 1881, pp. 381–394.
[4] Cady, W.G., *Piezoelectricity — An Introduction to the Theory and Applications of Electromechanical Phenomena in Crystals*, McGraw-Hill Book Company, Inc., 1946, 806 p.
[5] Tiersten, H.F., *Linear Piezoelectric Plate Vibrations — Elements of the Linear Theory of Piezoelectricity and the Vibrations of Piezoelectric Plates*, Springer Science + Business Media LLC, 1969, 211 p.
[6] Yaffe, B., Cook Jr, W.R. and Yaffe, H., *Piezoelectric Ceramics*, Academic Press, London, 1971, 328 p.
[7] van Randeraat, J., *Piezoelectric Ceramics*, Mullard Publications, Philips Application Book, 1974, 211 p.
[8] Ikeda, T., *Fundamentals of Piezoelectricity*, Oxford University Press, 1990, 263 p.
[9] Rosen, C.Z., Hiremath, B.V. and Newnham, R., *Piezoelectricity*, Key Papers in Physics, American Institute of Physics, 1992, 545 p.
[10] Katzir, S., *The Beginnings of Piezoelectricity — A Study in Mundane Physics*, Springer, Dordrecht, 2006, 273 p.
[11] Herbert, J.M., *Ferroelectric Transducers and Sensors*, Gordon and Breach Science Publishers, 1982, 439 p.
[12] Mattiat, O.E., (ed.), *Ultrasonic Transducer Materials*, Springer Science + Business Media, New York, 1971, 184 p.
[13] Zelenka, J., *Piezoelectric Resonators and Their Applications*, Vol. 24, Elsevier, 1986, 301 p.
[14] Tzou, H.S. and Anderson, G.L., (eds.), *Intelligent Structural Systems*, Springer Science + Business Media, Dordrecht, 1992, 454 p.
[15] Sashisa, T. and Kenjo, T., *An Introduction to Ultrasonic Motors*, Oxford Science Publications, Clarendon Press, 1993, 242 p.
[16] Uchino, K., *Piezoelectric Actuators and Ultrasonic Motors*, Kluwer Academic Publishers 1997, 351 p.
[17] Tzou, H.S., *Piezoelectric Shells — Distributed Sensing and Control of Continua*, Springer Science + Business Media, Dordrecht, 1993, 471 p.
[18] Rogacheva, N.N., *The Theory of Piezoelectric Shells and Plates*, CRC Press LLC, 1994, 253 p.
[19] Vinson, J.R., *Plate and Panel Structures of Isotropic, Composite and Piezoelectric Materials, Including Sandwich Construction*, Springer Dordrecht, 2005, 427 p.

[20] Gautschi, G., *Piezoelectric Sensorics — Force, Strain, Pressure, Acceleration and Acoustic Emission Sensors, Materials and Amplifiers*, Springer-Verlag, Berlin, 2002, 269 p.
[21] APC International, *Piezoelectric Ceramics: Principles and Applications*, APC International, 2002, 112 p.
[22] Yang, J., *An Introduction to the Theory of Piezoelectricity*, Springer Science +Business Media Inc., 2005, 298 p.
[23] Yang, J., *The Mechanics of Piezoelectric Structures*, World Scientific Publishing Co. Pte. Ltd., 2006, 313 p.
[24] Yang, J., *Analysis of Piezoelectric Devices*, World Scientific Publishing Co. Pte. Ltd., 2006, 379 p.
[25] Yang, J., editor, *Special Topics in the Theory of Piezoelectricity*, Springer Science + Business Media LLC, 2009, 329 p.
[26] Yang, J., *Antiplane Motions of Piezoceramics and Acoustic Wave Devices*, World Scientific Publishing Co. Pte. Ltd., 2010, 379 p.
[27] Moheimani, S.O. and Fleming, A.J., *Piezoelectric Transducers for Vibration Control and Damping*, Springer-Verlag, London, 2006, 270 p.
[28] Ye, Z.-G., *Handbook of Advanced Dielectric Piezoelectric and Ferroelectric Materials — Synthesis, Properties and Applications*, Woodhead Publishing Limited, 2008, 1060 p.
[29] Arnau, A., editor, *Piezoelectric Transducers and Applications*, Springer-Verlag, Berlin, 2008, 531 p.
[30] Safari, A. and Akdoğan, E.K., (eds.), *Piezoelectric and Acoustic Materials for Transducer Applications*, Springer Science + Business Media LLC 2008, 480 p.
[31] Heywang, W., Lubitz, K., and Wersing, W., *Piezoelectricity — Evolution and Future of a Technology*, Springer Series in Material Science, Vol. 114, Springer-Verlag, Berlin, 2008, 581 p.
[32] Brockman, T.H., *Theory of Adaptive Fiber Composites: From Piezoelectric Material behavior to Dynamics of Rotating Systems*, Springer Science + Business Media LLC, 2009, 218 p.
[33] Uchino, K., *Advanced Piezoelectric Materials: Science and Technology*, Woodhead Publishing, 2010, 678 p.
[34] Zhu, X., *Piezoelectric Ceramic Materials: Processing, Properties, Characterization and Applications*, Nova Science Publishers, 2010, 63 p.
[35] Nelson, W.G., *Piezoelectric Materials: Structure, Properties and Applications*, Nova Science Publishers, 2010, 259 p.
[36] Jalili, N., *Piezoelectric-Based Vibration Control — From Macro to Micro/Nano Scale Systems*, Springer Science + Business Media LLC, 2010, 516 p.
[37] Defaÿ, E., (ed.), *Integration of Ferroelectric and Piezoelectric Thin Films: Concepts and Applications for Microsystems*, ISTE Ltd. and John Wiley & Sons, Inc. 2011, 404 p.
[38] Erturk, A. and Inman, D.J., *Piezoelectric Energy Harvesting*, John Wiley & Sons, 2011, 416 p.
[39] Ilyas, M.A., *Piezoelectric Energy Harvesting: Methods, Progress and Challenges*, Momentum Press, 2017, 72 p.
[40] Ciofani, G. and Menciassi, A., *Piezoelectric Nanomaterials for Biomedical Applications*, Springer Science + Business Media LLC, 2012, 250 p.
[41] Yang, J., *Vibration of Piezoelectric Crystal Plates*, World Scientific Publishing Co. Pte. Ltd., 2013, 418 p.

[42] Qin, Q.-H., *Advanced Mechanics of Piezoelectricity*, Higher Education Press, Beijing and Springer-Verlag, Berlin, 2013, 331 p.

[43] Vijaya, M.S., *Piezoelectric Materials and Devices-Applications in Engineering and Medical Sciences*, CRC Press, Taylor & Francis Group LLC, 2013, 164 p.

[44] Dineva, P., Gross, D., Müller, R. and Rangelov, T., *Dynamic Fracture of Piezoelectric Materials: Solution of Time-Harmonic Problems via BIEM*, Springer International Publishing, Switzerland, 2014, 249 p.

[45] Levinzon, F., *Piezoelectric Accelerometers with Integral Electronics*, Springer International Publishing Switzerland, 2015, 168 p.

[46] Hehn, T. and Manoli, Y., *CMOS Circuits for Piezoelectric Energy Harvesters — Efficient Power Extraction, Interface Modeling and Loss Analysis*, Springer Science + Business Media, Dordrecht, 2015, 203 p.

[47] Ferrer, I., *Piezoelectric Materials and Devices,* NY Research Press, 2015, 176 p.

[48] Giurgiutiu, V., *Structural Health Monitoring with Piezoelectric Wafer Active Sensors*, Elsevier Science and Technology Books, 2017, 1024 p.

[49] Bhalla, S., Moharana, S., Kaur, N., and Talakokula, V., *Piezoelectric Materials: Applications in SHM, Energy Harvesting and Biomechanics*, John Wiley & Sons, 2017, 352 p.

[50] Chong, K.P., Liu, S.C. and Li, J.C., *Intelligent Structures*, Elsevier Science Publishers Ltd., 1990, 458 p.

[51] Gandhi, M.V. and Thompson, B.D., *Smart Materials and Structures*, Springer Science & Business Media, 1992, 312 p.

[52] Guran, A. and Inman, D.J., (eds.), *Wave Motion, Intelligent Structures and Nonlinear Mechanics*, World Scientific Publishing Co. Pte. Ltd., 1995, 301 p.

[53] Holnicki-Szulc, J. and Rodellar, J., (eds.), *Smart Structures-Requirements and Potential Applications in Mechanical and Civil Engineering*, NATO Science Series, Vol. 3. High Technology, Vol. 65, Kluwer Academic Publishers, 1999, 397 p.

[54] Srinivasan, A.V. and McFarland, D.M., *Smart Structures — Analysis and Design*, Cambridge University Press, 2001, 231 p.

[55] Gabbert, U. and Tzou, H.S., *Smart Structures and Structronic Systems, Solid mechanics and its Application, IUTAM Symposium*, Kluwer Academic Publishers, 2001, 362 p.

[56] Suleman, A., (ed.), *Smart Structures — Applications and Related Technologies,* Int. Centre for Mechanical Sciences, Courses and Lectures, No. 429, CISM, Udine, Italy, 2001, Springer, Wien, 358 p.

[57] Watanabe, K. and Ziegler, F., (eds.), *Dynamics of Advanced Materials and Smart Structures*, Springer Science & Business Media, 2003, 472 p.

[58] Cheng, F., Jiang, H., and Lou, K., *Smart Structures — Innovative Systems for Seismic Response Control*, CRC Press, Taylor & Francis Group, L.L.C., 2008, 652 p.

[59] Janocha, H., (ed.), *Adaptronics and Smart Structures: Basics, Materials, Design and Applications*, 2nd Revised edition, Springer-Verlag, Berlin, 2007, 543 p.

[60] Carrera, E., Brischetto, S. and Nali, P., *Plates and Shells for Smart Structures: Classical and Advanced Theories for Modelling and Analysis*, John Wiley & Sons, 2011, 322 p.

[61] Gaudenzi, P., *Smart Structures: Physical Behavior, Mathematical Modelling and Applications*, John Wiley & Sons, 2009, 194 p.

[62] Vepa, R., *Dynamics of Smart Structures*, John Wiley & Sons, 2010, 395 p.

[63] Wen, Y.K., (ed.), *Intelligent Structures — 2: Monitoring and Control*, Taylor & Francis, 2012, 354 p.

[64] Chopra, I. and Sirohi, J., *Smart Structures Theory*, Cambridge University Press, 2014, 905 p.

[65] Xu, Z.-D., Guo, Y.-Q., Zhu, J.-T., Xu, F.-H. and Zhang, X.-C., *Intelligent Vibration Control in Civil Engineering Structures*, Elsevier, 2016, 278 p.
[66] Abramovich, H., *Intelligent Materials and Structures*, De Gruyter, 2016, 378 p.
[67] Lopes Jr., V., Steffen Jr., V., Savi, M.A., (eds.), *Dynamics of Smart Systems and Structures: Concepts and Applications*, Springer International Publishing Switzerland, 2016, 342 p.
[68] Araujo, A.L. and Soares, C.A.M., (eds.), *Smart Structures and Materials, Selected Papers from the 7th ECCOMAS Thematic Conference on Smart Structures and Materials*, Springer International Publishing Switzerland, 2017, 292 p.
[69] Caliò, R., Rongala, U.B., Camboni, D., Milazzo, M., Stefanini, C., de Petris, G., Oddo, C.M., Piezoelectric energy harvesting solutions, *Sensors*, Vol. 14, 2014, pp. 4755–4790, doi:10.3390/s1403044755.
[70] Uchino, K., Piezoelectric actuator renaissance, *Energy Harvesting Syst.*, Vol. 1(1–2), 2014, pp. 45–56, doi:10.15151/ehs-2013-0021.
[71] Uchino, K., Glory of piezoelectric perovskites, *Sci. Technol. Adv. Mater.*, Vol. 16, 2015, Paper Id: 048001, 16 p, doi:10.1088/1468-6996/16/4/046001.
[72] Uchino, K., Piezoelectric energy harvesting systems-essentials to successful developments, *Energy Technol.*, Vol. 6, 2018, pp. 829–848, doi:10.1002/ente.201700785.
[73] Wei, C. and Jing, X., A comprehensive review on vibration energy harvesting: modelling and realization, *Renew. Sustainable Energy Rev.*, Vol. 74, 2017, pp. 1–18.
[74] Ahmed, R., Mir, F. and Banerjee, S., A review on energy harvesting approaches for renewable energies from ambient vibrations and acoustic waves using piezoelectricity, *Smart Mater. Struct.*, Vol. 26, 2017, Paper Id 085031, doi:10.1088/1361-665X/aa7bfb.
[75] Tran, N., Ghayesh, M.H. and Arjomandi, M., Ambient vibration energy harvesters: a review on nonlinear techniques for performance enhancement, *Int. J. Eng. Sci.*, Vol. 127, 2018, pp. 162–185.
[76] Priya, S., Song, H.-C., Zhou, Y., Varghese, R., Chopra, A., Kim, S.-G., Kanno, I., Wu, L., Ha, D.S., Ryu, J. and Polcawich, R.G., A review on piezoelectric energy harvesting: materials, methods and circuits, *Energy Harvesting Syst.*, Vol. 4(1), 2017, pp. 1–37, doi:10.15151/ehs-2016-0028.
[77] Avvari, P.V., Yang, Y.and Soh, C.K., Long-term fatigue behavior of a cantilever piezoelectric energy harvester, *Intell. Mater. Syst. Struct.*, Vol. 28(9), 2017, 23 p, doi:10.1177/1045389X16667552.
[78] Abramovich, H. and Har-nes, I., Analysis and experimental validation of a piezoelectric harvester with enhanced frequency bandwidth, *Materials*, Vol. 11, 2018, 1243; doi:10.3390/ma11071243.
[79] Todaro, M.T., Guido, F., Mastronardi, V., Desmaele, D., Epifani, G., Algieri, L., and De Vittorio, M., Piezoelectric MEMS vibrational energy harvesters: advances and outlook-a review article, *Microelectron. Eng.*, Vol. 183–184, 2017, pp. 23–36.
[80] Fang, H. and Wang, K.W., Piezoelectric vibration-driven locomotion systems-exploiting resonance and bistable dynamics, *J. Sound Vibration*, Vol. 391, 2017, pp. 153–169.
[81] Li, B., A short review of piezoelectricity-based vibration and acoustic energies harvesting. *Innovative Energy Res.*, Vol. 4(3), 2015, 100e113. doi:10.4172/2576-1463.1000e113.
[82] Narita, F. and Fox, M., A review on piezoelectric, magnetostrictive and magnetoelectric materials and device technologies for energy harvesting applications, *Adv. Eng. Mater.*, Vol. 20, 2018, Paper Id. 1700743, 22 p, doi:10.1002/adem.201700743.
[83] Yang, Z., Zhou, S., Zu, J. and Inman, D., High-performance piezoelectric energy harvesters and their applications-review, *Joule*, Vol. 2, 2018, pp. 642–697.

[84] Gafforelli, G., Xu, R., Corigliano, A. and Kim, S.-G., Modeling of a bridge shaped nonlinear piezoelectric energy harvester, *Energy Harvesting Syst.*, Vol. 1(3–4), 2014, pp. 179–187.

[85] Baur, C., Zhou, Y., Sipes, J., Priya, S. and Voit, W., Organic, flexible, polymer composites for high-temperature piezoelectric applications, *Energy Harvesting Syst.*, Vol. 1(3–4), 2014, pp. 167–177.

[86] Abdelkefi, A., Aeroelastic energy harvesting: A review, *Int. J. Eng. Sci.*, Vol. 100, 2016, pp. 112–135.

[87] Ramadan, K.S., Sameoto, D. and Evoy, S., A review of piezoelectric polymers as functional materials for electromechanical transducers, *Smart Materials and Structures*, Vol. 23, 2014, Paper Id 033001, doi:10.1088/0964-1726/23/3/03301.

[88] Wang, D., Yuan, G., Hao, G. and Wang, Y., All-inorganic flexible piezoelectric energy harvester enabled by two-dimensional mica, *Nano Energy*, Vol. 43, 2018, pp. 351–358.

[89] Tian, W., Ling, Z., Yu, W. and Shi, J., A review of MEMS scale piezoelectric energy harvester, *Appl. Sci., MDPI*, Vol. 8, 2018, Paper Id: 645, 20 p., doi:10.3390/app8040645.

[90] Kumar, A., Chhabra, D., Recent developments in the field of piezoelectric energy harvesting & advanced MEMS: An overview, *Int. J. Sci. Res. Develop.*, Vol. 4(4), 2016, pp. 1536–1540.

[91] Cha, Y., Chae, W., Kim, H., Walcott, H., Peterson, S.D. and Porfiri, M., Energy harvesting from a piezoelectric biomimetic fish tail, *Renew. Energy*, Vol. 86, 2016, pp. 449–458.

[92] Lee, K.Y., Gupta, M.K. and Kim, S.-W., Transparent flexible stretchable piezoelectric and triboelectric nanogenerators for power portable electronics, *Nano Energy*, Vol. 14, 2015, pp. 139–160.

[93] Hu, Y. and Wang, Z.L., Recent progress in piezoelectric nanogenerators as a sustainable power source in self-powered systems and active sensors, *Nano Energy*, Vol. 14, 2015, pp. 3–14.

[94] Briscoe, J. and Dunn, S., Piezoelectric nanogenerators — a review of nanostructured piezoelectric energy harvesters, *Nano Energy*, Vol. 14, 2015, pp. 15–29.

[95] Dagdeviren, C., Joe, P., Tuzman, O.L., Park, K.-I., Lee, K.J., Shi, Y., Huang, Y. and Rogers, J.A., Recent progress in flexible and stretchable piezoelectric devices for mechanical energy harvesting, sensing and actuation, *Extreme Mech. Lett.*, Vol. 9, 2016, pp. 269–281.

[96] Bera, B., Recent advances in piezoelectric nanogenerators in energy harvesting applications, *Imperial J. Interdisciplinary Res.*, Vol. 2(7), 2016, pp. 1274–1291.

[97] Jin, L., Ma, S., Deng, W., Yan, C., Yang Chu, X., Tian, G., Xiong, D., Lu, J. and Yang, W., Polarization-free high-crystallization β-PVDF piezoelectric nanogenerator toward self-powered 3D acceleration sensor, *Nano Energy*, Vol. 50, 2018, pp. 632–638.

[98] Rastegar, J., Pereira, C.M. and Feng, D., A review of piezoelectric-based electrical energy harvesting methods and devices for munitions, *Proc. SPIE*, Vol. 9799, 2016, 97992M (15 April 2016), doi:10.1117/12.2218747.

[99] Abramovich, H., Tsikhotsky, E. and Klein, G., An experimental determination of the maximal allowable stresses for high power piezoelectric generators, *J. Ceramic Sci. Technol.*, Vol. 4(3), 2013, pp. 131–136.

[100] Abramovich, H., Tsikhotsky, E. and Klein, G., An experimental investigation on PZT behavior under mechanical and cycling loading, *J. Mech. Behav. Mater.*, Vol. 22(3–4), 2013, pp. 129–136.

[101] Kim, S., Shen, J. and Ahad, M., Piezoelectric-based energy harvesting technology for road sustainability, *Int. J. Appl. Sci. Technol.*, Vol. 5(1), 2015, pp. 20–25.

[102] Duarte, F. and Ferreira, A., Energy harvesting on road pavements: state of the art, proceedings of the institution of civil engineers, *Energy*, Vol. 169(EN2), 2016, Paper 1500005, pp. 79–90.

[103] Kour, R. and Charif, A., Piezoelectric roads: energy harvesting method using piezoelectric technology, *Innovative Energy and Res.*, Vol. 5(1), 2016, Paper Id: 10000132, 6 p, doi:10.4172/2576-1463.1000132.

[104] Papagiannakis, A.T., Montoya, A., Dessouky, S. and Helffrich, J., Development and evaluation of piezoelectric prototypes for roadway energy harvesting, *J. Energy Eng., ASCE*, Vol. 143(5), 2017, Paper Id: 04017034, 7 p, doi:10.1061/(ASCE)EY.1943-7897.0000467.

[105] Yang, H., Wang, L., Hou, Y., Guo, M., Ye, Z., Tong, X. and Wang, D., Development in stacked-array-type piezoelectric energy harvester in asphalt pavement, *J. Mater. Civil Eng., ASCE*, Vol. 29(11), 2017, Paper Id: 04017224, 9 p.

[106] Qabur, A. and Alshammari, K., A systematic review of energy harvesting from roadways by using piezoelectric materials technology, *Innovative Energy Res.*, Vol. 7(1), 2018, Paper Id: 10000191, 6 p, doi:10.4172/2576-1463.1000191.

[107] Walubita, L.F., Djebou, D.C.S., Faruk, A.N.M., Lee, S.I., Dessouky, S. and Hu, X., Prospective of societal and environmental benefits of piezoelectric technology in road energy harvesting, *Sustainability*, Vol. 10, 2018, 383, 13 p, doi:10.3390/su10020383.

[108] Gripp, J.A.B. and Rade, D.A., Vibration and noise control using shunted piezoelectric transducers: A review, *Mech. Syst. Signal Process.*, Vol. 112, 2018, pp. 359–383.

[109] Yang, C.-S.W., Lai, Y.-A. and Kim, J.-Y., A piezoelectric brace for passive suppression of structural vibration and energy harvesting, *Smart Mater. Struct.*, Vol. 26, 2017, Paper Id 085005, doi:10.1088/1361-665X/aa76e98.

[110] Zheng, T., Wu, J., Xiao, D. and Zhu, J., Recent development in lead-free perovskite piezoelectric bulk materials, *Prog. Mater. Sci.*, Vol. 98, 2018, pp. 552–624.

[111] Quan, N.D., Bac, L.H., Thiet, D.V., Hung, V.N. and Dung, D.D., Current development in lead-free $Bi_{0.5}(Na,K)_{0.5}TiO_3$-based piezoelectric materials, *Adv. Mater. Sci. Eng.*, Vol. 2014, 2014, Paper Id: 365391, 13 p.

[112] Salim, M., Salim, D., Chandran, D., Aljibori, H.S. and Kherbeet, A.S., Review of nano piezoelectric devices in biomedicine applications, *J. Intell. Mater. Syst. Struct.*, Vol. 29(10), 2018, pp. 2105–2121.

[113] Pohanka, M., The piezoelectric biosensors: principles and applications-a review, *Int. J. Electromechanical Sci.*, Vol. 12, 2017, pp. 496–506, doi:10.20964/2017.01.04.

[114] Pohanka, M., Overview of piezoelectric biosensors, immunosensors and DNA sensors and their applications, *Materials, MDPI*, Vol. 11, 2018, Paper Id: 448, 13 p. doi:10.3390/ma11030448.

[115] Hennet, P., Piezoelectric bone surgery: a review of the literature and potential applications in veterinary oromaxillofacial surgery, *Frontiers Veterinary Sci.*, Vol. 2, Article 8, 2015, 7 p, doi:103389/fvets.2015.0008.

[116] Meller, C. and Hevas, T.E., Piezoelectric technology in otolaryngology and head and neck surgery: a review, *J. Laryngology Otology*, Vol. 131(S2), 2017, pp. S12–S18, doi:10.1017/S0022215117000767.

[117] Jacob, J., More, N., Kalia, K. and Kapusetti, G., Piezoelectric smart biomaterials for bone and cartilage tissue engineering, *Inflammation Regeneration*, Vol. 38(2), 2018, 11 p, doi:10.1186/s41232-018-0059-8.

[118] Tandon, B., Blaker, J.J. and Cartmell, S.H., Piezoelectric materials as stimulatory biomedical materials and scaffold for bone repair, *Acta Biomater.*, Vol. 73, 2018, pp. 1–20.

[119] Behera, B. and Nemade, H.B., Recent developments of piezoelectric motors with diverse operating principles, *J. Micro Smart Syst.*, Vol. 6, 2017, pp. 173–185, doi:10.1007/s41683-017-0015-x.

[120] Vinogradov, A.N., Ivanikin, I.A., Lubchenco, R.V., Matveev, Y.V. and Titov, P.A., Development and research of peristaltic multiphase piezoelectric micro-pump, *Int. J. Environ. Sci. Education*, Vol. 11(17), 2016, pp. 9723–9731.

[121] Jung, J., Lee, W., Kang, W., Shin, E., Ryu, J. and Choi, H., Review of piezoelectric micromachined ultrasonic transducers and their applications, *J. Micromech. Microeng.*, Vol. 27, 2017, Paper Id: 113001, 24 p.

[122] Khan, A., Abas, Z., Kim, H.S. and Oh, I.-K., Piezoelectric thin films: an integrated review of transducers and energy harvesting, *Smart Mater. Struct.*, Vol. 25, 2016, Paper Id 053002, doi:10.1088/0964-1726/25/5/053002.

[123] Liu, S. and Cohen, R.E., Origin of negative longitudinal piezoelectric effect, *Phys. Rev. Lett.*, Vol. 119, 2017, Paper Id: 207601, 5 p, doi:10.1103/PhysRevLett.119.207601.

[124] Calahorra, Y. and Kar-Narayan, S., Piezoelectricity in non-nitride III-V nanowires: challenges and opportunities, *J. Mater. Res.*, Vol. 33(6), 2018, pp. 611–624.

[125] Cui, C., Xue, F., Hu, W.-J. and Li, L.-J., Two-dimensional materials with piezoelectric and ferroelectric functionalities, NPJ/2D *Mater. Appl.*, Vol. 2(18), 2018, 14 p, doi:10.1038/s41699-018-0063-5.

[126] Garcia, Y., Ruiz-Blanco, Y.B., Marrero-Ponce, Y. and Sotomayor-Torres, C.M., Orthotropic piezoelectricity in 2D nanocellulose, *Sci. Rep.*, Vol. 6, 2016, Paper Id: 34616, doi:10.1038/srep34616.

[127] Tuukkanen, S. and Rajala, S., A survey of printable piezoelectric sensors, *IEEE Sensors*, Vol. 2015, 2015, pp. 1426–1429, doi:10.1109/ICSENS.2015.7370542.

[128] Marzeika, D., Vasiljev, P., Borodinas, S. and Yang, Y., Piezoelectric actuator based on two bending-type langevin transducers, *Energy Harvesting Syst.*, Vol. 2(3–4), 2015, pp. 193–199.

[129] Presas, A., Luo, Y., Wang, Z., Valentin, D. and Egusquiza, M., A review of PZT patches applications in submerged systems, *Sensors, MDPI*, Vol. 18, Paper Id: 2251, 21 p, doi:10.3390/s18072251.

Chapter 9

Introduction to Shape Memory Alloys

9.1. Basic Characteristics of Shape Memory Alloys

Shape Memory Alloys (SMAs) are metallic materials that demonstrate the ability to return to some previously defined shape or size when subjected to the appropriate thermal and/or mechanical loadings. Normally, these materials can be plastically deformed at some relatively low temperature, and upon exposure to some higher temperature will return to their shape prior to the deformation. It is custom to denote those materials that would present "shape memory" only due to heating as *one-way shape memory* alloys. Other materials would undergo a change in shape upon re-cooling having the property of a *two-way shape memory*. Those materials normally present another phenomenon called pseudoelasticity, which consists at displaying large elastic strains (6–7%) and their recovery when unloading, without leaving any plastic strains. Both phenomena, the shape memory and pseudoelasticity are enabled due to two metallurgical phases, the parent phase being *Austenite* (A) (at higher temperature) and at lower temperature, the second phase called *Martensite* (M) (see Fig. 9.1).

Historically, already in 1932, a Swedish researcher Arne Olander observed the shape and recovery ability of a gold–cadmium alloy (Au–Cd) and noted that it actually created motion [1]. In 1950, Chang and Read at Columbia University observed this unusual motion at the microscopic level by using x-rays to note the changes in crystal structure of Au–Cd [2]. Because of this study, other similar alloys were discovered including indium titanium. Only in 1963, Buehler and co-workers at the US Naval Ordinance Laboratory (NOL) [13, 14] observed the shape memory effect in a nickel and titanium alloy, today known as NITINOL (***Ni****ckel* ***Ti****tanium*-***N****aval*

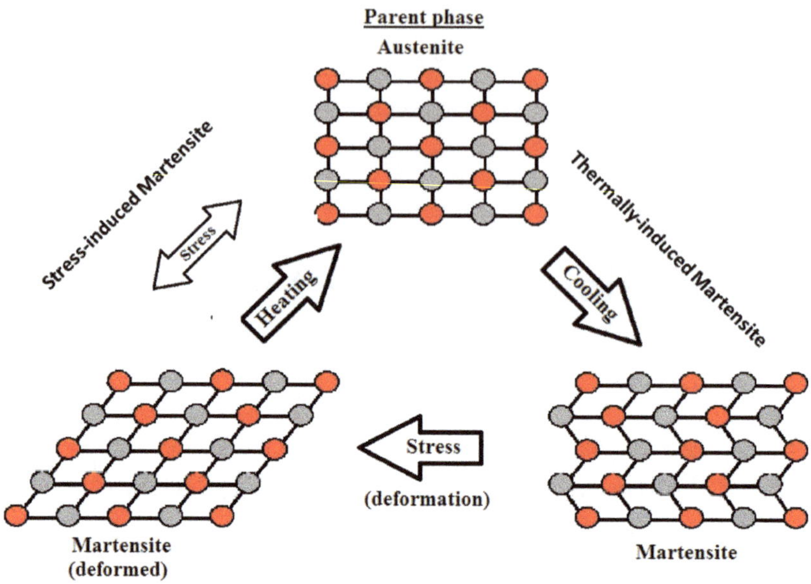

Fig. 9.1. Schematic illustration of shape memory effect.

O*rdnance* ***L****aboratory*) [3, 4]. Table 9.1 (from [5]) shows some of the SMA materials and their relevant properties.

As already stated before, the SMA's two phases are austenite and martensite, each having different crystal structure yielding different properties. The crystal structure of austenite is generally a cubic one, while the martensite's one can be tetragonal, orthorhombic or monoclinic. The transformation from parent metallurgical phase (austenite) to the martensite is made due to shear lattice distortion. This transformation is also named as *martensitic transformation*. Each formed martensitic crystal may have a different orientation direction, called a *variant* [6]. Usually, the assembly of martensitic variants may appear in two forms: the *twinned martensite* (M_t), which is formed by a combination of "self-accommodated" martensitic variants, and *detwinned* (reoriented) *martensite* (M_d), having a dominant specific variant. This reversible phase transformation from austenite (the parent phase) to martensite (the product phase) and vice versa, causes the unique behavior of SMAs.

For the simple case of only cooling the SMA material, namely, no mechanical loads are applied, the crystallographic structure would change from austenite to martensite, which is also called the *forward transformation*. This transformation causes the formation of several martensitic variants (for example up to 24 variants for Nitinol) and their space arrangement causes a

Table 9.1. Alloys with shape memory effect (from [5]).

Alloy	Composition	Transformation-temperature range °C	Transformation hysteresis Δ°C
Ag-Cd	44/49 at.% Cd[a]	−190 to −50	~15
Au–Cd	46.5/50 at.% Cd	30 to 100	~15
Cu–Al–Ni	14/14.5 wt% Al or 3/4.5 wt% Ni[b]	−140 to 100	~35
Cu–Sn	~15 at.% Sn	−120 to 30	
Cu–Zn	38.5/41.5 wt% Zn	−180 to −10	~10
Cu–Zn–X (X = Si, Sn, Al)	A few wt% X	−180 to 200	~10
In–Ti	18/23 at.% Ti	60 to 100	~4
Ni–Al	36/38 at.% Al	−180 to 100	~10
Ni–Ti	49/51 at.% Ni	−50 to 110	~30
Fe–Pt	~25 at.% Pt	~ −130	~4
Mn–Cu	5/35 at.% Cu	−250 to 180	~25
Fe–Mn–Si	32 wt% Mn, or 6 wt% Si	−200 to 150	~100

Note: [a]at.% = atomic percent.
[b]wt% = weight percent.

negligible average macroscopic shape change yielding a *twinned martensite*. Heating it from the martensitic phase transforms the crystal back to austenite phase in what it is called as the *reverse transformation*, again with no associated shape change.

The phase transformation from austenite phase to martensite phase and back has four characteristic temperatures associated with the phase transformation. For the forward transformation, in the absence of mechanical load, the temperature the material starts to transform to twinned martensite is called the martensitic start temperature (M_s), while the temperature at which this transformation is completed is named the martensitic finish temperature (M_f). At this temperature and below it, all the material is in the twinned martensitic phase. When performing the reverse transformation, and thus heating, the initiation of the austenite phase will start at the austenitic start temperature (A_s) with a complete transformation at the austenitic finish temperature (A_f). A schematic graph of the four transformation temperatures is presented in Fig. 9.2.

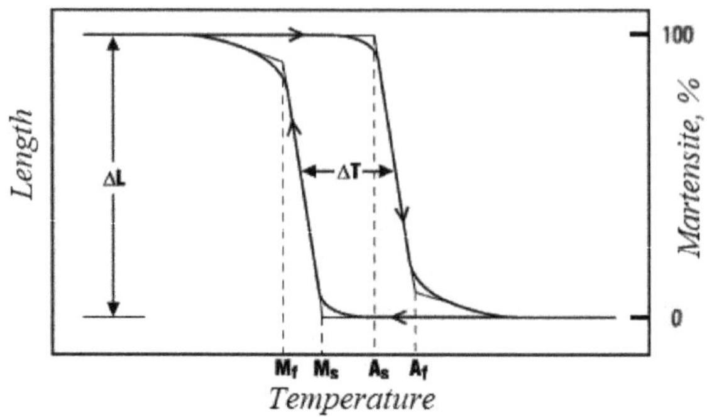

Fig. 9.2. Typical transformation vs. temperature curve for a SMA specimen.

Until now, we had talked about phase transformations in the absence of mechanical loads being applied on the SMA material. However, if such a mechanical load is being applied on the material, while being in its twinned martensitic phase, namely at a low temperature, it might be possible to induce a detwinning process by reorienting some of the variants. In contrast to the twinning process, which is associated with no change in the shape of the crystal, the detwinning process gives rise to a macroscopic shape change, yielding a deformed configuration with the load being released. Now without any mechanical load being applied to the deformed material, heating the SMA to a temperature above A_f will bring the detwinned martensite to austenite showing a complete shape recovery. Cooling back to a temperature below M_f causes this time the formation of twinned martensite showing no shape change. This process is sometimes referred to as the Shape Memory Effect (SME).

Note that the applied load to start the detwinning process must be above the minimum stress required for detwinning initiation, which is named the detwinning start stress (σ_s). The stress at which the complete process of detwinning of the martensite is achieved is termed the detwinning finish stress (σ_f). In addition, when the material is cooled in the presence of a stress larger than (σ_s), which is applied in the austenitic phase, the phase transformation will yield formation of a detwinned martensite, thus inducing a shape change. Reheating the material will again lead to a shape recovery while the stress is still being applied. One should note that the transformation temperature will increase with the increase of the applied stress (be it tension or compression).

Another interesting phenomenon is the capability to induce a phase transformation by only applying a sufficiently high mechanical stress to the material while being in its austenitic phase. This high stress will induce fully detwinned martensite created from austenite. If its temperature is above

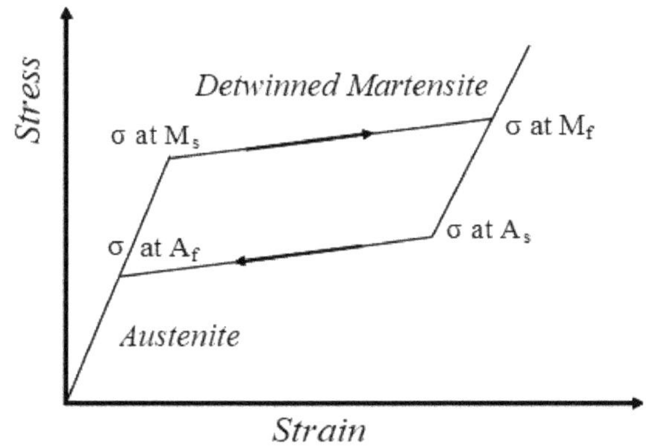

Fig. 9.3. Schematic stress–strain curve for the pseudoelasticity behavior.

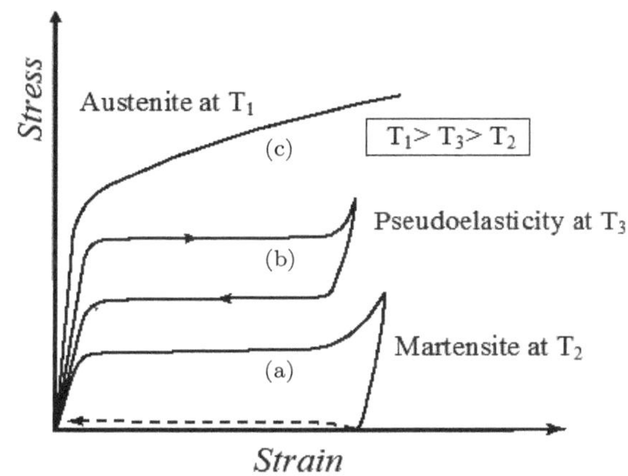

Fig. 9.4. Typical stress–strain curves at various temperatures: (a) martensite; (b) pseudoelasticity; and (c) austenite.

A_f, a complete shape recovery is obtained when unloading to its austenite phase. This behavior is usually named as the pseudoelasticity effect of SMA's (see Fig. 9.3). The loading stress at which the martensite transformation starts it is shown on Fig. 9.3 as σ at M_s, while σ at M_f denotes the completion of the process. For the unloading process of the SMA, the σ at A_s is the stress at which the austenite phase of the material is initiated, while σ at A_f is the stress at which the material completes its reverse transformation to austenite phase. Note that for the case of a material in its austenite phase tested above its M_s temperature, but below the A_f temperature, one can expect only a partial shape recovery. Figure 9.4 presents typical stress–strain curves for three cases, austenite and martensite phases and the pseudoelasticity behavior, at three different temperatures.

9.2. The Shape Memory Effect

As described above, the first effect to be shown by an SMA is, its shape memory effect (SME) when it is deformed in its twinned martensitic phase and then unloaded at a temperature below A_s. From this shape, heating the SMA above its A_f, the SMA will return to its original shape by transforming back into the parent austenitic phase. This property of the SMA is presented in Fig. 9.5 (from [6]) for a 3D curve portraying the experimental behavior of Nitinol under uniaxial loading in the form of axial stress σ (see Table 9.2 for the Nitinol properties).

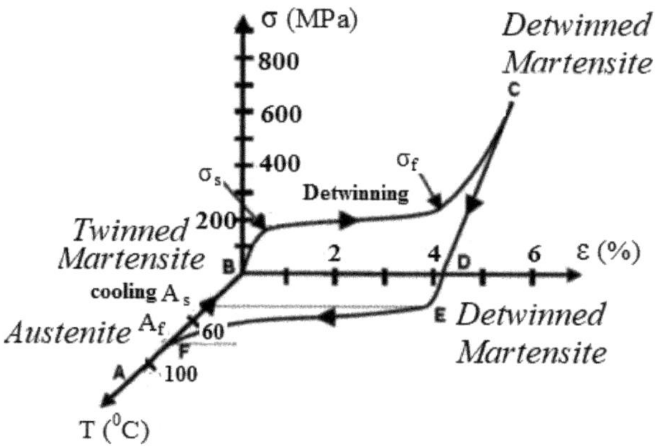

Fig. 9.5. A typical experimental Nitinol loading path exhibiting its shape memory effect (adapted from [6]).

Table 9.2. Nitinol's properties.[a]

Density	6.45 g/cm^3
Thermal conductivity	10 W/m^{0}K
Specific heat	322 J/kg^{0}K
Latent heat	24,200 J/kg
Ultimate tensile strength	750–960 MPa
Elongation to failure	15.5%
Yield strength (austenite)	560 MPa
Young's modulus (austenite)	75 GPa
Yield strength (martensite)	100 MPa

Note: [a]Taken from NiTi Aerospace, Inc., 2235 Polvorosa Ave, San Leandro, CA 94577, www.tiniaerospace.com.

Referring to Fig. 9.5, the process starts at point A (the parent phase — austenite). Cooling the material below M_s and M_f, in the absence of mechanical stress causes the formation of twinned martensite resulting, in point B on the graph. From this point, the twinned martensite is loaded by a stress exceeding its σ_s (start stress level), causing the starting of the reorientation process — the detwinning process. Note that the stress level is far lower than the permanent plastic yield stress of martensite. The detwinning process will be completed at σ_f (the end of the plateau in the σ–ε graph presented in Fig. 9.5). Elastically unloading the SMA material to zero stress, from point C to point D, without heating, keeps its detwinned martensite state. If heating starts, without mechanical stress the back transformation is initiated at A_s (point E) and finishes at A_f (point F). At any temperature above A_f, the material would be in its austenite phase and the original shape of the SMA is recovered (point A). The whole process described above is called *one-way shape memory effect* (OWSME), as the shape recovery was achieved due to the application of heating after the material had been detwinned due to an applied mechanical stress.

When an SMA is exposed to cyclic thermal loading in the absence of mechanical stresses, repeatable shape changes may occur. This behavior is called *two-way shape memory effect* (TWSME) or training. This process can induce changes in the microstructure, leading to macroscopically permanent changes in the material behavior. The training of an SMA is sometimes needed to obtain a material that its hysteretic response is stable and its inelastic strain saturates.

9.3. Pseudoelasticity

The second effect exhibited by an SMA is its pseudoelastic behavior, which is associated with stress-induced transformation, leading to large strain generation during loading which are reduced to zero upon unloading at temperatures above A_f. As pointed out by Kumar and Lagoudas [6], a pseudoelastic process can be either described by the path ($a \to b \to c \to d \to e \to a$) as presented in Fig. 9.6 as path 1, or by the path 2 in Fig. 9.6 which is performed at a nominally constant temperature above A_f.

Referring to Fig. 9.7, the process starts at zero stress at a temperature above A_f at point A. While the material is at its parent (austenite) phase, an elastic loading is applied bringing the curve to point B. The curve intersects the stress σ^{Ms}, namely the starting of transformation to the martensite phase diagram. This process is accompanied by the generation of large inelastic

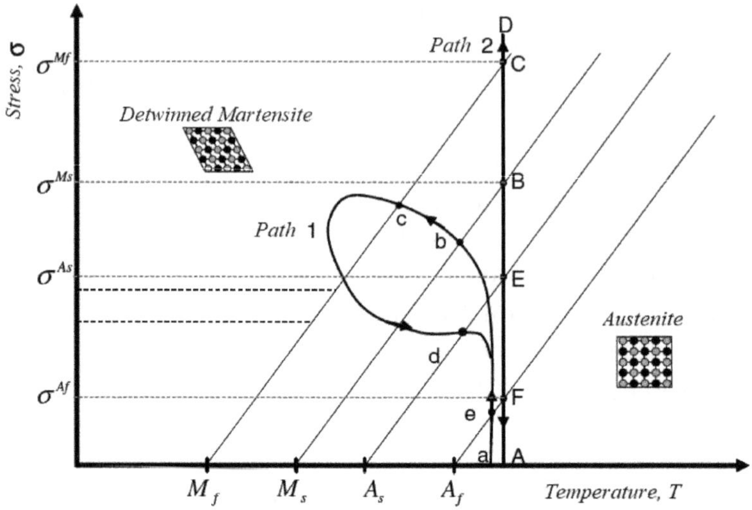

Fig. 9.6. A typical phase diagram with two pseudoelastic loading paths (taken from [6]).

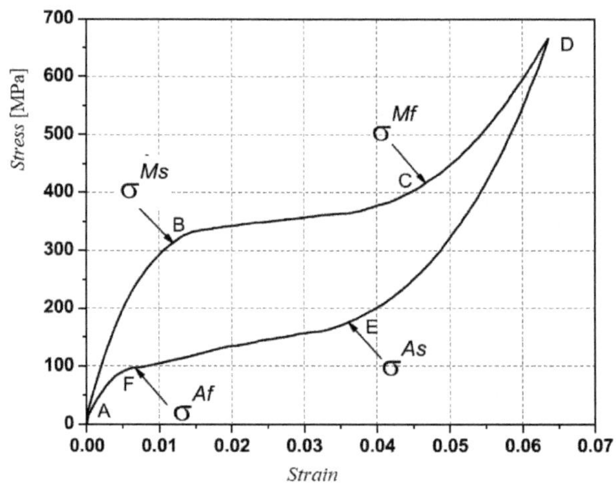

Fig. 9.7. A typical SMA pseudoelastic loading cycle (adapted from [6]).

strains (see Fig. 9.7). The curve continues to point C, characterized by the stress σ^{Mf}, at which the whole material is in its martensite phase.

At this point (C) a distinct change in slope on the σ–ε curve, associated with the elastic loading of the martensitic phase can be observed. Further increasing of the applied mechanical stress causes only elastic deformation of the martensite in its detwinned phase (the path from C to D). Point D is characterized by the largest stress and associated strain. Starting the unloading process from point D, the martensite phase is elastically unloaded up to the stress σ^{As} (point E) where it starts to revert back into its austenite phase. From point D to point E, the strain is recovered without permanent plastic strains. Point F, characterized by σ^{Af}, is the point where the curve meets the elastic region of the austenite phase. Further unloading the SMA to zero stress brings the material to the original point A without any residual strains.

One should note that passing the material a complete pseudoelastic cycle will form a hysteresis curve representing the energy being dissipated during this cycle. The size of the hysteresis would depend on the stress levels and the conditions of the testing.

9.4. Applications

The outstanding properties of SMAs are attracting technological interest in many fields of sciences and engineering, starting from medical up to aerospace applications. An interesting review of the SMA application in the medical field was written by Petrini and Migliavacca [7]. They claim that the shape memory effect and the pseudoelasticity effect together with good corrosion and bending resistance, biological and magnetic resonance compatibility, might explain the large market penetration, in the last 20 years, of SMA in the production of biomedical devices in dental, orthopedics, vascular, neurological, and surgical fields. Typical examples of such applications would be SMA orthodontic wires (Fig. 9.8), treatment of orthopedic issues using plates made of Nitinol (Fig. 9.9) and various uses of Nitinol devices like stents and filters in the vascular field (Figs. 9.10(a)–9.10(c)). The reader can find more applications of the SMA in the biomedical sector in Ref. [7].

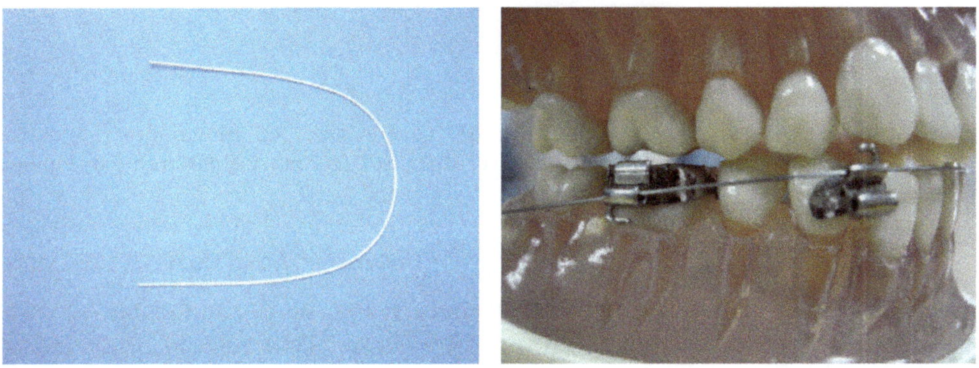

Fig. 9.8. Typical SMA orthodontic wires (adapted from [7]).

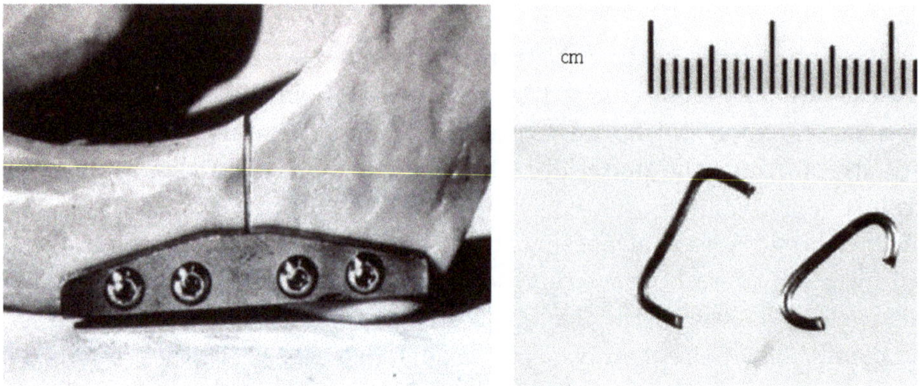

(a) (b)

Fig. 9.9. (a) NiTi plate for mandible fracture [8] and (b) staple before and after distraction [9].

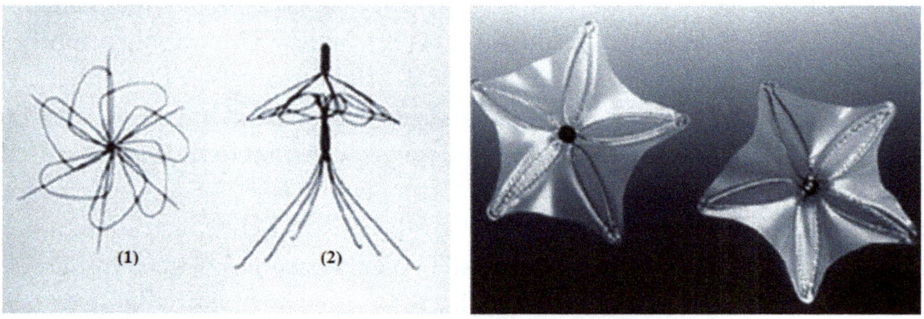

(a) (b)

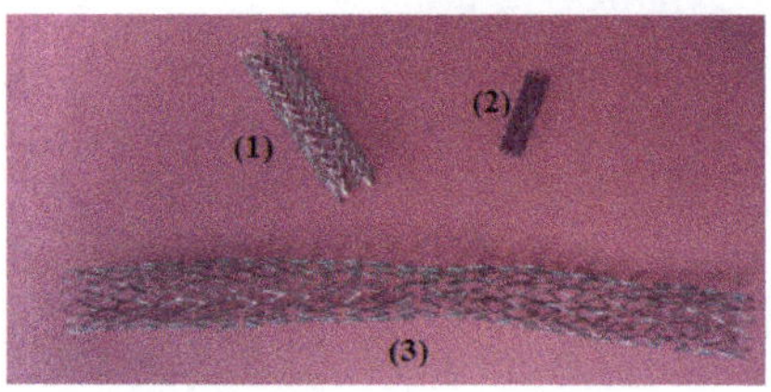

(c)

Fig. 9.10. Use of NiTi in the vascular field: (a1) Venous filter; (a2) Simon filter; (b) device to close ventricular septal defects; (c1) carotid stent; (c2) coronary stent; and (c3) femoral stent (adapted from [7]).

Although the biomedical applications of SMA material with an emphasis on Nitinol can be considered as the main usage of the material, one can find interesting and innovative designs involving SMA also in other industrial fields like mechanical, aerospace and civil engineering. Those applications will include fastening devices, actuators, SMA actuated valves, springs,

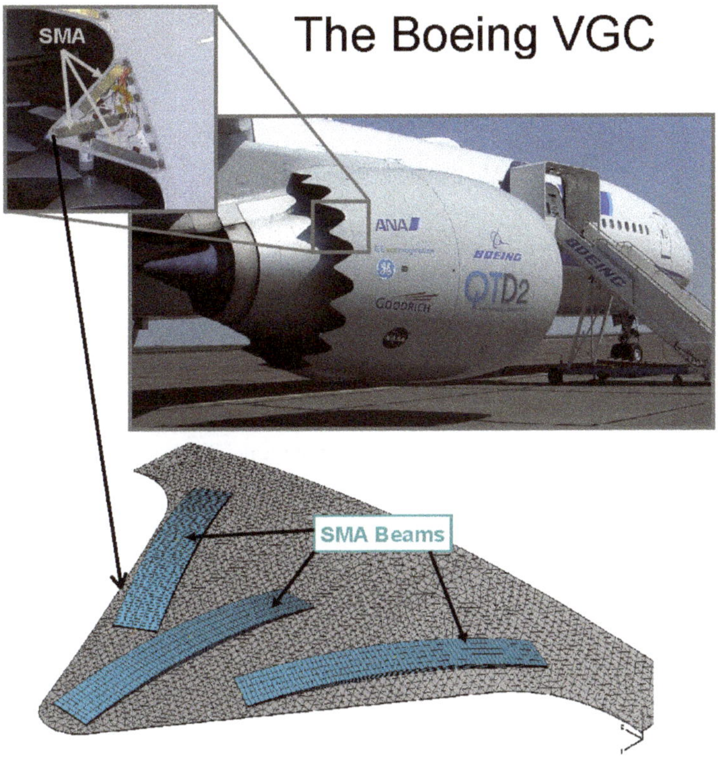

Fig. 9.11. Boeing's variable geometry chevron — VGC, a morphing aerospace structure for jet noise reduction [10].

damping devices, variable geometry chevrons, reconfigurable rotor blades, active hinge pin actuators, variable area nozzles, locking mechanisms, thermal switches and many other innovative designs to solve practical industrial problems. Figure 9.11 presents Boeing's variable geometry chevron — VGC, a morphing aerospace structure aimed at the reduction of noise emitted by the aircraft engine. CDI — Aerospace Engineering[a] company presents an interesting application of morphing airfoils using SMA material (see Fig. 9.12). A schematic drawing of plane wings with SMA wires which can change shape by applying voltages for heating the wires is presented in Fig. 9.13. This mechanics might replace the conventional hydraulic and electromechanical actuators existing today in aerospace industry. Another application of SMA wires was described by Pitt *et al.* [11] in their overview on the NASA lead SAMPSON smart inlet project. Various uses of SMA wires together with Flexskin or Conformal Moldline Technology (CMT) which is a

[a]www.continuum-dynamics.com.

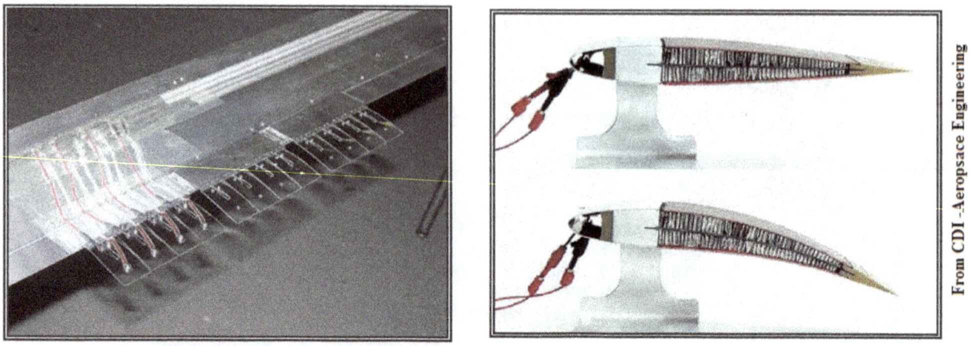

(a) (b)

Fig. 9.12. Morphing airfoils: (a) CDI's SMA — actuated tabs installed on a rotor blade and (b) CDI/Lockheed Martin continuously deformable SMA controlled airfoil section.

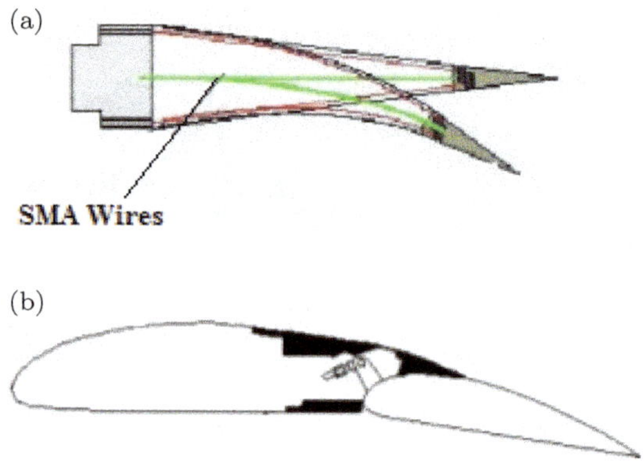

Fig. 9.13. Changing shape of a wing: and (a) SMA actuated flap, and (b) conventional flap.

Boeing developed technology designed to provide the structural flexibility or compliance needed for gross shape changes, while also providing structural stiffness and smooth surfaces. Flexskin is basically an elastomeric panel with structural rods running through. Some details of the integration of the SMA wires are presented in Fig. 9.14.

An actuator with SMA wires providing "latch" control and lockout until electronically released was developed by Autosplice[b] company which claims that the actuator can also provide "burst control" in actual applications (Fig. 9.15). S^3Lab[c] (Smart Structures and Systems Laboratory) affiliated with INHA University, Korea, presents other interesting application of SMA

[b]corp.autosplice.com.
[c]ssslab.com.

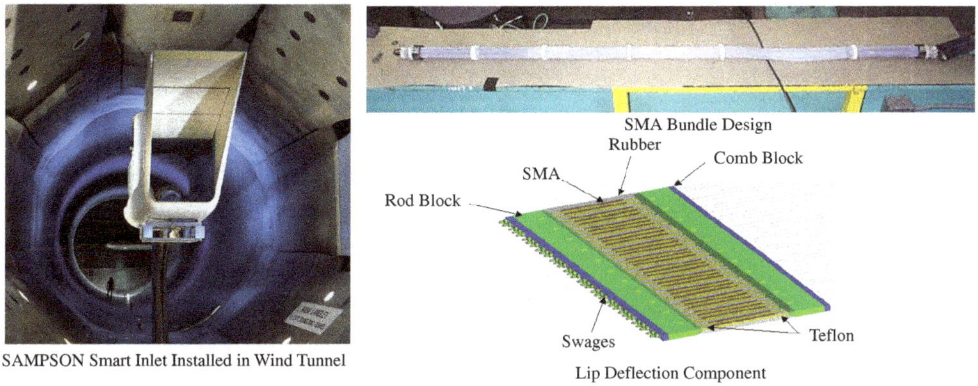

Fig. 9.14. NASA's SAMPSON smart inlet project with integrated CMT/SMA rods.

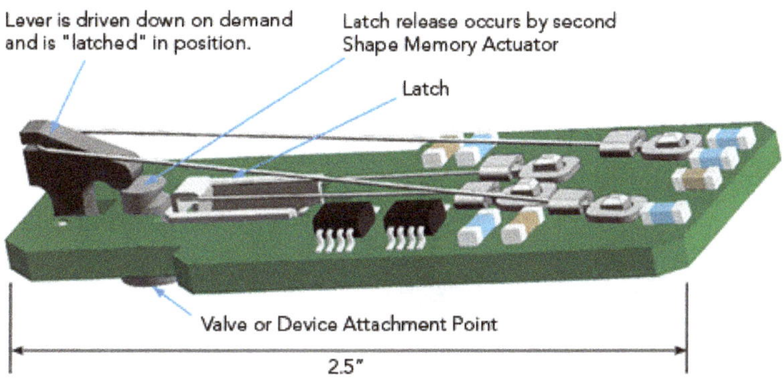

Fig. 9.15. SMA-based actuator (from Autosplice company).

material in the form of HDD (Hard Disk Drive) suspension and a robot micro-gripper as they are depicted in Figs. 9.16(a) and 9.16(b). The Computational Mechanics & Advanced Materials Group[d] affiliated with the department of civil engineering and architecture, University of Pavia, Italy is also active in designing and implementing SMA based devices as are presented in Fig. 9.17.

The Institute of Microstructure Technology (IMT) affiliate with the Karlsruhe Institute of Technology, Germany[e] presents a monostable SMA based microvalve (Fig. 9.18) made of a polymer housing, with an integrated fluid chamber, a membrane and an SMA microactuator being deflected by a microball. A zero current the microvalve is open, allowing the fluid to flow

---

[d]http://www-2.unipv.it/compmech/mat_const_mod.html.
[e]www.imt.kit.edu/english/1528.php.

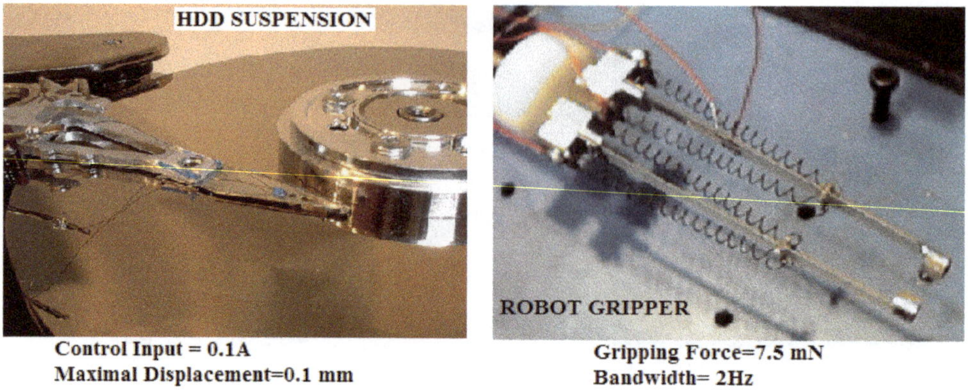

(a) (b)

Fig. 9.16. S^{3}Lab SMA-based devices: (a) HDD suspension; and (b) Robot-Gripper.

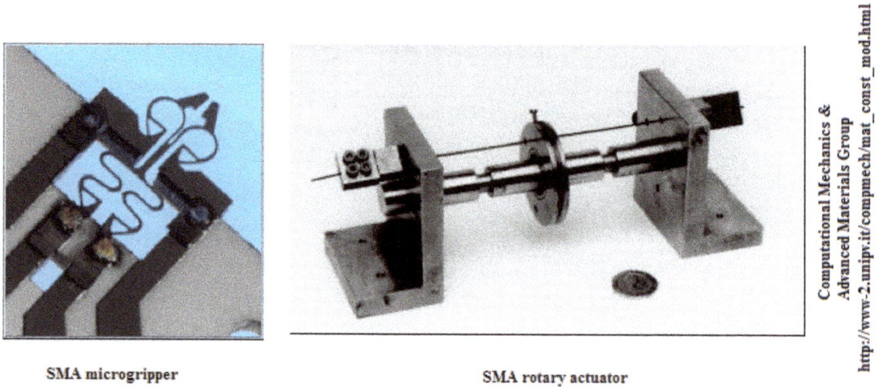

Fig. 9.17. Computational Mechanics & Advanced Materials Group SMA-based devices.

through the valve is a supply pressure is present. Electrically heating the microactuator it changes to its planar shape memory state and thus closing the microvalve.

Zuudee Holding Group[f], a well-known Chinese manufacturing group manufactures various SMA type springs including two ways NiTi SMA springs with a wire diameter between 0.1–10 mm for automobile and industrial uses (Fig. 9.19). Another application interesting application is the use of SMA Nitinol springs and Belleville washers as dampers within the NEESR-RC project led and performed by Georgia Tech[g] faculty, graduate and undergraduate students.

[f] http://zuudee.com/TiNi SMAAlloys/shape memory alloy NITI ASTM F2063-5.html.
[g] neesrcr.gatech.edu

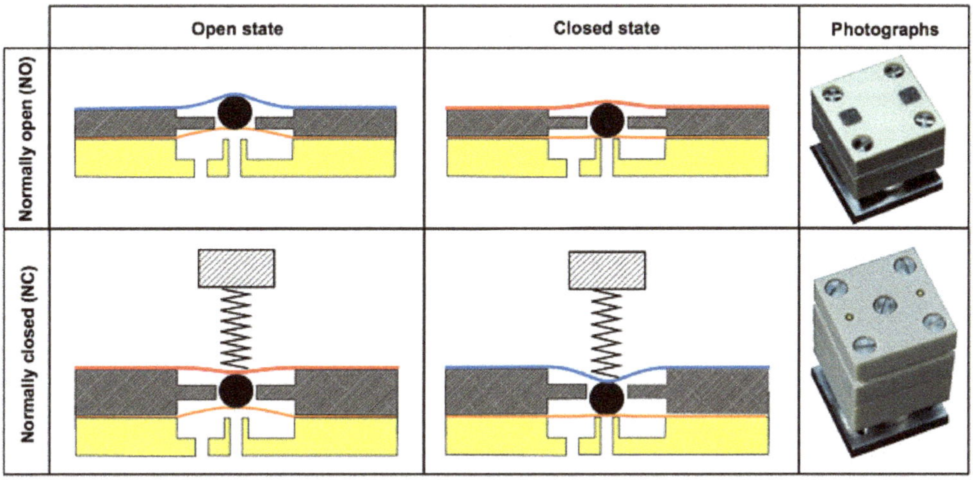

Fig. 9.18. The IMT KIT Germany monostable SMA based microvalve.

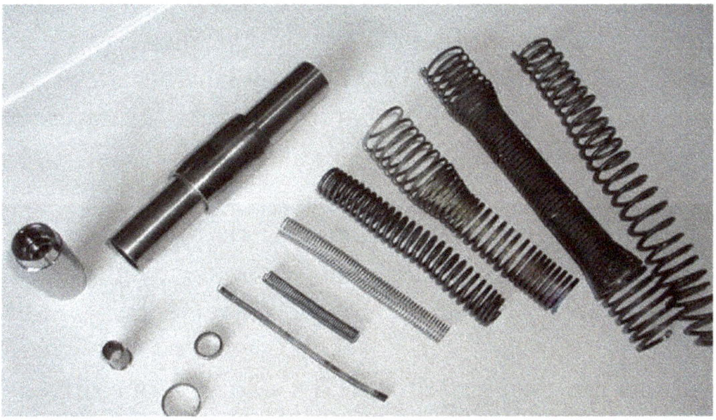

Fig. 9.19. The various NiTi SMA springs manufactured by Zuudee Holding Group.

The aim of the project is to retrofit existing houses and buildings with dampers (Fig. 9.20) built from SMA to allow a better seismic protection. The SMA-based system is considered to have the ductility and energy dissipation to prevent collapse and the ability to significantly reduce the residual deformations after an earthquake occurs. A tension/compression device was developed for applications as bracing elements in buildings. The device was designed to allow Nitinol helical springs or Belleville washers, to be used in compression. The results show that Nitinol helical springs produce good re-centering and damping behavior while the Nitinol Belleville washers show good potential to form the basis for a Nitinol damping device.

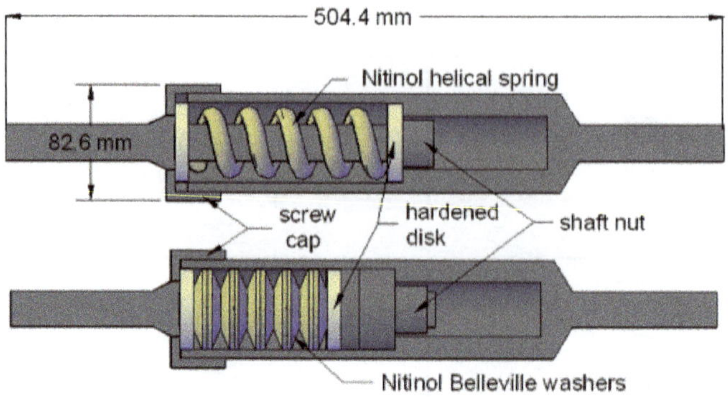

Fig. 9.20. The two SMA based dampers designed and manufactured within NEESR-RC project.

9.5. SMA Constitutive Equations

As described above, shape memory alloys (SMA) possess two distinct properties: the shape memory effect (SME) and its Pseudoelasticity effect (PE). Since the discovery of the SME property in Cu–Zn alloy by Greninger and Mooradian [12], and in Nitinol (NiTi) alloy by Buehler *et al.* [13, 14], Nitinol has been widely used as a smart material due to its large load capacity, high recovery strain (up to 8%), excellent fatigue performance and variable elasticity due to its phase transformation.

To simulate those specific properties, many constitutive laws had been proposed, such as in the form of phenomenological models [13–15], micromechanics models and 3D model for polycrystalline SMA based on a micro-plane theory [16].

To understand the behavior of an SMA alloy, an one-dimensional constitutive model will be presented and the various adaptations of this model to fit the experimental data will be outlined based on existing empirical methods published in the literature. Finally, a method to heat a SMA materials using electrical current will be described.

9.5.1. *1D constitutive equations for an SMA material*

Tanaka [13, 14] showed that a sufficient condition for the Clausius–Duhem inequality[h] to hold is the following equation:

$$\sigma(\varepsilon, T, \xi) = \rho_0 \frac{\partial \Phi}{\partial \varepsilon}, \tag{9.1}$$

[h]The Clausius–Duhem inequality is a way of expressing the *second law of thermodynamics* that is used in *continuum mechanics*. This inequality is particularly useful in determining whether the *constitutive relation* of a material is thermodynamically allowable.

where the stress σ is a function of the strain ε, the temperature T and the martensite fraction ξ and ρ_0 and Φ are the density of the material and the Helmholtz-free energy,[i], respectively.

Using differential calculus, Eq. (9.1) can be rewritten as

$$d\sigma = \frac{\partial \sigma}{\partial \varepsilon} d\varepsilon + \frac{\partial \sigma}{\partial T} dT + \frac{\partial \sigma}{\partial \xi} d\xi \equiv E d\varepsilon + \Theta dT + \Omega d\xi, \tag{9.2}$$

where $E = \rho_0 \frac{\partial^2 \Phi}{\partial \varepsilon^2}$ is the SMA modulus, $\Theta = \rho_0 \frac{\partial^2 \Phi}{\partial \varepsilon \partial T}$ is related to the thermal coefficient of expansion and $\Omega = \rho_0 \frac{\partial^2 \Phi}{\partial \varepsilon \partial \xi}$ is the transformation coefficient. One should note that according to this model the only state variables are the mono-axial strain ε, the temperature T and the volume fraction of the martensite phase, ξ.

Simplifying the constitutive model presented in Eq. (9.2) yields

$$(\sigma - \sigma_0) = E(\xi)(\varepsilon - \varepsilon_0) + \Theta(T - T_0) + \Omega(\xi)(\xi - \xi_0), \tag{9.3}$$

where the terms $\sigma_0, \varepsilon_0, T_0$ and ξ_0 stand for the initial state of the material. The modulus of elasticity $E(\xi)$ and the phase transformation coefficient $\Omega(\xi)$ are functions of the martensite volume fraction ξ and can be written as

$$E(\xi) = E_A + \xi(E_M - E_A) \quad \text{and} \quad \Omega(\xi) = -\varepsilon_L E(\xi) \tag{9.4}$$

where E_A and E_M represent the Young's moduli for the austenite and martensite phases, respectively and ε_L is the maximum recoverable strain. Tanaka's expressions for the martensite volume fraction were determined by a dissipation potential which depends both on the stress and the temperature and have an exponential form.

For the transformation from austenite to martensite (namely cooling) the equation is:

$$\xi_{A \to M} = 1 - e^{[\alpha_M (M_S - T) + \beta_M \sigma]}, \tag{9.5}$$

while the transformation from martensite to austenite (namely heating) the equation is

$$\xi_{M \to A} = e^{[\alpha_A (A_S - T) + \beta_A \sigma]} \tag{9.6}$$

The material constants appearing in Eqs. (4.25) and (4.26) are defined as:

$$\begin{aligned} \alpha_A &= \frac{\ln(0.01)}{(A_S - A_f)}, \quad \beta_A = \frac{\alpha_A}{C_A}, \\ \alpha_M &= \frac{\ln(0.01)}{(M_S - M_f)}, \quad \beta_M = \frac{\alpha_M}{C_M}. \end{aligned} \tag{9.7}$$

[i]In thermodynamics, the Helmholtz-free energy is a thermodynamic potential that measures the "useful" work obtainable from a closed thermodynamic system at a constant temperature.

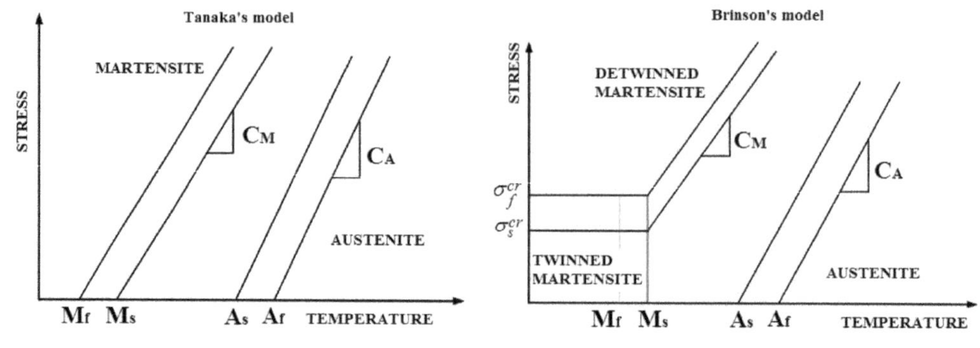

(a) (b)

Fig. 9.21. Stress–temperature curves: (a) Tanaka's model and (b) Brinson's model.

The coefficients E, Θ and Ω together with the other parameters like M_s, M_f, A_s, A_f, C_A and C_M are experimentally determined for each material.

The critical stress–temperature profile used for the Tanaka's model is presented in Fig. 9.21(a).

A further modification of the one-dimensional constitutive equation proposed by Tanaka was introduced by Liang and Rogers [18]. They presented the martensite volume fraction using a cosine formulation. Therefore, their formulation for the transformation from austenite to martensite (cooling) has the form:

$$\xi_{A\to M} = \frac{1-\xi_0}{2}\cos[a_M(T-M_f)+b_M\sigma]+\frac{1+\xi_0}{2}, \tag{9.8}$$

while for the inverse transformation from martensite to austenite (heating) their equation is

$$\xi_{M\to A} = \frac{\xi_0}{2}\cos[a_A(T-A_s)+b_A\sigma]+\frac{\xi_0}{2}. \tag{9.9}$$

The material constants appearing in Eqs. (9.8) and (9.9) are defined as

$$a_M = \frac{\pi}{(M_S-M_f)},\quad b_M = -\frac{a_M}{C_M},\quad a_A = \frac{\pi}{(A_f-A_s)},\quad b_A = -\frac{a_A}{C_A}. \tag{9.10}$$

where ξ_0 is the initial martensite volume fraction. Note that the argument of the cosine function can have a value between 0 and π leading to restrictions for the temperature and the stress, namely for the transformation from austenite to the martensite phase one would require that:

$$\begin{aligned} &M_f \le T \le M_s \\ &\text{and} \\ &C_M(T-M_f)-\frac{\pi}{|b_M|} \le \sigma \le C_M(T-M_f), \end{aligned} \tag{9.11}$$

while for the transformation from martensite to austenite (namely heating) the restrictions are

$$A_s \leq T \leq A_f$$

and

$$C_A(T - A_s) - \frac{\pi}{|b_A|} \leq \sigma \leq C_A(T - A_s). \quad (9.12)$$

One should note that both Tanaka and Liang and Rogers models can only correctly represent the phase transformation from martensite to austenite and its reverse transformation, namely the stress-induced martensitic transformation leading to the pseudoelasticity effect. Since the shape memory effect (SME) at lower temperatures is caused by the conversion between stress-induced martensite and temperature-induced martensite, these models cannot be implemented to the detwinning of martensite, which is responsible for the SME phenomena [19].

To answer this deficiency, a new model was proposed by Brinson [15, 20] in which the martensite volume fraction was divided into stress-induced fraction, ξ_S and temperature induced martensite fraction, ξ_T to yield:

$$\xi = \xi_S + \xi_T. \quad (9.13)$$

The original constitutive equation in the Brinson model was slightly modified from the Tanaka's one (Eq. (9.3)) and was written as

$$(\sigma - \sigma_0) = E(\xi)\varepsilon - E(\xi_0)\varepsilon_0 + \Theta(T - T_0) + \Omega(\xi)\xi - \Omega(\xi_0)\xi_0. \quad (9.14)$$

A simplified constitutive equation [20] has the following form:

$$\sigma = E(\xi)(\varepsilon - \varepsilon_L \xi_S) + \Theta(T - T_0). \quad (9.15)$$

To include the shape memory effect at temperatures below M_s, the transformation phase equations of Liang and Roger (see Eqs. (9.8) and (9.9)) were modified to include the two type of martensite volume fractions, the ξ_S and ξ_T [15]. A schematic presentation of the variation of the critical stresses with temperature for the transformation with the two types of volume fraction is shown in Fig. 9.21(b). The conversion to detwinned martensite is now written as

$$\xi_S = \frac{1 - \xi_{S_0}}{2} \cos\left[\frac{\pi}{\sigma_s^{cr} - \sigma_f^{cr}}[(\sigma - \sigma_f^{cr}) - C_M(T - M_s)]\right] + \frac{1 + \xi_{S_0}}{2},$$

$$\xi_T = \xi_{T_0} - \frac{\xi_{T_0}}{1 - \xi_{S_0}}(\xi_S - \xi_{S_0}) \quad (9.16)$$

provided the temperature and the stress are

$$T > M_s \text{ and}$$

$$\sigma_s^{cr} + C_M(T - M_s) < \sigma < \sigma_f^{cr} + C_M(T - M_s). \quad (9.17)$$

For the case of

$$T < M_s \quad \text{and}$$
$$\sigma_s^{cr} < \sigma < \sigma_f^{cr} \tag{9.18}$$

we will get the following expressions:

$$\xi_S = \frac{1-\xi_{S_0}}{2}\cos\left[\frac{\pi}{\sigma_s^{cr}-\sigma_f^{cr}}(\sigma-\sigma_f^{cr})\right] + \frac{1+\xi_{S_0}}{2},$$
$$\xi_T = \xi_{T_0} - \frac{\xi_{T_0}}{1-\xi_{S_0}}(\xi_S - \xi_{S_0}) + \Delta T_\xi. \tag{9.19}$$

The term ΔT_ξ will have the following values:

If $M_f < T < M_s$ and $T < T_0$ then:

$$\Delta T_\xi = \frac{1-\xi_{T_0}}{2}\{[\cos[a_M(T-M_f)]+1]\}; \text{ otherwise} : \Delta T_\xi = 0. \tag{9.20}$$

The back conversion to the parent phase, the austenite, is given by

$$\xi = \frac{\xi_0}{2}\left\{\cos\left[a_A(T - A_s - \frac{\sigma}{C_A})\right] + 1\right\},$$
$$\xi_S = \xi_{S_0} - \frac{\xi_{S_0}}{\xi_0}(\xi_0 - \xi) \quad \text{and} \quad \xi_T = \xi_{T_0} - \frac{\xi_{T_0}}{\xi_0}(\xi_0 - \xi). \tag{9.21}$$

One should note that the evolution kinetics as displayed by Brinson model might be incorrect for certain thermomechanical loadings yielding an inadmissible martensite fraction of $\xi > 1$. To overcome this problem, Chung *et al.* [23] represented the critical stress–temperature curve into eight bands (see Fig. 9.22) as follows:

- Band 1: Transformation from austenite or temperature-induced martensite to stress-induced martensite.
- Band 2: Mixture of temperature/stress-induced martensite and austenite (no transformation).
- Band 3: Transformation from martensite to austenite.
- Band 4: Pure austenite (no transformation).
- Band 5: Transformation from austenite or temperature-induced martensite to stress-induced martensite and transformation from austenite to temperature-induced martensite.
- Band 6: Transformation from austenite to temperature-induced martensite.
- Band 7: Transformation from temperature-induced martensite to stress-induced martensite.
- Band 8: Mixture of stress/temperature-induced martensite (no transformation).

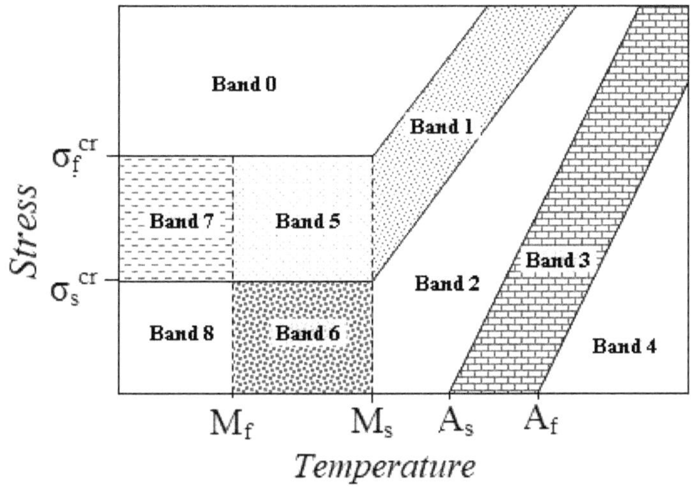

Fig. 9.22. The critical stress–temperature curve divided in eight bands (adapted from [12]).

Note that the Brinson model has a weakness at band 5. To answer this issue Chung *et al.* [21] proposed a modified model under the following five constraints:

(a) Under any circumstances, the total martensite fraction should comply with the following constraint: $\xi = \xi_S + \xi_T \leq 1$.
(b) The stress-induced martensite fraction must be 1 when $\sigma = \sigma_f^{cr}(\xi_S = 1)$.
(c) The total martensite fraction must be 1 when $T = M_f(\xi = \xi_S + \xi_T = 1)$.
(d) The transformation kinetics in band 5 must be continuous with band 1 (Fig. 9.22).
(e) The functions of the martensite volume fractions are all in cosine forms.

Note that Brinson's model satisfies all the above conditions when only the temperature is decreased or alternatively only the stress is increased. For the case when both the stress and the temperature are simultaneous altered of both the stress, the Brinson model is satisfying only conditions (d) and (e). To be consistent with the above five conditions, Eqs. (9.18)–(9.20) are modified and the formulation of the transformation are given in the following equations [21], provided the critical stresses are constant below M_s:

For the case of

$$T < M_s \quad \text{and} \quad \sigma_s^{cr} < \sigma < \sigma_f^{cr}, \tag{9.22}$$

we will get the following expressions

$$\xi_S = \frac{1-\xi_{S_0}}{2}\cos\left[\frac{\pi}{\sigma_s^{cr}-\sigma_f^{cr}}(\sigma-\sigma_f^{cr})\right]+\frac{1+\xi_{S_0}}{2},$$

$$\xi_T = \Delta T_\xi - \frac{\Delta T_\xi}{1-\xi_{S_0}}(\xi_S-\xi_{S_0}). \tag{9.23}$$

The term ΔT_ξ will have the following values:

If $M_f < T < M_s$ and $T < T_0$ then:

$$\Delta T_\xi = \frac{1-\xi_{S_0}-\xi_{T_0}}{2}\cos\left[\frac{\pi}{M_s-M_f}(T-M_f)\right]+\frac{1-\xi_{S_0}+\xi_{T_0}}{2},$$

otherwise $\Delta T_\xi = \xi_{T_0}$. (9.24)

The modification presented in Eqs. (9.21)–(9.24) answers the five conditions written above, and it can be shown that the modified Brinson's model will give consistent results for all initial conditions [21]. One should note that already in 1998, Bekker and Brinson [22] developed a consistent mathematical description of the evolution of the martensite fraction during the thermal thermoelastic phase transformation in a SMA induced by an general thermomechanical loading. This yielded a robust kinetics, which do not permit volume fractions to exceed unity; however the model additions of Chung *et al.* [21] to the original Brinson's model capture better the phenomena.

9.6. Heating an SMA Alloy Using Electrical Current

One of the ways to directly heat an SMA alloy having the shape of a wire is by connecting it to an electrical source. This type of heating is also known as the Joule heating, and its equation is given by [23]:

$$(\rho A)c_p\frac{dT(t)}{dt} = I^2R - h_c A_{\text{circ}}[T(t)-T_0], \tag{9.25}$$

where I is denoting the electrical current, R is the wire resistance per unit length, $T(t)$ is the temperature as a function of time, T_0 is the ambient temperature, $\rho A = m$, is the mass per unit length of the wire (ρ being its density and A the cross-sectional area of the wire), c_p is the specific heat, while the terms h_c and A_{circ} are the heat transfer coefficient and the circumferential area of the unit length of the wire, respectively. Assuming that both the current and the ambient temperature are constant and do not change with time, Eq. (9.25) can be solved to yield the following solution for the temperature as a function of time:

$$T(t) - T_0 = \frac{I^2 R}{h_c A_{\text{circ}}}(1 - e^{-t/t_{ht}}) + (T_{\text{start}} - T_0)e^{-t/t_{ht}},$$

$$\text{where} \quad t_{ht} = \frac{(\rho A)c_p}{h_c A_{\text{circ}}}, \tag{9.26}$$

where T_{start} is the temperature at the starting of the heating and t_{ht} is the time constant of the heat transfer process. For the case $T_{\text{start}} = T_0$, we get

$$T(t) - T_0 = \frac{I^2 R}{h_c A_{\text{circ}}}(1 - e^{-t/t_{ht}}). \tag{9.27}$$

Then the steady-state temperature to be measured on the wire will be:

$$T_{ss} = \frac{I^2 R}{h_c A_{\text{circ}}}(1 - e^{-t/t_{ht}}) + T_0. \tag{9.28}$$

Analyzing the exponential function presented by Eq. (9.27), one can calculate that at $t = 3t_{ht}$ the temperature will be approximately 95% of T_{ss}. Then the time, t_p, required to reach a predefined temperature T_p (thus heating the wire) will be:

$$t_p = -t_{ht} \ln \frac{T_{ss} - T_p}{T_{ss} - T_0}. \tag{9.29}$$

For the case of cooling the wire, we assume $I = 0$ in Eq. (9.26), yielding

$$T(t) - T_0 = (T_{\text{start}} - T_0)e^{-t/t_{ht}}. \tag{9.30}$$

The steady-state temperature will then be the ambient temperature and the time to reach a predefined temperature will be calculated according to the following equation:

$$t_p = -t_{ht} \ln \frac{T_p - T_0}{T_{\text{start}} - T_0}. \tag{9.31}$$

Once the time for heating and cooling can be estimated the next question to solve would be what power we need to apply to the Nitinol wire. The power for a given current I [*Ampere*] and a resistance $R[\Omega]$ is given by

$$P = I^2 R. \tag{9.32}$$

The resistance R can be calculated using the following equation:

$$R = \rho_r \frac{l}{A}, \tag{9.33}$$

where ρ_r (for Nitinol its value is $7.6 \times 10^{-5}\ \Omega/\text{cm}$) is the wire resistivity and l and A are its length and the cross-sectional area, respectively. Multiplying the power by time, t, will yield the required energy E, to heat a given wire from a starting temperature T_{start} to a predefined temperature T_p. This is given by

$$E = I^2 R \cdot t = c_p \cdot M \cdot (T_p - T_0) + L \cdot M, \tag{9.34}$$

where M being the mass of the wire (in grams), c_p [J/g∗C] and L [J/g] are the specific heat and the latent heat of transformation for the wire, respectively.[j] Substituting the expression for the resistance using Eq. (9.33) into Eq. (9.34) and solving for the current I yields

$$I = \sqrt{\frac{c_p \cdot M \cdot (T_p - T_0) + L \cdot M}{\frac{\rho_r l}{A} t}}. \tag{9.35}$$

Thus, for a given mass of wire, with a known length and cross-section, while the material properties c_p, L, and ρ_r are known, one can calculate the required current I to raise the temperature from T_0 to the predefined one T_p for a time t.

References

[1] Ölander, A., An electrochemical investigation of solid cadmium-gold alloys, *J. Am. Chem. Soc.*, Vol. 54(10), 1932, pp. 3819–3833, doi:10.1021/ja01349a004.

[2] Chang, L.C. and Read, T.A., Plastic deformation and diffusion less phase changes in metals — The gold-cadmium beta-phase, *Trans. AIME* Vol. 191, 1951, pp. 47–52.

[3] Buehler, W.J., Gilfrich, J.W. and Wiley, R.C., Effects of low-temperature phase changes on the mechanical properties of alloys near composition TiNi. *J. Appl. Phys.*, Vol. 34(5), 1963, pp. 1475–1477. doi:10.1063/1.1729603.

[4] Wang, F.E., Buehler, W.J. and Pickart, S.J., Crystal structure and a unique martensitic transition of TiNi, *J. Appl. Phys.*, Vol. 36(10), 1965, pp. 3232–3239. doi:10.1063/1.1702955.

[5] Shimizu, K. and Tadaki, T., *Shape Memory Alloys*, Funakubo, H. (ed.), Gordon and Breach Science Publishers, 1987.

[6] Kumar, P.K. and Lagoudas, D.C., Introduction to shape memory alloys, D.C. Lagoudas (ed.), in *Shape Memory Alloys*, doi:10.1007/978-0-387-47685-8 1, Springer Science +Business Media, LLC, 2008.

[7] Petrini, L. and Migliavacca, F., Biomedical applications of shape memory alloys, *J. Metallurgy*, Vol. 2011, 2011, Article ID 501483, 15 pages, doi:10.1155/2011/501483.

[8] Duerig, T.W., Melton, K.N., Stockel, D. and Wayman, C.M., *Engineering Aspects of Shape Memory Alloys*, Butterworth-Heinemann, London, UK, 1990, 512 p.

[9] Laster, Z., MacBean, A.D., Ayliffe, P.R. and Newlands, L.C., Fixation of a frontozygomatic fracture with a shape-memory staple, *British J. Oral Maxillofacial Surg.*, Vol. 39(4), 2001, pp. 324–325.

[10] Calkins, F.T., Mabe, J.H. and Butler, G.W., Boeing's variable geometry chevron: Morphing aerospace structures for jet noise reduction, *Proc. SPIE* Vol. 6171, 2006.

[j] For nitinol $c_p = 0.01$[J/g*C] and $L = 20$ [J/g].

[11] Pitt, D.M., Dunne, J.P. and White E.V., SAMPSON smart inlet design overview and wind tunnel test Part I — Design overview, *Proc. SPIE* Vol. 4698, 2002.

[12] Greninger, A.B. and Mooradian, V.G., Strain transformation in metastable beta copper-zinc and beta copper-tin alloys, *Trans. Metallurgical Soc. AIME*, Vol. 128, 1938, pp. 337–368.

[13] Buehler, W.J., Gilfrich, J.W. and Wiley, R.C., Effects of low-temperature phase changes on the mechanical properties of alloys near composition TiNi, *J. Appl. Phy.*, Vol. 34 (5), 1963, pp. 1475–1477, doi:10.1063/1.1729603.

[14] Wang, F.E. Buehler, W.J. and Pickart, S.J., Crystal structure and a unique martensitic transition of TiNi, *J. Appl. Phy.*, Vol. 36 (10), 1965, pp. 3232–3239, doi:10.1063/1.1702955.

[15] Tanaka, K. and Nagaki, S., A thermomechanical description of materials with internal variables in the process of phase transitions, *Ingenieur-Archiv*, Vol. 51(5), 1982, pp. 287–299.

[16] Tanaka, K., A thermomechanical sketch of shape memory effect: One-dimensional tensile behavior, *Res. Mechanica*, Vol. 18(3), 1986, pp. 251–263.

[17] Brinson, L.C., One-dimensional constitutive behavior of shape memory alloys: Thermomechanical derivation with non-constant material functions and redefined martensite internal variable, *J. Intelligent Mater. Syst. Struct.*, Vol. 4(2), 1993, pp. 229–242.

[18] Li, L., Li, Q. and Zhang, F., One-dimensional constitutive model of shape memory alloy with an empirical kinetics equation, *J. Metallurgy*, Vol. 2011, 2011, Article ID 563413, pp. 1–14, doi:10.1155/2011/563413.

[19] Barbarino, S., Saavedra Flores, E.I., Ajaj, R.M., Dayyani, I. and Friswell, M.I., A review on shape memory alloys with applications to morphing aircraft, *Smart Mater. Struct.*, Vol. 23, 2014, pp. 1–19, doi:10.1088/0964-1726/23/6/063001.

[20] Liang, C. and Rogers, C.A., One dimensional thermomechanical constitutive relations for shape memory material, *J. Intell. Mater. Syst. Struct.*, Vol. 1, 1990, pp. 207–234.

[21] Prahlad, H. and Chopra, I., Comparative evaluation of shape memory alloy constitutive models with experimental data, *J. Intell. Mater. Syst. Struct.*, Vol. 12, 2001, pp. 383–395.

[22] Brinson, L.C. and Huang, M.S., Simplifications and comparisons of shape memory alloy constitutive models, *J. Intell. Mater. Syst. Struct.*, Vol. 7, 1996, pp. 108–114.

[23] Chung, J., Heo, J. and Lee, J., Implementation strategy for the dual transformation region in the Brinson SMA constitutive model, *Smart Mater. Struct.*, Vol. 16, 2007, pp. N1–N5.

[24] Bekker, A. and Brinson, L.C., Phase diagram based description of the hysteresis behavior of shape memory alloys, *Acta Mater.*, Vol. 46, 1998, pp. 3649–3665.

[25] Song, H., Kubica, E. and Gorbet, R., Resistance modelling of SMA wire actuators, *Int. Workshop on Smart Materials, Structures & NDT in Aerospace Conference NDT*, 2–4 November 2011, Montreal, Quebec, Canada.

Chapter 10

Introduction to Fiber Optic

10.1. Basic Characteristics of a Fiber Optic

Fiber optic is defined as the science of light transmission through very fine glass or plastic fibers (see Fig. 10.1). Currently is used in more and more applications due to its advantages over copper conductors. These advantages include: large bandwidth (>10 GHz per km for single-mode fibers or between 200–600 MHz per km for multimode fibers compared to 10–25 MHz per km for electrical conductors), immunity to electromagnetic/radio frequency interference, easiness to install, requires less space and the weigh is 10–15 times lower than copper leading to reduced cost for installation. Moreover, the fibers are electrically insulated, do not emit sparks or cause short circuits and do not pose any shock hazards.

A fiber optic link (see Fig. 10.2) is composed of four main parts:

- Transmitter — This device converts electrical signals to optical ones. It contains a light source like a LED (Light Emitting Diode) or a LASER (Light Amplification by Stimulated Emission of Radiation) or a VCSEL (Vertical Cavity Emitting Laser). These light sources use an infrared band (approx. 850 nm[a]).
- Fiber — The optical fiber transmits the light using a special designed tube consisting of five different layers (see schematic view in Fig. 10.3)
  (a) Core — This is the central tube, in which the light is propagating and usually is made from either fused silica[b] (amorphous SO_2) or

[a] 1 nm = 10^{-9} m.

[b] Silica = silicon dioxide, SiO_2, found in the nature as quartz. Sand contains a large amount of silica.

Fig. 10.1. Various fiber optic cables.

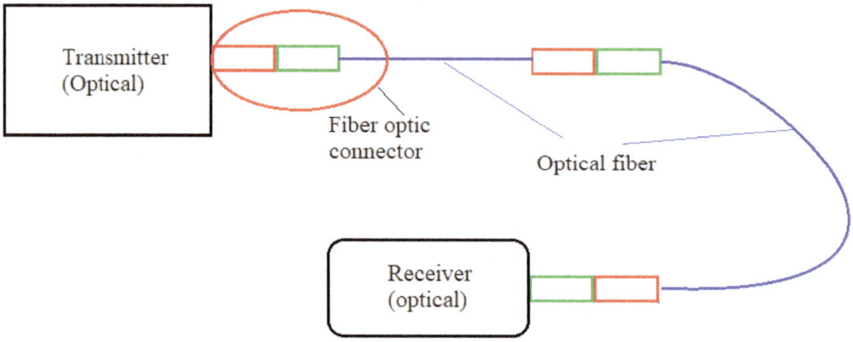

Fig. 10.2. Schematic fiber optic link.

doped silica (like Germania GeO_2 or phosphorous pentoxide P_2O_5 as dopers).

(b) Cladding is the second-layer engulfing the core. It is also made from doped silica (doped with Boria B_2O_3 or fluorine) to yield a refractive index (n-see Appendix A) lower than the refractive index of the core. This allows the light to be confined in the core by internal reflections at the core–cladding interface.

(c) Buffer coating protects the fiber during its installation and has a thickness of 900 microns.

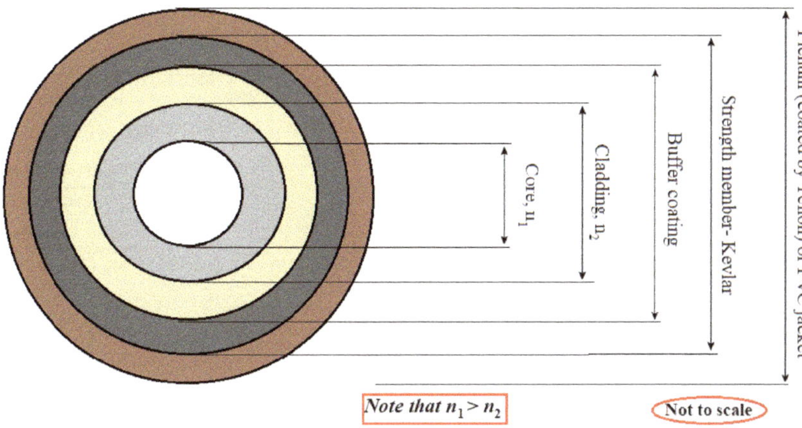

Fig. 10.3. Typical cross-section of a fiber optic cable.

(d) The fourth layer usually made of Kevlar[c] provides the required strength to the fiber cable.

(e) The external layer is a jacket made from PVC or plenum coated by Teflon[d] to prevent fire protection.

- Connectors — To operate smoothly connectors are used to link the fiber optic cable to the transmitter and the receiver. Two method exist:

(a) Direct connectivity — this method directly welds two fibers with an electric arc or what is known in the literature as *fusion splicer*. The good thing of this simple procedure is that there is almost no loss of light when passing from one fiber to the other, however the disadvantages like permanent connection, fragility and cost of the machine (fusion splicer) make the method less attractive to users.

(b) Physical connectors — this method involves connecting two fibers by physically connecting two fibers using two plugs. The connection can be mated and unmated yielding a robust connection with thousands of times of connections and disconnections. The disadvantages of this method are: higher loss of light as compared to the fusion splicer requires special tools and the process is longer.

One can find two types of connectors (see Table 10.1):

(1) Physical contact connectors — Use is made of a cap bonded to a polished fiber using epoxy and ensures alignment between the

[c]Kevlar® = created by Stephanie Kwolek at Du Pont™ is a para-amid synthetic fiber (hear-resistant and very strong). It is known for its high tensile strength-to-weight ratio.

[d]Teflon = a synthetic fluoropolymer of tetrafluoroethylene, or polythtrafluoroethylene having high melting point (327°C), thermal conductivity of 0.25 W/(m°K) and a chemical formula of $(C_2F_4)_n$.

Table 10.1. Fiber optic connectors (adapted from L-com® Global Connectivity).

Connector	Type	Coupling Type	# of fibers	Applications
	Connector LC	Snap on RJ45 style	1	Gigabit Ethernet,[a] Video multimedia
	Connector SC	Snap on	1	CATV,[b] Test equipment
	Connector ST	Twist on	1	LANs,[c] Military
	FC	Screw on	1	Datacom, Telecommunications
	MT-RJ	Snap on RJ45 style	2	Gigabit Ethernet ATM[d]
	MPO(MTP)	Push/Pull	6 or 12	Active device transceiver,[e] O/E,[f] modules interconnections QSFP,[g] transceivers

Notes: [a]Ethernet is a family of computer networking technologies commonly used in local area networks.
[b]CATV = Community Antenna Television, a cable television.
[c]LANs = Local Area Networks, a series of computers linked together to form a network in a prescribed location.
[d]ATM = Asynchronous Transmission mode.
[e]Transceiver is a device that is able to both transmit and receive.
[f]O/E = Opto-Electronic.
[g]QSFP = Quad Small Form-factor Pluggable is used at the interface of hardware to the fiber optic cable.

two fibers. The result is a robust connection with low light loss (approx. 0.3 dB) easy to be cleaned and cost effective.

(2) Expanded beam connectors — Lens are placed at the exit of each fiber to widen the ray and collimate the light passing through the fiber. The light loss is higher (0.8–2.5 dB) and the whole process is more complicated than the previous type of connectors, and therefore it is less found in the market.

- Receiver — the last part of the chain, aiming at converting the optical signals using a photodiode in electrical ones. The photodiodes can either be of type PIN (Positive Intrinsic Negative) or APD (Avalanche Photo Diode).

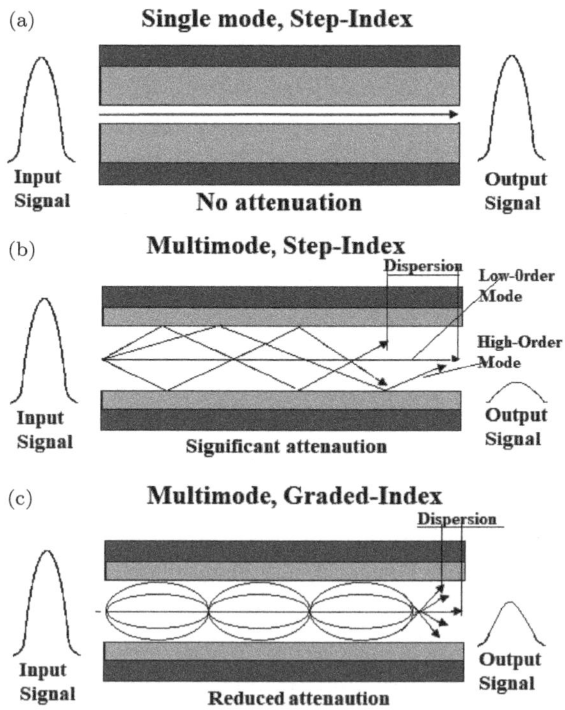

Fig. 10.4. The influence of the glass type on the fiber optic performance — a schematic drawing (adapted from L-com® Global Connectivity).

Another important property is the type of the glass used to manufacture the optic fiber. In today's high-speed networks, Graded Index Multimode fiber (the refractive index of the core is gradually decreased towards the cladding using often a parabolic law yielding a 'bend' of the light as it travels along the fiber) or Step Index Multimode or Single-Mode fiber (the light is reflected at the cladding-core interfaces due to the abrupt n in the material refractive index, n) is used to improve light transmission over long distances. Multimode fiber has a larger core and is typically used for short distances within buildings. Single-mode fiber has a smaller core and is used for long distances typically outside between buildings. As can be seen in Fig. 10.4, the signal is the same for the Step-index single-mode (a), while for the Step-index multimode, there is a considerable attenuation of the output signal compared with the input one and large dispersion (b). Using Graded-index multimode reduces the dispersion while improving the output signal (reduced attenuation) (c).

10.2. Fiber Optic — Historical Aspects

It is instructive to understand the evolution of fiber optic from its historical aspects. It is agreed, that the *Photophone*, a device capable of transmitting of sound on a light ray, created in 1880 by Alexander Graham Bell and his co-inventor Charles Sumner Tainter [1], as the starting point of fiber-optic communication. Although the test performed by Bell and Tainter on June 3rd, 1880, failed due to the atmosphere medium, it can be regarded as the starting point of the fiber-optic application to communication.

Later on, Harold Hopkins and Narinder Singh Kapany[e] (the latter also known as *Father of Fiber Optics*) at Imperial College, London, Great Britain, showed that even rolled fiberglass allows the transmission of light in contrast to the previous assumption that light can propagate only in a straight medium. Following the development of optical cladding by the Dutch scientist van Heel,[f] Kapany [2] was the first person to conceive the term 'fiber optics' in an article named Fiber Optics in *Scientific American* in November 1960.

A few years later, in 1966, Charles K. Kao and George Hockham proposed optical fibers while working at STC Laboratories (STL) at Harlow, UK. While Kao concluded that the fundamental limit on glass transparency is below 20 decibels per kilometer, which would be practical for communications. Hockham calculated that clad fibers should not radiate much light. They prepared together a paper proposing for the first time fiber-optic communications [3]. Kao is known in the literature as the Godfather of Broadband, the father of Fiber Optics, and the Father of Fiber Optic Communications, was awarded the 2009 Nobel Prize in physics for *groundbreaking achievements concerning the transmission of light in fibers for optical communication* and was knighted by Queen Elizabeth II *for services to fiber optic communications.*

One should note that ten years before, in 1958, Schawlow and Townes [4] published a paper, which is considered to be the first evidence for the laser operation but it wasn't until 1960 that the first light-emitting maser — better known as the laser — was constructed [5], enabling latter the extensive use of optical fibers for communications.

Interesting to mention that C. H. Townes shared with N. G. Basov and A.M. Prokhorov the 1964 Nobel Prize in physics *for fundamental work in the*

[e]Hopkins, H.H. and Kapany, N.S., A flexible fibrescope, using static scanning, *Nature*, Vol. 173, 1954, pp. 39–41.

[f]Simons, C.A.J., A.C.S. van Heel: teacher and inspirator of technical optics, *Proc. SPIE,* 3190, *Fifth International Topical Meeting on Education and Training in Optics* (8 December 1997); https://doi.org/10.1117/12.294379

Table 10.2. Transmission capability vs. years.

Generation	Year realized	Bit rate	Repeater[a] spacing	Operating wave-length
1st (graded-index fibers)	1980	45 Mb/s	10 km	0.8 μm
2nd (single-mode fibers)	1985	100 Mb/s up to 1.7 Gb/s	50 km	1.3 μm
3rd (single-mode lasers)	1990	10 Gb/s	100 km	1.55 μm
4th (optical amplifiers)	1996	10 Tb/s	>10,000 km	1.45–1.62 μm
5th (Raman amplification)	2002	40–160 Gb/s	24,000–35,000 km	1.53–1.57 μm

Note: [a]Repeater — a wireless network device that repeats wireless signals to extend range. It is not connected with cable to a router/modem or users.

field of quantum electronics, which has led to the construction of oscillators and amplifiers based on the maser–laser principle.

Corning Glass Works successfully developed optical fibers in 1970, with a low attenuation of about 20dB/km for communications purposes. At the same time GaAs compact semiconductor lasers were developed by researchers (Zhores Alferov's group) at the Ioffe Institute in Leningrad (today St. Petersburg) , Russia and in parallel and independently by Panish & Hayashi at Bell Labs, NJ, USA, suitable for transmitting light through fiber optic cables for long distances.

Since 1970 and up today, the transmission capability had enormously improved as can be seen in Table 10.2 and Fig. 10.5.

Note that the straight line in Fig. 10.5 denotes that the BL term will double each year.

It is worth to remember the capability of wireless networks and their dramatic advances through the years,[g] like the 1st generation (1G in 1980s), 2nd generation (2G in 1990s), 3rd generation (3G in 2000s), 4th generation (4G in 2010s) and the 5th generation (5G in 2020s). The 5G is expected to display enhanced mobile broadband (EMBB), massive machine-type communications (MMTC) and ultra-reliable and low-latency communication (URLLC) [1].

[g]Bit — a basic unit of information having the value of either 0 or 1, used in computing and digital communications.

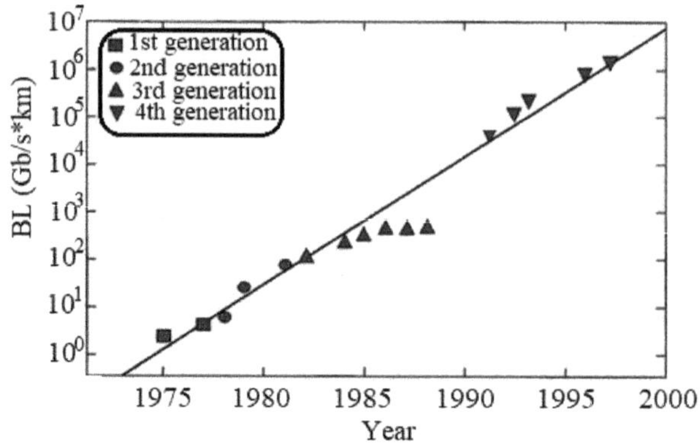

Fig. 10.5. BL (Bit 15*length) vs. year for the first four generations (adapted from [1]).

Surely, it can be stated that the topic of fiber optic is today well established and documented in many books (for instance [6–16]) and thousands of manuscripts. In what follows, we shall restrict ourselves only to Bragg gratings and their applications.

10.3. Fiber Bragg Grating

Fiber Bragg Grating (FBG) is an optical fiber sensor capable of measuring accurately mechanical strains and temperature induced strains. The name Bragg was given to the device following the Bragg law.[h] The FBG discovery is attributed to Kenneth Hill and his co-investigators, who discovered it already in 1978 while working at Communication Research Center, Canada [17, 18]. Besides sensing strain and temperature, it can detect also pressure, stress and refractive index of glass. FBG can be found in many industries, like aerospace, civil and maritime as well as in biochemical, health and biomedical devices.

The schematic structure of a typical FBG is presented in Fig. 10.6.

The manufacturing process to induce grating into the core of the fiber is named *inscription* or *writing*. There are a few ways to introduce the grating [19]:

(a) An interferometric procedure — in which the photo-sensible region of the core is exposed to an interferometric pattern to yield the whole grating.

[h]Lawrence Sir William Bragg proposed in 1912 what is known as the Bragg Law of X-ray diffraction. The law is used for the determination and investigation of crystals and thin films. Sir Bragg and his father William Henry Bragg won the Nobel prize in Physics in 1915, *for their services in the analysis of crystal structures by means of X-rays.*

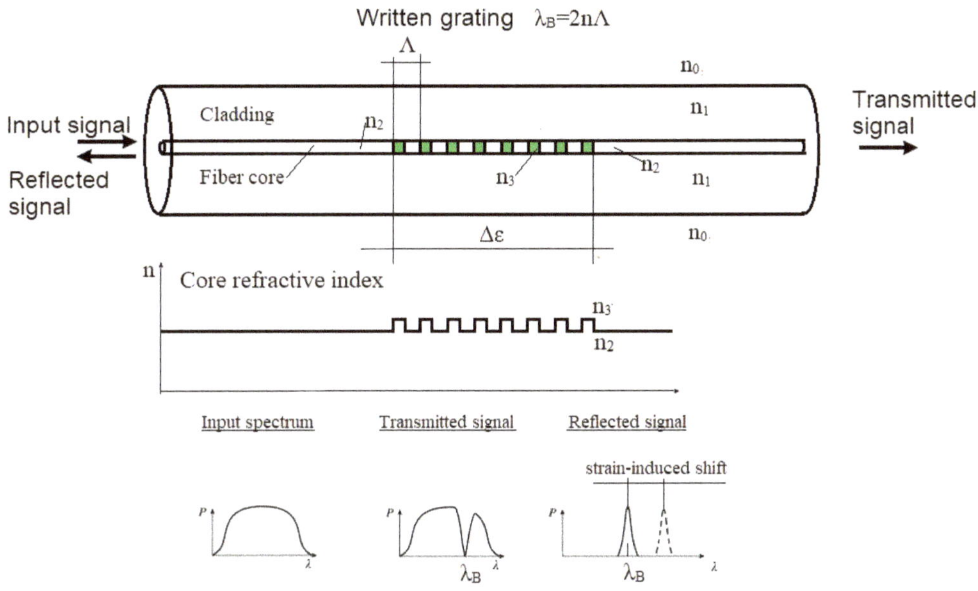

Fig. 10.6. A schematic view of the transmitted and reflection spectra for an FBG.

This is done by illuminating a proper mask having the FBG period. Note that this approach procedure can be done also without a mask.

(b) The direct point-by-point method — a laser is used to write the grating, while controlling its parameter and the nonlinear absorption of the laser pulse. A translational platform having the fiber on it is moving and the laser is writing one point after the other to form the grating.

(c) The continuous core-scanning procedure — in this method the only parameters to be controlled are connected to the moving of the translational platform on which a fiber is attached.

As reported in [20], besides the uniform grating described above, there are also chirped type grating, long period grating, tilted grating, phase-shifted grating and supper-structure FBG grating.

For a simple FBG with periodic, equally spaced perturbations (see Fig. 10.6) the strength of the modulation is only positive [19]. The refractive index along the grating will vary as a square wave having a duty cycle of half-period Λ of the grating and can be written as

$$\begin{aligned} n_{\text{core}} &= n_{\text{core}\,0} & 0 < x < \Lambda/2, \\ n_{\text{core}} &= n_{\text{core}\,0} + \Delta n, & \Lambda/2 < x < \Lambda, \end{aligned} \tag{10.1}$$

where Δn is the modulation strength and n_{core0} is the refractive index at the untouched core.

The operating system is as follows: part of the light traveling forward through the core of the fiber is reflected back due to the grating and the rest will pass over. This causes the light beam to have two components that passes in two opposite directions. At certain wavelengths decided by the grating parameters, the Bragg condition is satisfied yielding a constructive interference of the reflected light. The spectrum of the reflected light would present a peak at the Bragg wavelength, λ_B, while the transmitted light will show a notch filter (like a transfer function — see Fig. 10.6). Those light components far from λ_B will travel unperturbed.

Remembering that the Bragg law is given by

$$2d \sin\theta = n\lambda, \quad n = 1, 2, 3, 4, \ldots, \tag{10.2}$$

where d is the distance between two adjacent crystalline planes, θ is the incidence angle and λ is the wavelength, one can obtain the expression for λ_B. The angle θ is 90° and d is the distance between adjacent peaks (see Appendix A), and for $n = 1$ we get $\lambda = 2d$, and one has to take into account the refraction index of the fiber (as the original Bragg law was derived for vacuum) leading to

$$\lambda_B = 2n_{\text{eff}}\Lambda. \tag{10.3}$$

The Bragg is therefore a function of the effective refractive index of the fiber, n_{eff}, and the grating pitch, Λ.

As described in [19], application of the coupled-mode theory, (see [21]), enables the calculation of the reflectivity and transmissivity spectrum. Defining the following parameters

$$\alpha \equiv \frac{2\pi}{\lambda}\overline{\Delta n_{\text{eff}}}, \quad \kappa \equiv \frac{\pi}{\lambda}\upsilon\overline{\Delta n_{\text{eff}}}, \quad L \equiv N\Lambda, \tag{10.4}$$

where α is the DC self-coupling coefficient, $\overline{\Delta n_{\text{eff}}}$ is the DC index change spatially averaged over a grating period, κ is the AC coupling coefficient, υ is the FBG order (unit for an uniform FBG) and N is the number of periods leads to the reflected amplitude

$$R_{\text{am.}} = -\frac{\kappa \sinh(\sqrt{\kappa^2 - \alpha^2 L})}{\alpha^2 \sinh(\sqrt{\kappa^2 - \alpha^2 L}) + i(\sqrt{\kappa^2 - \alpha^2})\cosh(\sqrt{\kappa^2 - \alpha^2 L})}, \tag{10.5}$$

and power

$$\text{Power} = |R^2_{\text{am.}}| = -\frac{\sinh^2(\sqrt{\kappa^2 - \alpha^2 L})}{\cosh^2(\sqrt{\kappa^2 - \alpha^2 L}) - \frac{\alpha^2}{\kappa^2}}. \tag{10.6}$$

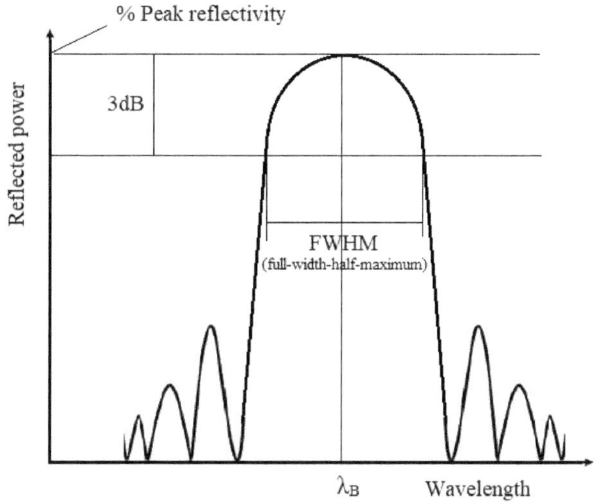

Fig. 10.7. Schematic reflectivity of an FBG.

The full-width-half-maximum (FWHM) or the bandwidth (see Fig. 10.7) is written as

$$\mathrm{FWHM}_{\lambda} = \lambda_B \beta \sqrt{\left(\frac{\Delta n}{2n_{c0}}\right)^2 + \left(\frac{1}{N}\right)^2},$$

$$\text{where} \quad \beta \approx \begin{cases} 1 & \text{for high reflectivity grating,} \\ 0.5 & \text{for weak reflectivity grating.} \end{cases} \qquad (10.7)$$

As can be seen in Fig. 10.7, the spectrum is composed from a central peak having a variable bandwidth according to the given parameters of the FBG, embraced from both side-by-side lobes.

For different uses [20, 21], other grating profiles can be considered like chirped or phased-shifted grating which might reduce the side lobes. Another interesting profile named an apodized grating profile, which consists in a non-uniform strength of the modulation along the grating (like for raised-cosine or Gaussian profile), is shown to drastically reduce the amplitudes of the side lobes.

10.4. Fiber Bragg Grating Used as Strain Sensor

As described in Fig. 10.6, strain or temperature will induce a shift in the Bragg wavelength. This shift can be calculated as follows: partial differentiating Eq. (10.3) yields

$$\Delta\lambda_B = 2\left(\Lambda\frac{\partial n_{\text{eff}}}{\partial\varepsilon} + n_{\text{eff}}\frac{\partial\Lambda}{\partial\varepsilon}\right)\Delta\varepsilon,$$

$$+2\left(\Lambda\frac{\partial n_{\text{eff}}}{\partial T} + n_{\text{eff}}\frac{\partial\Lambda}{\partial T}\right)\Delta T = (\Delta\lambda_B)_{\text{strain}} + (\Delta\lambda_B)_{\text{temp.}} \quad (10.8)$$

The first part of the left-hand side of Eq. (10.8) presents the shift in Bragg wavelength due to strain, while the second part is due to temperature.

Equation (10.8) can be rewritten in another useful form to yield [19]:

$$\frac{\Delta\lambda_B}{\lambda_B} = (1-\rho_E)\varepsilon + (\beta+\chi)\Delta T, \quad (10.9)$$

where ρ_E is the photo-elastic coefficient stemming from the Pockel's stress-optic tensor [23], ε is the strain in the direction of the fiber, β is the thermal expansion of the material of the fiber (like silica) and χ is the thermo-optic coefficient of the material [22] (the temperature dependence of the refractive index, dn_{eff}/dT). The authors in [22] present the following data (fiber silica with germanium doped core) for the constants in Eq. (10.9): $\rho_E = 0.22$, $\beta = 0.55\times10^{-6}/°\text{C}$ and $\chi = 8.6\times10^{-6}/°\text{C}$ leading to $\Delta\lambda_B/\Delta\varepsilon = 1.2pm/\mu s$ and $\Delta\lambda_B/\Delta T = 14.18pm/°\text{C}$. Note that in [24] a different value is presented for the relationship strain-Bragg shift: $\Delta\lambda_B/\lambda_B = 0.79\varepsilon$.

One should note that using FBG as strain sensor, the strain output would normally include the added temperature strain. A compensation is needed to subtract the influence of the temperature so that the output will provide only mechanical strains. An interesting method is presented in [22], in which two parallel mounted FBGs, one bonded while the other is un-bonded and provides the temperature induced strain, which is subtracted from the reading of the first FBG to yield the net strain encountered at the particular part of the structure (on which the FBGs are attached). Typical studies on the use of FBG to sense strain are presented in [25–32].

10.5. Interrogators for Fiber Bragg Grating

The information gathered by the use of FBG should be decoded to enable its use. This process is named demodulation technique or interrogation. There are several techniques for FBG based sensors [19, 20]: Bulk optics, filtering (having either passive edge or active bandpass filtering), interferometric, laser sensing and others. According to [19] the most used interrogation techniques are the filtering and interferometric methods. The filtering method converts the wavelength shift of the FBG output into a change in an electrical signal using a photo-diode which collects the optical signal of the FBG reflected spectrum (instead of using a conventional optical spectrum analyzer — OSA, which was found to be slow and expensive to use [19]). The interferometric methods allocate an optical phase difference to the Bragg wavelength shift. An unbalanced asymmetric Mach Zehnder interferometer is normally used, to receive the reflected spectrum of the sensing grating and passed to a phase modulator [24] located on one of the arms of the interferometer. The change in the measured intensity can be transformed into a phase difference once the interferometer phase and its associated intensity output is known. Then the required wavelength shift is obtained using an adequate signal processing applied on the photodiode voltage, leading to resolutions in the range of $pm/\sqrt{Hz}$ [19].

More on interrogators can be found in [19, 20] or in books dealing with optical fibers, like [6–16].

References

[1] Liu, X., Evolution of fiber-optic transmission and networking toward the 5G era, *Rev. Science*, Vol. 22, 2019, pp. 489–506.

[2] Kapany, N.S., *Fiber Optics Principles and Applications*, Academic Press, 1967, 429 p.

[3] Kao, K.C. and Hockham, G.A. (1966). Dielectric fibre surface waveguides for optical frequencies, *Proc. IEEE*, Vol. 113, 1966, pp. 1151–1158.

[4] Schawlow, A.L. and Townes, C.H., Infrared and optical masers, *Phys. Rev.*, Vol. 112(6), 1958, pp. 1940–1949.

[5] Maiman, T., Stimulated optical radiation in Ruby, *Nature*, Vol. 187, 1960, pp. 493–494.

[6] Saleh, B.E.A. and Teich, M.C., *Fundamentals of Photonics*, Chapter 8: Fiber optics, John Wiley & Sons, 1991, 947 p.

[7] Ansari, F., editor, *Applications of Fiber Optic Sensors in Engineering Mechanics* American Society of Civil Engineers, 1993, 230 p.

[8] Reese, R.T. and Kawahara, W.A., (eds.), *Handbook on Structural Testing* Society for Experimental Mechanics (US), Fairmont Press, 1993, 402 p.

[9] Udd, E., (ed.), *Fiber Optic Smart Structures*, John Wiley & Sons, 1995, 688 p.

[10] Culshaw, B., *Smart Structures and Materials*, Artech House, 1996, 207 p.

[11] Van Steenkiste, R.J. and Springer, G.S., *Strain and Temperature Measurement with Fiber Optic Sensors*, Technomic Publishing Co, 1997, 294 p.

[12] Ansari, F., (ed.), *Fiber Optic Sensors for Construction Materials and Bridges*, Technomic Publishing Co., 1998, 267 p.

[13] Ghatak, A. and Thyagarajan, K., *Introduction to Fiber Optics*, Cambridge University Press, 1998, 581 p.

[14] Measures, R.M., *Structural Monitoring with Fiber Optic Technology*, Academic Press, 2001, 716 p.

[15] Lopez-Higuera, J.M. (ed.), *Handbook of Optical Fibre Sensing Technology,* John Wiley & Sons, 2002, 828 p.

[16] Al-Amri, M., El-Gomati,M. and Zubairy, M.S., *Optics in Our Time*, Chapter 4: Applications, Springer Open, 2016, 504 p.
[17] Hill, K.O., Fujii, Y., Johnson, D.C. and Kawasaki, B.S., Photosensitivity in optical fiber waveguides: application to reflection fiber fabrication, *Appl. Phys. Lett.*, Vol. 32(10), 1978, pp. 647–649.
[18] Hill, K.O. and Meltz, G., Bragg grating technology fundamentals and overview, *J. Lightwave Technol.*, Vol. 15(8), 1997, pp. 1263–1276.
[19] Campanella, C.E., Cuccovillo, A., Campanella, C., Yurt, A. and Passaro, V.M.N., Fibre Bragg grating based strain sensors: review of technology and applications, *Sensors*, Vol. 18, 2018, Paper Id 3115, 27 p., doi:10.3390/s18093115.
[20] Sahota, J.K., Gupta, N. and Dhawan, D., Fiber Bragg grating sensors for monitoring of physical parameters: a comprehensive review, *Optical Eng.*, Vol. 59(6), 2020, Paper Id 060901.
[21] Pierce, J.R., Coupling modes of propagation, *J. Appl. Phys.*, Vol. 25, 1954, pp. 179–183.
[22] Werneck, M.M., Allil, R.C.S.B., Ribeiro, B.A. and de Nazaré, F.V.B., A guide to fiber Bragg grating sensors, *INTECH Open Science Book*, Chapter 1, 2013, 25 p, http://dx.doi.org/10.5772/54682.
[23] Nye, J.F., *Physical Properties of Crystals: Their Representation by Tensors and Matrices*, Oxford University Press, 1957, 324 p.
[24] Gagliardi, G., Salza, M., Avino, S., Ferraro, P. and De Natale, P., Probing the ultimate limit of fiber-optic strain sensing, *Science*, Vol. 330, 2010, pp. 1081–1084.
[25] Fidanboylo, K. and Efendioğlu, H.S., Fiber optic sensors and their applications, *5th International Advanced Technologies Symposium (IATS'09)*, May 13–15, 2009, Karabuk, Turkey.
[26] Pevec, S. and Donlagić, D., Multiparameter fiber-optic sensors: a review, *Optical Eng.*, Vol. 58(7), Paper Id 072009, 2019, 26 p., doi:10.1117/1.OE.58.7.072009.
[27] Guemes, A., Fernandez-Lopez, A. and Soller, B.J., Optical fiber distributed sensing-physical principles and applications, *Struct. Health Monitoring*, Vol. 9(3), 2010, pp. 233–245, doi:10.1177/1475921710365263.
[28] Sánchez, D.M., Gresil, M. and Soutis, C., Distributed internal strain measurement during composite manufacturing using optical fibre sensors, *Composite Sci. Technol.*, Vol. 120, 2015, pp. 49–57, doi:10.1016/j.compscitech.2015.09.023.
[29] Chandarana, N., Sánchez, D.M., Soutis, C. and Gresil, M., Early damage detection in composites by distributed strain and acoustic event monitoring, *Procedia Eng.*, Vol. 188, 2017, pp. 88–95, doi:10.1016/j.proeng.2017.04.515.
[30] Caucheteur, C., Guo, T. and Albert, J., Polarization-assisted fiber Bragg grating sensors, Tutorial and review, *J. Lightwave Technol.*, Vol. 35(16), 2027, pp. 3311–3322.
[31] Sano, Y. and Yoshino, T., Fast optical wavelength interrogator employing arrayed waveguide grating for distributed fiber Bragg grating sensors, *J. Lightwave Technol.*, Vol. 21(1), 2003, pp. 132–139.
[32] Freydin, M., Ratner, M.K. and Raveh, E. D., Fiber-optics-based aeroelastic shape sensing, *AIAA J.*, Vol. 57(12), 2019, pp. 5094–5103.

Appendix A: Geometrical Optics — Basic Concepts

Geometrical optics is a model of optics that describes light propagation in terms of rays. The ray in geometric optics is useful for approximating the paths along which light propagates.

One should remember that the speed of light in vacuum and its associated frequency are

$$c = 2.99792458 \times 10^8\ \text{m/s},$$
$$f = c/\lambda \cong 10^{15}\ \text{Hz}.$$

When the light is traveling in a material, like water for instance, it will slow down, leading to a term called index of refraction, $n = c/v$, where v is the speed of the light through that material (see Table A.1 for various values of n).

Remembering that the wave speed (v) is related to the frequency (f) and wavelength (λ), $v = f/\lambda$ and that the light frequency does not change with the medium it travels through, we obtain the following relation between the index of refraction and the wavelength of the light

$$\begin{aligned} &\lambda_0 = c/f && \text{@vacuum,} \\ &\lambda = v/f && \text{@other medium} \\ &\Rightarrow \frac{\lambda_0}{\lambda} = \frac{c}{v} = n. \end{aligned} \tag{A.1}$$

Reflection of light

When the light hits glossy surfaces, like for example mirrors, it will be reflected back, as shown schematically in Fig. A.1. The law of reflection states that the incident ray, the normal and the reflected ray lay in the same plane and $\theta_i = \theta_r$.

Table A.1. Index of refraction for common materials (from *CRC Handbook of Chemistry and Physics*).

Material	Air	Water	quartz	Fused oil	Whale glass	Crown Salt	Asphalt	Diamond	Lead
n	1.0003	1.33	1.4585	1.46	1.52	1.54	1.635	2.42	2.6

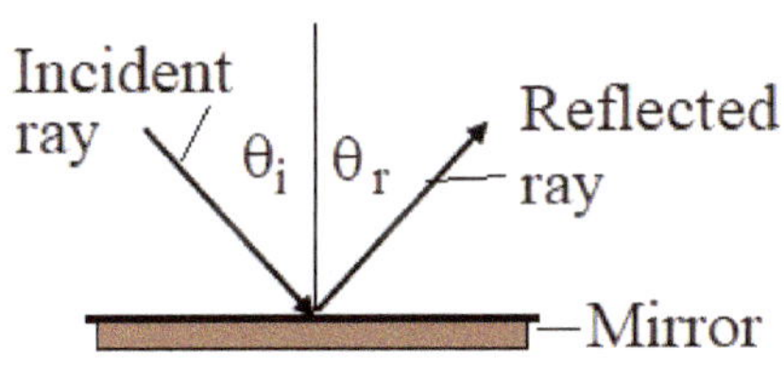

Fig. A.1. Schematic view of a reflected ray.

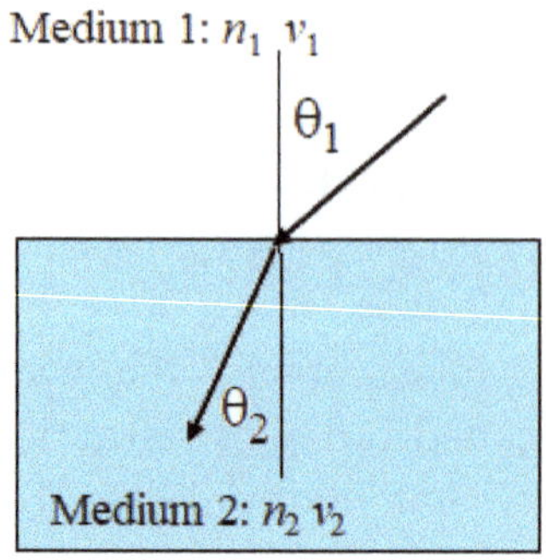

Fig. A.2. Schematic view of a refracted ray — Snell's law.

Snell's law — refractions

When light travels from one medium with index of refraction n_1 to a second medium having an index of refraction n_2 it changes its direction, as shown schematically in Fig. A.2. Snell's law[i] can be written as

$$n_1 \sin\theta_1 = n_2 \sin\theta_2 \Rightarrow v_1 \sin\theta_2 = v_2 \sin\theta_1. \tag{A.2}$$

Critical angle and total internal reflection

When light travels from a medium with high value of n to another medium having lower n, the application of the Snell's law might require that the sine of the angle would be larger that unity. This of course cannot happened and therefore for such cases the light will be totally reflected, which is known in literature as *total internal reflection*. The *critical angle* is defined as the largest possible angle of incidence, which still allows refraction. This can be written as

$$\begin{aligned}
&n_1 \sin\theta_1 = n_2 \sin\theta_2,\\
&\theta_{\text{crit}} = \arcsin\left(\frac{n_2}{n_1}\sin\theta_2\right)\\
&\text{but } \sin\theta_2 = 1\\
&\Rightarrow \theta_{\text{crit}} = \arcsin\left(\frac{n_2}{n_1}\right).
\end{aligned} \tag{A.3}$$

Numerical aperture and acceptance angle of a fiber

Numerical Aperture (NA) is a non-dimensional quantity related the acceptance angle of a given angle. The term is written as

$$NA = n_o \sin\theta_a, \tag{A.4}$$

[i]Snell is the English family name version of Willebrord Snellius (13 June 1580–30 Oct. 1626) a Dutch astronomer and mathematician.

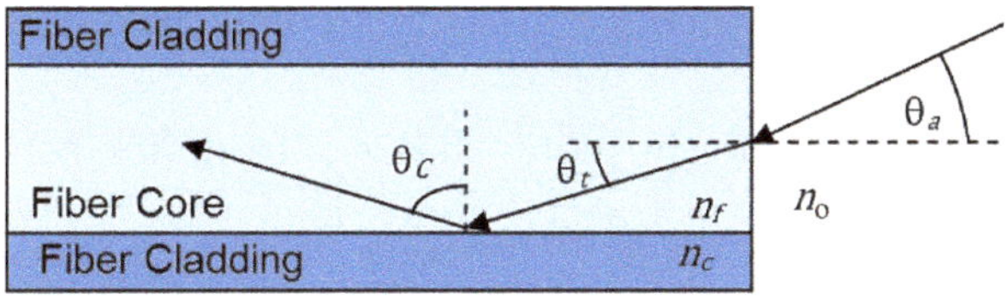

Fig. A.3. A light ray impinging the fiber at half its acceptance angle.

where θ_a is the maximum half of the acceptance angle of a fiber, and n_o is the index of refraction of the material outside the fiber. Typically the outside material is air, and therefore $n_o = 1$, yielding $NA = \sin\theta_a$. Another way of defining the NA is using the schematic drawing presented in Fig. A.3. A ray at the maximum half acceptance angle propagates into a fiber. It will hit the interface cladding–core of the finer and it will be reflected. According to Snell's law, one can write the following expression:

$$
\begin{aligned}
&n_o \sin\theta_a = n_f \sin\theta_t \quad \text{(air core)}\\
&n_f \sin\theta_c = n_c \sin\theta_r \quad \text{(core cladding)}\\
&\text{but} \quad \sin\theta_{\text{crit},c} = \left(\frac{n_c}{n_f}\sin\theta_r\right) \quad \text{and} \quad \sin\theta_r = 1\\
&\Rightarrow \sin\theta_{\text{crit},c} = \left(\frac{n_c}{n_f}\right) \text{ however } \quad \theta_{\text{crit},c} + \theta_t = 90^\circ \Rightarrow \sin\theta_{\text{crit},c} = \cos\theta_t\\
&\Rightarrow \sin\theta_{\text{crit},c} = \left(\frac{n_c}{n_f}\right) = \cos\theta_t = \sqrt{1-\sin^2\theta_t}\\
&\text{or} \quad \sin\theta_t = \sqrt{1-\frac{n_c^2}{n_f^2}}.
\end{aligned}
\tag{A.5}
$$

Using the first row in Eq. (A.5) and substituting the equation in the last row of Eq. (A.5) yields

$$
\begin{aligned}
&n_o \sin\theta_a = n_f \sin\theta_t \quad \text{(air core)}\\
&\text{but} \quad \sin\theta_t = \sqrt{1-\frac{n_c^2}{n_f^2}}\\
&\Rightarrow n_o \sin\theta_a = n_f\sqrt{1-\frac{n_c^2}{n_f^2}} = \sqrt{n_f^2 - n_c^2} = NA\\
&\text{or} \quad \sin\theta_{a,\max} = \sqrt{\frac{n_f^2-n_c^2}{n_o^2}} \Rightarrow \theta_{a,\max} = \arcsin\left(\sqrt{\frac{n_f^2-n_c^2}{n_o^2}}\right)
\end{aligned}
\tag{A.6}
$$

The total angle of acceptance for the light conus would be twice the last expression of Eq. (A.6).

The refractive index of the core changes from its center towards the cladding. The profile of the refractive index is generally given as

$$n^2(r) = n_f^2\left[1 - 2\Delta\left(\frac{r}{0.5d}\right)^{\alpha}\right] \quad r < 0.5d \text{ (@core)}$$

$$n^2(r) = n_c^2 = \text{constant} \quad r > 0.5d \quad \text{(@cladding)}$$

$$\text{where} \quad \Delta = \frac{n_f^2 - n_c^2}{2n_f^2}$$

and α for special cases is :

$\alpha = 1$ for triangular profile

$\alpha = 2$ for parabolic profile

$\alpha \to \infty$ for Step-Index profile. (A.7)

Another parameter is the normalized frequency, V and is given as

$$V = 2\pi\frac{0.5d}{\lambda}NA = \pi\frac{d}{\lambda}NA, \text{ where}$$

$$NA = n_f\sqrt{2\Delta}, \tag{A.8a}$$

where d is the core diameter and λ is the light wavelength. The V parameter determines the fiber-operating regime: mono-mode propagation or multi-mode propagation.

The number of modes guided through the core is a function of the above profile, namely

$$N \approx \frac{V^2}{2}\frac{\alpha}{\alpha+2}. \tag{A.8b}$$

For a step-index profile if $V < 2.405$, then only a single mode will propagate. This is also called the fundamental mode LP_{01} with a cut-off frequency of $V_c = 0$.

For a graded-index profile ($\alpha = 2$), we get

$$V_c \approx 2.405 \cdot \sqrt{2} \approx 3.4. \tag{A.9}$$

Interference

When two rays have the same wavelength and are in phase (see Fig. A.4), they will add, yielding a wave having the sum of the peak-to-peak amplitudes of the input rays thus leading to what it is called a *Constructive Interference*.

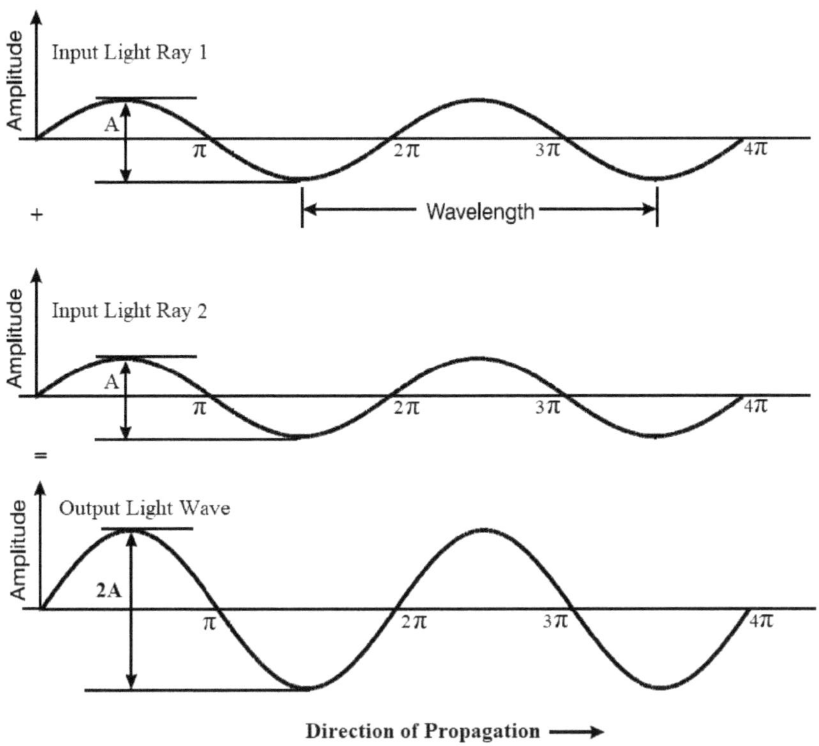

Fig. A.4. Constructive interference — a schematic view.

The opposite case leading to *Destructive Interference* (see Fig. A.5) is due to the "out of phase" state of one of the rays compared to the other. If the phase is π, then the output light peak-to-peak amplitude would vanish.

Diffraction grating for light waves

Diffraction grating or simply a grating is a thin film made of glass or even plastic material that has a large number of lines per mm etched on it. The density of the grating can be 250 lines/mm up to 3000 lines/mm and even higher. When light emanating from a bright and small source would pass through the grating it will generate a large number of independent sources (see Fig. A.6). Those sources are in-phase ones or *coherent sources*. Each source would send rays in all directions. Placing a screen at a D distance from the grating would reveal bright fringes on it (Fig. A.6). The principal maxima would appear on both sides of the central maxima. Note that as all the sources are in-phase at fringe P_0, the brightest one, and as not all the waves are in phase at P_3, it is much dimmer than P_0.

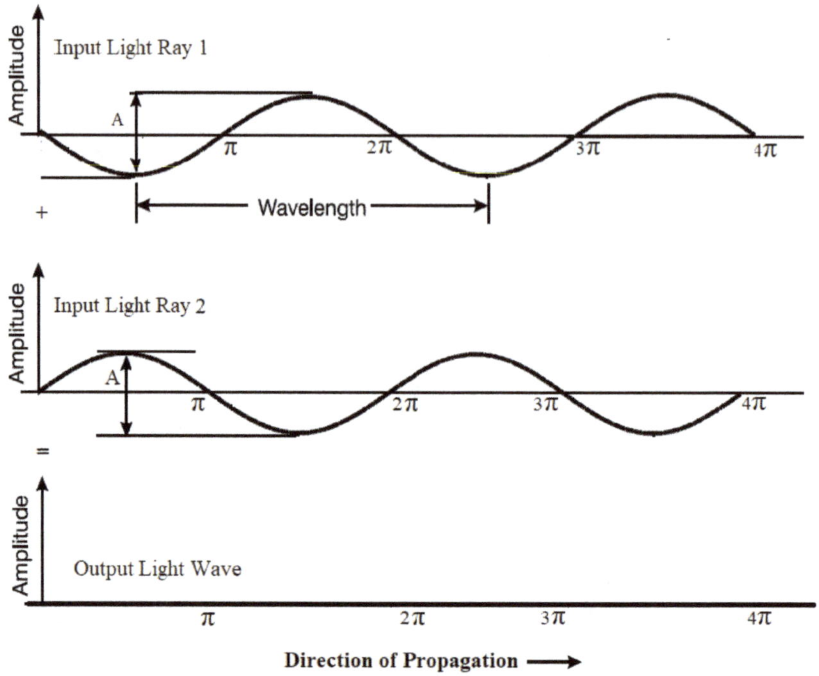

Fig. A.5. Destructive interference (out-of-phase = π) — a schematic view.

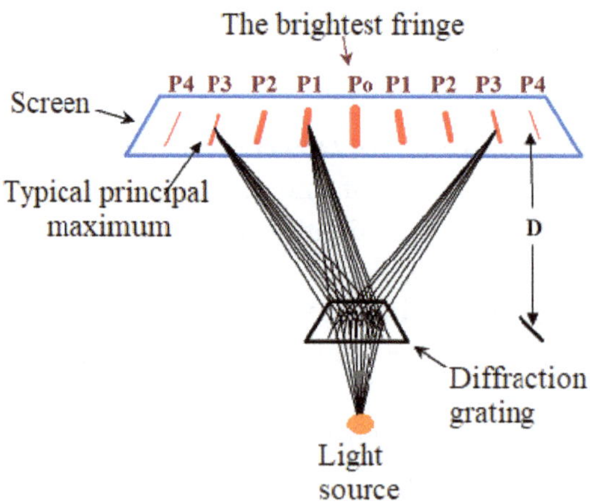

Fig. A.6. Schematic view of a diffraction grating and the associated screen (not to scale).

Assuming that the distance between two adjacent lines to be d and N being the number of lines per mm of grating one can write

$$d = \frac{1}{N} \quad \text{or} \quad N = \frac{1}{d}. \tag{A.10}$$

Then the expression to find the principal maxima for a diffraction grating would be

$$d \sin\theta_n = n\lambda, \quad n = 1, 2, 3, \ldots,$$

where

$$\lambda = \text{wavelength}. \tag{A.11}$$

Chapter 11

Static and Dynamic Behavior of Nano-beams*

11.1. Introduction

Nano-beam is one of the morphologies of nano-scale structures. As a beam, the nano-beam has one dimension (length) larger in scale than the other dimensions, e.g. width and height in a square section or the radius in a round section. The true scale in such a beam is the nano scale, which is a one of a billion of a meter (10^{-9} [m]).

These kinds of beams may be synthesized from metals or semiconducting materials. In Fig. 11.1, an SEM (Scanning Electron Microscope) image of a single-electron transistor is presented. The source and drain electrodes are bridged by C_{60}–Au–C_{60} (Laser-ablated gold atoms reacting with C_{60}) nano-particles.

Figure 11.2 a silicon nano-beam with cavities is presented in order to investigate the dependence of the quality on the initial nano-beam with the change of the holes sizes.

Due to exceptionally good physical, mechanical and electrical properties as well as display technologies, as the reflectivity of the rods can be altered by changing their orientation with an applied electric field, nano-sized structures have attracted much investment to develop innovatory applications in a wide range of disciplines.

Nanotechnology has opened a new area in engineering and materials science. Besides experimental studies at the nano-scale, which may be

*The content of the present chapter is based on the master thesis written by Mr. Oran Buch, under my supervision, carrying the name: *Dynamic buckling of nano-beams* at the Faculty of Aerospace Engineering, Technion, I.I.T., 32000 Haifa, Israel, Feb. 2015, 205p. (in English).

Fig. 11.1. SEM image of a single-electron transistor.
Source: http://www.phys.sinica.edu.tw/~quela/images/sample/10.jpg.

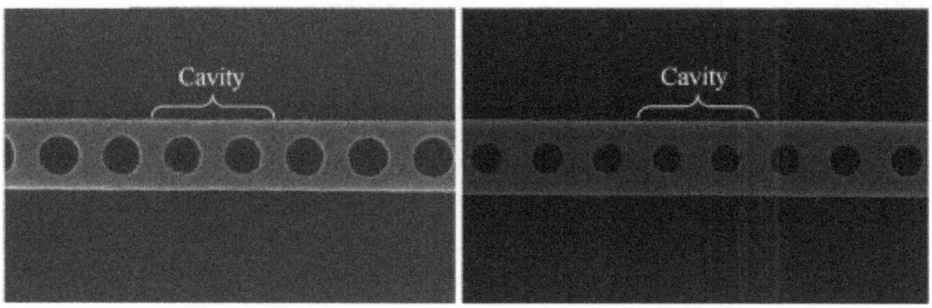

Fig. 11.2. SEM image of the cavity region of a silicon nano-beam.
Source: http://imagebank.osa.org/getImage.xqy?img=OG0kcC5sYXJnZSxqb3NhYi0yOS0yLUE1NS1nMDAy.

expensive and complicated, there are three main theoretical approaches to investigate modeling of nanostructures [1]:

- Atomistic modeling;
- Hybrid atomistic–continuum mechanics;
- Continuum mechanics.

Atomic Modeling includes techniques such as [1]:

- Classical molecular dynamics (MD);
- Tight-binding molecular dynamics (TBMD);
- Density functional theory (DFT).

Atomic and molecular models, while, certainly conceptually valid for small length scales, are difficult to formulate accurately and are usually computationally intensive [2]. Hybrid atomistic–continuum mechanics allows the user to directly incorporate interatomic potential into the continuum analysis [1]. Continuum mechanics includes classical (or local) beam, plate and shell theories that are practical for analyzing nanostructures for large-scale systems, i.e. a structure not in nano-scale built out of basic elements with characteristic size in the nano scale [1].

Hybrid atomistic–continuum formulation allows the simulation of complex phenomena at the nano- and micro-scales without the prohibitive cost of fully atomistic approach (atomistic modeling). Atomistic modeling is limited to the region of interest, hybrid atomistic formulation allows calculations in the entire domain by dividing it to computational regions and continuum mechanics allows analyzing the entire region as a whole.[a]

Continuum mechanics approach is less computationally expensive than the former two approaches and its formulations are relatively simple. These advantages make continuum modeling employed as an alternative way to simulate some phenomena in nanostructures such as buckling, wave propagation and free vibration [1].

A consensus was reached that the refined non-local models, based on the theory developed by Eringen, are superior to their local counterparts [1]. This conclusion was bound to be accepted since the continuum mechanics is based on the continuous assumption. However, since the theory also belongs to the category of continuum mechanics, the applicability of the non-local theories needs to be justified i.e. identify the magnitude of the scale coefficient (small-scale parameter) e_0a which is the key in revealing scale effect in models and assuring the proper use of continuum mechanics equations (local) in nanotechnology (non-local).

Therefore, this continuum theory on one hand is suitable for modeling sub-micro or nano-sized structures, while on the other hand it avoids enormous computational efforts when compared with discrete atomistic or molecular dynamics simulations [3].

[a]http://link.springer.com/chapter/10.1007%2F978-1-4020-3286-8_133#page-1.

Arash and Wang [1] conducted a comparison between several methods of investigating beams in nano-scale. More comparison analyses were performed in [2, 3], in which the lead conclusion is the advantage of the non-local method by Eringen [7] then any other methods mentioned in their investigation. A short list of typical references is present in what follows.

Wang and Feng [4] examined surface effects on the axial buckling and the transverse vibrations of nano-wires by using the Timoshenko beam theory, while Wang *et al.* [5, 6] investigated the elastic buckling analysis of micro- and nano-rods/tubes based on Eringen's non-local elasticity theory [7] and Eringen and Edelen [8] by using the Timoshenko beam theory. Later on, Eringen [9] published another study in which she showed that the linear nonlocal elasticity theory can be reduced to singular partial differential equations for a special class of physically admissible kernels instead of integro-partial differential equations. Solutions for screw dislocations and surface waves are presented, with experimental observations and atomic lattice dynamics supporting nicely the theoretical results. Thai [10] investigated and compared Eringen's theory with the non-local Bernoulli–Euler, non-local Timoshenko and Reddy beam theories. Yan and Jiang [11] investigated the influence of surface effects, including residual surface stress, surface elasticity and surface piezoelectricity, on the vibrational and buckling behavior of piezoelectric nano-beams by using the Bernoulli–Euler beam. Sears and Batra [12] investigated carbon single and multi-layered nano-tubes under axial loads and Zhang *et al.* [13] analyzed hybrid non-local Bernoulli–Euler beam model for the bending, buckling, and vibration analyzes of micro and nano-beams. Other researchers like, Wang and Feng [14] tested the influence of surface elasticity and residual surface tension on nanowires, while Nishio *et al.* [15] investigated multi-layered nano-tubes buckling with constant outer diameter and as a function of thickness under axial load. Wang *et al.* [16] studied the buckling of familiar carbon nano-tubes under the influence of main factors such as dimensions, boundary conditions, temperature, strain rate, and chirality. They provided the present state-of-the-art on the buckling of carbon nano-tubes. Lee *et al.* [16] studied the buckling behavior of a non-uniform nano-wire under axial compression using the non-local elasticity theory. Wang and Yang [18] addressed the issue of bending stability of multiwalled carbon nanotubes.

11.2. The Static Non-local Nano-beam Model and Buckling

Figure 11.3 presents the schematic nano-beam model. First, the Timoshenko model will be used which was described in Chapter 2 as FSDT model [19].

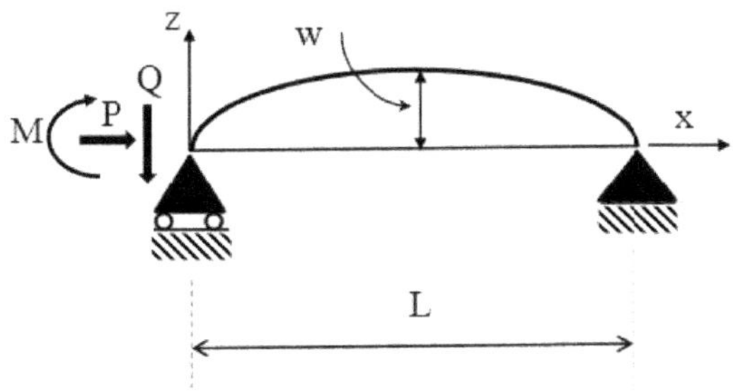

Fig. 11.3. A schematic nano-beam subjected to axial compressive load, P, shear force Q and moment M.

Accordingly, the static equilibrium equations can be written as

$$-\frac{dM}{dx}+Q=0,$$

$$-\frac{dQ}{dx}+P\frac{d^2w}{dx^2}=0, \quad \text{where}$$

$$M=\int_A \sigma_{xx} z dA, \quad Q=\kappa\int_A \sigma_{xz} dA, \tag{11.1}$$

where κ is the Timoshenko shear correction factor (usually taken as 5/6 for a rectangular cross section[b]) and σ_{xx} and σ_{xz} are the normal and shear stresses on the cross section. To deal with nano-beams, we shall introduce the term *non-local* which is represented by $e_0 a$, a scale coefficient that incorporates the small scale effect, as suggested by Eringen [7]. One should note that $\boldsymbol{a}$ is the internal characteristic length (like a lattice parameter or a C–C[c] bond length) and e_0 is experimentally determined or approximated, by matching the dispersion curves of plane waves with those of atomic lattice dynamics [6, 10].

While *non-local* term implies that the stress at a point depends on the strains at all other points on the body, the term *local* associated to larger scale bodies, assumes that the stress at a given point depends only on the strain at that point.

Therefore using the Timoshenko model with the non-local Eringen model [7] (see Appendix A) leads to the following moment and shear force

[b]The accurate value is $\kappa=\frac{10(1+\upsilon)}{12+11\upsilon}$ for a rectangular cross-section and $\kappa=\frac{6(1+\upsilon)}{7+6\upsilon}$ for a solid circular cross-section.
[c]C–C=Carbon–Carbon.

equations

$$M - (e_0 a)^2 \frac{d^2 M}{dx^2} = EI \frac{d\phi}{dx},$$
$$Q = \kappa GA \left(\phi + \frac{dw}{dx} \right), \tag{11.2}$$

where G is the shear modulus of the beam, A = the area of the beam's cross-section and ϕ is the rotation due to shear.

Using Eqs. (11.2), one can write the two-coupled governing equations for the Timoshenko non-local nano-beam

$$-\frac{dM}{dx} + Q = 0 \Rightarrow \frac{dM}{dx} - Q = 0 \Rightarrow (e_0 a)^2 P \frac{d^3 w}{dx^3} + EI \frac{d^2 \phi}{dx^2}$$
$$- \kappa GA \left(\phi + \frac{dw}{dx} \right) = 0,$$
$$-\frac{dQ}{dx} + P \frac{d^2 w}{dx^2} = 0 \frac{dQ}{dx} - P \frac{d^2 w}{dx^2} = 0 \Rightarrow \kappa GA \left(\frac{d\phi}{dx} + \frac{d^2 w}{dx^2} \right) - P \frac{d^2 w}{dx^2} = 0. \tag{11.3}$$

Some boundary conditions associated with Eqs. (11.3) would be

$$\text{Pinned:} \quad w = 0 \quad \text{and} \quad M = (e_0 a)^2 P \frac{d^3 w}{dx^3} + EI \frac{d^2 \phi}{dx^2} = 0,$$
$$\text{Clamped:} \quad w = 0 \quad \text{and} \quad \phi = 0,$$
$$\text{Free:} \quad M = (e_0 a)^2 P \frac{d^3 w}{dx^3} + EI \frac{d^2 \phi}{dx^2} = 0 \quad \text{and}$$
$$Q - P \frac{dw}{dx} = \kappa GA \left(\phi + \frac{dw}{dx} \right) - P \frac{dw}{dx} = 0. \tag{11.4}$$

Note that the *pinned* boundary conditions are also called *simply supported* boundary conditions in the literature.

Equations (11.3) can be decoupled to yield

$$\frac{d^4 \widetilde{w}}{d\widetilde{x}^4} + \lambda^2 \frac{d^2 \widetilde{w}}{d\widetilde{x}^2} = 0,$$
$$\frac{d^3 \phi}{d\widetilde{x}^3} + \lambda^2 \frac{d\phi}{d\widetilde{x}} = 0, \quad \text{where}$$
$$\widetilde{w} = w/L, \quad \widetilde{x} = x/L,$$
$$\lambda^2 = \frac{\frac{PL^2}{EI}}{\left[1 - \frac{P}{\kappa GA} - (e_0 a)^2 \frac{P}{EI}\right]}. \tag{11.5}$$

The solutions of Eqs. (11.5) have the following form:

$$\widetilde{w}(\widetilde{x}) = C_1 \sin\lambda\widetilde{x} + C_2 \cos\lambda\widetilde{x} + C_3\widetilde{x} + C_4,$$
$$\phi(\widetilde{x}) = -C_1\lambda\left(1 - \frac{P}{\kappa GA}\right)\cos\lambda\widetilde{x} + C_2\lambda\left(1 - \frac{P}{\kappa GA}\right)\sin\lambda\widetilde{x} - C_3. \quad (11.6)$$

The four constants, C_1, C_2, C_3 and C_4 are determined by imposing the adequate boundary conditions. For pinned-pinned boundary conditions we get

$$\begin{bmatrix} 0 & 0 & 0 & 1 \\ \sin\lambda & \cos\lambda & 1 & 1 \\ \Gamma\sin\lambda & \Gamma\cos\lambda & 0 & 0 \\ 0 & \Gamma & 0 & 0 \end{bmatrix} \begin{Bmatrix} C_1 \\ C_2 \\ C_3 \\ C_4 \end{Bmatrix} = \begin{Bmatrix} 0 \\ 0 \\ 0 \\ 0 \end{Bmatrix}, \quad \text{where}$$
$$\Gamma = 1 - \frac{P}{\kappa GA} - (e_0a)^2\frac{P}{EI}. \quad (11.7)$$

For the clamped–clamped case, we obtain

$$\begin{bmatrix} 0 & 1 & 0 & 1 \\ \sin\lambda & \cos\lambda & 1 & 1 \\ \Lambda\lambda & 0 & 1 & 0 \\ -\Lambda\lambda\cos\lambda & \Lambda\lambda\sin\lambda & -1 & 0 \end{bmatrix} \begin{Bmatrix} C_1 \\ C_2 \\ C_3 \\ C_4 \end{Bmatrix} = \begin{Bmatrix} 0 \\ 0 \\ 0 \\ 0 \end{Bmatrix}, \quad \text{where}$$
$$\Lambda = 1 - \frac{P}{\kappa GA}. \quad (11.8)$$

For a cantilevered beam (clamped–free), the expression will be

$$\begin{bmatrix} 0 & 1 & 0 & 1 \\ -\Lambda\lambda & 0 & -1 & 0 \\ \Gamma\sin\lambda & \Gamma\cos\lambda & 0 & 0 \\ 0 & 0 & -P & 0 \end{bmatrix} \begin{Bmatrix} C_1 \\ C_2 \\ C_3 \\ C_4 \end{Bmatrix} = \begin{Bmatrix} 0 \\ 0 \\ 0 \\ 0 \end{Bmatrix}, \quad (11.9)$$

while for a clamped–pinned case we get

$$\begin{bmatrix} 0 & 1 & 0 & 1 \\ \sin\lambda & \cos\lambda & 1 & 1 \\ \Lambda\lambda & 0 & 1 & 0 \\ \Gamma\sin\lambda & \Gamma\cos\lambda & 0 & 0 \end{bmatrix} \begin{Bmatrix} C_1 \\ C_2 \\ C_3 \\ C_4 \end{Bmatrix} = \begin{Bmatrix} 0 \\ 0 \\ 0 \\ 0 \end{Bmatrix}. \quad (11.10)$$

Table 11.1. Characteristic equations for various boundary conditions using Timoshenko non-local theory.

Boundary conditions	Characteristic equation	Euler buckling load P_{local}
Pinned–pinned	$\sin\lambda = 0$	$\frac{\pi^2 EI}{L^2}$
Clamped–clamped	$\sin\lambda/2 = 0$	$4\frac{\pi^2 EI}{L^2}$
Clamped–free	$\sin\lambda = 0$	$\frac{\pi^2 EI}{4L^2}$
Clamped–pinned	$\Lambda\lambda = \tan\lambda$	$2.0457\frac{\pi^2 EI}{L^2}$

Table 11.2. Properties of the nano-beams.

Property	Value
Young's modulus, $E\,\text{N/m}^2$	$7.0 \cdot 10^{10}$
Poisson's ratio, υ	0.3
Shear modulus, $G\,\text{N/m}^2$	$2.692 \cdot 10^{10}$
Density, $\rho\,\text{kg/m}^3$	2700
Timoshenko's shear factor, κ	5/6
Diameter of a characteristic nano-beam, d m	$1.0 \cdot 10^{-9}$
Cross-sectional area of the nano-beam, $A\,\text{m}^2$	$7.845 \cdot 10^{-19}$
Second moment of area, $I\,\text{m}^4$	$4.9087 \cdot 10^{-38}$

For each type of boundary conditions, we demand that the determinant of the coefficients matrix would vanish, leading to a characteristic equation from which an eigenvalue (the buckling load of the given nano-beam) is calculated. The results are summarized in Table 11.1.

The nano-beam buckling load using the Timoshenko non-local theory can be written as

$$P_{\text{Tim.non-local}} = \frac{P_{\text{local}}}{1 + \frac{P_{\text{local}}}{\kappa GA} + (e_0 a)^2 \frac{P_{\text{local}}}{EI}} \tag{11.11}$$

One can see from Eq. (11.11), that for a regular-scaled beam, $(e_0 a)^2 = 0$, the Timoshenko buckling load is lower that the Bernoulli–Euler buckling load. For a nano-beam, $(e_0 a)^2 \neq 0$ the buckling load would be even lower than the Bernoulli–Euler one.

Typical results are next presented. The properties for the nano-beams are presented in Table 11.2.

The computational results are presented in Figs. 11.4 and 11.5 for pinned–pinned and clamped–clamped boundary conditions, respectively using Timoshenko local and Timoshenko non-local theories.

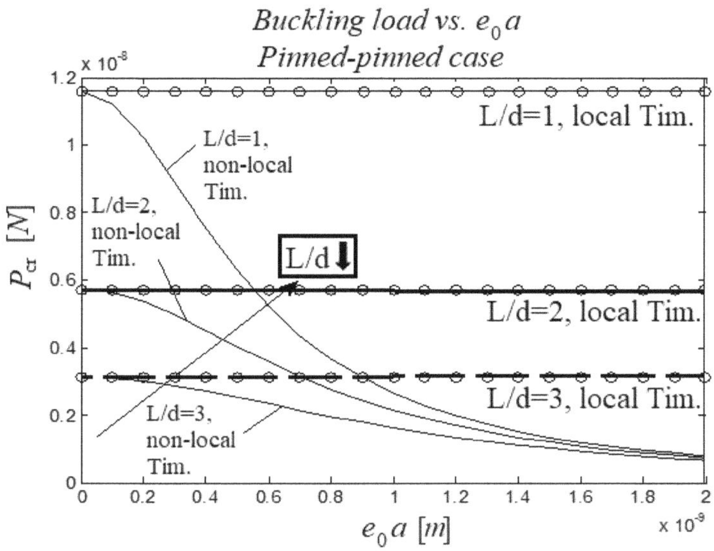

Fig. 11.4. Buckling load vs. (e_0a) scale coefficient (pinned–pinned case).

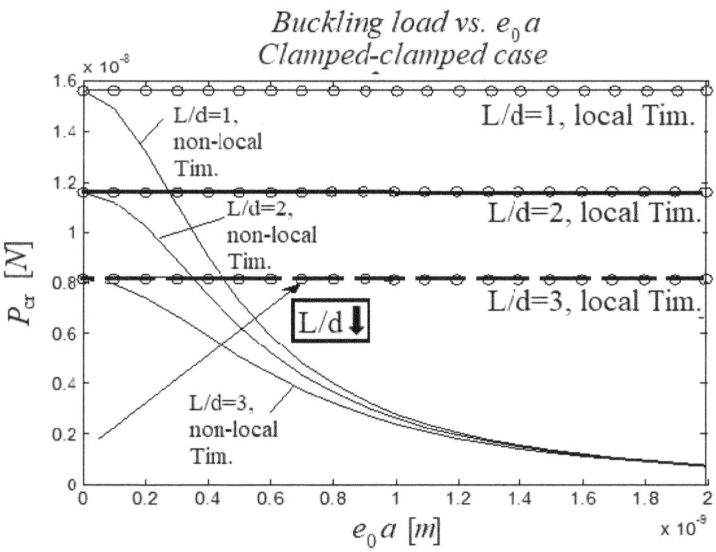

Fig. 11.5. Buckling load vs. (e_0a) scale coefficient (clamped–clamped case).

One should note that while for a local Euler–Bernoulli case $P_{cr,\text{C–C}} = 4P_{cr,p\text{–}p}$ for a local Timoshenko case this ratio is less than 4 and depends on the Timoshenko's shear factor, κ (Table 11.2) yielding the following ratio $4(1+\widehat{\alpha})/(1+4\widehat{\alpha})$ where $\widehat{\alpha} = (\pi^2 EI)/(\kappa GAL^2)$. This is expected also for the non-local Timoshenko case.

11.3. The Dynamic Non-local Nano-beam Model — Vibrations and Natural Frequencies

Adding to Eqs. (11.3) the dynamic terms and assuming $w(x,t) = W(x)e^{i\omega t}$ and $\phi(x,t) = \Phi(x)e^{i\omega t}$leads to the following two equations of motion:

$$-\frac{dM}{dx} + Q = -\rho I \frac{d^2\phi}{dt^2} \Rightarrow -\frac{dM}{dx} + Q = -I_3 \frac{d^2\phi}{dt^2} = \omega^2 I_3 \Phi(x),$$

$$-\frac{dQ}{dx} + P\frac{d^2W}{dx^2} = -\rho A \frac{d^2w}{dt^2} \Rightarrow -\frac{dQ}{dx} + P\frac{d^2W}{dx^2} = -I_1 \frac{d^2w}{dt^2} = \omega^2 I_1 W(x),$$

where

$$I_1 = \rho A, I_3 = \rho I, \rho = \text{mass density}. \tag{11.12}$$

Writing the expressions for the shear force and the moment for the dynamic case yields

$$Q = \kappa GA\left(\Phi + \frac{dW}{dx}\right), \frac{dM}{dx} = Q - I_3\omega^2\Phi \Rightarrow \frac{d^2M}{dx^2} = \frac{dQ}{dx} - I_3\omega^2\frac{d\Phi}{dx},$$

$$\frac{dQ}{dx} = P\frac{d^2W}{dx^2} - I_1\omega^2 W,$$

$$M = EI\frac{d\Phi}{dx} + (e_0 a)^2\frac{d^2M}{dx^2} = EI\frac{d\Phi}{dx} + (e_0 a)^2\left(\frac{dQ}{dx} - I_3\omega^2\frac{d\Phi}{dx}\right)$$

$$= EI\frac{d\Phi}{dx} + (e_0 a)^2\left(P\frac{d^2W}{dx^2} - I_1\omega^2 W - I_3\omega^2\frac{d\Phi}{dx}\right). \tag{11.13}$$

Substituting the expressions for Q and Min Eqs. (11.12) provides the two coupled equations of motions in terms of W and Φ

$$-\frac{dM}{dx} + Q = \omega^2 I_3 \Phi(x) \Rightarrow EI\frac{d^2\Phi}{dx^2} - (e_0 a)^2$$

$$\left(I_1\omega^2\frac{dW}{dx} + I_3\omega^2\frac{d\Phi}{dx} - P\frac{d^3W}{dx^3}\right),$$

$$-\kappa GA\left(\Phi + \frac{dW}{dx}\right) + I_3\omega^2\Phi = 0,$$

$$-\frac{dQ}{dx} + P\frac{d^2W}{dx^2} = \omega^2 I_1 W(x) \Rightarrow \kappa GA\left(\frac{d\Phi}{dx} + \frac{d^2W}{dx^2}\right)$$

$$-P\frac{d^2W}{dx^2} + \omega^2 I_1 W = 0. \tag{11.14}$$

To facilitate the calculation of the natural frequencies of nano-beams the above equations will be non-dimensionalized to yield

$$\chi\left(1-\frac{\alpha^2\delta^2}{\varsigma^2}\right)\frac{d^2\Phi}{d\widetilde{x}^2}+\left(\frac{\delta^2\chi}{\varsigma^2}-1\right)\Phi+\alpha^2\lambda^2\frac{d^3\widetilde{W}}{d\widetilde{x}^3}-(1+\alpha^2\delta^2\chi)\frac{d\widetilde{W}}{d\widetilde{x}}=0,$$

$$\frac{d\Phi}{d\widetilde{x}}+(1-\lambda^2)\frac{d^2\widetilde{W}}{d\widetilde{x}^2}+\delta^2\chi\widetilde{W}=0,$$

where

$$\lambda^2=\frac{P}{\kappa GA},\delta^2=\omega^2\frac{I_1L^4}{EI}=\omega^2\frac{\rho AL^4}{EI},\chi=\frac{EI}{\kappa GAL^2},$$

$$\alpha=\frac{e_0a}{L},\varsigma=\frac{L\sqrt{A}}{\sqrt{I}},\widetilde{x}=\frac{x}{L}. \tag{11.15}$$

The decoupling process of the two coupled equations appearing in Eq. (11.15) is as follows.

First the unknowns $\widetilde{W}$and Φ are assumed to have the following forms $\widetilde{W}(\widetilde{x})=\overline{W}(\widetilde{x})e^{\beta\widetilde{x}},\Phi(\widetilde{x})=\overline{\Phi}(\widetilde{x})e^{\beta\widetilde{x}}$ and then substituted in Eqs. (11.15). This results in the following matrix notation:

$$\begin{bmatrix}M_{11} & M_{12}\\ M_{21} & M_{22}\end{bmatrix}\begin{Bmatrix}\overline{W}\\ \overline{\Phi}\end{Bmatrix}=\begin{Bmatrix}0\\0\end{Bmatrix}. \tag{11.16}$$

The determinant of the matrix in Eq. (11.16) is must vanish to obtain a unique solution. Therefore

$$\det[M]=M_{11}M_{22}-M_{12}M_{21}=\overline{a}\beta^4+\overline{b}\beta^2+\overline{c}=0,$$

where

$$\overline{a}=\lambda^2\alpha^2+(\lambda^2-1)\left(1-\frac{\delta^2\alpha^2}{\varsigma^2}\right)\chi,$$

$$\overline{b}=-\lambda^2-\delta^2\chi\left(\chi+\frac{\lambda^2-1+\alpha^2\delta^2\chi}{\varsigma^2}+\alpha^2\right),$$

$$\overline{c}=\delta^2\chi-\frac{\delta^4\chi^2}{\varsigma^2}=\delta^2\chi\left(1-\frac{\delta^2\chi}{\varsigma^2}\right). \tag{11.17}$$

The solutions for Eq. (11.17) are

$$\beta_1\equiv s_1=\sqrt{\frac{-\overline{b}^2+\sqrt{\overline{b}^2-4\overline{a}\overline{c}}}{2\overline{a}}},$$

$$\beta_2\equiv s_2=\sqrt{\frac{-\overline{b}^2-\sqrt{\overline{b}^2-4\overline{a}\overline{c}}}{2\overline{a}}}. \tag{11.18}$$

The general solutions for Eqs. (11.17) can be written as (see also Chapter 4)

$$\overline{W}(\overline{x}) = A_1 \cosh(s_1\overline{x}) + A_2 \sinh(s_1\overline{x}) + A_3 \cos(s_2\overline{x}) + A_4 \sin(s_2\overline{x}), \tag{11.19}$$

$$\overline{\Phi}(x) = B_1 \cosh(s_1 x) + B_2 \sinh(s_1 x) + B_3 \cos(s_2 x) + B_4 \sin(s_2 x). \tag{11.20}$$

Substituting back into Eqs. (11.17) one obtains the relations between A_1, A_2, A_3 and A_4 and B_1, B_2, B_3 and B_4 and the solutions can be casted into

$$\overline{W}(x) = A_1 \cosh(s_1 x) + A_2 \sinh(s_1 x) + A_3 \cos(s_2 x) + A_4 \sin(s_2 x), \tag{11.21}$$

$$\begin{aligned}\overline{\Phi}(\overline{x}) &= A_2\alpha_1 \cosh(s_1\overline{x}) + A_1\alpha_1 \sinh(s_1\overline{x}) \\ &\quad + A_4\alpha_2 \cos(s_2\overline{x}) - A_3\alpha_2 \sin(s_2\overline{x}), \quad \text{where} \\ \alpha_1 &= \frac{(1-\lambda^2)s_1^2 + \delta^2\chi}{s_1}, \quad \alpha_2 = \frac{(1-\lambda^2)s_2^2 + \delta^2\chi}{s_2}\end{aligned} \tag{11.22a}$$

Substituting the boundary conditions of the problem in Eqs. (11.21) and (11.22) yield a demand for the coefficients determinant to vanish from which natural frequencies and their associated mode shapes are calculated. For pinned–pinned boundary conditions we have

$$\begin{aligned}&\sinh(s_1)\sin(s_2)[-s_2\alpha_2 - s_1\alpha_1]^2 = 0 \Rightarrow \sin(s_2) = 0 \rightarrow s_2 = n\pi, \\ &n = 1, 2, 3, \ldots.\end{aligned} \tag{11.22b}$$

As one can see, the solution for the pinned–pinned case leads to a closed form solution while for the case of clamped–clamped case we get

$$\sinh(s_1)\sin(s_2)[\alpha_2^2 - \alpha_2^2]^2 + 2[\alpha_1\alpha_2 \cos(s_2)\cosh(s_1) - 1] = 0, \tag{11.23}$$

which is not a closed-form solution and the natural frequencies will be obtained numerically.

Some numerical results are shown in Figs. 11.6 and 11.7 for the first frequency of a beam as a function of the scale factor (e_0a) at 0% and 50% of the critical buckling load, for $L/d = 10$.

For all the scale coefficients (e_0a) calculated within the present parametric investigation and for all axial compression loads, the beam transverse displacement has the same modal shape, half-wave sine.

The first frequency squared is plotted as a function of the axial compression and displayed in Fig. 11.8, while the non-dimensional version is shown in Fig. 11.9. As for local theories, the present Timoshenko non-local theory

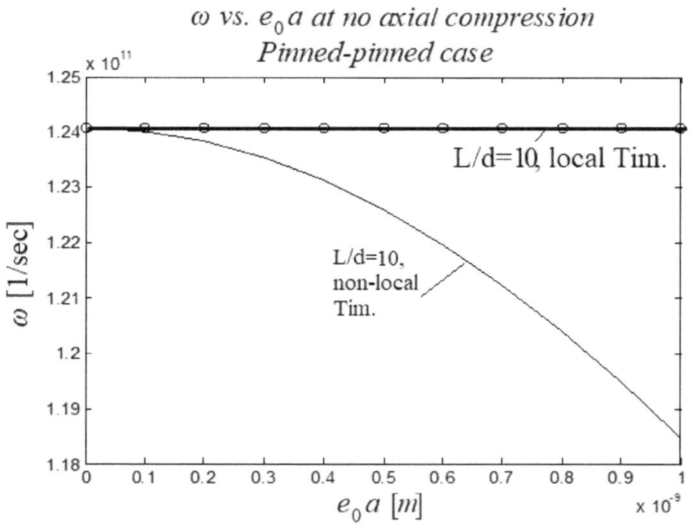

Fig. 11.6. First frequency as a function of the scale coefficient e_0a, @ $P = 0 * P_{cr}$ (pinned–pinned case).

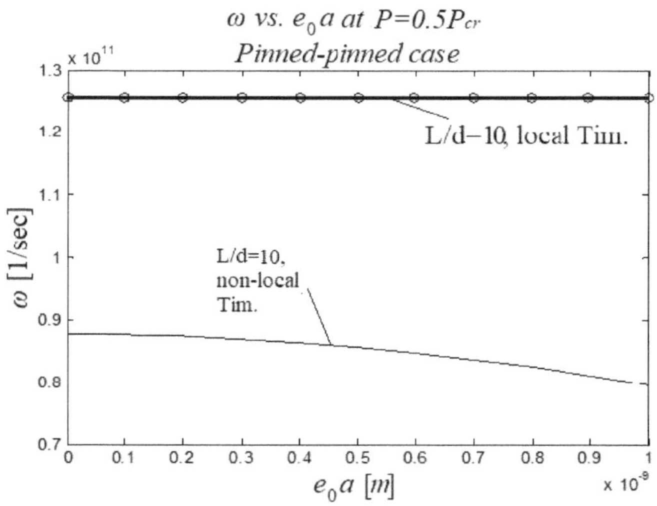

Fig. 11.7. First frequency as a function of the scale coefficient e_0a, @ $P = 0.5P_{cr}$ (pinned–pinned case).

shows an inverse linear dependency of the first frequency squared with the axial compression (see more details in Chapter 5 of the present book).

Results for the clamped–clamped case are presented in Figs. 11.10–11.12.

While for the pinned–pinned case, the mode shape for the first frequency was not changed due to the scaling coefficient e_0a, Fig. 11.12 shows a slight modification for the clamped–clamped shape mode at unloaded case and

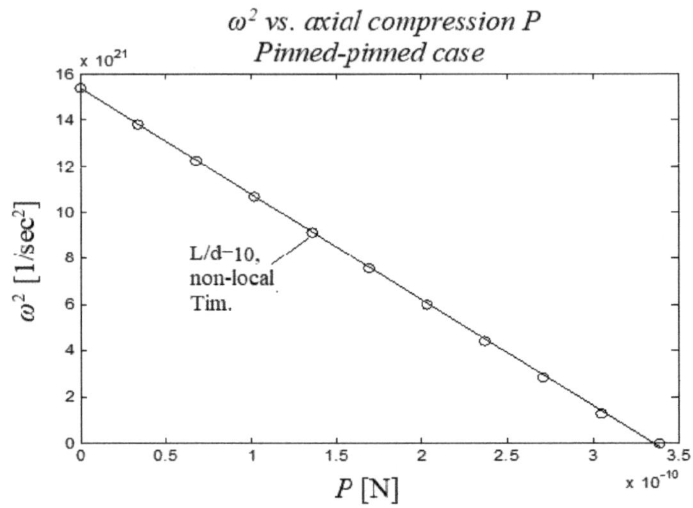

Fig. 11.8. First frequency squared ω^2 as a function of the axial compression P (pinned–pinned case).

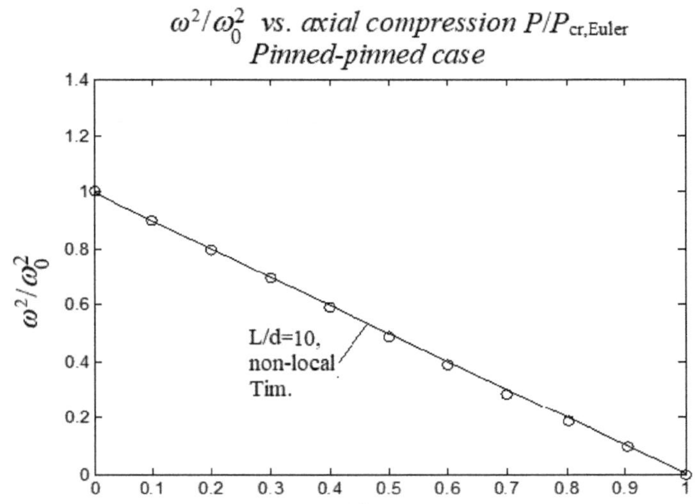

Fig. 11.9. Non-dimensional first frequency squared ω^2/ω_0^2 as a function of the nondimensional axial compression $P/P_{\mathrm{cr,Euler}}$ (pinned–pinned case).

a larger modification for loaded cases. The dependency of the first natural frequency squared with the axial compression is almost inverse linear similar to the one presented in Figs. 11.8 and 11.9. This behavior is again identical to the behavior presented for local beam theories in Chapter 5 of the present book.

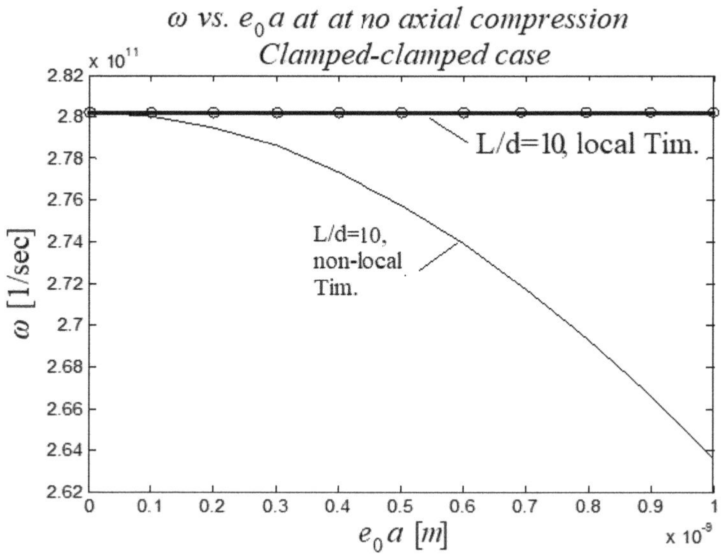

Fig. 11.10. First frequency as a function of the scale coefficient e_0a, @ $P = 0 * P_{cr}$ (clamped-clamped case).

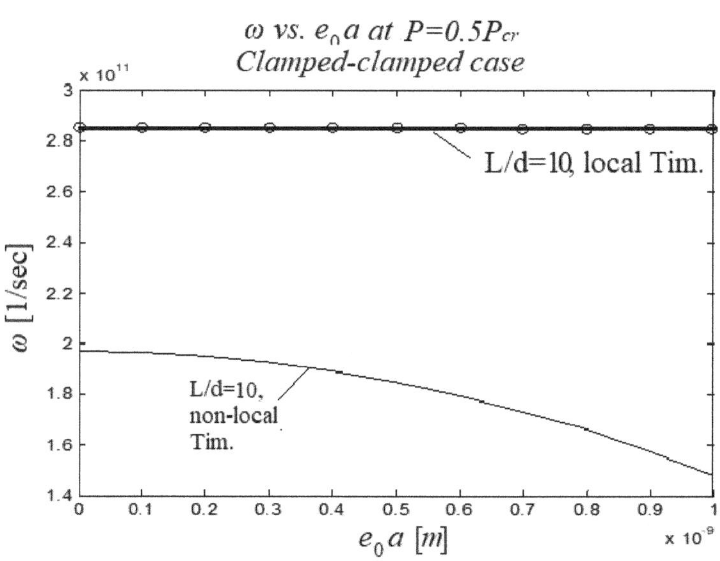

Fig. 11.11. First frequency as a function of the scale coefficient e_0a, @ $P = 0.5 * P_{cr}$ (clamped–clamped case).

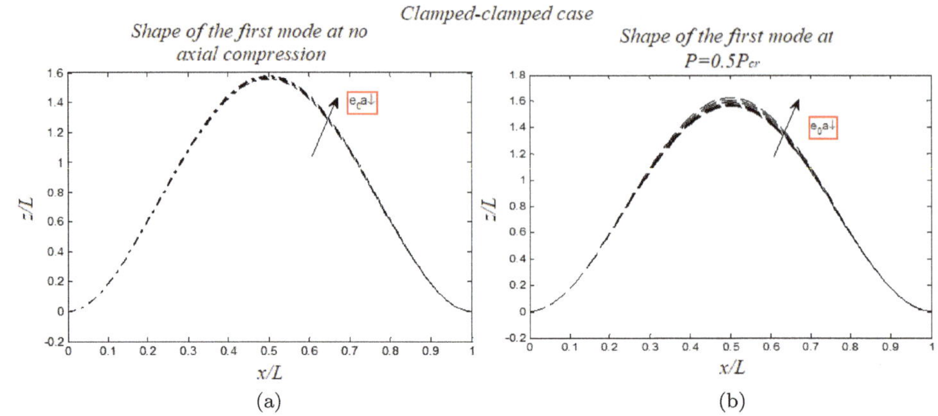

Fig. 11.12. Shape of the first mode for decreasing values of the scale coefficient, e_0a: (a) at zero axial compression and (b) at $P = 0.5 * P_{cr}$ (clamped–clamped case).

11.4. The Dynamic Non-local Nano-beam Model — Dynamic Buckling

The differential equations of motion for a beam under time-dependent axial compression loading $N_o(t)$ (see Fig. 11.13) for the two variables, w (lateral deflection) and u (axial deflection) can be written as (see Chapter 7)

$$N_{x,x} = I_0\ddot{u} - I_1\ddot{w}_{,x}, \tag{11.24}$$

$$[M_{x,x} + N_x \cdot w_{,x} + I_2 \cdot \ddot{w}_{,x}]_{,x} = I_0\ddot{w} + I_1\ddot{u}_{,x}, \tag{11.25}$$

where

$$(I_0;\ I_1;\ I_2) = \rho\int_{-h/2}^{h/2}(1;\ z;\ z^2)dz, \tag{11.26}$$

and $()_{,x}$ is the first partial differentiation with respect to x, $\ddot{u}, \ddot{w}$ are the second partial differentiation of the axial and lateral displacements with respect to time, t, h is the total thickness of the column and ρ is the density. For $I_1 = 0$ Eq. (11.24) simplifies to yield

$$N_{x,x} = I_0\ddot{u},$$

$$[M_{x,x} + N_x w_{,x} + I_2\ddot{w}_{,x}]_{,x} = I_0\ddot{w}. \tag{11.27}$$

The axial compression load for a non-local case has the following expression:

$$N_x - (e_0a)^2 N_{x,xx} = EA\varepsilon_{xx} = EA\left\{u_{,x} + \frac{1}{2}[w_{,x}^2 - w_{0,x}^2]\right\}$$

but $N_{x,x} = I_0\ddot{u} \rightarrow N_{x,xx} = I_0\ddot{u}_{,x}$

$$\Rightarrow N_x = (e_0a)^2 I_o\ddot{u}_{,x} + EA\left\{u_{,x} + \frac{1}{2}[w_{,x}^2 - w_{0,x}^2]\right\}. \tag{11.28}$$

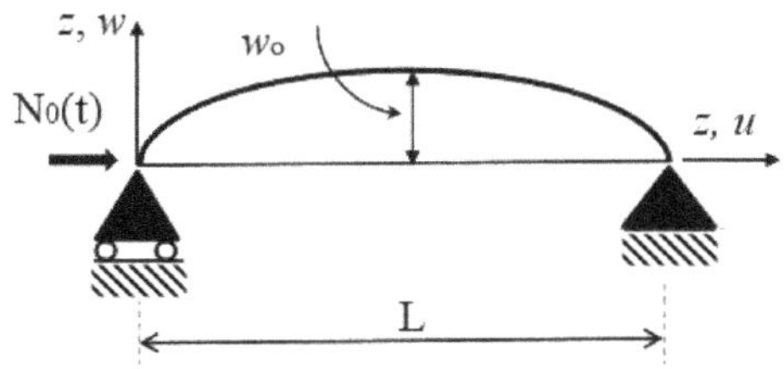

Fig. 11.13. A beam with initial geometric imperfection, w_0 subjected to time-dependent axial compression loading — a schematic view.

Deriving Eq. (11.28) once with respect to x, and inserting in the first equation in Eqs. (11.27) yields

$$(e_0a)^2 I_o \ddot{u}_{,xx} + EA\left\{u_{,x} + \frac{1}{2}[w_{,x}^2 - w_{0,x}^2]\right\}_{,x} = I_o\ddot{u}. \tag{11.29}$$

Performing the required derivation with respect to x in Eq. (11.29) leads to the final differential equation for the axial displacement u

$$(e_0a)^2 I_o \ddot{u}_{,xx} + EA\{u_{,xx} + w_{,x}w_{,xx} - w_{0,x}w_{0,xx}\}_{,x} = I_o\ddot{u} \tag{11.30}$$

The bending moment expressed by the lateral deflection for a non-local case has the following expression:

$$\begin{aligned} &M_x - (e_0a)^2 M_{x,xx} = -EI(w - w_0)_{xx} \\ &\text{but } {}_{x,xx} = I_0\ddot{w} - (N_x w_{,x})_{,x} - I_2\ddot{w}_{,xx} \\ &\Rightarrow M_x = (e_0a)^2[I_0\ddot{w} - (N_x w_{,x})_{,x} - I_2\ddot{w}_{,xx}] - EI(w - w_0)_{xx} \end{aligned} \tag{11.31}$$

Deriving Eq. (11.28) twice with respect to x, and inserting in the second equation in Eqs. (11.27) yields

$$\begin{aligned} &M_{x,xx} = (e_0a)^2[I_0\ddot{w} - (N_x w_{,x})_{,x} - I_2\ddot{w}_{,xx}]_{,xx} - EI(w - w_0)_{,xxxx} \\ &\text{but } [M_{x,x} + N_x w_{,x} + I_2\ddot{w}_{,x}]_{,x} = I_0\ddot{w} \\ &\Rightarrow (e_0a)^2[I_0\ddot{w} - (N_x w_{,x})_{,x} - I_2\ddot{w}_{,xx}]_{,xx} - EI(w - w_0)_{,xxxx} + (N_x w_{,x})_{,x} \\ &\quad + I_2\ddot{w}_{,xx} = I_0\ddot{w}. \end{aligned} \tag{11.32}$$

Performing the various derivations in Eq. (11.32) and rearranging yields

$$\begin{aligned} &(e_0a)^2 I_0\ddot{w}_{,xx} - (e_0a)^2(N_{x,xxx}w_{,x} + 3N_{x,xx}w_{,xx} + 3N_{x,x}w_{,xxx} + N_x w_{,xxxx}) \\ &\quad - (e_0a)^2 I_2\ddot{w}_{,xxxx} - EI(w - w_0)_{,xxxx} \\ &\quad + (N_{x,x}w_{,x} + N_x w_{,xx}) + I_2\ddot{w}_{,xx} = I_0\ddot{w}. \end{aligned} \tag{11.33}$$

Equations (11.30) and (11.33) together with the following boundary conditions (pinned-pinned case) will provide the necessary equations to solve the problem.

$$
\begin{aligned}
&w(0,t) = w(L,t) = 0,\\
&M_x(0,t) = M_x(L,t) = 0,\\
&N_x(0,t) = -N_0(t),\\
&u(L,t) = 0.
\end{aligned} \tag{11.34}
$$

The initial conditions for the dynamic problem are

$$
\begin{aligned}
&w(x,t=0) = w_0, \quad \dot{w}(x,t=0) = 0,\\
&u(x,t=0) = 0, \quad \dot{u}(x,t=0) = 0.
\end{aligned} \tag{11.35}
$$

The solution of the non-linear Eqs. (11.30) and (11.33) is done using the Galerkin method with the following variables:

$$
\begin{aligned}
&w(x,t) = A(t)\sin\frac{\pi x}{L}, \quad w_0(x,t) = A_0 \sin\frac{\pi x}{L},\\
&u(x,t) = B(t)\cos\frac{\pi x}{2L}, \quad N_0(t) = N_0 \sin\frac{\pi t}{T}.
\end{aligned}
$$

For more details of the solution procedure, see [20].

Typical results are next presented using the properties presented in Table 11.3.

The behavior of the lateral displacement, w, of the impacted beam, as a function of the axial compressed load can be separated into two regions.

Table 11.3. Properties of single-walled nano tubes (SWNT).

Property	Value
Young's modulus, E N/m^2	5.5×10^{10}
Poisson's ratio, v	0.3
Density, ρ kg/m^3	2300
Outer diameter of a characteristic beam, d_0 m	2.0×10^{-9}
Wall thickness, $t = 0.1d_0$ m	2.0×10^{-10}
Inner diameter of a characteristic beam, $d_i = (d_0 - 2t)$ m	1.6×10^{-9}
Cross-sectional area of the beam, A m^2	1.131×10^{-18}
Second moment of area, I m^4	4.9087×10^{-38}
Characteristic length, $L = 10d_0$ m	2.0×10^{-8}
Scale coefficients, e_0a m	$0, d_0, 2d_0$
Natural frequency, ω_0 1/sec	7.7259×10^{11}
Natural time period, T_0 sec	8.1326×10^{-12}
Buckling load (local Euler), P_{cr} N	6.2927×10^{-8}
Initial geometric imperfection, $w_0 = 0.1d_0$ $(w_0/t = 1)$ m	2.0×10^{-10}

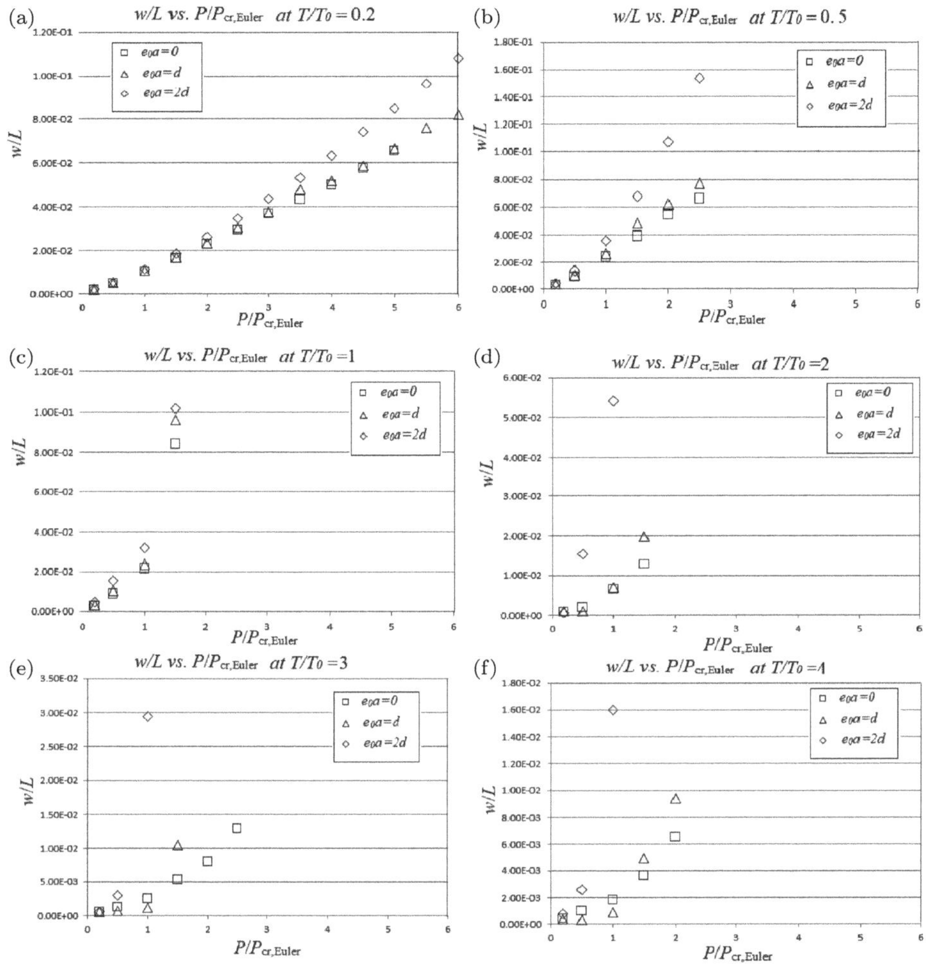

Fig. 11.14. Non-dimensional lateral amplitude vs. $P/p_{\text{cr.Euler}}$ with $e_0a = 0$, d, $2d$ at: (a) $T/T_0 = 0.2$; (b) $T/T_0 = 0.5$; (c) $T/T_0 = 1$; (d) $T/T_0 = 2$; (e) $T/T_0 = 3$; and (f) $T/T_0 = 4$.

In the first region, the slope of the curve would be quite moderate, while in the second region a small increase in the dynamic compression load would result in a much larger increase in the lateral displacement (bending) response which is known as in the literature as the Budiansky and Hutchinson dynamic buckling criterion (see Chapter 7). Figures 11.14(a)–11.14(f) present some of the results obtained for ramp type dynamic axial compression load.

From Fig. 11.14, one can see that the ratio between the non-dimensional lateral amplitude and the dynamic axial compression load increases with the increase of the scale coefficient e_0a per each period.

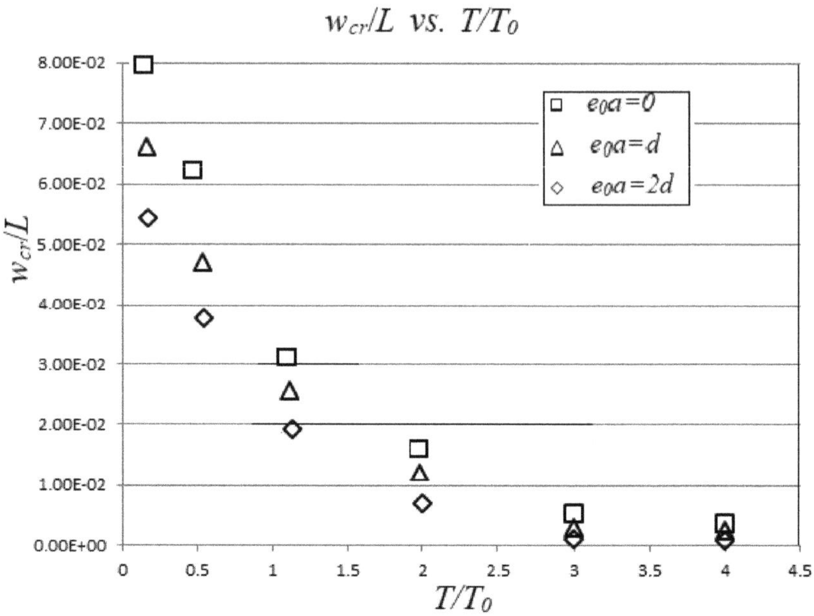

Fig. 11.15. Non-dimensional buckling lateral amplitude as a function of non-dimensional period time for different scale coefficients e_0a/d, (pinned–pinned case).

The lateral amplitude of the beam increases as the dynamic axial compression load increases per each period.

The point where the lateral amplitude increase largely per small increase in the dynamic axial compression load is going towards the point where the dynamic axial compression load is close to the static critical buckling load with the increase of the period or with the decrease of the frequency (and becoming more static like behavior).

Figure 11.15 presents a graph for the lateral non-dimensional at buckling vs. the non-dimensional period time, for three different scale coefficients e_0a. It is clearly seen that the lateral displacement at buckling would reach maximal values for short time durations.

11.5. Conclusions

Based on the study performed on nano-beams using the Timoshenko non-local theory one can outline the following conclusions:

(a) Using either Timoshenko or Bernoulli–Euler theories to calculate the buckling loads of nano-beams tend to overestimate the buckling.

(b) Using either Timoshenko or Bernoulli–Euler theories to calculate the natural frequencies of Nano-beams tend to overestimate the frequencies.
(c) The non-local theory (Eringen's theory) shows that the $(e_0a)^2$ parameter has a large influence on both the buckling and the natural frequencies of nano-beams.
(d) Nano-beams might also buckle dynamically, with the parameter $(e_0a)^2$ largely influencing the behavior of the beam.
(e) For fast load applications, the buckling load is much higher than the static critical buckling load (like for regular beams).
(f) For slow load applications, the buckling load is close to the static critical buckling load (like for regular beams).

References

[1] Arash, B. and Wang, Q., A review on the application of nonlocal elastic models in modeling of carbon nanotubes and graphenes, *Comput. Mater. Sci.*, Vol. 51, 2011, pp. 303–313.
[2] Peddieson, J., Buchanan, G.R. and McNitt, R.P., Application of nonlocal continuum models to nanotechnology, *Int. J. Eng. Sci.*, Vol. 41, 2003, pp. 305–312, PII: S0020-7225(02)00210-0.
[3] Lu, P., Zhang, P.Q., Lee, H.P., Wang, C.M. and Reddy, J.N., Non-local elastic plate theories, *Proc. R. Soc. A*, Vol. 463, 2007, pp. 3225–3240.
[4] Wang, G.F. and Feng, X.Q., Timoshenko beam model for buckling and vibration of nanowires with surface effect, *J. Phys. D: Appl. Phys.*, Vol. 42, 2009, Paper Id. 155411, 5p.
[5] Wang, C.M., Zhang, Y.Y., Ramesh, S.S. and Kitipornchai, S., Buckling analysis of micro-and nano-rods/tubes based on nonlocal Timoshenko beam theory, *J. Phys. D: Appl. Phys.*, Vol. 39, 2006, pp. 3904–3909.
[6] Wang, C.M., Zhang, Y.Y. and He, X.Q., Vibration of nonlocal Timoshenko beams, *Nanotechnology*, Vol. 18, 2007, Paper Id.105401, 9 p.
[7] Eringen, A.C., Nonlocal polar elastic continua, *Int. J. Eng. Sci.*, Vol. 10, 1972, pp. 1–16.
[8] Eringen, A.C. and Edelen, D.G.B., On nonlocal elasticity, *Int. J. Eng. Sci.*, Vol. 10, 1972, pp. 233–248.
[9] Eringen, A.C., On differential equations of nonlocal elasticity and solutions of screw dislocation and surface waves, *J. Appl. Phys.*, Vol. 54, 1983, pp. 4703–4710.
[10] Thai, H.T., A nonlocal beam theory for bending, buckling and vibration of nanobeams, *Int. J. Eng. Sci.*, Vol. 52, 2012, pp. 56–64.
[11] Yan, Z. and Jiang, L.Y., The vibrational and buckling behaviors of piezoelectric nanobeams with surface effects, *Nanotechnology*, Vol. 22, 2011, Paper Id. 245703, 7 p.
[12] Sears, A. and Batra, R.C., Buckling of multiwalled carbon nanotubes under axial compression, *Phys. Rev. B*, Vol. 73(8), 2006, Paper Id. 085410, 11 p.
[13] Zhang, Y.Y., Wang, C.M. and Challamel, N., Bending, buckling and vibration of micro/nano beams by hybrid nonlocal beam model, *J. Eng. Mech., ASCE*, Vol. 136(5), 2010, pp. 562–574.

[14] Wang, G.F. and Feng, X.Q., Surface effects on buckling of nanowires under uniaxial compression, *Appl. Phys. Lett.*, Vol. 94, 2009, Paper Id. 141913, 3 p.
[15] Nishio, M., Akita, S. and Nakayama, Y., Buckling test under axial compression for multiwall carbon nanotubes, *Japanese J. Appl. Phys.*, 44(34), 2005, pp. L1097–L1099.
[16] Wang, C.M. Zhang, Y.Y., Xiang Y. and Reddy, J.N., Recent studies on buckling of carbon nanotubes, *Appl. Mech. Rev.*, Vol. 16, No. 3, 2010, Paper Id. 030804, 18 p.
[17] Lee, H.L., Chang, R.P. and Chang, W.J., Buckling analysis of nonuniform nanowires under axial compression, *Proc. World Congr. Engineering* 2011, *WCE 2011*, July 6–8, 2011, London, UK.
[18] Wang, X. and Yang, H.K., Bending stability of multiwalled carbon nanotubes, *Phys. Rev. B*, Vol. 73(8), 2006, Paper Id. 0854109, 8 p.
[19] Abramovich, H., Natural frequencies of Timoshenko beams under compressive axial loads, *J. Sound Vibration*, Vol. 157(3), 1992, pp. 183–189.
[20] Buch, O.,Dynamic buckling of nano-beams, Master thesis at the Faculty of Aerospace Engineering, Technion, I.I.T., 32000 Haifa, Israel, Feb. 2015, 205 p. (in English).

Appendix A: The Non-local Model

The constitutive equation of classical elasticity is an algebraic relationship between the stress and strain tensors while that of *Eringen's non-local* elasticity (see [1–3]) involves spatial integrals which represent weighted averages of the contributions of strain tensors of all the points in the body to the stress tensor at the given point.

Though it is mathematically difficult to obtain the solution for *non-local* elasticity problems due to the spatial integrals in *Eringen's* constitutive equations, these integro-partial constitutive differential equations can be converted to equivalent differential constitutive equations under certain conditions.

The stress according to the *non-local Eringen* theory [5] can be simplified and finally be expressed as an ordinary differential equation.

The equation of motion for non-local linear elastic solids is

$$t_{ij,j} + f_i = \rho \ddot{u}_i, \tag{A.1}$$

where t_{ij} is the stress tensor of the non-local elasticity, f_i is the body (and/or applied) forces, ρ is the mass density and u_i the displacement vector. The stress tensor t_{ij} is defined as

$$t_{ij}(x) = \int_v \alpha(|x' - x|)\sigma_{ij}(x')dv, \tag{A.2}$$

where v is the body volume, x is a reference point, $\alpha(|x' - x|)$ is the non-local kernel function and σ_{ij} is the local stress tensor of classical elasticity theory at any point x' in the body. The stress satisfies the constitutive relations as follows:

$$\sigma_{ij} = C_{ijkl}\varepsilon_{kl}, \quad \varepsilon_{kl} = \frac{u_{k,l} + u_{l,k}}{2}, \tag{A.3}$$

where C_{ijkl} is the elastic modulus components with the symmetry properties and ε_{kl} is the strain tensor.

According to the properties of the non-local kernel, and as discussed in detail in [9], when the characteristic length ratio tends to zero, $\alpha(|x|)$ must revert to the Dirac delta measure so that the classical elasticity limit is included in the limit of vanishing internal characteristic length.

When $\alpha(|x|)$ takes on a Green's function of a linear differential operator $\mathcal{L}$, one gets

$$\mathcal{L}[\alpha(|x' - x|)] = \delta(|x' - x|). \tag{A.4}$$

where $\delta(|x|)$ is the Dirac delta. Then the non-local constitutive relation in Eq. (A.2) is reduced to the differential equation as follows:

$$\mathcal{L}[t_{ij}] = \sigma_{ij} \tag{A.5}$$

and the integro-partial differential equation of t_{ij} in Eq. (A.1) is correspondingly reduced to the partial differential equation:

$$\sigma_{ij,j} + \mathcal{L}(f_i - \rho\ddot{u}_i) = 0. \tag{A.6}$$

Eringen proposed a non-local model with a linear differential operator $\mathcal{L}$ defined as

$$\begin{gathered} \mathcal{L} \equiv 1 - (e_0 a)^2\nabla^2, \quad \text{where} \\ \nabla^2 = \left(\frac{\partial}{\partial x} + \frac{\partial}{\partial y} + \frac{\partial}{\partial z}\right)^2. \end{gathered} \tag{A.7}$$

Therefore, according to Eq. (A.3) and Eqs. (A.5–A.7), the constitutive relations with this kernel function may be simplified to:

$$[1 - (e_0 a)^2\nabla^2]t_{ij} = C_{ijkl}\varepsilon_{kl}. \tag{A.8}$$

Thus, the simplified *non-local* constitutive equation for the normal stress and strain in a one-dimensional (x) direction is given by [6]

$$\sigma_{xx} - (e_0 a)^2\frac{d^2\sigma_{xx}}{dx^2} = E\varepsilon_{xx}, \tag{A.9}$$

where E is the material Young's modulus and $e_0 a$ is the scale coefficient that incorporates the small-scale effect.

References

[1] Eringen, A.C., Nonlocal polar elastic continua, *Int. J. Eng. Sci.*, Vol. 10, 1972, pp. 1–16.
[2] Eringen, A.C. and Edelen, D.G.B., On nonlocal elasticity, *Int. J. Eng. Sci.*, Vol. 10, 1972, pp. 233–248.
[3] Eringen, A.C., On differential equations of nonlocal elasticity and solutions of screw dislocation and surface waves, *J. Appl. Phys.*, Vol. 54, 1983, pp. 4703–4710.
[4] Wang, C.M., Zhang, Y.Y., Ramesh, S.S. and Kitipornchai, S., Buckling analysis of micro-and nano-rods/tubes based on nonlocal Timoshenko beam theory, *J. Phys. D: Appl. Phys.*, Vol. 39, 2006, pp. 3904–3909.
[5] Wang, C.M., Zhang, Y.Y. and He, X.Q., Vibration of nonlocal Timoshenko beams, *Nanotechnology*, Vol. 18, 2007, Paper Id.105401, 9 p.

Index

B

H

I

M

N

Q

R

S

Lightning Source UK Ltd.
Milton Keynes UK
UKHW050634110621
385329UK00003B/328

9 789811 214240